The Warming Papers

The Warming Papers

*The Scientific Foundation
for the Climate Change Forecast*

Edited by

David Archer

and

Raymond Pierrehumbert

(W) **WILEY-BLACKWELL**

A John Wiley & Sons, Ltd., Publication

This edition first published 2011 © 2011 by Blackwell Publishing Ltd

Blackwell Publishing was acquired by John Wiley & Sons in February 2007. Blackwell's publishing program has been merged with Wiley's global Scientific, Technical and Medical business to form Wiley-Blackwell.

Registered Office
John Wiley & Sons Ltd, The Atrium, Southern Gate, Chichester, West Sussex, PO19 8SQ, UK

Editorial Office
9600 Garsington Road, Oxford, OX4 2DQ, UK

For details of our global editorial offices, for customer services and for information about how to apply for permission to reuse the copyright material in this book please see our website at www.wiley.com/wiley-blackwell.

The right of the author to be identified as the author of this work has been asserted in accordance with the UK Copyright, Designs and Patents Act 1988.

Library of Congress Cataloging-in-Publication Data

The warming papers : the scientific foundation for the climate change forecast / edited by David Archer
and Raymond Pierrehumbert.
      p.   cm.
   Includes bibliographical references and index.
   ISBN 978-1-4051-9616-1 (pbk.) – ISBN 978-1-4051-9617-8 (hardcover)
 1. Greenhouse effect, Atmospheric.   2. Greenhouse gases.   3. Global temperature changes.   I. Archer, David, 1960–
II. Pierrehumbert, Raymond T.
   QC912.3.W37 2011
   551.5–dc22

                                     2010040516

A catalogue record for this book is available from the British Library.

Set in 10.5/13pt Minion by SPi Publisher Services, Pondicherry, India
Printed and bound in Malaysia by Vivar Printing Sdn Bhd

1   2011

Contents

Preface

Global warming is arguably the defining scientific issue of modern times, but it is not widely appreciated that the foundations of our understanding are almost two centuries old. The sensitivity of climate to changes in atmospheric CO_2 was first estimated about one century ago, and the rise in atmospheric CO_2 concentration was discovered half a century ago. The fundamentals of the science underlying the forecast for human-induced climate change were being published and debated long before it started to appear in the newspapers.

The aim of this book is to gather together the classic scientific papers that are the scientific foundation for the forecast of global warming and its consequences. These are not necessarily the latest in the state of play; there can be subsequent quantitative revision. But these papers are the big ideas. Some of the good old good ones can be heavy going, it must be admitted, so we will try to guide the reader with some verbage of our own, unworthy though it may be. We summarize the results for you, and provide the latest revisions from the ongoing literature, how strong the water vapor feedback turned out to be, for example. We will fill in the context, the personalities, and the aftermath of the ideas in the papers. We'll also presume to provide short comments where they occur to us in boxes throughout the papers, signposts to help guide the casual reader.

Part I

Climate Physics

The Greenhouse Effect

Fourier, J. (1827). Mémoire sur les Températures du Globe Terrestre et des Espaces Planétaires. *Mémoires de l'Académie Royale des Sciences*, **7**, 569–604. 25 pages.

Joseph Fourier (1768–1830) is generally credited with the discovery of what is now known as the greenhouse effect. In fact, his contribution to the study of planetary temperature is even more profound than that. Fourier introduced the problem of planetary temperature as a proper object of study in physics, and established a largely correct physical framework for attacking the problem. His work set the stage for most of the further developments in this area over the remainder of the nineteenth century. Indeed, it was only toward the end of that century that physics had caught up to the point that the first quantitative estimates of the Earth's temperature based on Fourier's concepts could be attempted.

If much of Fourier's reasoning in this paper seems qualitative, it should be recognized that most of the areas of physics that Fourier needs to call on were in their infancy in Fourier's day. Infrared radiation (called "dark heat" or "dark radiation" at the time) had been discovered in 1800 by the astronomer Sir Frederick William Herschel, and it was the subject of intense inquiry. Infrared was the "dark energy" of its day and it was perhaps no less mysterious to physicists of Fourier's day than is the dark energy talked about by today's physicists. There was some understanding from the work of Fourier's contemporaries, Dulong and Petit, that the rate of heat loss by infrared radiation increases with temperature, and it was known that infrared could carry heat through a near-vacuum. There was, however, only a limited ability to do quantitative calculations involving infrared heat transfer. Thermodynamics was in its infancy. The very nature of heat was still being hotly debated; the landmark energy conservation experiments of Joule that showed the equivalence of mechanical work and heat would not be carried out until 1843. Against this context, the general correctness of Fourier's great leap of intuition seems all the more remarkable.

In his 1827 paper, Fourier introduces five key concepts:

1. The temperature of the Earth, or indeed any planet, is determined by a balance between the rate at which the energy is received and the rate at which the energy is lost. There is therefore a need to determine the sources and sinks of a planet's energy.
2. There are three possible sources of heat: Sunlight, heat diffusing from the hot interior of the planet, and heat communicated from the general "temperature of space." Of these, the amount of heat leaking out of the Earth's interior is too small to play a significant role in the Earth's surface temperature.
3. Emission of infrared radiation is the only means by which a planet loses heat. Since the rate of energy loss by infrared radiation increases with the temperature of the body, the planet can come into equilibrium by heating up until the rate at which it loses energy by infrared emission equals the rate at which it gains energy from its energy sources.
4. Visible light is converted into infrared light when it is absorbed at a solid or liquid surface.
5. The atmosphere has an asymmetric effect on the incoming sunlight and the outgoing infrared, because the atmosphere is largely transparent to sunlight but is relatively opaque

to infrared. This retards the rate at which the planet loses energy, for any given temperature. The result is that the atmosphere keeps the planet warmer than it would have been if the atmosphere had been transparent to infrared radiation.

Fourier's inferences concerning the minimal influence of the Earth's interior heat on climate are drawn from observations of the way temperature varies with depth below the Earth's surface. Of all Fourier's claims in the 1827 paper, this is the one that is most backed up by quantitative reasoning, though the actual mathematical analysis appears in Fourier's other papers and is not reproduced in the 1827 essay. Fourier's greatest work as a mathematical physicist was the formulation of the partial differential equation describing the diffusion of heat within a body, and the development of the mathematical techniques required to solve it. The full range of these developments were engaged in Fourier's interpretation of the Earth's subsurface temperature variations. Indeed, Fourier states that the problem of planetary temperatures provided the main impetus for his formulation of the analytical theory of heat. His theory of heat was applied to the problem in two basic ways. First, since the rate of heat flow is proportional to the temperature gradient, the measured increase of time-mean temperature with depth itself shows that the interior of the Earth is hotter than the surface, and gives an estimate of the heat flux, provided that one can estimate the thermal conductivity of the Earth. The flux Fourier arrived at using this procedure was an overestimate compared to modern calculations because he used the thermal conductivity of iron, but his calculation nonetheless showed the diffusion of heat from the interior to be an insignificant factor in surface temperature. The second kind of problem Fourier did was to impose the observed time-periodic daily and seasonal fluctuations of temperature at the surface as a boundary condition, and then calculate what the subsurface temperature fluctuations should look like. It was this kind of calculation that led Fourier to develop what we now call Fourier series, so as to decompose the complex time-periodic boundary condition into a sum of simple sines and cosines for which the problem is analytically tractable. This calculation correctly predicts that the diurnal variation of temperature should decay rapidly with depth and the annual variation more slowly. The calculation also gives an estimate of the amount of heat that flows into and out of the surface from sunlight in the course of the diurnal and seasonal cycle, and thus provides an additional check on the importance of solar energy in determining the Earth's surface temperature.

It takes away nothing from Fourier's brilliance to point out the one stupendous blunder in his paper. Fourier thought that the heat the Earth receives from the general temperature of interplanetary space was a crucial factor in the Earth's climate, on a par with energy received from the Sun. He thought the temperature of space to be somewhat below the minimum temperatures observed in Winter in the Arctic – roughly 200 K in modern terms. He viewed this as one of his principle discoveries, and claimed that without this source of heat, the Earth would become infinitely cold at night and in the winter, and that no life would be possible. In essence, Fourier's view was that 200 K was the natural temperature that all Solar System planets would relax to if there were no absorption of sunlight. Conceptually, he was not entirely wrong, though the correct number for the "temperature of space" in this sense would be more nearly 5 K than 200 K, but Fourier's estimate of the temperature of space was based on highly dubious reasoning that did not justify his level of certainty by any means. The assumption that Arctic night temperatures represent the temperature of space neglects the role played by the long time required for the ocean to cool down ("thermal inertia") and by the ability of air and ocean currents to transport heat from warmer parts of the planet to the poles. Fourier knew about these effects, and even mentions them explicitly elsewhere in the essay. Evidently, he thought they were too ineffective to account for the observed winter and night-time temperature, though his reasons for preferring the more exotic solution of a high temperature of space remain obscure.

In any event, Fourier's misconception about the temperature of space was corrected by Claude Pouillet in 1838. Pouillet's main contribution to science was a largely correct measurement of the

Solar Constant, though his estimate of the corresponding temperature of the Sun was in error because of shortcomings of then-current representations of blackbody radiation. In the course of these measurements, Pouillet found that the temperature of space was far below the value supposed by Fourier, and nothing more was heard thereafter about the role of the temperature in space in climate.

de Saussure's Hot Boxes

In thinking about the effect of the atmosphere on the Earth's energy balance, Fourier drew on the behavior of a simple device invented by the Swiss Alpinist Horace-Bénédict de Saussure (1740–99). This device, called a *heliothermometer*, consisted of a wooden box insulated with cork and wool, with a lid consisting of one or more panes of transparent glass (Fig. 1). The interior walls were painted black so as to absorb nearly all the sunlight entering the box, and a thermometer was placed in the box so that its temperature could be determined. de Saussure devised this instrument as a means of measuring the intensity of sunlight, so that he could test the hypothesis that it is colder atop mountains because the sunlight is weaker there. The idea was to trap the energy of sunlight inside the box, and keep the interior isolated from the surrounding so that the temperature in the box would be responsive to the intensity of the sunlight rather than the temperature of the surroundings. Using the heliothermometer, de Saussure correctly concluded that sunlight becomes, if anything, more intense at higher elevations, so that some other physical process must come into play. "Hot-Boxes" such as de Saussure's were popular toys among scientists throughout the nineteenth century, and many succumbed to the temptation to use them as solar cookers. de Saussure writes that "Fruits ... exposed to this heat were cooked and became juicy." Herschel himself took a hot-box with him to South Africa in 1830, and reported: "As these temperatures [up to 240°F] far surpass that of boiling water, some amusing experiments were made by exposing eggs, meat, etc. [to the heat inside the box], all of which, after a moderate length of exposure, were found perfectly cooked. ... [On] one occasion a very respectable stew of meat was prepared and eaten with no small relish by the entertained bystanders."

Neither de Saussure nor Fourier hit on the correct explanation of the decline of temperature with altitude, which involves the cooling of air parcels as they are lifted and expand. Nonetheless, the behavior of the heliothermometer provoked a lot of useful thinking about the energy carried by sunlight. Fourier's use of the analogy was to show that if one keeps the rate of energy *input* by sunlight the same, but retards the rate of energy *loss* by putting on a pane of glass, then when the system comes into equilibrium its temperature will be greater than it would have been without the glass in place. Fourier knew that the glass was transparent to sunlight and largely opaque to infrared, but he also knew that in the typical experiment the glass retards heat loss, in part, by simply trapping warm air in the box and keeping it from blowing away. He alludes to the fact that the experiment would still yield an elevation of temperature even if performed in a vacuum, but his use of the subjunctive in the original French suggests that this is a thought experiment, rather than one he actually carried out.

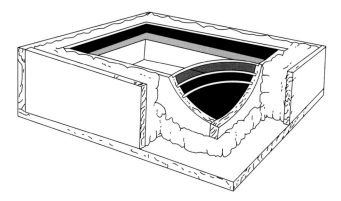

Fig. 1.1 **Artist's conception of the Saussure's improved hot box.**

What Fourier Did Not Do

One thing Fourier did not do was coin the term "greenhouse effect," though his use of de Saussure's heliothermometer as an analogue could be considered similar to a greenhouse analogue. de Saussure's box is indeed a kind of miniature greenhouse. In any event, Fourier showed a clear awareness of the imperfection of the analogy, stating explicitly that the temperature in the hot box was influenced by turbulent heat transfers that have no proper counterpart in the planetary temperature problem.

Further, Fourier did not compute the temperature of the Earth in the absence of an atmosphere and concluded that it was colder than the observed temperature. In fact, he never actually computed the Earth's temperature based on a balance between incoming sunlight and outgoing infrared, though he could have attempted this using the Dulong–Petit radiation law. It is not clear why Fourier thought the atmosphere had to have a warming role. Rather than this being demanded by too cold temperatures in the absence of an atmosphere, Fourier seems to be inferring that the atmosphere ought to act like a pane of glass in being transparent to sunlight but opaque to infrared; he shows awareness of the downward infrared radiated by the atmosphere, but it is not clear what the basis of Fourier's leap of intuition about the atmosphere was. In any event, he was right, and his work stimulated a great deal of further research on the effect of the atmosphere on infrared, and ultimately Tyndall's definitive experiments to be discussed next.

On the Temperatures of the Terrestrial Sphere and Interplanetary Space

JEAN-BAPTISTE JOSEPH FOURIER

Translator's note. This is a translation of Jean-Baptiste Joseph Fourier's "Mémoire sur les Températures du Globe Terrestre et des Espaces Planétaires," which originally appeared in *Mémoires d l'Académie Royale des Sciences de l'Institute de France* **VII** 570–604 1827. The original text is most readily accessible in the 1890 edition of Fourier's collected *Oeuvres*, Volume 2, edited by M. Gaston Darboux (Gauthier-Villars et Fils:Paris). This work is available online from the Bibliothèque Nationale de France (search catalogue.bnf.fr for author "Fourier, Jean-Baptiste-Joseph"). In the version reprinted in the *Oeuvres*, it is noted that a very slightly different version of the essay also appeared in the *Annales de Chimie et de Physique*, vol XXVII, pp 136–167; 1824, under the title "Remarques générales sur les températures du globe terrestre et des espaces planétaires."

An English translation of Fourier's article has not been available in print for more than a century. Although the article is widely cited, it is my experience that its actual contents are not well known in the Anglophone community (and they are hardly better known among Francophones). My object in doing a new translation is to help rectify this situation, while using some of my own knowledge of physics of climate to help put Fourier's arguments in the clearest possible light. I have put a premium on readability rather than literal translation, and in some cases I have taken the liberty of rephrasing some sentences so as to make Fourier's reasoning more evident; I do not think that in doing so I have read more into the text than Fourier himself put there, but readers seeking the finer nuances of Fourier's meaning will of course have to read the original. I have not consulted any of the existing translations in carrying out the present one, though I can recommend to the reader's attention the annotated translation by W. M. Connolley, available online only at www.wmc.care4free.net/sci/fourier_1827.

I have provided some commentary in the form of footnotes, which are marked by my initials.

Note that for variety, Fourier often uses *globe terrestre* for "Earth," This also serves to remind the reader of the connection with Fourier's earlier idealized work on heat diffusion in a sphere. In the title, I have preserved this sense, but for the most part the phrase has simply been translated as "Earth" in the text.

R. T. Pierrehumbert
1 September, 2004
Chicago, IL, USA

The question of the Earth's temperature distribution, one of the most important and most difficult of all Natural Philosophy, is made up of rather diverse elements that must be considered from a general point of view. It has occurred to me that it would be useful to unite in a single work the principle consequences of this theory; the analytical details that have been omitted here can for the most part be found in the Works which I have already published. Above all, I wish to present to physicists, in a broader picture, the collection of pertinent phenomena and the mathematical relations amongst them.

It is first necessary to distinguish the three sources from which the Earth derives its heat:

(1) The Earth is heated by solar radiation, the unequal distribution of which produces the diversity of climates;

(2) It participates in the common temperature of interplanetary space, being exposed to irradiation by countless stars which surround all parts of the solar system;

(3) The Earth has conserved in the interior of its mass, a part of the primordial heat which it had when the planets originally formed.

By considering each of these three causes and the phenomena which it produces, we will come to understand as clearly as possible, within the limitations of the current state of science, the principal characteristics of these phenomena. In order

Fourier, J.-B. F. (1827) On the temperatures of the terrestrial sphere and interplanetary space. *Mémoires de l'Académie Royale des Sciences* 7: 569–604.

to provide an overview of this grand question, and to give a first indication of the results of our investigations, we shall present them first in summary form. This summary, in a manner of speaking, serves as an annotated table of contents to my work on the subject.

Our solar system is located in a region of the universe of which all points have a common and constant temperature, determined by the light rays and the heat sent by all the surrounding stars. This cold temperature of the interplanetary sky is slightly below that of the Earth's polar regions. The Earth would have none other than this same temperature of the Sky, were it not for two causes which act together to further heat it. The first is the interior heat which the globe possessed when the planetary bodies were formed, and of which only a part has escaped through the surface. The second cause is the continual action of solar radiation, which has penetrated the whole mass of the Earth and which leads at the surface to the difference in climates from one place to another.

The primordial heat of the globe no longer has any significant effect at the surface, but it can still be immense in the interior of the Earth. The temperature of the surface does not exceed by more than a thirtieth of a degree the value that it will eventually achieve after a long time has passed: At first, it diminished very rapidly; however, at present the diminution continues only exceedingly slowly.

The observations collected so far indicate that the points of a vertical line continued into the solid earth become warmer with increasing depth, and this rate of increase has been estimated at 1 degree for each 30 to 40 meters. Such a result implies a very high temperature for the interior of the Earth; it can not arise from the action of solar radiation: rather, it is naturally explained by the heat the Earth has retained from the time of its origin.

This rate of increase, on the order of 1 degree per 32 m, will not always remain the same: It will diminish progressively; however, a great many centuries (much more than 30,000 years) will pass before it will be reduced to half of its present value.

It is possible that other yet-unknown causes can explain the same facts, and that there are other general or incidental sources of terrestrial heat. If so, one will discover them through comparison of the results of the present theory against observations.

The heat rays which the Sun incessantly sends to the Earth produce two very distinct effects there:

The first is periodic and affects the outer envelope of the planet, while the other is constant; one observes it in deep places, for example at 30 m below the surface. The temperature of these locations is subject to hardly any change in the course of the year, it is fixed; however the deep temperature varies substantially from one climatic zone to another: it results from the perpetual action of solar radiation and the inequal exposure of the surface to these rays, from the equator to the poles. One can determine the time which had to pass in order for the solar radiation to produce the diversity of climates observed today. All these results are in accord with dynamical theories which have led us to recognize the stability of the Earth's axis of rotation.

The periodic effect of solar heating consists of both diurnal and annual variations. Observations of this type are reproduced exactly and in all details by the theory. The comparison of results with observations can be used to measure the thermal conductivity of the material of which the crust of the Earth is formed.

The presence of the atmosphere and surface waters has the effect of rendering the distribution of heat more uniform. In the Ocean and in lakes, the most cold molecules – or more precisely, those with the greatest density – direct themselves continually towards lower regions, and the transport of heat due to this cause is much more rapid than that which can be accomplished in solid bodies by means of thermal conductivity. Mathematical examination of the former effect will require numerous and exact observations: they will serve to clarify how these internal fluid motions keep the internal heat of the globe from having a notable effect in the depths of the waters.[1] Liquids conduct heat very poorly; but they have, as do gaseous materials, the the property of being able to transport it rapidly in certain directions through fluid motions. It is this same property which, in combination with centrifugal force, displaces and mixes all parts of the atmosphere and those of the Ocean; it involves organized and immense currents.

The interposition of air greatly modifies the effects of heat at the surface of the globe. The rays of the Sun, in traversing the layers of the atmosphere compressed by their own weight, heats them very inequally: Those which are the most tenuous are also the most cold, because they attenuate and absorb a lesser quantity of these rays.[2] The heat of the Sun, arriving in the form of visible light, has the ability to penetrate transparent solid or liquid substances, but loses this ability almost completely when it is converted, by its interaction with the terrestrial body, into dark radiant heat.

This distinction between luminous heat and dark heat explains the increase of temperature caused by transparent bodies. The body of water which covers a great part of the globe and the polar ice pose less of an obstacle to the incident luminous heat than to the dark heat, which returns in the opposite sense to exterior space.[3] The presence of the atmosphere produces an effect of the same sort, but which, in the present state of theory and owing further to lack of observations with which theory may be compared, cannot yet be exactly defined. However great the effect may be, one would not suppose that the temperature caused by the incidence of the rays of the Sun on an extremely large solid body would greatly exceed that which one would observe on exposing a thermometer to the light of that star.

The radiation from the highest layers of the atmosphere, whose temperature is very cold and nearly constant, influences all meteorological features which we observe: this radiation can be rendered more easily detectible by means of reflection from concave mirrors. The presence of clouds, which intercept these rays, tempers the cold of the nights.[4]

One thus sees that the surface of the Earth is located between one solid mass, whose central heat may surpass that of incandescent matter, and an immense region whose temperature is below the freezing point of mercury.

All the preceding considerations apply equally well to other planetary bodies. One can consider them as being placed in an environment whose common temperature is constant and somewhat below that of the terrestrial polar regions. This temperature – the temperature of the heavens – is the temperature that would be found at the surface of the most distant planets, for the Solar radiation would be too weak, even augmented by the state of the surface, to have a significant effect; From the state of the Earth we know further, that on other planets (whose formation could hardly have been much later than that of the Earth) the interior remanent heat no longer causes any significant elevation of surface temperature.

It is similarly likely that, for most of the planets, the polar temperature is only slightly greater than that of interplanetary space. As for the mean temperature caused by the action of the Sun on each of these bodies, we are in a state of ignorance, because it can depend on the presence of an atmosphere and the state of the surface. One can only assign, in a very imprecise manner, the mean temperature which the Earth would acquire if it were transported to the same position as the planet in question.

After this discussion, we will treat in succession the various parts of the question. First we must set forth a remark the significance of which bears on all these parts, because it is founded on the nature of the differential equations governing the movement of heat. Namely, we make use of the fact that the effects which arise from each of the three causes which we have discussed above can be calculated separately, as if each of these causes existed in isolation. It suffices then to combine the partial effects; they can be freely superposed, just as for the problem of final oscillations of bodies.[5]

We shall describe first the principal results caused by the prolonged action of solar rays on the Earth.

If one places a thermometer at a considerable depth below the surface of the solid Earth, for example at 40 meters, this instrument indicates a fixed temperature.[6]

One observes this fact at all points of the globe. This deep subsurface temperature is constant for any given location; however, it is not the same in all climates. Generally speaking, it decreases as one moves towards the poles.

If one observes the temperature of points much closer to the surface, for example at 1 m or 5 m or 10 m of depth, one notices very different behavior. The temperature varies during the course of a day or a year; however, we will for the moment idealize the problem by supposing that the skin of the Earth wherein such temperature variations occur is eliminated. We then consider the fixed temperatures of the new surface of the globe.

One can imagine that the state of the mass has varied continually in accord with the heat received from the heat source. This variable temperature state gradually alters, and more and more approaches a final state which no longer varies in time. At that time, each point of the solid sphere has acquired – and conserves – a fixed temperature, which depends only on the position of the point in question.

The final state of the mass, of which the heat has penetrated through all parts, is precisely analogous to that of a vessel which receives, through its upper opening, a liquid which furnishes a constant source, and which allows liquid to escape at a precisely equal rate through one or more openings.

Thus, the solar heat accumulates in the interior of the globe and is continually renewed. It penetrates the portions of the surface near the equator, and escapes through the polar regions. The first question of this type which has been subjected to calculation can be found in a dissertation which I read at the Institute of France at the end of 1807, article 115, p. 167.[7] This work has been deposited in the Archives of the Academy of Sciences. At the time, I took up this first question in order to offer a remarkable example of the application of the new theory presented in the article, and to show how analysis reveals the routes followed by solar heat in the interior of the globe.

Let us now restore the upper envelope of the Earth, for which the points are not sufficiently deep for their temperatures to be time-independent. One then notices a more intricate range of phenomena, which can be completely accounted for by our analysis. At a moderate depth, such as 3 m to 4 m, the temperature observed does not vary in the course of the day; however, it changes very noticably in the course of a year; it alternately rises and falls. The amplitude of these variations, that is to say the difference between the *maximum* and the *minimum* of temperature, is not the same at all depths; it becomes less as the distance from the surface becomes greater. The points lying on a vertical line do not all achieve their extremes of temperature at the same time. The amplitude of the variations, and the time of year at which the highest, mean and lowest temperatures are achieved, change with the position of the point on the vertical. The same applies to the quantities of heat which alternately descend and rise: all these quantities have very definite relations amongst each other, which are indicated by experiment and which analysis expresses very distinctly. The observations conform to the results furnished by the theory; there is not any natural effect more completely explained than this. The mean annual temperature of any given point of the vertical, that is, the mean of all values observed at this point in the course of a year, is independent of depth. It is the same for all points of the vertical, and in consequence, the same as that observed immediately below the surface: it is the invariable temperature of deep places.

It is obvious that, in the statement of this proposition, we have idealized away the interior heat of the globe, and with greater reason the accessory effects which could modify this result in any given place. Our principle object is to bring to light the general nature of the phenomena.

We have said above that the diverse effects can be considered separately. It should also be noted that all of the numerical evaluations given in this article are presented only as examples of how the calculation may be performed. The meteorological observations needed to reveal the heat capacity and permeability of the materials which make up the globe are too uncertain and limited to permit the calculation of precise results; nonetheless, we present these numbers in order to show how the formulae are applied. However inexact these evaluations may be, they serve to give a more correct idea of the phenomena than would general mathematical expressions bereft of numerical application.

In the portions of the envelope closest to the surface, a thermometer would rise and fall in the course of each day. These diurnal variations become insignificant at a depth of 2 m to 3 m. Below these depths, one observes only annual variations, which themselves disappear at yet greater depths.

If the speed of rotation of the Earth about its axis were to become incomparably greater, and if the same were to occur for the movement of the planet about the Sun, one would no longer find diurnal and annual temperature variations of the sort described above; the points of the surface would attain and conserve the fixed deep-Earth temperature. In general, the depth to which one must go in order for the variations to be significant has a very simple relation with the length of the period with which the effects repeat at the surface. This depth is exactly proportional to the square root of the period. It is for this reason that the diurnal variations penetrate to a depth nineteen times less than that at which one can still detect annual variations.

The question of the periodic movement of solar heat was treated for the first time and solved in a separate writing which I submitted to the Institute of France in October 1809. I reproduced this solution in a piece sent at the end of 1811, which was printed in our Collected Works.

The same theory provides the means of measuring the total quantity of heat which, in the course of a year, determines the alternation of the seasons. Our goal in choosing this example of the application of the formulae is to show that there exists a necessary relation between the law of periodic variations and the total quantity of heat transfer which accompanies this oscillation; once this law is known from observations of one given climate, one can deduce the quantity of heat which is introduced into the Earth and which later returns to the air.

By considering a law similar to that which holds in the interior of the globe, one finds the following results. One eighth of a year before the temperature of the surface rises to its mean value, the Earth begins to accumulate heat; the rays of the Sun penetrate the Earth for six months. Then, the movement of the Earth's heat reverses direction; it exits and expands through the air and outer space: the quantity of heat exchanged in these oscillations over the course of a year is expressed by the calculation. If the terrestrial envelope were formed of a metallic substance, such as wrought iron (a substance which I chose as an example because its thermal coefficients have been measured), the heat which produces the alternation of the seasons would, for the climate of Paris and for each square meter of surface, be equivalent to that required to melt a cylindrical column of ice with a base of one square meter and a height of about 3 m.[8] Though the value of the thermal coefficients specific to the material of which the globe is formed have not been measured, one sees easily that they would give a result much less that that which I have just indicated. The result is proportional to the square root of the product of the heat capacity per unit volume and the thermal conductivity.

Let us now consider the second cause of the terrestrial heat, which resides, according to us, in interplanetary space. The temperature of this space is that which a thermometer would show if the Sun and all planetary bodies which accompany it were to cease to exist, assuming the instrument to be placed anywhere in the region of the heavens presently occupied by the solar system.

We shall now indicate the principle facts which have led us to recognize the existence of this characteristic temperature of interplanetary space, independent of the presence of the Sun, and independent of the primitive heat that the globe has been able to retain. To obtain knowledge of this remarkable phenomenon, one must consider what the temperature of the Earth would be if it received only the heat of the Sun; further, to render the problem more tractable, one can at first neglect the effect of the atmosphere. Then, if there were no agency maintaining a common and constant temperature in interplanetary space, that is to say if the Earth and all bodies forming the solar system were located in a region deprived of all heat, one would observe effects completely contrary to those which which we are familiar. The polar regions would be subject to intense cold,

and the decrease of temperature from equator to pole would be incomparably more rapid and more extreme than is observed.[9]

Under this hypothesis of absolutely cold space, if such a thing is possible to conceive of, all effects of heat, such as we observe at the surface of the globe, would be due solely to the presence of the Sun. The least variation of the distance of the Earth from this star would lead to considerable changes in the temperature, and the eccentricity of the Earth's orbit would give rise to new forms of seasonal variations.

The alternation of day and night would produce effects both sudden and totally different from those we observe. The surface of bodies would be exposed all of a sudden, at the beginning of night, to an infinitely intense cold. The living world, both animal and vegetable, could not survive such a rapid and strong action, which repeats in the opposite sense at sunrise.

The primitive heat conserved in the interior of the terrestrial mass cannot supplant the exterior temperature of space, and would not prevent any of the effects we have just described; for we know with certainty, by theory and observations, that this central heat has long ago become insensible at the surface, notwithstanding that it can be very great at a moderate depth.

From these various remarks we conclude, and principally from the mathematical examination of the question, that there must be a physical cause which is always present, which moderates the temperatures of the surface of the globe, and which gives this planet a fundamental heat independent of the action of the Sun and of the primitive heat retained in the interior of the planet. This fixed temperature which the Earth receives from space differs little from that which one measures at the Earth's poles. Of necessity, the temperature of space is below the temperature characterizing the coldest lands; however, in making this comparison, one must admit only selected observations, and not consider episodes of extremely intense cold caused by accidental effects such as evaporation, violent winds or extraordinary expansion of the air.

Having recognized the existence of this fundamental temperature of space without which the observed pattern of temperature at the Earth's surface would be inexplicable, we note that the origin of this phenomenon is obvious. It is due to the radiation of all the bodies of the universe, whose light and heat can reach us. The stars which we can see with our own eyes, the countless multitude of stars visible by telescope, or the dark bodies which fill the universe, the atmospheres which surround these immense bodies, the tenuous material strewn through various parts of space, act together to form these rays which penetrate all parts of the planetary regions. One cannot conceive of the existence of such an assemblage of luminous or heated bodies, without admitting also that any given point of the space containing it must acquire a definite temperature.[10]

The immense number of bodies compensates for the inequality of their temperatures, and renders the radiation essentially uniform.

This temperature of space is not the same in all parts of the universe; however, it does not vary much over the dimensions of a planetary system, since this size is incomparably smaller than the distance separating the system from the radiant bodies. Thus, the Earth finds the same temperature of the heavens at all parts of its orbit.

The same applies to the other planets of our system; they all participate equally in the communal temperature, which is more or less augmented by the incidence of the rays of the Sun, according to the distance of the planet from this star. As for the problem of assigning the temperature that each planet is expected to attain, the principles which furnish an exact theory are as follows. The intensity and distribution of heat at the surface of these bodies depends on the distance from the Sun, the inclination of the axis of rotation, and the state of the surface. The temperature is very different, even in the mean, from that which an isolated thermometer placed at the location of the planet would measure, for the solidity of the planet, its great size and doubtless also the presence of the atmosphere and the nature of the surface act together to determine the mean temperature.

The original heat conserved in the interior of the mass has long ago stopped having any noticable effect at the surface; the present state of the terrestrial envelope allows us to know with certainty that the primitive heat of the surface has almost entirely dissipated. We regard it as very likely, given the construction of our solar system, that the polar temperatures of each planet, or at least most of them, is little different from that of space. This polar temperature is essentially the same for all bodies, despite the fact that their distances from the Sun differ greatly.

One can determine with reasonable precision the amount of heat which the Earth would acquire if it were substituted for each of the planets; however, the temperature of the planet itself cannot be assigned, because one would need to know the state of its surface and of its atmosphere. This difficulty no longer applies for the bodies situated at the extremities of the solar system, such as the planet discovered by Herschel. The exposure of this planet to sunlight is insignificant. Its surface temperature is therefore little different from that of interplanetary space. We have stated this result in a public discourse delivered recently in the presence of the Academy. One sees that this result applies only to the most distant planets. We do not know any means of assigning the mean temperature of the other planets with any precision.

The movements of the air and the waters, the extent of the oceans, the elevation and form of the surface, the effects of human industry and all the accidental changes of the Earth's surface modify the temperature of each climate. The basic character of phenomena arising from fundamental causes survives, but the thermal effects observed at the surface are different from those which would be seen without the influence of these accessory causes.

The mobility of water and air tends to moderate the effects of heat and cold, and renders the temperature distribution more uniform; however it would be impossible for the action of the atmosphere to supplant the universal cause arising from the communal temperature of interplanetary space; if this cause did not exist, one would observe, despite the action of the atmosphere and the oceans, enormous differences between the polar and equatorial temperature.

It is difficult to know just to what extent the atmosphere affects the mean temperature of the globe, and here the guidance of rigorous mathematical theory ceases. One is indebted to the celebrated explorer M. de Saussure[11] for an experiment which appears to be well suited to clarifying this question. The experiment consists of exposing a vessel covered by one or more sheets of highly transparent glass (placed at some distance from each other) to the rays of the Sun. The interior of the vessel is covered with an thick layer of blackened cork, suited to absorb and retain the heat. The heated air is contained in all parts of the apparatus, either in the interior of the box or in each gap between two plates of glass. Thermometers placed in this vessel and in the spaces between the plates register the degree of heat acquired in these cavities. This instrument was exposed to the Sun at or near noontime, and it has been observed, in various experiments, that the thermometer in the vessel raises to 70°, 80°, 100°, 110° or even higher (octogesimal[12] division). Thermometers placed within the gaps between the sheets of glass indicate a much lower degree of heat acquired, decreasing steadily from the bottom of the box up to the top gap.

The effect of solar heat on air contained within a transparent enclosure has been known for a long time. The apparatus which we have just described is designed for the purpose of maximizing the acquired heat, and above all with the purpose of comparing the solar effect on a very high mountain with that on the plain below. This observation is remarkable by virtue of the accurate and extensive conclusions the inventor of the apparatus has drawn: he has repeated the experiments several times at Paris and at Edinburgh, and found consistent results.

The theory of this instrument is easy to formulate. It suffices to remark that: (1) the heat acquired is concentrated, because it is not dissipated immediately by exchange of air with the surroundings; (2) the heat emanated by the Sun has properties different from those of dark heat. The rays of this star are for the most part transmitted through the glass without attenuation and reach the bottom of the box. They heat the air and the surfaces which contain it: the heat communicated in this way ceases to be luminous, and takes on the properties of dark radiant heat. In this state, the heat cannot freely traverse the layers of glass which cover the vessel; it accumulates more and more in the cavity enclosed by materials which conduct heat poorly, and the temperature rises to the point at which the incident heat is exactly balanced by the dissipated heat. One could verify this explanation, and render the consequences more evident, if one were to vary the conditions of the experiment by employing colored or darkened glass, and by making the cavities containing the thermometers empty of air. When one examines this effect by quantitative calculations, one finds results which conform entirely to those which the observations have yielded[13] It is necessary to consider this range of observations and the results of the calculations very carefully if one is to understand the the influence of the atmosphere and the waters on the thermometric state of the Earth.

In effect, if all the layers of air of which the atmosphere is formed were to retain their density and transparency, but lose only the mobility which they in fact possess, this mass of air would become solid, and being exposed to the rays of the Sun, would produce an effect of the same type as that which we have just described. The heat, arriving in the form of light as far as the solid surface of the Earth, suddenly and almost entirely loses its ability to pass through transparent solids; it will accumulate in the lower layers of the atmosphere, which will therefore acquire elevated temperatures. One will observe at the same time a diminution of the degree of heat acquired as one moves away from the surface of the Earth.[14] The mobility of the air, which is displaced rapidly in all directions and which rises when it is heated, and the irradiation by dark heat in the air diminishes the intensity of the effects which would take place in a transparent and solid atmosphere, but it does not completely eliminate these effects. The reduction of heat in elevated regions of the air does not fail to take place; it is thus that the temperature is augmented by the interposition of the atmosphere, because the heat has less trouble penetrating the air when it is in the form of light, than it has exiting back through the air after it has been converted to dark heat.

We will now consider the heat of the Earth itself, which it possessed at epochs when the planets were formed, and which continues to dissipate at the surface, under the influence of the low temperature of interplanetary space.

The notion of an interior fire, as a perpetual cause of several grand phenomena, has recurred in all the ages of Philosophy. The goal which I have set myself is to know exactly the laws by which a solid sphere, heated by long immersion in a medium, loses its primitive heat once it is transported to a space with constant temperature lower than that of the first medium. This difficult question, not treatable by mathematical techniques formerly known, was resolved by a new method of calculation which is also applicable to a variety of other phenomena.

The form of the terrestrial sphere, the regular disposition of interior layers made manifest by experiments with pendula, their growing density with depth, and various other considerations concur to prove that a very intense heat once penetrated all parts of the globe. This heat dissipates by radiation into the surrounding space, whose temperature is much below the freezing point of water. Now, the mathematical expression of the law of cooling shows that the primitive heat contained in a spherical mass of dimension as big as the Earth diminishes much more rapidly at the surface than at parts situated at great depth. The latter retain almost all of their heat for an immense time; there is no doubt about the truth of the conclusions, because I have calculated this time for metallic substances having much greater thermal conductivity than the materials making up the globe.

However, it is obvious that the theory alone can teach us only about the laws to which the phenomena are subject. It remains to examine if, in the layers of the globe we are able to penetrate, one finds some evidence of this central heat. One must verify, for example, that, below the surface, at distances where diurnal and annual variations cease entirely, temperatures increase with depth along a vertical extended into the interior of the solid earth: Now all the facts which have been gathered and discussed by the most experienced observers have taught us the magnitude of this increase: it has been estimated at 1° for each 30 m to 40 m of depth.

The object of the mathematical question is to discover the definite conclusions which one can deduce from this fact alone, considering it as given by direct observation, and to prove that it determines: (1) the location of the source of heat; (2) the temperature excess remaining at the surface.

It is easy to conclude, as a result of exact analysis, that the increase of temperature with depth cannot be produced by prolonged action of the rays of the Sun. The heat emanating from this star does accumulate in the interior of the globe, but the accumulation has long since ceased; further, if the accumulation were still continuing, one would observe a temperature increase in precisely the opposite sense as that which is observed.

The cause which gives greater temperature to layers located at greater depth is therefore a constant or variable interior source of heat, placed somewhere below the points of the globe which it has been possible to penetrate. This cause raises the temperature of the Earth's surface above the value that it would have under the action of the Sun alone. However, this excess of surface temperature has become almost imperceptible; we can be assured of this because there exists a mathematical relation between the value of temperature increase per meter and the amount by which the surface temperature still exceeds the value it would have if there were no interior heat source. For us, measuring the rate of increase of temperature with depth is the same thing as measuring the temperature excess of the surface.

For a globe made of iron, a rate of increase of a thirtieth of a degree per meter would yield only a quarter of a centessimal degree of excess surface

temperature at the present. This elevation is in direct ratio to the conductivity of the material of which the envelope is formed, all other things being equal. Thus, the surface temperature excess of the actual Earth caused by the interior heat source is very small; it is certainly less than a thirtieth of a centessimal degree. It should be noted that this last conclusion applies regardless of the supposition which one may make about the nature of the internal heat source, whether it be regarded as local or universal, constant or variable.

When one carefully examines all the observations relating to the shape of the Earth, according to the principles of dynamical theory, one cannot doubt that the planet received a very high temperature at its origin; further, the present distribution of heat in the terrestrial envelope is that which would be observed if the globe had been formed in a medium of a very high temperature, whereafter the globe cooled continually.

The question of terrestrial temperatures has always appeared to me to be one of the greatest objects of cosmological study, and I have had this subject principally in view in establishing the mathematical theory of heat. I first determined the time-varying state of of a solid globe which, after having been kept for a long time in a heated medium, has been transported to a cold space. I also considered the time-varying state of a sphere which, having been plunged successively in two or more media of varying temperature, is subjected to a final cooling in a space having constant temperature. After having remarked on the general consequences of the solution to this problem, I examined more specifically the case where the primitive temperature acquired in the heated medium became constant throughout the mass; further, supposing the sphere to be extremely large, I investigated the progressive diminution of temperature in layers sufficiently close to the surface. If one applies the results of this analysis to the terrestrial globe in order to know what would be the successive effects of an initial formation similar to that which we have just considered, one sees that the increase of a thirtieth of a degree per meter, considered as the result of interior heat, was once much greater. One sees further that this temperature gradient now varies extremely slowly. As for the temperature excess of the surface, it varies according to the same law; the secular diminution or the quantity by which it reduces in the course of a century is equal to the present value divided by twice the number of centuries that have flown by since the beginning of the cooling. The age of historical monuments provides us with a lower limit to this number, whence one concludes that from the time of the Greek school of Alexandria up to our time, the surface temperature has not diminished (by this cause) by three hundredths of a degree. Here one again encounters the stable character presented by all great phenomena of the universe. This stability is, by the way, a necessary result independent of the initial state of the mass, because the present temperature excess is extremely small and can do nothing else but continue to diminish for an indefinitely prolonged time.

The effect of the primitive heat which the globe has retained has therefore become essentially imperceptible at the Earth's surface; however it is still manifest at accessible depths, because the temperature of lower layers increases with distance from the surface. This increase, measured in fixed units, would not have the same value at much greater depths: it diminishes with depth; however the same theory shows us that the temperature excess, which is nearly zero at the surface, can be enormous at a distance of several tens of kilometers; it follows that the heat of intermediate depth layers could far surpass the that of incandescent matter.

The passage of centuries will bring great changes in these interior temperatures; at the surface, however, these changes are essentially done, and the continual loss of primitive heat cannot result in any cooling of the climate.

It is important to observe that the accessory causes can cause temperature variations at any given place which are incomparably more significant than those arising from the secular cooling of the globe.

The establishment and progress of human societies, and the action of natural forces, can notably change the state of the ground surface over vast regions, as well as the distribution of waters and the great movements of the air. Such effects have the ability to make the mean degree of heat vary over the course of several centuries, for the analytic expressions contain coefficients which depend on the state of the surface, and which greatly influence the temperature.

Though the effect of the interior heat is no longer perceptible at the surface of the Earth, the total quantity of this heat which dissipates in a given amount of time, such as a year or a century, is measurable, and we have determined it; that which traverses one square meter of surface during a century and expands into celestial space

could melt a column of ice having this square meter as its base, and a height of about 3 m.[15]

This conclusion derives from a fundamental proposition which belongs to all questions regarding the movement of heat, and which applies above all to those of the terrestrial temperature: I speak of the differential equation which expresses for each instant the state of the surface. This equation, whose truth is palpable and easy to demonstrate, establishes a simple relation between the temperature of an element of the surface and the movement of heat in the direction of the normal to the surface. What renders this theoretical result very important, and more suitable than any other to clarify the questions which are the subject of this Article, is that the relation applies independent of the form and the dimensions of the body, and regardless of the nature of the substances – homogeneous or diverse – of which the interior mass is composed. Hence, the consequences which one deduces from this equation are absolute; they hold equally, whatever might have been the material constitution or original state of the globe.

We have published, in the course of the year 1820, a summary of an Article on the secular cooling of the terrestrial globe (*Bulletin des Sciences, Societé philomathique*, 1820, p. 48 ff). One has reported there the principal formulae, and notably those which express the time-varying state of an extremely large solid body uniformly heated up to a given depth. If the initial temperature, instead of being the same up to a very great distance from the surface, results from a successive immersion in several media with different temperatures, the consequences are neither less simple nor less remarkable. When all is said and done, this case and several others which we have considered are included as special cases of the general expressions which have been indicated.

The reading of this extract gives me the occasion to note that formulae (1) and (2) reported there have not been correctly transcribed. I will make up for this omission afterwards. In any case, the error affects neither the other formulae nor the conclusions contained in the extract.

In order to describe the principal thermometric effects which arise from the presence of the oceans, let us suppose for the moment that the water of the Ocean is drained from the basins which contain it, leaving behind immense cavities in the solid Earth. If this state of the Earth's surface, deprived of the atmosphere and the waters, were to persist for a great many centuries, the solar heat would produce alternations of temperature similar to those which we observe on the continents, and subject to the same laws. The diurnal or annual variations cease at certain depths, and a temporally invariable state would form in the interior layers which continually transports equatorial heat toward the polar regions.

At the same time, as the original heat of the globe dissipates through the exterior surface of the basins, one would observe there, as in all other parts of the surface, an increase of temperature with depth along a line normal to the surface of the bottom.

It is necessary to remark here that the increase of temperature due to the original heat depends principally on the normal distance from the surface. If the exterior surface were horizontal, one would find equal temperatures along horizontal lower layers; however if the surface of the solid Earth is convex, these layers of equal temperature would not be at all horizontal, and they would differ from level surfaces. They follow the sinuous form of the surface: it is for this reason that, in the interior of mountains, the central heat can penetrate up to a great height. This is a complex effect, which one can determine by mathematical analysis keeping in mind the form and the absolute elevation of the masses.

If the surface were concave, one would observe an analogous effect in the opposite sense, and this case applies to the hypothetical water-free oceans which we are considering. The layers of equal temperature would be concave, and this state would be found if the Earth were not covered by waters.

Let us suppose now that, after this same state has lasted a great many centuries, one re-establishes the waters at the bottom of the oceans and lakes, and that they remain exposed to the alternation of the seasons. When the temperature of the upper layers of the liquid becomes less than that of the lower parts, though surpassing the freezing point of water by only a few degrees, the density of these upper layers increases; they will descend more and more, and come to occupy the bottom of the basins which they will then cool by their contact: at the same time, the warmer and lighter waters rise to replace the upper waters, whence infinitely varied movements are established in the liquid masses, whose general effect will be to transport heat toward upper regions.

These phenomena are more complex in the interior of the great oceans, because the inhomogeneity of temperature there occasions currents in the opposite sense and thus displaces the waters of far-removed regions.

The continual action of these causes is modified by another property of water, that which limits the growth of density and causes it to reverse when the temperature falls to near the freezing point. The solid bottom of the oceans is therefore subject to a special action which sustains itself forever, and which has perpetually cooled the bottom since time immemorial, by the contact with a liquid having a temperature exceeding by only a few degrees that of melting ice. One finds in consequence that the temperature of the waters decreases with depth; this deep temperature is on the order of $4°$ at the bottom of most lakes in our climate. In general, if one observes the temperature of the ocean at ever greater depths, one approaches this limit which corresponds to the greatest density; however one must, in questions of this type, keep in mind the nature of the waters, and above all the communication established by the currents: this last cause can totally change the results.

The increase of temperature, which we observe in Europe when carrying a thermometer into the interior of the solid globe at great depths, therefore cannot survive in the interior of the oceans, and more generally the order of temperature variations must be the reverse.

As for the portions located immediately below the bottom of the oceans, the law of increase of heat is not that which applies in continental lands. These temperatures are determined by a peculiar cooling action, the vessel being exposed, as we have said, to perpetual contact with a liquid which retains the same temperature at all times. It is to clarify this part of the problem of terrestrial temperatures that I determined, in the *Analytic Theory of Heat* (Chapter IX, p 427 ff), the expression for the time-variable state of a solid primitively heated in some manner, and for which the surface is kept during an indefinite time at a constant temperature. The analysis of this problem allows us to know precisely the law by which the exterior influence causes the temperature of the solid to vary. In general, after having established the fundamental equations of movement of heat, and the method of calculation which serves to integrate them, I turned to the solution of the questions pertinent to the study of terrestrial temperatures, and made known the relations of this study to the systematic behavior of the world.

After having explained separately the principles governing terrestrial temperatures, one must bring together all the effects we have just described into a general point of view, and from there form a correct idea of the operation of the full range of phenomena.

The Earth receives the rays of the Sun, which penetrate its mass and are converted there into dark heat; the Earth also possesses heat of its own which it retains from its origin, and which dissipates continually at the surface; finally this planet receives rays of light and heat from the countless stars among which the solar system is located. These are the three general causes which determine terrestrial temperatures. The third, that is to say the influence of the stars, is equivalent to the presence of an immense region closed in all parts, whose constant temperature is little inferior to that which we observe in polar lands.

One could without doubt suppose that radiant heat has properties as yet unknown, which could take the place in some way of this fundamental temperature which we attribute to space; however, in the present state of physical science, and without recourse to properties other than those which derive from observations, all the known facts can be explained naturally. It suffices to posit that the planetary bodies are in a space whose temperature is constant. We have therefore investigated the question of what this temperature must be in order for the thermometric effects to be similar to what we observe; now, the predicted effects differ entirely from observations if one supposes that space is absolutely cold; however, if one progressively increases the common temperature of the environment which encloses this space, the results come to approach the observations. One can affirm that the present phenomena are those which would be produced if the irradiation by the stars gives each point of planetary space a temperature of about 40° below zero (octogesimal division).

The primitive interior heat which is still not at all completely dissipated produces only a very small effect at the surface of the Earth; the primitive heat is more evidently manifest by the augmentation of temperature in deep layers of the Earth. At the greatest distances from the surface, the temperature can surpass the highest temperatures ever measured to date.

The effect of the solar rays is periodic in the upper layers of the terrestrial envelope; it is fixed in all the deeper places. This fixed temperature of the lower portions is not the same for all of them; it depends principally on the latitude of the place.

The solar heat accumulates in the interior of the globe, whose state becomes time-independent.

That which penetrates in equatorial regins is exactly compensated by the heat which flows out through the polar regions. Thus the Earth returns to celestial space all the heat which it receives from the Sun, and adds to it a part which derives from its own primitive heat.

All the terrestrial effects of the Sun's heat are modified by the interposition of the atmosphere and by the presence of the waters. The grand movements of these fluids renders the temperature distribution more uniform.

The transparency of the waters and that of the air act together to augment the degree of heat acquired, because incident luminous heat penetrates easily to the interior of the mass, but the dark heat exits with more difficulty when following the contrary route.

The alternations of the seasons are accompanies by an immense quantity of solar heat which oscillates in the terrestrial envelope, passing under the surface for six months, and returning from the Earth to the air during the other half of the year. Nothing can shed better light on this question than the experiments which have as their object the precise measurement of the effect produced by the rays of the Sun on the terrestrial surface.

I have summarized, in this Article, all the principle elements of the analysis of the problem of terrestrial temperatures. It is made up of several results of my research, which have been published long ago. When I first endeavored to treat this type of question, there was no mathematical theory of heat, and one could even doubt such a theory to be possible. The Articles and Works which I have set forth contain the exact solution of fundamental questions; they have been submitted and communicated publicly, or printed and analyzed in scientific collections over the past several years.

In the present writing, I have set myself another goal, that of calling attention to one of the greatest objects of Natural Philosophy, and to set forth an overview of the general conclusions. I have hoped that the geometers will not see these researches only as a question of calculation, but that they will consider also the importance of the subject. One cannot at present resolve all the uncertainties in such a vast subject, which embraces, besides the results of a novel and difficult mathematical analysis, exceedingly varied physical concepts. For the future, it remains to take many more precise observations; one will also study the movement of heat in liquids and air. Possibly, additional properties of radiant heat will be discovered, as well as further processes which can modify the temperature distribution of the globe. However, all the principle laws governing the movement of heat are already known; this theory, which rests on invariable foundations, forms a new branch of mathematical Science: it consists at present of the differential equations for the movement of heat in solids and liquids, solutions of these first equations, and theorems relating to the equilibrium properties of radiant heat.

One of the principle features of the analysis which expresses the distribution of heat in solid bodies is the ability to superpose simple solutions in order to build the solution of more complex problems. This property derives from the nature of the differential equations for the movement of heat, and applies also the the problem of the long-term oscillation of bodies; however, the superposition property belongs more particularly to the theory of heat, since the most complex effects can truly be resolved into simple movements. This proposition does not express a law of nature, and I do not mean to imply anything of this sort; it expresses an enduring property, and not a cause. One one would find the same result in dynamical questions wherein one considers resistive forces which cause a rapid cessation of the effect produced.

The applications of the theory of heat have demanded prolonged analytical research, and it was first necessary to formulate the method of calculation, regarding as constant the specific coefficients which enter into the equations; for, this condition establishes itself spontaneously, and endures for an infinite time once the differences in temperature become sufficiently small, as one observes in the problem of terrestrial temperatures. Moreover, in this question (which is the most important application of the theory of heat), the demonstration of the principle results is independent of the homogeneity and the nature of the interior layers of the Earth.

The analytic theory of heat can be extended as required to treat the most varied applications. The list of principles which serve to generalize the theory is as follows:

- Suppose that the coefficients are subject to very small variations, which have been fixed by observation. One can then determine, by the method of successive substitutions, the corrections which go beyond the results of the the first calculation.
- We have demonstrated several general theorems which are not at all dependent on the form of

the body, or on its homogeneity. The general equation relating to area is a proposition of this type. One finds another very remarkable example if one compares the movement of heat in similar bodies, whatever may be the nature of these bodies.

- While the complete solution of these differential equations depends on expressions which are difficult to discover, or on tables which have not yet been created, one can nonetheless determine the limits between which the unknown quantities are necessarily bounded. One arrives thus at definite conclusions regarding the object in question.
- In the research on the temperature distribution of the Earth, the large size of the planet allows one to adopt a simplified form of the equations, and allows for much easier interpretation. Though the nature of the interior masses and their thermal properties are unknown, one can deduce solely from observations made at accessible depths conclusions of the greatest importance regarding the stability of climate, the present excess surface temperature due to the primitive heat, and the secular variation of temperature growth with depth. It is in this fashion that we have been able to demonstrate that this increase, which is on the order of $1°$ per $32\,\mathrm{m}$ in diverse European locations, once had a much larger value. At present its rate of diminution is so slow as to be imperceptible, and it will take more than thirty thousand years before the temperature gradient is reduced to half its present value. This conclusion is not at all uncertain, despite the lack of knowledge of the interior state of the globe, for the interior masses, whatever their state and temperature may be, communicate only an insignificant quantity of heat to the surface over immense stretches of time. For example, I wished to know what would be the effect of an extremely heated mass of the same size of the Earth, placed some leagues below the surface. Here is the result of this inquiry.

If, below a depth of 12 leagues, one were to replace the terrestrial mass down to the center of the globe by a matter whose temperature is five hundred times that of boiling water, the heat communicated by this mass to the neighborhood of the surface would remain imperceptible for a very long time; certainly more than two hundred thousand years would pass before one could observe a single degree of temperature increase at the surface. Heat penetrates solid masses – and especially those of which the terrestrial envelope is formed – so slowly that a separation of only a very few leagues suffices to render it inappreciable during twenty centuries application of the most intense heat.

A careful examination of the conditions to which the planetary system are subject leads to the conclusion that these bodies were made from the mass of the Sun, and it can be said that there is no observed phenomenon which fails to buttress this opinion. We do not know how the interior of the Earth has lost this original heat; one can only affirm that at the surface the excess of heat due to this cause has become essentially undetectable; the thermometric state of the globe no longer varies but with extreme lassitude; and, if one were to imagine that the portion a few leagues below the surface were replaced by either ice or the very substance of the Sun having the same temperature of that star, a great number of centuries would flow by before one observed any appreciable change in the surface temperature. The mathematical theory of heat furnishes several other consequences of this type, whose certainty is independent of all hypotheses regarding the state of the interior of the terrestrial globe.

These theories have an extensive and fertile future ahead of them, and nothing will contribute more to their perfection than a numerous set of precise experiments; for, mathematical analysis (if we may be permitted to reiterate this reflection here)[16] can deduce general phenomena and lend simple form to the expression of the laws of nature; however, the application of these laws to very complex effects demands a long series of exact observations.

NOTES

1   Here, Fourier is evidently referring to the fact that temperature decreases with depth in the ocean whereas it increases with depth in the solid crust. The latter is explained easily by Fourier's diffusion equation, whereas the former requires a quite different explanation. *RTP*

2   Here Fourier is attempting to explain the fact that the atmospheric temperature decreases with height. He seeks to explain this by the effect of density on solar absorption, whereas the correct explanation involves the joint action of convection in lifting air parcels with the cooling resulting from expansion of the parcels. Nonetheless, the rest of the paragraph makes clear that

Fourier understands that the atmosphere is mostly transparent to solar radiation. *RTP*

3   Fourier seems to imply that the ocean has a greenhouse effect similar to that of the atmosphere. This is a puzzling, since Fourier knows that the ocean gets colder with depth rather than warmer. It is true that water is more transparent to visible light than it is to infrared, and therefore would seem to have the properties necessary to produce a greenhouse effect. The main reason that the Ocean has no greenhouse effect is that the sunlight is absorbed mostly in the top 100 m, and that a well mixed state of water is isothermal, rather than having a temperature decrease with height as is the case for a compressible substance like air. The ocean in fact causes an *anti-greenhouse* effect, in that the temperature of the bottom of the ocean is lower than what it would be if the water were removed. *RTP*

4   This paragraph refers to the warming of the surface by downwelling infrared radiation coming from the atmosphere. Fourier's many articles on infrared radiation make reference to observations documenting the presence of this radiation. *RTP*

5   Fourier here refers to the linearity of the equations of heat diffusion. He is evidently unaware that other parts of the physics to which he refers (notably the intensity of infrared radiation, as described by the yet-to-be-discovered Stefan Boltzman law) are not linear. *RTP*

6   i.e. independent of time of day or time of year. *RTP*

7   see p 3–28 of the *Oeuvres*, vol. 2. *RTP*

8   Equivalent to a mean flux of about 60 W/m^2 into the surface for one half of the year, followed by the same amount out of the surface for the other half. This is considerably in excess of most estimates of the surface energy imbalance over land, probably because Fourier used the conductivity of iron in his estimate. *RTP*

9   It is strange that Fourier neglects the effect of thermal inertia and atmosphere-ocean heat transports, which easily account for the moderation of polar and nighttime cooling. Fourier mentions these effects further along, but dismisses them without having any quantitative reason for doing so. *RTP*

10   This argument is qualitatively right, but quantitatively wrong. The actual "temperature of space," which may be identified with the microwave background radiation, is more like 4 degrees Kelvin than 200 degrees Kelvin, as Fourier supposed. *RTP*

11   Horace Bénédict de Saussure, 1740–1799, a scientist and mountaineer who was primarily interested in the factors governing weather and climate on mountains. He is widely regarded as the first mountain meteorologist, and is known also as the grandfather of the celebrated linguist Ferdinand de Saussure.

12   The octogesimal temperature scale, also known as the Reaumur scale, divides the temperature range between the freezing and boiling points of water into 80 equally spaced degrees. A comparison with de Saussure's data suggests that Fourier may have actually converted the values to centigrade here, but erroneously continued to refer to them as octogesimal. *RTP*

13   Fourier refers to the existence such calculations, but I have not located them anywhere in his published works. In his discussion of variations on de Saussure's experiment, Fourier is probably describing his

expectation of what the results of such experiments would be rather than referring to experiments which have actually been carried out and reported. This is underscored by his use of the conditional tense in the original. In any event de Saussure could not have performed experiments with an evacuated box, given the technology available to him. On the other hand, many other investigators did reproduce de Saussure's results, so it is not out of the question that Fourier had actual knowledge of some results from experiments such as he describes. *RTP*

14 This reasoning is partially correct for an atmosphere which does not move, but fails to capture the true reason that atmospheric temperature decreases with height. In fact, buoyancy driven motion greatly enhances the vertical decay of temperature, through the cooling of lifted air parcels as they expand. It is clear that Fourier understood that air cools as it expands (see his remark about episodic bouts of intense cold), but he doesn't seem to have connected this effect with the general decrease of atmospheric temperature with height. He also fails to identify the important role this temperature decrease plays in limiting radiation of infrared to space, via reducing the temperature of the "radiating surface." *RTP*

15 Equivalent to 318 mW/m^2, which is 3–4 times modern estimates of geothermal heat flux. As Fourier implies, the overestimate arises from using the conductivity of iron.

16 Discours Préliminaire of *Théorie Analytique de la Chaleur*

2

Wagging the Dog

Tyndall, J. (1861). On the absorption and radiation of heat by gases and vapours, and on the physical connexion of radiation, absorption, and conduction. *Philosophical Magazine Series 4*, **22**, 169–194, 273–285. 12 pages.

Technical Innovation

John Tyndall labored long and hard, he would have us be assured, to improve the technology for measuring the interactions of gas molecules with infrared radiation, what he called "radiant heat." He measured the absorption of IR by the cool gas, and the emission of IR by the gas when it is heated. Gases in his study that absorb also emit a principal now known as Kirkhoff's law. Tyndall also demonstrated that when a gas is very dilute or at low enough pressure, the absorption increases proportionately with the concentration of the gas, but at a high enough concentration the gas absorbs all of the IR, a phenomenon now called the band saturation effect.

Figure 2.1 shows results from Tyndall's Tables I and II, demonstrating the difference between a saturated and an unsaturated gas. When the gas is unsaturated, the absorption of the radiation by the gas varies linearly with the amount of gas in the cell. At high gas concentration, further increases in the amount of the gas have little effect on the IR absorption.

The basic idea behind the measurement of infrared energy flux or intensity is still used today, for example, in electronic thermometers that peer into a baby's ear (much recommended over the other method). The incoming radiation warms up one pole of a device called a thermopile. A thermopile has two poles, and it produces a measurable voltage that is proportional to the temperature difference between the poles. A thermopile is a collection of thermocouples, consisting of junctions of dissimilar metals at each pole, wired together to produce a stronger, more easily measured voltage.

Tyndall measured the signal from the thermopile using a galvanometer, which measures electrical current, a device now usually called an ammeter. Presumably, the voltage from the thermopile was driven through a resistor to produce an electric current. The galvanometer consists of a rotating part on springs, and a fixed part. A set of permanent magnets are mounted on the fixed part, surrounding the rotating part. The electricity flows through a coil on the rotating part, inducing a magnetic field and causing the rotating part to twist slightly, straining against the springs. A needle mounted on the twisting part allowed a measurement to be recorded, which Tyndall expressed in units of degrees of deflection.

Many of the technical improvements that Tyndall developed have to do with improving the sensitivity of the galvanometer. In addition to improving the intrinsic sensitivity of the device by replacing the dye in the silk used to insulate the coil (he claimed it had a magnetic field), Tyndall figured out that if he balanced the two sides of the thermopile against each other, he could obtain greater sensitivity of measurement. Let us suppose that a gas absorbs 1% of an incoming IR beam, and you use the thermopile to compare the intensity of the IR beam to the intensity of IR from the cool surroundings, just the lab walls shining in. To measure the impact of the absorbing gas,

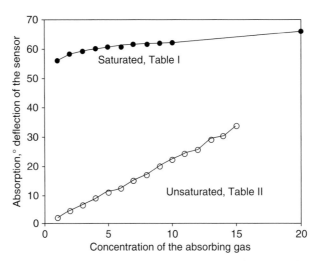

Fig. 2.1 **Tyndall's measurements demonstrated the band saturation effect, which causes the absorption of light to be insensitive to the amount of gas that the light traverses, if the gas concentration is high enough. If virtually all the light is absorbed, then adding more gas has only a small effect. The difference between the data in Tables I and II is that the concentrations of gases in Table I are higher, and therefore saturated. Tyndall also recognized that the atmosphere could be either saturated or unsaturated.**

you would have to be able to detect a small signal change of only 1%. Tyndall's idea was to balance the two sides of the thermopile against each other, by setting up an IR source to the reference side that is the same intensity as the beam on the sample side, before the gas has been introduced. The 1% change in the absolute intensity of the sample beam produces a much larger relative signal against the balanced reference beam of the thermopile. Tyndall needed this increased sensitivity to detect the IR absorption of dilute mixtures of greenhouse gases, such as in air.

Conceptual Breakthroughs

Tyndall's laboratory labors were well informed, driven even, by questions of the Earth's climate and the greenhouse effect. The most jaw-dropping implication of his study was that most (or, as it subsequently turned out, essentially all) of the greenhouse activity of the atmosphere is due to a few trace gases such as water vapor and carbon dioxide. Tyndall realized that this discovery opened the door to an easy way to change the climate of the Earth through time. Instead of waiting for the entire size or mass of the atmosphere to change, all that needs to change is the concentration of a few trace gases. When it comes to the trace greenhouse gases, a little goes a long way.

Scientists are still working out the factors that control the CO_2 concentration of the atmosphere. Tyndall nods to CO_2 as a potential agent of climate change, but points out in particular the variability of the humidity of the air, and speculates about the role that water vapor as a greenhouse gas might play in climate. As it turns out, the Earth's temperature is affected by greenhouse forcing from humidity, but the averaging of the temperatures over wide areas and around the year tends to eliminate much simple correlation between temperature and day-to-day humidity variations. Tyndall was correct in his conclusion that water vapor is the most powerful greenhouse gas in the atmosphere, but as it turns out water vapor is not considered to be a primary climate forcing, because the average water vapor concentration of the atmosphere is now thought to be closely controlled by the hydrologic cycle, in that if the air gets too humid, it rains. For this reason running a garden sprinkler does not lead to global warming in the way that running a gasoline-powered leaf blower does (by emitting CO_2). However, water vapor acts as a positive feedback in the climate system, amplifying a temperature change driven by any other factor such as rising CO_2 concentration.

Tyndall took a few steps in the direction of working out why some gases interact with IR light and some do not. The IR-transparent ones are all simple single-element gases like O_2, N_2, and H_2. The molecular formulas of the gases were not available to Tyndall, so he assumed these to be simple atoms, as opposed to molecules comprised of multiple copies of the same element. The IR behavior of these gases was very different from that of the compound gases like H_2O, CO_2, and ethylene (C_2H_4), which he found to be IR active. Tyndall interpreted his results as evidence for chemical bonds, as in, for example, ammonia as opposed to a mixture of nitrogen and hydrogen gases. The flaw in his conclusion is that diatomic molecules such as N_2 and O_2 have chemical bonds also, it turns out, but are not greenhouse gases because their electronic symmetry when they vibrate does not present an electric dipole to the electromagnetic field, and thus does not produce light. Tyndall phrased it that the compound gases "present a broader side" to the mysterious, gelatinous, gooey substance known as the ether, the medium within which light was thought to propagate, while "the simple atoms do not, – that in consequence of these differences the ether must swell into billows when the former are moved, while it merely trembles into ripples when the latter are agitated. ..." It is difficult to read this sentence today without a bit of mirthful joy, but if we substitute "electromagnetic field" for "ether," and "electromagnetic radiation" for the billows and ripples, Tyndall's intuition was clearly in the right direction.

On the Absorption and Radiation of Heat by Gases and Vapours, and on the Physical Connexion of Radiation, Absorption, and Conduction

JOHN TYNDALL ESQ., F.R.S. & C.*

§ 1. The researches on glaciers which I have had the honour of submitting from time to time to the notice of the Royal Society, directed my attention in a special manner to the observations and speculations of De Saussure, Fourier, M. Pouillet, and Mr. Hopkins, on the transmission of solar and terrestrial heat through the earth's atmosphere. This gave practical effect to a desire which I had previously entertained to make the mutual action of radiant heat and gases of all kinds the subject of an experimental inquiry.

Our acquaintance with this department of Physics is exceedingly limited. So far as my knowledge extends, the literature of the subject may be stated in a few words.

From experiments with his admirable thermo-electric apparatus, Melloni inferred that for a distance of 18 or 20 feet the absorption of radiant heat by atmospheric air is perfectly insensible.[1]

With a delicate apparatus of the same kind, Dr. Franz of Berlin found that the air contained in a tube 3 feet long absorbed 3·54 per cent. of the heat sent through it from an Argand lamp; that is to say, calling the number of rays which passed through the exhausted tube 100, the number which passed when the tube was filled with air was only 96·46.[2]

In the sequel I shall refer to circumstances which induce me, to conclude that the result obtained by Dr. Franz is due to an inadvertence in his mode of observation. These are the only experiments of this nature with which I am acquainted, and they leave the field of inquiry now before us perfectly unbroken ground.

§ 2. At an early stage of the investigation, I experienced the need of a first-class galvanometer. My instrument was constructed by that excellent workman, Sauerwald of Berlin. The needles are suspended independently of the shade; the latter is constructed so as to enclose the smallest possible amount of air, the disturbance of aërial currents being thereby practically avoided. The plane glass plate, which forms the cover of the instrument, is close to the needle; so that the position of the latter can be read off with ease and accuracy either by the naked eye or by a magnifying lens.

The wire of the coil belonging to this instrument was drawn from copper obtained from a galvano-plastic manufactory in the Prussian Capital; but it was not free from the magnetic metals.

In consequence of its impurity in this respect, when the needles were perfectly astatic they deviated as much as 30° right and left of the neutral line. To neutralize this, a "compensator" was made use of, by which the needle was gently drawn to zero in opposition to the magnetism of the coil.

But the instrument suffered much in point of delicacy from this arrangement, and accurate quantitative determinations with it were unattainable. I therefore sought to replace the Berlin coil by a less magnetic one. Mr. Becker first supplied me with a coil which reduced the lateral deflection from 30° to 3°.

But even this small residue was a source of great annoyance to me; and for a time I almost despaired of obtaining pure copper wire. I knew that Professor Magnus had succeeded in obtaining it for his galvanometer, but the labour of doing so was immense.[3] Previous to undertaking a similar task, the thought occurred to me, that for my purpose a magnet furnished an immediate and perfect test as to the quality of the wire. Pure copper is *diamagnetic*; hence its repulsion or attraction by the magnet would at once declare its fitness or unfitness for the purpose which I had in view.

Fragments of the wire first furnished to me by M. Sauerwald were strongly attracted by the magnet. The wire furnished by Mr. Becker, when covered with its green silk, was also attracted, though in a much feebler degree.

* Tyndall, J. (1861) On the absorption and radiation of heat by gases and vapours, and on the physical connexion of radiation, absorption, and conduction. *Philosophical Magazine* Series 4, Vol. 22: 169–94, 273–85.

I then removed the green silk covering from the latter and tested the naked wire. *It was repelled.* The whole annoyance was thus fastened on the green silk; some iron compound had been used in the dyeing of it; and to this the deviation of my needle from zero was manifestly due.

I had the green coating removed and the wire overspun with white silk, clean hands being used in the process. A perfect galvanometer is the result. The needle, when released from the action of a current, returns accurately to zero, and is perfectly free from all magnetic action on the part of the coil. In fact while we have been devising agate plates and other elaborate methods to get rid of the great nuisance of a magnetic coil,[4] the means of doing so are at hand. Nothing is more easy to be found than diamagnetic copper wire. Out of eleven specimens, four of which were furnished by Mr. Becker, and seven taken at random from our laboratory, nine were found diamagnetic and only two magnetic.

Perhaps the only defect of those fine instruments with which Du Bois Raymond conducts his admirable researches in animal electricity is that above alluded to. The needle never comes to zero, but is drawn to it by a minute magnet. This defect may be completely removed. By the substitution of clean white silk for green, however large the coil may be, the compensator may be dispensed with, and a great augmentation of delicacy secured. The instrument will be rendered suitable for quantitative measurements; effects which are now beyond the reach of experiment will be rendered manifest; while the important results hitherto established will be obtained with a fraction of the length of wire now in use.[5]

§ 3. Our present knowledge of the deportment of liquids and solids, would lead to the inference that, if gases and vapours exercised any appreciable absorptive power on radiant heat, the absorption would make itself most manifest on heat emanating from an obscure source. But an experimental difficulty occurs at the outset in dealing with such heat. How must we close the receiver containing the gases through which the calorific rays are to be sent? Melloni found that a glass plate one-tenth of an inch in thickness intercepted all the rays emanating from a source of the temperature of boiling water, and fully 94 per cent. of the rays from a source of 400° Centigrade. Hence a tube closed with glass plates would be scarcely more suitable for the purpose now under consideration, than if its ends were stopped by plates of metal.

Rock-salt immediately suggests itself as the proper substance; but to obtain plates of suitable size and transparency was exceedingly difficult. Indeed, had I been less efficiently seconded, the obstacles thus arising might have been insuperable. To the Trustees of the British Museum I am indebted for the material of one good plate of salt; to Mr. Harlin for another; while Mr. Lettsom, at the instance of Mr. Darker,[6] brought me a piece of salt from Germany from which two fair plates were taken. To Lady Murchison, Sir Emerson Tennant, Sir Philip Egerton, and Mr. Pattison my best thanks are also due for their friendly assistance.

The first experiments were made with a tube of tin polished inside, 4 feet long and 2·4 inches in diameter, the ends of which were furnished with brass appendages to receive the plates of rock-salt. Each plate was pressed firmly against a flange by means of a bayonet joint, being separated from the flange by a suitable washer. Various descriptions of leather washers were tried for this purpose and rejected. The substance finally chosen was vulcanized india-rubber very lightly smeared with a mixture of bees-wax and spermaceti. A T-piece was attached to the tube, communicating on one side with a good air-pump, and on the other with the external air, or with a vessel containing the proper gas.

The tube being mounted horizontally, a Leslie's cube containing hot water was placed close to one of its ends, while an excellent thermo-electric pile, connected with its galvanometer, was presented to the other. The tube being exhausted, the calorific rays sent through it fell upon the pile, a permanent deflection of 30° being the consequence. The temperature of the water was in the first instance purposely so arranged as to produce this deflection.

Dry air was now admitted into the tube, while the needle of the galvanometer was observed with all possible care. Even by the aid of a magnifying lens I could not detect the slightest change of position. Oxygen, hydrogen, and nitrogen, subjected to the same test, gave the same negative result. The temperature of the water was subsequently lowered so as to produce a deflection of 20° and 10° in succession, and then heightened till the deflection amounted to 40°, 50°, 60° and 70°; but in no case did the admission of air, or any of the above gases into the exhausted tube, produce any sensible change in the position of the needle.

It is a well-known peculiarity of the galvanometer, that its higher and lower degrees represent different amounts of calorific action. In my instrument, for example, the quantity of heat necessary to move the needle from 60° to 61° is about twenty times that required to move it from 11° to 12°. Now in the case of the small deflections above referred to, the needle was, it is true, in a sensitive position; but then the total amount of heat passing through the tube was so inconsiderable that a small per-centage of it, even if absorbed, might well escape detection. In the case of the large deflections, on the other hand, though the total amount of heat was large, and though the quantity absorbed might be proportionate, the needle was in such a position as to require a very considerable abstraction of heat to produce any sensible change in its position. Hence arose the thought of operating, if possible, with large quantities of heat, while the needle intended to reveal its absorption should continue to occupy its position of maximum delicacy.

The first attempt at solving this problem was as follows:—My galvanometer is a differential one—the coil being composed of two wires wound side by side, so that a current could be sent through either of them independent of the other. The thermo-electric pile was placed at one end of the tin tube, and the ends of one of the galvanometer wires connected with it. A copper ball heated to low redness being placed at the other end of the tube, the needle of the galvanometer was propelled to its stops near 90°. The ends of the second wire were now so attached to a second pile that when the latter was caused to approach the copper ball, the current thus excited passed through the coil in a direction opposed to the first one: Gradually, as the second pile was brought nearer to the source of heat, the needle descended from the stops, and when the two currents were nearly equal the position of the needle was close to zero.

Here then we had a powerful flux of heat through the tube; and if a column of gas four feet long exercised any sensible absorption, the needle was in the position best calculated to reveal it. In the first experiment made in this way, the neutralization of one current by the other occurred when the tube was filled with air; and after the exhaustion of the tube had commenced, the needle started suddenly off in a direction which indicated that a *less* amount of heat passed through the partially exhausted tube, than through the tube filled with air. The needle, however, soon stopped, turned, descended quickly to zero, and passed on to the other side, where its deflection became permanent. The air made use of in this experiment came

direct from the laboratory, and the first impulsion of the needle was probably due to the aqueous vapour precipitated as a cloud by the sudden exhaustion of the tube. When, previous to its admission, the air was passed over chloride of calcium, or pumice-stone moistened with sulphuric acid, no such effect was observed. The needle moved steadily in one direction until its maximum deflection was attained, and this deflection showed that in all cases radiant heat was absorbed by the air within the tube.

These experiments were commenced in the spring of 1859, and continued without intermission for seven weeks. The course of the inquiry during this whole period was an incessant struggle with experimental difficulties. Approximate results were easily obtainable; but I aimed at exact measurements, which could not be made with a varying source of heat like the copper ball. I resorted to copper cubes containing fusible metal, or oil, raised to a high temperature; but was not satisfied with their action. I finally had a lamp constructed which poured a sheet of gas-flame along a plate of copper; and to keep the flame constant, a gas regulator specially constructed for me by Mr. Hulet was made use of. It was also arranged that the radiating plate should form one of the walls of a chamber which could be connected with the air-pump and exhausted, so that the heat emitted by the copper plate might cross a vacuum before entering the experimental tube. With this apparatus I determined approximately the absorption of nine gases and twenty vapours during the summer of 1859. The results would furnish materials for a long memoir; but increased experience and improved methods have enabled me to substitute for them others of greater value; I shall therefore pass over the work of these seven weeks without further allusion to it.

On the 9th of September of the present year (1860) I resumed the inquiry. For three weeks I worked with the plate of copper as my source of heat, but finally rejected it on the score of insufficient constancy. I again resorted to the cube of hot oil, and continued to work with it up to Monday the 29th of October. During the seven weeks just referred to, I experimented from eight to ten hours daily; but these experiments, though more accurate, must unhappily share the fate of the former ones. In fact the period was one of discipline—a continued struggle against the difficulties of the subject and the defects of the locality in which the inquiry was conducted.

My reason for making use of the high sources of heat above referred to was, that the absorptive power of some of the gases which I had examined was so small that, to make it clearly evident, a high temperature was essential. For other gases, and for *all* the vapours that had come under my notice, a source of lower temperature would have been not only sufficient, but far preferable. I was finally induced to resort to boiling water, which, though it gave greatly diminished effects, was capable of being preserved at so constant a temperature that deflections which, with the other sources, would be masked by the errors of observation, became with it true quantitative measures of absorption.

§ 4. The entire apparatus made use of in the experiments on absorption is figured on Plate III. S S′ is the *experimental tube*, composed of brass, polished within, and connected, as shown in the figure, with the air-pump, A A. At S and S′ are the plates of rock-salt which close the tube air-tight. The length from S to S′ is 4 feet. C is a cube containing boiling water, in which is immersed the thermometer *t*. The cube is of cast copper, and on one of its faces a projecting ring was cast to which a brass tube of the same diameter as S S′, and capable of being connected air-tight with the latter, was carefully soldered. The face of the cube within the ring is the radiating plate, which is coated with lampblack. Thus between the cube C and the first plate of rock-salt there is *a front chamber* F, connected with the air-pump by the flexible tube D D, and capable of being exhausted independently of S S′. To prevent the heat of conduction from reaching the plate of rock-salt S, the tube F is caused to pass through a vessel V, being soldered to the latter where it enters it and issues from it. This vessel is supplied with a continuous flow of cold water through the influx tube *i i*, which dips to the bottom of the vessel; the water escapes through the efflux tube *e e*, and the continued circulation of the cold liquid completely intercepts the heat that would otherwise reach the plate S.

The cube C is heated by the gas-lamp L. P is the thermoelectric pile placed on its stand at the end of the experimental tube, and furnished with two conical reflectors, as shown in the figure. C′ is the *compensating cube*, used to neutralize by its radiation[7] the effect of the rays passing through S S′. The regulation of this neutralization was an operation of some delicacy; to effect it the double screen H was connected with a winch and screw arrangement, by which it could be advanced or withdrawn through extremely minute spaces. For this most useful adjunct I am indebted to the kindness of my friend

Mr. Gassiot. N N is the galvanometer, with perfectly astatic needles and perfectly non-magnetic coil; it is connected with the pile P by the wires *w w*; Y Y is a system of six chloride-of-calcium tubes, each 32 inches long; R is a U-tube containing fragments of pumice-stone, moistened with strong caustic potash; and Z is a second similar tube, containing fragments of pumice-stone wetted with strong sulphuric acid. When *drying* only was aimed at, the potash tube was suppressed. When, on the contrary, as in the case of atmospheric air, both moisture and carbonic acid were to be removed, the potash tube was included. G G is a holder from which the gas to be experimented with was sent through the drying-tubes, and thence through the pipe *pp* into the experimental tube S S′. The appendage at M and the arrangement at O O may for the present be disregarded; I shall refer to them particularly by and by.

The mode of proceeding was as follows: — The tube SS′ and the chamber F being exhausted as perfectly as possible, the connexion between them was intercepted by shutting off the cocks *m, m′*. The rays from the interior blackened surface of the cube C passed first across the vacuum F, then through the plate of rock-salt S, traversed the experimental tube, crossed the second plate S′, and being concentrated by the anterior conical reflector, impinged upon the adjacent face of the pile P. Meanwhile the rays from the hot cube C′ fell upon the opposite face of the pile, and the position of the galvanometer needle declared at once which source was predominant. A movement of the screen H back or forward with the hand sufficed to establish an approximate equality; but to make the radiations perfectly equal, and thus bring the needle exactly to 0°, the fine motion of the screw above referred to was necessary. The needle being at 0°, the gas to be examined was admitted into the tube; passing, in the first place, through the drying apparatus. Any required quantity of the gas may be admitted; and here experiments on gases and vapours enjoy an advantage over those with liquids and solids, namely, the capability of changing the density at pleasure. When the required quantity of gas had been admitted, the galvanometer was observed, and from the deflection of its needle the absorption was accurately determined.

Up to about its 36th degree, the degrees of my galvanometer are all equal in value; that is to say, it requires the same amount of heat to move the needle from 1° to 2° as to move it from 35° to 36°. Beyond this limit the degrees are equivalent to larger amounts of heat. The instrument was accurately calibrated by the method recommended by Melloni (*Thermochrose*, p. 59); so that the precise value of its larger deflections are at once obtained by reference to a table. Up to the 36th degree, therefore, the simple deflections may be regarded as the expression of the absorption ; but beyond this the absorption equivalent to any deflection is obtained from the table of calibration.

§ 5. The air of the laboratory, freed from its moisture and carbonic acid, and permitted to enter until the tube was filled, produced a deflection of about .... 1°.

Oxygen obtained from chlorate of potash and peroxide of manganese produced a deflection of about .... 1°.

One specimen of nitrogen, obtained from the decomposition of nitrate of potash, produced a deflection of about .... 1°.

Hydrogen from zinc and sulphuric acid produced a deflection of about .... 1°.

Hydrogen obtained from the electrolysis of water produced a deflection of about .... 1°.

Oxygen obtained from the electrolysis of water, and sent through a series of eight bulbs containing a strong solution of iodide of potassium, produced a deflection of about .... 1°.

In the last experiment the electrolytic oxygen was freed from its ozone. The iodide of potassium was afterwards suppressed, and the oxygen, plus its ozone, admitted into the tube; the deflection produced was .... 4°.

Hence the small quantity of ozone which accompanied the oxygen in this case trebled the absorption of the oxygen itself.[8]

I have repeated this experiment many times, employing different sources of heat. With sources of high temperature the difference between the ozone and the ordinary oxygen comes out very strikingly. By careful decomposition a much larger amount of ozone might be obtained, and a corresponding large effect on radiant heat produced.

In obtaining the electrolytic oxygen, I made use of two different vessels. To diminish the resistance of the acidulated water to the passage of the current, I placed in one vessel a pair of very large platinum plates, between which the current from a battery of ten of Grove's cells was transmitted. The oxygen bubbles liberated on so large a surface were extremely minute, and the gas thus generated, on being sent through iodide of potassium, scarcely coloured the liquid; the characteristic odour of ozone was also almost entirely absent. In the second vessel smaller plates were

used. The bubbles of oxygen were much larger, and did not come into such intimate contact with either the platinum or the water. The oxygen thus obtained showed the characteristic reactions of ozone; and with it the above result was obtained.

The total amount of heat transmitted through the tube in these experiments produced a deflection of .... 71°·5.

Taking as unit of heat the quantity necessary to cause the needle to move from 0° to 1°, the number of units expressed by the above deflection is .... 308.

Hence the absorption by the above gases amounted to about 0·33 per cent.

I am unable at the present moment to range with certainty oxygen, hydrogen, nitrogen, and atmospheric air in the order of their absorptive powers, though I have made several hundred experiments with the view of doing so. Their proper action is so small that the slightest foreign impurity gives one a predominance over the other. In preparing the gases, I have resorted to the methods which I found recommended in chemical treatises, but as yet only to discover the defects incidental to these methods. Augmented experience and the assistance of my friends will, I trust, enable me to solve this point by and by. An examination of the whole of the experiments induces me to regard hydrogen as the gas which exercises the lowest absorptive power.

We have here the cases of minimum gaseous absorption. It will be interesting to place in juxtaposition with the above results some of those obtained with olefiant gas—the most highly absorbent permanent gas that I have hitherto examined. I select for this purpose an experiment made on the 21st of November.

The needle being steady at zero in consequence of the equality of the actions on the opposite faces of the pile, the admission of olefiant gas gave a permanent deflection of .... 70°·3.

The gas being completely removed, and the equilibrium reestablished, a plate of polished metal was interposed between one of the faces of the pile and the source of heat adjacent. The total amount of heat passing through the exhausted tube was thus found to produce a deflection of ......75°.

Now a deflection of 70°·3 is equivalent to 290 units, and a deflection of 75° is equivalent to 360 units; hence more than seven-ninths of the total heat was cut off by the olefiant gas, or about 81 per cent.

The extraordinary energy with which the needle was deflected when the olefiant gas was admitted into the tube, was such as might occur had the plates of rock-salt become suddenly covered with an opake layer. To test whether any such action occurred, I polished a plate carefully, and projected against it for a considerable time a stream of the gas; there was no dimness produced. The plates of rock-salt, moreover, which were removed daily from the tube, usually appeared as bright when taken out as when they were put in.

The gas in these experiments issued from its holder, and had there been in contact with cold water. To test whether it had chilled the plates of rock-salt, and thus produced the effect, I filled a similar holder with atmospheric air and allowed it to attain the temperature of the water; but its action was not thereby sensibly augmented.

In order to subject the gas to ocular examination, I had a glass tube constructed and connected with the air-pump. On permitting olefiant gas to enter it, not the slightest dimness or opacity was observed. To remove the last trace of doubt as to the possible action of the gas on the plates of rock-salt, the tin tube referred to at the commencement was perforated at its centre and a cock inserted into it; the source of heat was at one end of the tube, and the thermo-electric pile at some distance from the other. The plates of salt were entirely abandoned, the tube being open at its ends and consequently full of air. On allowing the olefiant gas to stream for a second or two into the tube through the central cock, the needle flew off and struck against its stops. It was held steadily for a considerable time between 80° and 90°.

A slow current of air sent through the tube gradually removed the gas, and the needle returned accurately to zero.

The gas within the holder being under a pressure of about 12 inches of water, the cock attached to the cube was turned quickly on and off; the quantity of gas which entered the tube in this brief interval was sufficient to cause the needle to be driven to the stops, and steadily held between 60° and 70°.

The gas being again removed, the cock was turned once half round as quickly as possible. The needle was driven in the first instance through an arc of 60°, and was held permanently at 50°.

The quantity of gas which produced this last effect, on being admitted into a graduated tube, was found not to exceed one-sixth of a cubic inch in volume.

The tube was now taken away, and both sources of heat allowed to act from some distance on the thermo-electric pile. When the needle was at zero, olefiant gas was allowed to issue from a common argand burner into the air between one of the sources of heat and the pile. The gas was invisible, nothing was seen in the air, but the needle immediately declared its presence, being driven through an arc of 41°. In the four experiments last described, the source of heat was a cube of oil heated to 250° Centigrade, the compensation cube being filled with boiling water.[9]

Those who like myself have been taught to regard transparent gases as almost perfectly diathermanous, will probably share the astonishment with which I witnessed the foregoing effects. I was indeed slow to believe it possible that a body so constituted, and so transparent to light as olefiant gas, could be so densely opake to any kind of calorific rays; and to secure myself against error, I made several hundred experiments with this single substance. By citing them at greater length, however, I do not think I could add to the conclusiveness of the proofs just furnished, that the case is one of true calorific absorption.[10]

§ 6. Having thus established in a general way the absorptive power of olefiant gas, the question arises, "What is the relation which subsists between the density of the gas and the quantity of heat extinguished ?"

I sought at first to answer this question in the following way:— An ordinary mercurial gauge was attached to the air-pump; the experimental tube being exhausted, and the needle of the galvanometer at zero, olefiant gas was admitted until it depressed the mercurial column 1 inch, the consequent deflection being noted; the gas was then admitted until a depression of 2 inches was observed, and thus the absorption effected by gas of 1, 2, 3, and more inches tension was determined. In the following Table the first column contains the tensions in inches, the second the deflections, and the third the absorption equivalent to each deflection.

Table I. Olefiant Gas.

Tensions in inches	Deflections	Absorption
1	56°	90
2	58·2	123
3	59·3	142
4	60·0	157
5	60·5	168
6	61·0	177
7	61·4	182
8	61·7	186
9	62·0	190
10	62·2	192
20	66·0	227

No definite relation between the density of the gas and its absorption is here exhibited. We see that an augmentation of the density *seven times* about *doubles* the amount of the absorption ; while gas of 20 inches tension effects only $2\frac{1}{2}$ times the absorption of gas possessing 1 inch of tension.

But here the following reflections suggest themselves:—It is evident that olefiant gas of 1 inch tension, producing so large a deflection as 56°, must extinguish a large proportion of the rays which are capable of being absorbed by the gas, and hence the succeeding measures having a less and less amount of heat to act upon must produce a continually smaller effect. But supposing the quantity of gas first introduced to be so inconsiderable that the number of rays extinguished by it is a vanishing quantity compared with the total number capable of absorption, we might reasonably expect that in this case a double quantity of gas would produce a double effect, a treble quantity a treble effect, or in general terms, that the absorption would, for a time, be proportional to the density

To test this idea, a portion of the apparatus, which was purposely omitted in the description already given, was made use of. O O, Plate III., is a graduated glass tube, the end of which dips into the basin of water B. The tube can be stopped above by means of the stopcock *r*; *d d* is a tube containing fragments of chloride of calcium. The tube O O being first filled with water to the cock *r*, had this water displaced by olefiant gas; and afterwards the tube S S', and the entire space between the cock *r* and the experimental tube, was exhausted. The cock *n* being now closed and *r'* left open, the cock *r* at the top of the tube O O was carefully turned on and the gas permitted to enter the tube S S' with extreme slowness. The water rose in O O, each of whose smallest divisions represents a volume of $\frac{1}{50}$ th of a cubic inch. Successive measures of this capacity were admitted into the tube and the absorption in each case determined.

In the following Table the first column contains the quantity of gas admitted into the tube; the second contains the corresponding deflection, which, within the limits of the Table, expresses the absorption; the third column contains the absorption, calculated on the supposition that it is proportional to the density.

This Table shows the correctness of the foregoing surmise, and proves that for small quantities of gas the absorption is exactly proportional to the density.

Table II. Olefiant Gas. Unit-measure $\frac{1}{50}$ th of a cubic inch.

Measures of gas	Absorption	
	Observed	Calculated
1	2·2	2·2
2	4·5	4·4
3	6·6	6·6
4	8·8	8·8
5	11·0	11·0
6	12·0	13·2
7	14·8	15·4
8	16·8	17·6
9	19·8	19·8
10	22·0	22·0
11	24·0	24·2
12	25·4	26·4
13	29·0	28·6
14	30·2	29·8
15	33·5	33·0

Let us now estimate the tensions of the quantities of gas with which we have here operated. The length of the experimental tube is 48 inches, and its diameter 2·4 inches; its volume is therefore 218 cubic inches. Adding to this the contents of the cocks and other conduits which led to the tube, we may assume that each fiftieth of a cubic inch of the gas had to diffuse itself through a space of 220 cubic inches. The tension, therefore, of a single measure of the gas thus diffused would be $\frac{1}{11,000}$ th of an atmosphere,—a tension capable of depressing the mercurial column connected with the pump $\frac{1}{367}$ th of an inch, or about $\frac{1}{15}$ th of a millimetre!

But the absorptive energy of olefiant gas, extraordinary as it is shown to be by the above experiments, is far exceeded by that of some of the vapours of volatile liquids. A glass flask was provided with a brass cap furnished with an interior thread, by means of which a stopcock could be screwed air-tight on to the flask. Sulphuric ether being placed in the latter, the space above the liquid was completely freed of air by means of a second air-pump. The flask, with its closed stopcock, was now attached to the experimental tube; the latter was exhausted and the needle brought to zero. The cock was then turned on so that the ether-vapour slowly entered the experimental tube. An assistant observed the gauge of the air-pump, and when it had sunk an inch, the stopcock was promptly closed. The galvanometric deflection consequent on the partial cutting off of the calorific rays was then noted; a second quantity of the vapour, sufficient to depress the gauge another inch, was then admitted, and in this way the absorptions of five successive measures, each possessing within the tube 1 inch of tension, were determined.

In the following Table the first column contains the tensions in inches, the second the deflection due to each, and the third the amount of heat absorbed, expressed in the units already referred to. For the purpose of comparison I have placed the corresponding absorption of olefiant gas in the fourth column.

Table III. Sulphuric Ether.

Tensions in inches	Deflections	Absorption	Corresponding absorption by olefiant gas
1	64·8	214	90
2	70·0	282	123
3	72·0	315	142
4	73·0	330	154
5	73·0	330	163

For these tensions the absorption of radiant heat by the vapour of sulphuric ether is more than twice the absorption of olefiant gas. We also observe that in the case of the former the successive absorptions approximate more quickly to a ratio of equality. In fact the absorption produced by 4 inches of the vapour was sensibly the same as that produced by 5.

But reflections similar to those which we have already applied to olefiant gas are also applicable to ether. Supposing we make our unit-measure small enough, the number of rays first destroyed will vanish in comparison with the total number, and for a time the fact will probably manifest itself that the absorption is directly proportional to the density. To examine whether this is the case, the other portion of the apparatus, omitted in the general description, was made use of. K is a small flask with a brass cap, which is closely screwed on to the stopcock c'. Between the cocks c' and c, which latter is connected with the experimental tube, is the chamber M, the capacity of which was accurately determined. The flask k was partially filled with ether, and the air above the liquid removed. The stopcock c' being shut off and c turned on, the tube S S' and the chamber M are exhausted. The cock c is now shut off, and c' being turned on, the chamber M becomes filled with pure ether vapour. By turning c' off and c on, this quantity of vapour is allowed to diffuse itself through the experimental tube, and its absorption determined; successive measures are

thus sent into the tube, and the effect produced by each is noted. Measures of various capacities were made use of, according to the requirements of the vapours examined.

In the first series of experiments made with this apparatus, I omitted to remove the air from the space above the liquid; each measure therefore sent in to the tube was a mixture of vapour and air. This diminished the effect of the former; but the proportionality, for small quantities, of density to absorption exhibits itself so decidedly as to induce me to give the observations. The first column, as usual, contains the measures of vapour, the second the observed absorption, and the third the calculated absorption. The galvanometric deflections are omitted, their equivalents being contained in the second column. In fact as far as the eighth observation, the absorptions are merely the record of the deflections.

Table IV. Mixture of Ether Vapour and Air. Unit-measure $\frac{1}{50}$th of a cubic inch.

Measures	Absorption	
	Observed	Calculated
1	4·5	4·5
2	9·2	9·0
3	13·5	13·5
4	18·0	18·0
5	22·8	23·5
6	27·0	27·0
7	31·8	31·5
8	36·0	36·0
9	39·7	40·0
10	45·0	45·0
20	81·0	90·0
21	82·8	95·0
22	84·0	99·0
23	87·0	104·0
24	88·0	108·0
25	90·0	113·0
26	93·0	117·0
27	94·0	122·0
28	95·0	126·0
29	98·0	131·0
30	100·0	135·0

Up to the 10th measure we find that density and absorption augment in precisely the same ratio. While the former varies from 1 to 10, the latter varies from 4·5 to 45. At the 20th measure, however, a deviation from proportionality is apparent, and the divergence gradually augments from 20 to 30. In fact 20 measures tell upon the rays capable of

Table V. Sulphuric Ether. Unit-measure $\frac{1}{100}$ th of a cubic inch.

Measures	Absorption		Measures	Absorption	
	Observed	Calculated		Observed	Calculated
1	5·0	4·6	17	65·5	77·2
2	10·3	9·2	18	68·0	83·0
4	19·2	18·4	19	70·0	87·4
5	24·5	23·0	20	72·0	92·0
6	29·5	27·0	21	73·0	96·7
7	34·5	32·2	22	73·0	101·2
8	38·0	36·8	23	73·0	105·8
9	44·0	41·4	24	77·0	110·4
10	46·2	46·2	25	78·0	115·0
11	50·0	50·6	26	78·0	119·6
12	52·8	55·2	27	80·0	124·2
13	55·0	59·8	28	80·5	128·8
14	57·2	64·4	29	81·0	133·4
15	59·4	69·0	30	81·0	138·0
16	62·5	73·6			

being absorbed,—the quantity destroyed becoming so considerable, that every additional measure encounters a smaller number of such rays, and hence produces a diminished effect.

With ether vapour alone, the results recorded in the following Table were obtained. Wishing to determine the absorption exercised by vapour of very low tension, the capacity of the unit-measure was reduced to $\frac{1}{100}$ th of a cubic inch.

We here find that the proportion between density and absorption holds sensibly good for the first eleven measures, after which the deviation gradually augments.

I have examined some specimens of ether which acted still more energetically on the thermal rays than those above recorded. No doubt for smaller measures than $\frac{1}{100}$ th of a cubic inch the above law holds still more rigidly true; and in a suitable locality it would be easy to determine with perfect accuracy $\frac{1}{10}$ th of the absorption produced by the first measure; this would correspond to $\frac{1}{1000}$ th of a cubic inch of vapour. But on entering the tube the vapour had only the tension due to the temperature of the laboratory, namely 12 inches. This would require to be multiplied by 2·5 to bring it up to that of the atmosphere. Hence the $\frac{1}{1000}$ th of a cubic inch, the absorption of which I have affirmed to be capable of measurement, would, on being diffused through a tube possessing a capacity of 220 cubic inches, have a tension of $\frac{1}{220} \times \frac{1}{2.5} \times \frac{1}{1000} = \frac{1}{500,000}$ th part of an atmosphere!

I have now to record the results obtained with thirteen other vapours. The method of experiment was in all cases the same as that just employed in the case of ether, the only variable element being the size of the unit-measure; for with many substances no sensible effect could be obtained with a unit volume so small as that used in the experiments last recorded. With bisulphide of carbon, for example, it was necessary to augment the unit-measure 50 times to render the measurements satisfactory.

Table VI. Bisulphide of Carbon. Unit-measure $\frac{1}{2}$ a cubic inch.

Measures	Absorption	
	Observed	Calculated
1	2·2	2·2
2	4·9	4·4
3	6·5	6·6
4	8·8	8·8
5	10·7	11·0
6	12·5	13·0
7	13·8	15·4
8	14·5	17·6
9	15·0	19·0
10	15·6	22·0
11	16·2	24·2
12	16·8	26·4
13	17·5	28·6
14	18·2	30·8
15	19·0	33·0
16	20·0	35·2
17	20·0	37·4
18	20·2	39·6
19	21·0	41·8
20	21·0	44·0

As far as the sixth measure the absorption is proportional to the density ; after which the effect of each successive measure diminishes. Comparing the absorption effected by a quantity of vapour which depressed the mercury column half an inch, with that effected by vapour possessing one inch of tension, the same deviation from proportionality is observed.

By mercurial gauge

Tension	Absorption
$\frac{1}{2}$ inch	14·8
1 inch	18·8

These numbers simply express the galvanometric deflections, which, as already stated, are strictly proportional to the absorption as far as 36° or 37°. Did the law of proportion hold good, the absorption due to 1 inch of tension ought of course to be 29·6 instead of 18·8.

Whether for equal volumes of the vapours at their maximum density, or for equal tensions as measured by the depression of the mercurial column, bisulphide of carbon exercises the lowest absorptive power of all the vapours which I have hitherto examined. For very small quantities, a volume of sulphuric ether vapour, at its maximum density in the measure, and expanded thence into the tube, absorbs 100 times the quantity of radiant heat intercepted by an equal volume of bisulphide of carbon vapour at its maximum density. These are the extreme limits of the scale, as far as my inquiries have hitherto proceeded. The action of every other vapour is less than that of sulphuric ether, and greater than that of bisulphide of carbon.

A very singular phenomenon was repeatedly observed during the experiments with bisulphide of carbon. After determining the absorption of the vapour, the tube was exhausted as perfectly as possible, the trace of vapour left behind being exceedingly minute. Dry air was then admitted to cleanse the tube. On again exhausting, after the first few strokes of the pump a jar was felt and a kind of explosion heard, while dense volumes of blue smoke immediately issued from the cylinders. The action was confined to the latter, and never propagated backwards into the experimental tube.

It is only with bisulphide of carbon that this effect has been observed. It may, I think, be explained in the following manner:—To open

the valve of the piston, the gas beneath it must have a certain tension, and the compression necessary to produce this appears sufficient to cause the combination of the constituents of the bisulphide of carbon with the oxygen of the air. Such a combination certainly takes place, for the odour of sulphurous acid is unmistakeable amid the fumes.

Table VII. Amylene. Unit-measure $\frac{1}{10}$ th of a cubic inch.

Measures	Absorption	
	Observed	Calculated
1	3·4	4·3
2	8·4	8·6
3	12·0	12·9
4	16·5	17·2
5	21·6	21·5
6	26·5	25·8
7	30·6	30·1
8	35·3	34·4
9	39·0	38·7
10	44·0	43·0

To test this idea I tried the effect of compression in the air-syringe. A bit of tow or cotton wool moistened with bisulphide of carbon, and placed in the syringe, emitted a bright flash when the air was compressed. By blowing out the fumes with a glass tube, this experiment may be repeated twenty times with the same bit of cotton.

It is not necessary even to let the moistened cotton remain in the syringe. If the bit of tow or cotton be thrown into it, and out again as quickly as it can be ejected, on compressing the air the luminous flash is seen. Pure oxygen produces a brighter flash than atmospheric air. These facts are in harmony with the above explanation.

For these quantities the absorption is proportional to the density, but for large quantities the usual deviation is observed, as shown by the following observations:—

By mercurial gauge

Tension	Deflection	Absorption
$\frac{1}{2}$ inch	60°	157
1 inch	65	216

Did the proportion hold good, the absorption for an inch of tension ought of course to be 314 instead of 216.

Table VIII. Iodide of Ethyle.
Unit-measure $\frac{1}{10}$ th of a cubic inch.

Measures	Absorption	
	Observed	Calculated
1	5·4	5·1
2	10·3	10·2
3	16·8	15·3
4	22·2	20·4
5	26·6	25·5
6	31·8	30·6
7	35·6	35·9
8	40·0	40·8
9	44·0	45·9
10	47·5	51·0

By mercurial gauge

Tension	Deflection	Absorption
$\frac{1}{2}$ inch	56°·3	94
1 inch	58·2	120

The deflections here are very small; the substance, however, possesses so feeble a volatility, that the tension of a measure of its vapour, when diffused through the experimental tube, must be infinitesimal. With the specimen which I examined, it was not practicable to obtain a tension sufficient to depress the mercury gauge $\frac{1}{2}$ an inch; hence no observations of this kind are recorded.

Table IX. Iodide of Methyle.
Unit-measure $\frac{1}{10}$ th of a cubic inch.

Measures	Absorption	
	Observed	Calculated
1	3·5	3·4
2	7·0	6·8
3	10·3	10·2
4	15·0	13·6
5	17·5	17·0
6	20·5	20·4
7	24·0	23·8
8	26·3	27·2
9	30·0	30·6
10	32·3	34·0

By mercurial gauge

Tension	Deflection	Absorption
$\frac{1}{2}$ inch	48°·5	60
1 inch	56·5	96

Table X. Iodide of Amyle.
Unit-measure $\frac{1}{10}$ th of a cubic inch.

Measures	Absorption	
	Observed	Calculated
1	0·6	0·57
2	1·0	1·1
3	1·4	1·7
4	2·0	2·3
5	3·0	2·9
6	3·8	3·4
7	4·5	4·0
8	5·0	4·6
9	5·0	5·1
10	5·8	5·7

Up to the 10th measure, or thereabouts, the proportion between density and absorption holds good, from which onwards the deviation from the law gradually augments.

Subsequent observations lead me to believe that the absorption by chloroform is a little higher than that given in the below Table.

The difference between the measurements when equal *tensions* and when equal *volumes* at the maximum density are made use of is here strikingly exhibited.

Table XI. Chloride of Amyle.
Unit-measure $\frac{1}{10}$ th of a cubic inch.

Measures	Absorption	
	Observed	Calculated
1	1·3	1·3
2	3·0	2·6
3	3·8	3·9
4	5·1	5·2
5	6·8	6·5
6	8·5	7·8
7	9·0	9·1
8	10·9	10·4
9	11·3	11·7
10	12·3	13·0

By mercurial gauge

Tension	Deflection	Absorption
$\frac{1}{2}$ inch	59°	137
1 inch	not practicable.	

Table XII. Benzole.
Unit-measure $\frac{1}{10}$ th of a cubic inch.

| Measures | Absorption | |
	Observed	Calculated
1	4·5	4·5
2	9·5	9·0
3	14·0	13·5
4	18·5	18·0
5	22·5	22·5
6	27·5	27·0
7	31·6	31·5
8	35·5	36·0
9	39·0	40·0
10	44·0	45·0
11	47·0	49·0
12	49·0	54·0
13	51·0	58·5
14	54·0	63·0
15	56·0	67·5
16	59·0	72·0
17	63·0	76·5
18	67·0	81·0
19	69·0	85·5
20	72·0	90·0

By mercurial gauge

Tension	Deflection	Absorption
$\frac{1}{2}$ inch	54°	78
1 inch	57	103

Table XIII. Methylic Alcohol.
Unit-measure $\frac{1}{10}$ th of a cubic inch.

| Measures | Absorption | |
	Observed	Calculated
1	10·0	10·0
2	20·0	20·0
3	30·0	30·0
4	40·5	40·0
5	49·0	50·0
6	53·5	60·0
7	59·2	70·0
8	71·5	80·0
9	78·0	90·0
10	84·0	100·0

By mercurial gauge

Tension	Deflection	Absorption
$\frac{1}{2}$ inch	58°·8	133
1 inch	60·5	168

Table XIV. Formic Ether.
Unit-measure $\frac{1}{10}$ th of a cubic inch.

| Measures | Absorption | |
	Observed	Calculated
1	8·0	7·5
2	16·0	15·0
3	22·5	22·5
4	30·0	30·0
5	35·2	37·5
6	39·5	45·0
7	45·0	52·5
8	48·0	60·0
9	50·2	67·5
10	53·5	75·0

By mercurial gauge

Tension	Deflection	Absorption
$\frac{1}{2}$ inch	58°·8	133
1 inch	62·5	193

Table XV. Propionate of Ethyle.
Unit-measure $\frac{1}{10}$ th of a cubic inch.

| Measures | Absorption | |
	Observed	Calculated
1	7·0	7·0
2	14·0	14·0
3	21·8	21·0
4	28·8	28·0
5	34·4	35·0
6	38·8	42·0
7	41·0	49·0
8	42·5	56·0
9	44·8	63·0
10	46·5	70·0

By mercurial gauge

Tension	Deflection	Absorption
$\frac{1}{2}$ inch	60°·5	168
1 inch	not practicable.	

In the case of alcohol I was obliged to resort to a unit-measure of $\frac{1}{2}$ a cubic inch to obtain an effect about equal to that produced by benzole with a measure possessing only $\frac{1}{10}$ th of a cubic inch in capacity; and yet for equal tensions of 0·5 of an inch, alcohol cuts off precisely twice as much heat as benzole. There is also an enormous difference between alcohol and sulphuric ether when equal measures at the maximum density are compared; but to bring the alcohol and ether vapours up to a

Table XVI. Chloroform.
Unit-measure $\frac{1}{10}$ th of a cubic inch.

| Measures | Absorption | |
	Observed	Calculated
1	4·5	4·5
2	9·0	9·0
3	13·8	13·5
4	18·2	18·0
5	22·3	22·5
6	27·0	27·0
7	31·2	31·5
8	35·0	36·0
9	39·0	40·5
10	40·0	45·0

common tension, the density of the former must be many times augmented. Hence it follows that when *equal tensions* of these two substances are compared, the difference between them diminishes considerably. Similar observations apply to many of the substances whose deportment is recorded in the foregoing Tables; to the iodide and chloride of amyle, for example, and to the propionate of ethyle. Indeed it is not unlikely that with equal tensions the vapour of a perfectly pure specimen of the substance last mentioned would be found to possess a higher absorptive power than that of ether itself.

Table XVII. Alcohol.
Unit-measure $\frac{1}{2}$ th of a cubic inch.

| Measures | Absorption | |
	Observed	Calculated
1	4·0	4·0
2	7·2	8·0
3	10·5	12·0
4	14·0	16·0
5	19·0	20·0
6	23·0	24·0
7	28·5	28·0
8	32·0	32·0
9	37·5	36·0
10	41·5	40·0
11	45·8	44·0
12	48·0	48·0
13	50·4	52·0
14	53·5	56·0
15	55·8	60·0

By mercurial gauge

Tension	Deflection	Absorption
$\frac{1}{2}$ inch	60°	157
1 inch	not practicable.	

It has been already stated that the tube made use of in these experiments was of brass polished within, for the purpose of bringing into clearer light the action of the feebler gases and vapours. Once, however, I wished to try the effect of chlorine, and with this view admitted a quantity of the gas into the experimental tube. The needle was deflected with prompt energy; but on pumping out[11], it refused to return to zero. To cleanse the tube, dry air was introduced into it ten times in succession ; but the needle pointed persistently to the 40th degree from zero. The cause of this was easily surmised: the chlorine had attacked the metal and partially destroyed its reflecting power; thus the absorption by the sides of the tube itself cut off an amount of heat competent to produce the deflection mentioned above. For subsequent experiments the interior of the tube had to be repolished.

Though no other vapour with which I had experimented produced a permanent effect of this kind, it was necessary to be perfectly satisfied that this source of error had not vitiated the experiments. To check the results, therefore, I had a length of 2 feet of similar brass tube coated carefully on the inside with lampblack, and determined by means of it the absorptions of all the vapours which I had previously examined, at a common tension of 0·3 of an inch. A general corroboration was all I sought, and I am satisfied that the few discrepancies which the measurements exhibit would disappear, or be accounted for, in a more careful examination.

In the following Table the results obtained with the blackened and with the bright tubes are placed side by side, the tension in the former being three-tenths, and in the latter five-tenths of an inch.

Table XVIII.

| Vapour | Absorption | | |
	Bright tube, 0·5 tension	Blackened tube, 0·3 tension	Absorption with bright tube proportional to
Bisulphide of Carbon..................	5·0	21	23
Iodide of Methyle	15·8	60	71
Benzole	17·5	78	79
Chloroform...........	17·5	89	79
Iodide of Ethyle ...	21·5	94	97
Wood-spirit..........	26·5	123	120
Methylic Alcohol.	29·0	133	131
Chloride of Amyle	30·0	137	135
Amylene..............	31·8	157	143

The order of absorption is here shown to be the same in both tubes, and the quantity absorbed in the bright tube is, in general, about $4\frac{1}{2}$ times that absorbed in the black one. In the third column, indeed, I have placed the products of the numbers contained in the first column by 4·5. These results completely dissipate the suspicion that the effects observed with the bright tube could be due to a change of the reflecting power of its inner surface by the contact of the vapours.

With the blackened tube the order of absorption of the following substances, commencing with the lowest, stood thus:—

> Alcohol,
> Sulphuric ether,
> Formic ether,
> Propionate of ethyle;

whereas with the bright tube they stood thus:—

> Formic ether,
> Alcohol,
> Propionate of ethyle,
> Sulphuric ether.

As already stated, these differences would in all probability disappear, or be accounted for on re-examination. Indeed very slight differences in the purity of the specimens used would be more than sufficient to produce the observed differences of absorption.[12]

§7. *ACTION of permanent Gases on Radiant Heat.*—The deportment of oxygen, nitrogen, hydrogen, atmospheric air, and olefiant gas has been already recorded. Besides these I have examined carbonic oxide, carbonic acid, sulphuretted hydrogen, and nitrous oxide. The action of these gases is so much feebler than that of any of the vapours referred to in the last section, that, in examining the relationship between absorption and density, the measures used with the vapours were abandoned, and the quantities of gas admitted were measured by the depression of the mercurial gauge.

Table XIX. Carbonic Oxide.

Tension in inches	Absorption	
	Observed	Calculated
0·5	2·5	2·5
1·0	5·6	5·0
1·5	8·0	7·5
2·0	10·0	10·0
2·5	12·0	12·5
3·0	15·0	15·0
3·5	17·5	17·5

Up to a tension of $3\frac{1}{2}$ inches the absorption by carbonic oxide is proportional to the density of the gas. But this proportion does not obtain with large quantities of the gas, as shown by the following Table:—

Tension in inches	Deflection	Absorption
5	18°·0	18
10	32·5	32·5
15	41·0	45

Table XX. Carbonic Acid.

Tension in inches	Absorption	
	Observed	Calculated
0·5	5·0	3·5
1·0	7·5	7·0
1·5	10·5	10·5
2·0	14·0	14·0
2·5	17·8	17·5
3·0	21·8	21·0
3·5	24·5	24·5

Here we have the proportion exhibited, but not so with larger quantities.

Tension in inches	Deflection	Absorption
5	25·0°	25
10	36·0	36
15	42·5	48

Table XXI. Sulphuretted Hydrogen.

Tension in inches	Absorption	
	Observed	Calculated
0·5	7·8	6
1·0	12·5	12
1·5	18·0	18
2·0	24·0	24
2·5	30·0	30
3·0	34·5	36
3·5	36·0	42
4·0	36·5	48
4·5	38·0	54
5·0	40·0	60

The proportion here holds good up to a tension of 2·5 inches, when the deviation from it commences and gradually augments.

Though these measurements were made with all possible care, I should like to repeat them.

Dense fumes issued from the cylinders of the air-pump on exhausting the tube of this gas, and I am not at present able to state with confidence that a trace of such in a very diffuse form within the tube did not interfere with the purity of the results.

Table XXII. Nitrous Oxide.

Tension in inches	Absorption	
	Observed	Calculated
0·5	14·5	14·5
1·0	23·5	29·0
1·5	30·0	43·5
2·0	35·5	58·0
2·5	41·0	71·5
3·0	45·0	87·0
3·5	47·7	101·5
4·0	49·0	116·0
4·5	51·5	130·5
5·0	54·0	145·0

Here the divergence from proportionality makes itself manifest from the commencement.

I promised at the first page of this memoir to allude to the results of Dr. Franz, and I will now do so. With a tube 3 feet long and blackened within, an absorption of 3·54 per cent. by atmospheric air was observed in his experiments. In my experiments, however, with a tube 4 feet long and polished within, which makes the distance traversed by the reflected rays more than 4 feet, the absorption is only one-tenth of the above amount. In the experiments of Dr. Franz, carbonic acid appears as a feebler absorber than oxygen. According to my experiments, for small quantities the absorptive power of the former is about 150 times that of the latter; and for atmospheric tensions, carbonic acid probably absorbs nearly 100 times as much as oxygen.

The differences between Dr. Franz and myself admit, perhaps, of the following explanation. His source of heat was an argand lamp, and the ends of his experimental tube were stopped with plates of glass. Now Melloni has shown that fully 61 per cent. of the heat-rays emanating from a Locatelli lamp are absorbed by a plate of glass one-tenth of an inch in thickness. Hence in all probability the greater portion of the rays issuing from the lamp of Dr. Franz was expended in heating the two glass ends of his experimental tube. These ends thus became secondary sources of heat which radiated against his pile. On admitting air into the tube, the partial withdrawal by conduction and convection of the heat of the glass plates would produce an effect exactly the same as that of true absorption. By allowing the air in my tube to come into contact with the radiating plate, I have often obtained a deflection of twenty or thirty degrees,—the effect being due to the cooling of the plate, and not to absorption. It is also certain that had I used heat from a luminous source, I should have found the absorption of 0·33 per cent. considerably diminished.

§ 8. I have now to refer briefly to a point of considerable interest as regards the effect of our atmosphere on solar and terrestrial heat. In examining the separate effects of the air, carbonic acid, and aqueous vapour of the atmosphere, on the 20th of last November, the following results were obtained:—

Air sent through the system of drying-tubes and through the caustic-potash tube produced an absorption of about ....1.

Air direct from the laboratory, containing therefore its carbonic acid[13] and aqueous vapour, produced an absorption of ....15.

Deducting the effect of the gaseous acids, it was found that the quantity of aqueous vapour diffused through the atmosphere on the day in question, produced an absorption at least equal to thirteen times that of the atmosphere itself.

It is my intention to repeat and extend these experiments on a future occasion;[14] but even at present conclusions of great importance may be drawn from them. It is exceedingly probable that the absorption of the solar rays by the atmosphere, as established by M. Pouillet, is mainly due to the watery vapour contained in the air. The vast difference between the temperature of the sun at midday and in the evening, is also probably due in the main to that comparatively shallow stratum of aqueous vapour which lies close to the earth. At noon the depth of it pierced by the sunbeams is very small; in the evening very great in comparison.

The intense heat of the sun's direct rays on high mountains is not, I believe, due to his beams having to penetrate only a small depth of air, but to the comparative absence of aqueous vapour at those great elevations.

But this aqueous vapour, which exercises such a destructive action on the obscure rays, is comparatively transparent to the rays of light. Hence the differential action, as regards the heat coming from the sun to the earth and that radiated from the earth into space, is vastly augmented by the aqueous vapour of the atmosphere.

De Saussure, Fourier, M. Pouillet, and Mr. Hopkins regard this interception of the terrestrial rays as exercising the most important influence on climate. Now if, as the above experiments indicate, the chief influence be exercised by the aqueous vapour, every variation of this constituent must produce a change of climate. Similar remarks would apply to the carbonic acid diffused through the air, while an almost inappreciable admixture of any of the hydrocarbon vapours would produce great effects on the terrestrial rays and produce corresponding *changes of climate*. It is not, therefore, necessary to assume alterations in the density and height of the atmosphere to account for different amounts of heat being preserved to the earth at different times; a slight change in its variable constituents would suffice for this. Such changes in fact *may have produced all the mutations of climate which the researches of geologists reveal*. However this may be, the facts above cited remain; they constitute true causes, the *extent* alone of the operation remaining doubtful.

The measurements recorded in the foregoing pages constitute only a fraction of those actually made; but they fulfil the object of the present portion of the inquiry. They establish the existence of enormous differences among colourless gases and vapours as to their action upon radiant heat; and they also show that, when the quantities are sufficiently small, the absorption in the case of each particular vapour is exactly proportional to the density.

These experiments furnish us with purer cases of molecular action than have been hitherto attained in experiments of this nature. In both solids and liquids the cohesion of the particles is implicated; they mutually control and limit each other. A certain action, over and above that which belongs to them separately, comes into play and embarrasses our conceptions. But in the cases above recorded the molecules are perfectly free, and we fix upon them individually the effects which the experiments exhibit; thus the mind's eye is directed more firmly than ever on those distinctive physical qualities whereby a ray of heat is stopped by one molecule and unimpeded by another.

§ 9. *Radiation of Heat by Gases.*—It is known that the quantity of light emitted by a flame depends chiefly on the incandescence of solid matter,—the brightness of an ignited jet of ordinary gas, for example, being chiefly due to the solid particles of carbon liberated in the flame.

Melloni drew a parallel between this action and that of radiant heat. He found the radiation from his alcohol lamp greatly augmented by plunging a spiral of platinum wire into the flame. He also found that a bundle of wire placed in the current of hot air ascending from an argand chimney gave a copious radiation, while when the wire was withdrawn no trace of radiant heat could be detected by his apparatus. He concluded from this experiment that air possesses the power of radiation in so feeble a degree, that our best thermoscopic instruments fail to detect this power.[15]

These are the only experiments hitherto published upon this subject; and I have now to record those which have been made in connexion with the present inquiry. The pile furnished with its conical reflector was placed upon a stand, with a screen of polished tin in front of it. An alcohol lamp was placed behind the screen so that its flame was entirely hidden by the latter; on rising above the screen, the gaseous column radiated its heat against the pile and produced a considerable deflection. The same effect was produced when a candle or an ordinary jet of gas was substituted for the alcohol lamp.

The heated products of combustion acted on the pile in the above experiments, but the radiation from pure air was easily demonstrated by placing a heated iron spatula or metal sphere behind the screen. A deflection was thus obtained which, when the spatula was raised to a red heat, amounted to more than sixty degrees. This action was due solely to the radiation of the air; no radiation from the spatula to the pile was possible, and no portion of the heated air itself approached the pile so as to communicate its warmth by contact to the latter. These effects are so easily produced that I am at a loss to account for the inability of so excellent an experimenter as Melloni to obtain them.

My next care was to examine whether different gases possessed different powers of radiation; and for this purpose the following arrangement was devised. P (fig. 1) represents the thermo-electric pile with its two conical reflectors; S is a double screen of polished tin; A is an argand burner consisting of two concentric rings perforated with orifices for the escape of the gas; C is a heated copper ball; the tube *t t* leads to a gas-holder containing the gas to be examined. When the ball C is placed on the argand burner, it of course heats the air in contact with it; an ascending current is established, which acts on the pile as in the experiments last described. It was found necessary to neutralize this radiation from the heated air, and for this purpose a large Leslie's cube L, filled with water a few degrees above the temperature of the air, was allowed to act on the opposite face of the pile.

Fig. 1.

When the needle was thus brought to zero, the cock of the gas-holder was turned on; the gas passed through the burner, came into contact with the ball, and ascended afterwards in a heated column in front of the pile. The galvanometer was now observed, and the limit of the arc through which its needle was urged was noted. It is needless to remark that the ball was entirely hidden by the screen from the thermo-electric pile, and that, even were this not the case, the mode of neutralization adopted would still give us the pure action of the gas.

The results of the experiments are given in the following Table, the figure appended to the name of each gas marking the number of degrees through which the radiation from the latter urged the needle of the galvanometer[16]:—

> Air .... 0°
> Oxygen 0
> Nitrogen 0
> Hydrogen 0
> Carbonic oxide ... 12
> Carbonic acid 18
> Nitrous oxide 29
> Olefiant gas 53

The radiation from air, it will be remembered, was neutralized by the large Leslie's cube, and hence the 0° attached to it merely denotes that the propulsion of air from the gas-holder through the argand burner did not augment the effect. Oxygen, hydrogen, and nitrogen, sent in a similar manner over the ball, were equally ineffective. The other gases, however, not only exhibit a marked action, but also marked differences of action. Their radiative powers follow precisely the same order as their powers of absorption. In fact, the deflections actually produced by their respective absorptions at 5 inches tension are as follow:—

> Air .... A fraction of a degree
> Oxygen .... ”   ”
> Nitrogen .... ”   ”
> Hydrogen .... ”   ”
> Carbonic oxide .... 18°
> Carbonic acid .... 25°
> Nitrous oxide .... 44°
> Olefiant gas .... 61°

It would be easy to give these experiments a more elegant form, and to arrive at greater accuracy, which I intend to do on a future occasion; but my object now is simply to establish the general order of their radiative powers. An interesting way of exhibiting both radiation and absorption is as follows:—When the polished face of a Leslie's cube is turned towards a thermoelectric pile the effect produced is inconsiderable, but it is greatly augmented when a coat of varnish is laid upon the polished surface. Instead of the coat of varnish, a film of gas may be made use of. Such a cube, containing boiling water, had its polished face turned towards the pile, and its effect on the galvanometer neutralized in the usual manner. The needle being at 0°, a film of olefiant gas, issuing from a narrow slit, was passed over the metal. The increase of radiation produced a deflection of 45°. When the gas was cut off, the needle returned accurately to 0°.

The absorption by a film may be shown by filling the cube with cold water, but not so cold as to produce the precipitation of the aqueous vapour of the atmosphere. A gilt copper ball, cooled in a freezing mixture, was placed in front of the pile, and its effect was neutralized by presenting a beaker containing a little iced water to the opposite face of the pile. A film of olefiant gas was sent over the ball, but the consequent deflection proved that the absorption, instead of being greater, was less than before. The ball, in fact, had been coated by a crust of ice, which is one of the best absorbers of radiant heat. The olefiant gas, being warmer than the ice, partially neutralized its absorption.

When, however, the temperature of the ball was only a few degrees lower than that of the atmosphere, and its surface quite dry, the film of gas was found to act as a film of varnish; it augmented the absorption.

A remarkable effect, which contributed at first to the complexity of the experiments, can now be explained. Conceive the experimental tube exhausted and the needle at zero ; conceive a small quantity of alcohol or ether vapour admitted; it cuts off a portion of the heat from one source, and the opposite source triumphs. Let the consequent deflection be 45°. If dry air be now admitted till the tube is filled, its effect of course will be slightly to augment the absorption and make the above deflection greater. But the following action is really observed:— when the air first enters, the needle, instead of ascending, descends; it falls to 26°, as if a portion of the heat originally cut off had been restored. At 26°, however, the needle stops, turns, moves quickly upwards, and takes up a permanent position a little higher than 45°. Let the tube now be exhausted, the withdrawal of the mixed air and vapour ought of course to restore the equilibrium with which we started; but the following effects are observed:— When the exhaustion commences, the needle moves upwards from 45° to 54°; it then halts, turns, and descends speedily to 0°, where it permanently remains.

After many trials to account for the anomaly, I proceeded thus:—A thermo-electric couple was soldered to the external surface of the experimental tube, and its ends connected with a galvanometer. When air was admitted, a deflection was produced, which showed that the air, on entering the vacuum, was heated. On exhausting, the needle was also deflected, showing that the interior of the tube was chilled. These are indeed known effects; but I was desirous to make myself perfectly sure of them. I subsequently had the tube perforated and thermometers screwed into it air-tight. On filling the tube the thermometric columns rose, on exhausting it they sank, the range between the maximum and minimum amounting in the case of air to 5° Fahr.

Hence the following explanation of the above singular effects. The absorptive power of the vapour referred to is very great, and its radiative power is equally so. The heat generated by the air on its entrance is communicated to the vapour, which thus becomes a temporary source of radiant heat, and diminishes the deflection produced in the first instance by its presence.

The reverse occurs when the tube is exhausted; the vapour is chilled, its great absorptive action on the heat radiated from the adjacent face of the pile comes more into play, and the original effect is augmented. In both cases, however, the action is transient; the vapour soon loses the heat communicated to it, and soon gains the heat which it has lost, and matters then take their normal course.

§ 10. *On the Physical Connexion of Radiation, Absorption, and Conduction.*—Notwithstanding the great accessions of late years to our knowledge of the nature of heat, we are as yet, I believe, quite ignorant of the atomic conditions on which radiation, absorption, and conduction depend. What are the specific qualities which cause one body to radiate copiously and another feebly ? Why, on theoretic grounds, must the equivalence of radiation and absorption exist? Why should a highly diathermanous body, as shown by Mr. Balfour Stewart, be a bad radiator, and an adiathermanous body a good radiator? How is heat conducted? and what is the strict physical meaning of good conduction and bad conduction ? Why should good conductors be, in general, bad radiators, and bad conductors good radiators ? These, and other questions, referring to facts more or less established, have still to receive their complete answers. It is less with a hope of furnishing such than of shadowing forth the possibility of uniting these various effects by a common bond, that I submit the following reflections to the notice of the Royal Society.

In the experiments recorded in the foregoing pages, we have dealt with *free* atoms, both simple and compound, and it has been found that in all cases absorption takes place. The meaning of this, according to the dynamical theory of heat, is that no atom is capable of existing in vibrating ether without accepting a portion of its motion. We may, if we wish, imagine a certain roughness of the surface of the atoms which enables the ether to *bite* them and carry the atom along with it. But no matter what the quality may be which enables any atom to accept motion from the agitated ether, the same quality must enable it to impart motion to still ether when it is plunged in the latter and agitated. It is only necessary to imagine the case of a body immersed in water to see that this must be the case. There is a polarity here as rigid as that of magnetism. From the existence of absorption, we may on theoretic grounds infallibly infer a capacity for radiation; from the existence of radiation, we may with equal certainty infer a capacity for absorption; and each of them must be regarded as the measure of the other[17].

This reasoning, founded simply on the mechanical relations of the ether and the atoms immersed in it, is completely verified by experiment. Great differences have been shown to exist among gases as to their powers of absorption, and precisely similar differences as regards their powers of radiation. But what specific property is it which makes one free molecule a strong absorber, while another offers scarcely any impediment to the passage of radiant heat ? I think the experiments throw some light upon this question. If we inspect the results above recorded, we shall find that the *elementary* gases hydrogen, oxygen, nitrogen, and the *mixture* atmospheric air, possess absorptive and radiative powers beyond comparison less than those of the *compound* gases. Uniting the atomic theory with the conception of an ether, this result appears to be exactly what ought to be expected. Taking Dalton's idea of an elementary body as a single sphere, and supposing such a sphere to be set in motion in still ether, or placed without motion in moving ether, the communication of motion by the atom in the first instance, and the acceptance of it in the second, must be less than when a number of such atoms are grouped together and move as a system. Thus we see that hydrogen and nitrogen, which, when *mixed* together, produce a small effect, when *chemically united* to form ammonia, produce an enormous effect. Thus oxygen and hydrogen, which, when mixed in their electrolytic proportions, show a scarcely sensible action, when chemically combined to form aqueous vapour exert a powerful action. So also with oxygen and nitrogen, which, when mixed, as in our atmosphere, both absorb and radiate feebly, when united to form oscillating systems, as in nitrous oxide, have their powers vastly augmented. Pure atmospheric air, of 5 inches tension, does not effect an absorption equivalent to more than the one-fifth of a degree, while nitrous oxide of the same tension effects an absorption equivalent to fifty-one such degrees. Hence the absorption by nitrous oxide at this tension is about 250 times that of air. No fact in chemistry carries the same conviction to my mind, that air is a *mixture* and not a *compound*, as that just cited. In like manner, the absorption by carbonic oxide of this tension is nearly 100 times that of oxygen alone; the absorption by carbonic acid is about 150 times that of oxygen; while the absorption by olefiant

gas of this tension is 1000 times that of its constituent hydrogen. Even the enormous action last mentioned is surpassed by the vapours of many of the volatile liquids, in which the atomic groups are known to attain their highest degree of complexity.

I have hitherto limited myself to the consideration, that the compound molecules present broad sides to the ether, while the simple atoms with which we have operated do not,—that in consequence of these differences the ether must swell into billows when the former are moved, while it merely trembles into ripples when the latter are agitated,—that, in the interception of motion also, the former, other things being equal, must be far more influential than the latter; but another important consideration remains. All the gases and vapours whose deportment we have examined are transparent to light; that is to say, the waves of the visible spectrum pass among them without sensible absorption. Hence it is plain that their absorptive power depends on the periodicity of the undulations which strike them. At this point the present inquiry connects itself with the experiments of Nièpce, the observation of Foucault, the surmises of Ångström, Stokes, and Thomson, and those splendid researches of Kirchhoff and Bunsen which so immeasurably extend our experimental range. By Kirchhoff it has been conclusively shown that every atom absorbs in a special degree those waves which are synchronous with its own periods of vibration. Now, besides presenting broader sides to the ether, the association of simple atoms to form groups must, as a general rule, render their motions through the ether more sluggish, and tend to bring the periods of oscillation into isochronism with the slow undulations of obscure heat, thus enabling the molecules to absorb more effectually such rays as have been made use of in our experiments.

Let me here state briefly the grounds which induce me to conclude that an agreement in period alone is not sufficient to cause powerful absorption and radiation—that in addition to this the molecules must be so constituted as to furnish *points d'appui* to the ether. The heat of contact is accepted with extreme freedom by rock-salt, but a plate of the substance once heated requires a great length of time to cool. This surprised me when I first noticed it. But the effect is explained by the experiments of Mr. Balfour Stewart, by which it is proved that the radiative power of heated rock-salt is extremely feeble. Periodicity can have no influence here, for the

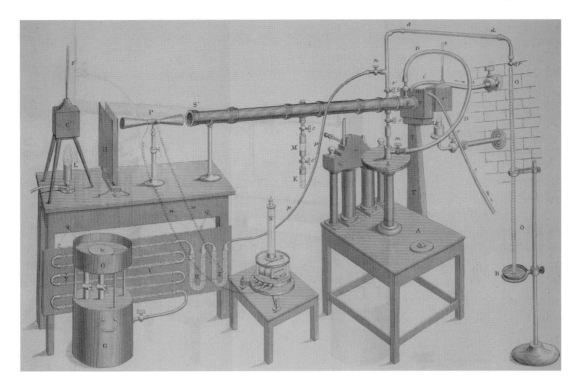

ether is capable of accepting and transmitting impulses of all periods; and the fact that rock-salt requires more time to cool than alum, simply proves that the molecules of the former glide through the ether with comparatively small resistance, and thus continue moving for a longer time; while those of the latter presenting broad sides to the ether, speedily communicate to it the motion which we call heat. This power of gliding through still ether possessed by the rock-salt molecules, must of course enable the moving ether to glide round them, and no coincidence of period could, I think, make such a body a powerful absorber.

Many chemists, I believe, are disposed to reject the idea of an atom, and to adhere to that of equivalent proportions merely. They figure the act of combination as a kind of interpenetration of one substance by another. But this is a mere masking of the fundamental phenomenon. The value of the atomic theory consists in its furnishing the physical explanation of the law of equivalents: assuming the one, the other follows; and assuming the act of chemical union as Dalton figured it, we see that it blends harmoniously with the perfectly independent conception of an ether, and enables us to reduce the phenomena of radiation and absorption to the simplest mechanical principles.

Considerations similar to the above may, I think, be applied to the phenomena of *conduction*. In the Philosophical Magazine for August 1853, I have described an instrument used in examining the transmission of heat through cubes of wood and other substances. When engaged with this instrument, I had also cubes of various crystals prepared, and determined with it their powers of conduction. With one exception, I found that the conductivity augmented with the diathermancy. The exception was furnished by a cube of very perfect rock-crystal, which conducted slightly better than my cube of rock-salt. The latter, however, had a very high conductive power; in fact rock-salt, calcareous spar, glass, selenite, and alum stood in my experiments, as regards conductivity, exactly in their order of diathermancy in the experiments of Melloni. I have already adduced considerations which show that the molecules of rock-salt glide with facility through the ether; but the ease of motion which these molecules enjoy must facilitate their mutual collision. Their motion, instead of being expended on the ether which exists between them, and communicated by it to the external ether, is in great part transferred directly from particle to particle, or in other words, is freely conducted. When a molecule of alum, on the contrary, approaches a neighbour molecule, it produces a swell in the intervening ether, which swell is in part transmitted, not to the molecules, but to the general ether of space, and thus lost as regards conduction. This lateral waste prevents the motion from penetrating the alum to any great extent, and the substance is what we call a bad conductor[18].

Such considerations as these could hardly occur without carrying the mind to the kindred question of electric conduction; but the speculations have been pursued sufficiently far for the present, and must now abide the judgment of those competent to decide whether they are the mere emanations of fancy, or a fair application of principles which are acknowledged to be secure.

The present paper, I may remark, embraces only the first section of these researches.

NOTES

1   *La Thermochrose*, p. 136.
2   Pogg. *Ann.* vol. xciv. p. 342.
3   Pogg. *Ann.* vol. lxxxiii. p. 489; and Phil. Mag. 1852, vol. iii. p. 82.
4   See Melloni upon this subject, *Thermochrose*, pp. 31–33.
5   Mr. Becker, to whose skill and intelligence I have been greatly indebted, furnished me with several specimens of wire of the same fineness as that used by Du Bois Raymond, some covered with green silk and others with white. The former were invariably attracted, the latter invariably repelled. In all cases the *naked* wire was repelled.
6   During the course of the inquiry, I have often had occasion to avail myself of the assistance of this excellent mechanician.
7   It will be seen that in this arrangement I have abandoned the use of the differential galvanometer, and made the thermo-electric pile the differential instrument.
8   It will be seen further on that this result is in harmony with the supposition that ozone, obtained in the manner described, is a *compound* body.
9   With a cube containing boiling water I have since made this experiment visible to a large audience.
10 It is evident that the old mode of experiment might be applied to this gas. Indeed, several of the solids examined by Melloni are inferior to it in absorptive power. Had time permitted, I should have checked my results by experiments made in the usual way; this I intend to do on a future occasion.
11 Dense dark fumes rose from the cylinders on this occasion ; a similar effect was produced by sulphuretted hydrogen.
12 In illustration of this I may state, that of two specimens of methylic alcohol with which I was furnished by two of my chemical friends, one gave an absorption of 84 and the other of 203. The former specimen had been purified with great care, but the latter was not pure. Both specimens, however, went under the common name of methylic alcohol. I have had a special apparatus constructed with a view to examine the influence of ozone on the interior of the experimental tube.
13 And a portion of sulphurous acid produced by the two gas-lamps used to heat the cubes.
14 The peculiarities of the locality in which this experiment was made render its repetition under other circumstances necessary.
15 La Thermochrose, p. 94.
16 I have also rendered these experiments on radiation visible to a large audience. They may be readily introduced in lectures on radiant heat.
17 This was written long before Kirchhoff's admirable papers on the relation of emission to absorption were known to me.
18 In the above considerations regarding conduction, I have limited myself to the illustration furnished by two compound bodies; but the elementary atoms also differ among themselves as regards their powers of accepting motion from the ether and of communicating motion to it. I should infer, for example, that the atoms of platinum encounter more resistance in moving through the ether than the atoms of silver. It is needless to say that the physical texture of a substance also has a great influence.

3

By the Light of the Silvery Moon

Arrhenius, S. (April 1896). On the influence of carbonic acid in the air upon the temperature of the ground. *The London, Edinburgh, and Dublin Philosophical Magazine and Journal of Science Series 5*, **41** (251). 39 pages.

In Arrhenius' 1896 paper we witness the birth of modern climate science. Working with incomplete theoretical basis and a few beams of moonlight, Arrhenius calculated the warming that would result from doubling the CO_2 concentration of the atmosphere, a quantity that modern climate scientists call the climate sensitivity. Granted he may have gotten lucky to get what is essentially the right answer, but more importantly his approach was well guided, and brilliantly creative. Along the way, Arrhenius described the water vapor feedback, which about doubles the impact of changing CO_2, and the ice albedo feedback, which is largely responsible for the intensified warming in high latitudes. Although Arrhenius is best known for the Arrhenius equation, which describes the effect of temperature on the rates of chemical reactions, his 1896 paper contribution stands squarely at the foundation of Earth science.

Fourier described the greenhouse effect resulting from an atmosphere that selectively passes incoming visible light, and absorbs outgoing infrared radiation. Tyndall showed that the capacity to absorb and emit infrared radiation is shared by only a few trace gases in the atmosphere, notably water vapor and CO_2. However, technology was not available in Tyndall's time or Arrhenius' to measure the strength of the absorption by the gases, which as it turns out varies wildly depending on the exact wavelength of the radiation. Putting matters even further out of Arrhenius' reach, the detailed absorption spectrum depends on the pressure of the gas, because of interactions between molecules that alter the vibrational frequencies that the molecules can undergo. The CO_2 absorption spectrum consists of a collection of very narrow peaks, which broaden and coalesce with increasing pressure. Even today, a detailed calculation of the absorption and emission of IR by a column of atmosphere is not trivial; it can be done by computer models known as line-by-line codes, based on megabytes of detailed spectral information for the various greenhouse gases, but these calculations are too computationally expensive, that is to say slow, to be done in the full climate models that are used to predict things such as, say, the climate sensitivity or global warming forecasts. Climate models use approximate codes to calculate the balance of radiation energy. Under these circumstances, what should we suppose were the odds of Arrhenius doing this calculation by hand and getting the right answer?

The basis of Arrhenius' scheme is measurements of the IR intensity of moonlight made by Samuel Pierpont Langley. Langley was trying to estimate the temperature of the moon based on the knowledge that the intensity of IR emission goes up as the temperature of the emitting object rises. He invented and used a device called a bolometer to measure the IR intensity, manifested as a change in temperature of a piece of metal coated with an absorbing layer of soot that was exposed to the light beam, relative to another that was not. Langley selected particular wavelengths by sending the IR through a prism made of salt, which was known to be one of the few solids that is transparent to IR radiation (Fig. 3.1).

Although the measurements were not intended for this purpose, Arrhenius' idea was to use the data to calculate the absorption of IR by the entire atmospheric column of CO_2 and water vapor.

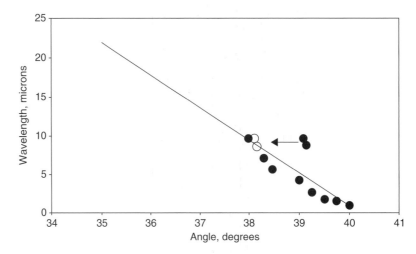

Fig. 3.1   **This is the relationship between the deflection angle of Langley's infrared beams and the wavelengths of the light as reported by Arrhenius. The two outliers are probably a misprint; if the angles of 39.xx° are replaced by 38.xx°, they fall on the same relation as the rest of the values quoted. At the time Arrhenius was writing, experimental data relating wavelength to refraction angle through Langley's salt prism was not available for angles below 38.x°, so Arrhenius estimated the relationship using a linear fit. For Arrhenius' purposes, the exact form of the relationship does not actually matter, since it was only the absorption summed up over all wavelengths that counts for the greenhouse calculation. You can do this sum equally well using the deflection angle or the wavelength. However, where the relationship matters is in comparing the spectral data Arrhenius deduced to modern accurate absorption data for water vapor and carbon dioxide. Jean-Louis Dufresne, in his Habilitation thesis (a sort of super PhD thesis done by advanced researchers in the French academic system) found that the nonlinearities in the wavelength–angle relation are very important to doing the comparison correctly.**

The measurements were made over many nights, under differing weather conditions, which meant different amounts of water vapor to absorb the IR. The moon was also at different elevations in the sky on the different nights, sending the moonbeams through the atmosphere either vertically or obliquely, through differing inventories of CO_2. Regressing the IR intensity data against the varying inventories of CO_2 and water vapor, Arrhenius calculated the apparent absorption coefficients of the greenhouse gases in the atmosphere. One of the many formidable challenges Arrhenius faced was disentangling the water vapor and CO_2 absorption effects. In some parts of the spectrum, this is not really possible using the kind of data available to Arrhenius, but science proceeds by making the best use possible of whatever data there is, and that is what Arrhenius did. Up until recently, the standard wisdom was that Arrhenius did quite well at getting the absorption properties more or less right, but Jean-Louis Dufresne's discovery of the importance of the nonlinearity in the relationship between refraction angle and wavelength has changed that picture. In Fig. 3.2, it is argued that Langley's data was sufficient to allow Arrhenius to do a fairly good job of the water vapor absorption properties near 6.5-micron wavelength, but did not extend far enough into the long-wave infrared to pick up the CO_2 absorption feature near 15 micron, which is the most important one for global warming.

Arrhenius was aware that any substance that can absorb IR will also emit its intensity dependent upon its temperature. Some of the IR light that they measured may have come from emission by gases in the atmosphere rather than coming directly from the moon. This physics passed without comment in Arrhenius' derivation of the absorption coefficients, although it is clearly accounted for in the climate modeling in the second part of the paper. Perhaps Arrhenius was assuming implicitly that the moon is warm enough that any IR emission coming from our own atmosphere would be negligible, or at least that IR emission from our atmosphere would be

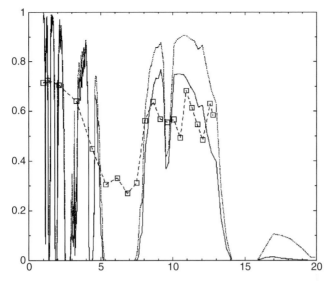

Fig. 3.2a **The spectral properties of CO_2 and water vapor in the atmosphere as deduced by Arrhenius compared with results from a modern atmospheric radiation code. Each figure shows the transmission (vertical axis) as a function of the wave number in microns (horizontal axis). A transmission of 1 corresponds to a transparent atmosphere, whereas a transmission of 0 corresponds to a completely absorbing atmosphere. The dashed lines with symbols in each figure give the results of the calculation by Arrhenius using Langley's lunar infrared transmission data. The dotted and solid lines are the results of a modern radiation code, with two different assumptions about the water vapor content of the atmosphere. Figure 3.2a compares results calculated with both water vapor and CO_2.**

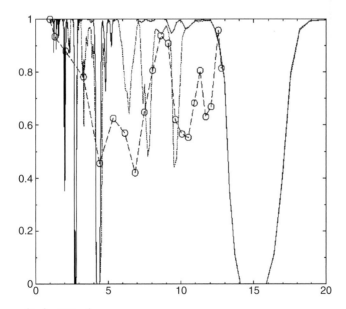

Fig. 3.2b **Shows results for CO_2 alone.**

reflected in the absorption coefficients that he derived from the data. If the air in the atmosphere were all the same temperature as the surface of the Earth and the moon, then presumably the amount of gas intercepting the moonbeam would have no impact on IR intensity, and the absorption coefficients Arrhenius derived would be small, indicating correctly that the greenhouse

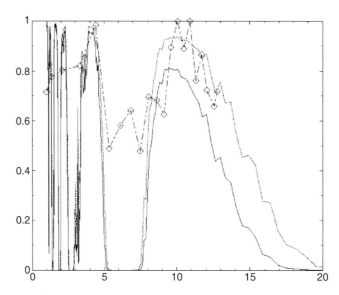

Fig. 3.2c   **Shows results for water vapor alone. The calculation by Arrhenius does a qualitatively good job of reproducing the main absorption features of water vapor around 5 micron, but the CO_2 spectroscopy bears little resemblance to the correct pattern based on accurate modern laboratory measurements. In particular it completely misses the 15 micron absorption region, which is principally responsible for anthropogenic global warming. There is nothing essentially wrong with the technique used by Arrhenius. The main challenge is to separate water vapor from CO_2 effects, and if Langley's data had extended to the 15 micron region, where the CO_2 effects dominate, the masterful analysis technique Arrhenius employed would have worked very well. These calculations were carried out by Jean-Louis Dufresne (*L'effet de serre: sa découverte, son analyse par la méthode des puissances nettes échangées et les effets de ses variations récentes et futures sur le climat terrestre.* Habilitation thesis, Université Pierre et Marie Curie, 2009) and are used with the kind permission of the author.**

effect is weak. This benefit of the doubt is undermined somewhat by the statement that it is "a physical absurdity" for the IR intensity to increase with passage through the gas (page 245), or that "it is not permissible to assume that the radiation could be strengthened by its passage" through the gas (page 252), and the calculation of the near-complete extinction of the radiation intensity by very high gas inventories following Beer's law in Table III (page 251). The main problem with the spectroscopy inferred by Arrhenius was not the effect of atmospheric thermal emission. It was the fact that Langley's measurements did not extend to sufficiently long wavelengths to pick up the principle CO_2 absorption feature near 15 micron (see Fig. 3.2).

Arrhenius was also aware of the band saturation effect, carefully described by Tyndall, in which the absorption depends linearly on temperature for low concentrations or pressures of the gas, but as the absorbing gas inventory increases, eventually all the incoming IR is absorbed, and further increases in gas concentration have only a small impact on the IR flux. One of the chief objections, raised by Angstrom in particular, to the proposition that CO_2 can affect the climate was that the absorption bands of CO_2 were already saturated. This would have been an impossible determination to make in the laboratory at this time, because CO_2 at differing altitudes in the atmosphere has different spectra and therefore differing effects on the IR absorption. Arrhenius' effective absorption coefficients, based on the change in IR intensity with a change in the inventories of the greenhouse gases, actually included the effect of band saturation. Had the gases in the atmosphere been more saturated than they are, the effect of humidity and lunar zenith angle would have been smaller than Langley measured, and Arrhenius would have gotten smaller absorption coefficients. O clever, clever man!

So, Arrhenius obtained the best estimate of the CO_2 and water vapor spectroscopy the methods of his day would permit, but then how to turn that into a calculation of temperature? Basically, what Arrhenius had to do was invent the field of climate modeling, which had never before been done quantitatively for planets with atmospheres. He wrote a model taking into account the radiative energy exchanges between the atmosphere and space, between the atmosphere and the ground, and between the ground and space (owing to transmission of infrared through the atmosphere). The model took into account the fact that the water vapor content of the atmosphere is expected to increase as the atmosphere gets warmer. He used his spectroscopic data to calculate the atmospheric infrared absorption and emission that appears in the energy balance equations. To keep the calculation manageable (remember, he had to do all of this by hand), he wrote a model in which the entire atmosphere was characterized by a single temperature. Arrhenius was not implying that the atmosphere really was isothermal – he was only assuming that the atmosphere's radiative properties could be characterized by a single vertically aver-aged temperature of some sort. His discussion shows that he was well aware of the inaccuracies inherent in a one-layer model of the atmosphere, but given the tedium of the calculations, who could fault him from stopping at one layer, when the results already looked interesting? The one-layer model, and its comparison to modern multilayer models, is described in the Box, **The Arrhenius one-layer model of the greenhouse effect.**

Arrhenius' calculation of the impact of changing CO_2 concentration was done in a gridded calculation space, just like modern climate models. The grid in Arrhenius' case was by latitude and season. Arrhenius recognized that the radiation energy budget does not necessarily balance at any given location on the Earth. More energy comes in at the equator from the intense sun-shine than leaves locally by infrared. The difference is carried to higher latitudes by winds and ocean currents, and is ultimately balanced by excess outgoing radiation over incoming at higher latitudes. Arrhenius assumed that the heat fluxes that he could diagnose today from his gridded model would also apply in an altered, double-CO_2 world. This assumption was not precisely cor-rect, but it was certainly the correct place to start.

The impact of clouds on climate is difficult to model even to this day. Clouds scatter and reflect incoming sunlight, and they also act as blackbodies in the atmosphere, absorbing essentially all of the IR radiation that hits them, and re-radiating at an intensity dependent on the temperature of the cloud droplets. Arrhenius does not mention the IR effects of clouds, but he is well aware of the vis-ible (albedo) effect. His solution, as for the heat fluxes, is to assume that the distribution and impacts of clouds will be the same in the altered double-CO_2 world. Again, this was an excellent place to start, given the limited options available. Clouds remain the main uncertainty in modern climate models, but there is no indication that the behavior of clouds has a large biasing effect on the temperature impact of changes in atmospheric CO_2 concentration.

Arrhenius constructed the water vapor feedback by power of pure reason. As the temperature of the Earth rises in response to a doubling of atmospheric CO_2, the saturation vapor pressure of the atmosphere rises following the Clausius–Clapeyron relationship. If the air can hold more water vapor, then presumably it will. Arrhenius assumed that the relative humidity, the expres-sion of water vapor content relative to the saturation concentration, would remain the same with a change in climate. If the natural world has an average relative humidity of 80%, then the warmer world will also be 80% saturated, but since the warmer air can hold more water vapor, it holds 80% of a higher concentration, the absolute humidity rising proportionally to the saturation value. The effect of water vapor on climate is therefore to amplify a temperature change, driven in this case by doubling atmospheric CO_2.

Arrhenius also recognized the cooling effect of ice and snow at the Earth's surface when it reflects incoming visible light, and the fact that a warming world can thaw the ice, leaving a less reflective ground or ocean surface to absorb the incoming sunlight. Ice therefore acts as another positive amplifying feedback in the climate system, leading to a stronger temperature change from increasing CO_2 concentration, in particular in high latitudes where the ice is. The ice albedo feedback, as it is now called, is among the reasons why high latitudes warm more than low latitudes, an effect we see today in the Arctic, and in the forecast for the future, both from modern climate models and in Arrhenius' own results. (The Antarctic, it must be noted, is something of a special case, with cooling in the interior of the continent and little loss of sea ice over the past decades, for dynamical reasons having to do with the intensity of the circumpolar jet in the atmosphere and the loss of stratospheric ozone, a greenhouse gas, known as the ozone hole.)

In the end, Arrhenius predicted that doubling the CO_2 concentration of the atmosphere would raise the temperature of the surface of the Earth by about 6°C. This is hauntingly similar to the climate sensitivity found today. With the benefit of over a century of conceptual advances and an explosion of computer power that would have seemed like magic to Arrhenius, we now expect that the Earth would warm by about 2.5–4°C. However, it must be said that with regard to the specific number he came up with, Arrhenius was more lucky than right. There are two sources of error in his calculation, which were inevitable products of the state of the art at the time. The first source of error is the limitation in the accuracy with which he could estimate the true absorption spectrum of CO_2. The second source of error is in the use of a one-layer model of the atmosphere to compute the greenhouse effect. The one-layer model, used with correct spectroscopy, leads to an underestimate of the true climate sensitivity, mainly because with only one layer water vapor excessively masks the effects of carbon dioxide because a one-layer model lacks the high, cold dry parts of the atmosphere where CO_2 packs the most punch (see Box: The Arrhenius one-layer model of the greenhouse effect.). The inaccuracies in the spectroscopy Arrhenius used, however, bias the sensitivity to the high side, which more than compensated for the low bias of the one-layer model. Thus, if Arrhenius had had correct spectroscopy in his one-layer model, he would have predicted a modest (though still significant) climate sensitivity. If he had used a modern multilevel model with his inaccurate spectroscopy, he would have found an extraordinarily high climate sensitivity, well in excess of the high end of the IPCC range.

Would history have been different in either of those cases? We can only speculate, but given that even the rather alarming climate sensitivity he came up with was insufficient to stir much sense of concern for decades after Arrhenius' seminal paper, it seems unlikely. It is unseemly to dwell too much on the specific number Arrhenius came up with, which was a product of unavoidable technical shortcomings of his day. The genius in the work of Arrhenius is that he turned Fourier's rather amorphous and unquantified notion of planetary temperature into exactly the correct conceptual framework, even going so far as getting the notion of water vapor feedback right. Most importantly, he correctly identified the importance of satisfying the energy balance both at the top of the atmosphere and at the surface. Conceptual errors regarding this point plagued the subject long after the spectroscopy had improved. If the climate theorists and spectroscopists of the next few decades had only fully understood Arrhenius' paper, many false steps could have been avoided. As it stands, correct spectroscopy was not brought together with a correct conceptual framework in a multilevel model until the seminal work of Manabe in the early 1960s. It is rather fortuitous that the number we now have for climate sensitivity is similar to the one that Arrhenius came up with, but what is not fortuitous is that nothing that has come up in the intervening century or more has shaken the basic conceptual foundation of the greenhouse effect that Arrhenius laid down. Not even a little. In this, Arrhenius was prescient and 100% right. While we can now compute the effects of CO_2 on climate at a level of detail and confidence that Arrhenius could hardly have dreamed of, we are basically doing the same energy book-keeping as Arrhenius taught us how to do, but only in vastly elaborated detail with vastly better fundamental spectroscopic data.

Arrhenius was mostly interested in the cause of the ice ages, and he predicted that the CO_2 concentration of the atmosphere during glacial time might have been 150 ppm (we now know it was 180–200 ppm). He recognized the possibility that humans could alter the climate of the Earth, by "evaporating the coal mines into the atmosphere," but estimated that it would take 1000 years to double the CO_2 concentration. This was actually a reasonable conclusion at the time. CO_2 emissions have grown exponentially since then, and it takes real courage of conviction, or wild-eyed alarmism, to extrapolate a present-day trend far into the future based on exponential growth. Anyway, as a Swede he felt that perhaps a bit of warming might be pleasant. Not an attitude, we hasten to add, that Swedes today would generally concur with. They seem to like their broad sweeps of Northern tundra and short, sweet summers with berries in the pastures. To say nothing of a century of investment in hydropower designed to make use of the climate of the past century, and not that of the scary new world global warming is taking us into.

Box: The Arrhenius One-Layer Model of the Greenhouse Effect

The one-layer model of Arrhenius represents the atmosphere by a single layer with temperature T_a having emissivity e. According to the laws of radiation physics, the emissivity also gives the absorption, so that the transmission is $(1 - e)$. The emissivity is a function of temperature, because of the temperature dependence of the atmosphere's water vapor content. Arrhenius solved for the atmospheric temperature assuming only infrared radiative exchanges between the atmosphere and the ground, and between the atmosphere and space. He then used the result to compute the infrared leaving the top of the atmosphere – the *Outgoing Longwave Radiation*, or OLR. In equilibrium, this must balance the absorbed solar radiation. The calculation is laid out in Fig. 3.3.

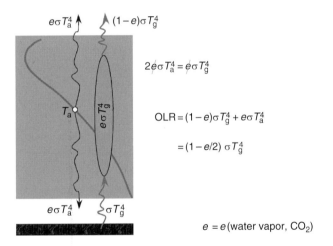

$$e\sigma T_a^4 \qquad (1-e)\sigma T_g^4$$

$$2e\sigma T_a^4 = e\sigma T_g^4$$

$$OLR = (1-e)\sigma T_g^4 + e\sigma T_a^4$$

$$= (1 - e/2)\, \sigma T_g^4$$

$$e\sigma T_a^4 \qquad \sigma T_g^4$$

$$e = e(\text{water vapor}, CO_2)$$

Fig. 3.3 **Schematic of Arrhenius' one-layer atmosphere model.**

A modern multilayer model looks like Fig. 3.4, in which a radiation model is used to compute the mean altitude from which infrared escapes to space. The temperature there is T_{rad}, and it gets colder relative to the ground as more greenhouse gas is added to the atmosphere. Adding a greenhouse gas warms the ground because in equilibrium T_{rad} has to stay fixed so as to balance the absorbed solar radiation, but it occurs at a higher altitude, so you have to follow the temperature gradient a longer distance before you hit the ground. Another refinement taken into account in modern models is that heat exchange with the ground is not just radiative, but also contains contributions from turbulent exchange of heat and moisture. The effect of these is to keep the ground temperature close to the overlying air temperature, and so the details of the turbulent transfer are relatively inconsequential.

It is straightforward to insert modern spectroscopy into the Arrhenius one-layer model. All you have to do is to use a modern radiation code to compute the emissivity e of a slab, taking into account the temperature-dependent water vapor content of the slab. Increasing the CO_2 makes the emissivity closer to 1, and therefore warms the surface. In Fig. 3.5, the calculation has been organized graphically.

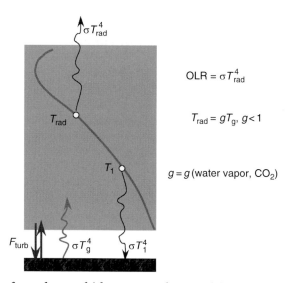

Fig. 3.4   **Schematic of a modern multi-layer atmosphere model.**

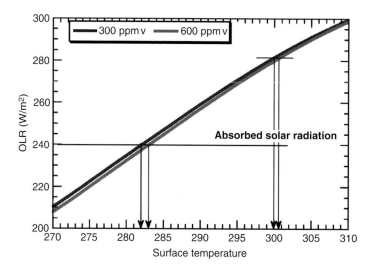

Fig. 3.5   **One-layer model with water vapor feedback, using modern spectroscopy.**

We specify a surface temperature, use the Arrhenius balance equations to compute the corresponding atmospheric temperature (using the temperature and CO_2-dependent emissivity), and then use both the surface and atmospheric temperature to compute the infrared emission out of the top of the atmosphere – the OLR. Where the straight line corresponding to the absorbed solar radiation intersects the OLR curve, we read off the equilibrium surface temperature. All climate theory, in essence, amounts to some variant on computing this curve with ever-greater sophistication. The results show that where the absorbed solar radiation yields an equilibrium temperature around 280 K at 300 ppmv CO_2, doubling the CO_2 would yield about a degree of warming. The warming is somewhat less if the base case corresponds to a typical tropical temperature. These results include the one-layer version of the water vapor feedback. They are smaller than what a multilayer model would give, because the isothermal model of the atmosphere puts too much water vapor high up, and therefore underestimates the increase in the emissivity caused by a doubling of CO_2.

Box: The Saturation Fallacy

Knut Ångström (1857–1910) was the son of the Swedish physicist Anders Ångström who lent his name to the unit of length widely used in spectroscopy and atomic physics. The younger Ångström, like his father, was very interested in the properties of radiation. If there were a prize for papers that set back the study of global warming, surely Knut's paper on the absorption of infrared by CO_2 and water vapor would be a strong contender. This paper was among the first to introduce what might be called the *Saturation Fallacy* – the idea that at its present concentration CO_2 absorbs as much infrared as it possibly can, so introduction of more CO_2 cannot change the climate. If you put enough infrared absorber into a layer of air, it absorbs everything. You cannot absorb more than 100% of the radiation. Moreover, laws of radiation physics that were well established at the time said that such a perfect absorber would also radiate like an ideal black-body, meaning that addition of more absorber could not change the emission, if you held the temperature of the layer fixed. As a variant on this argument, Ångström also claimed that even without this *saturation* of the CO_2 effect, the absorption by water vapor would be so strong that it would accomplish the same thing, leaving little role for CO_2 changes to affect climate. Such was the force of Ångström's claim that it may well have been instrumental in keeping most atmospheric scientists from taking Arrhenius seriously for nearly a half century. Even today, one sometimes hears the saturation argument used by some less scrupulous global warming deniers.

But Ångström was wrong, and he was wrong on many counts. First, his laboratory measurements of the absorption properties of CO_2 were inaccurate. We know today, from precise laboratory measurements, that CO_2 is not anywhere close to saturated in the Earth's conditions. In fact, it is not even completely saturated for the atmosphere of Venus, which has 300 000 times as much CO_2 in it as the Earth's atmosphere! Second, it turns out that even if CO_2 and water vapor *were* saturated in the sense claimed by Ångström, it would not prevent addition of more CO_2 from warming the climate. The reason is that the temperature and density of the atmosphere decline with altitude, and so there is always some region up there that is tenuous enough and dry enough (by virtue of being cold) that it is unsaturated – and it is from *this* region that infrared escapes to space when the lower atmosphere is saturated. The "thinning, cooling, and drying" argument could have been made using the physics known at the time of Arrhenius, but strangely enough it did not become appreciated until the much later work of Plass and Manabe. The fact that reduced pressure also limits gaseous absorption – and hence saturation aloft – adds to the effect, but is in no way crucial to the "thinning, cooling, and drying" argument. Let us first take a closer look at the absorption properties of CO_2, using modern laboratory data.

Suppose we were to sit at sea level and shine an infrared flashlight with an output of one Watt upward into the sky. If all the light from the beam were then collected by an orbiting astronaut with a sufficiently large lens, what fraction of a Watt would that be? The question of saturation amounts to the following question: How would that fraction change if we increased the amount of CO_2 in the atmosphere? Saturation refers to the condition where increasing the amount of CO_2 fails to increase the absorption, because the CO_2 was already absorbing – essentially all there is to absorb at the wavelengths where it absorbs at all. Think of a conveyor belt with red, blue, and green M&M candies going past. You have one guy who only eats red M&Ms, and he can eat them fast enough to eat half of the M&Ms going past him. Thus, he reduces the M&M flux by half. If you put another guy next to him who can eat at the same rate, he will eat all the remaining red M&Ms. Then, if you put a third guy in the line, it will not result in any further decrease in the M&M flux, because all the M&Ms that they like to eat are already gone. You would need an eater of green M&Ms to make further reductions in the flux.

Ångström and his followers believed that the situation with CO_2 and infrared was like the situation with the red M&Ms. To understand how wrong they were, we need to look at modern measurements of the rate of absorption of infrared light by CO_2. The rate of absorption is a very intricately varying function of the wavelength of the light. At any given wavelength, the amount of light surviving goes down like the exponential of the number of molecules of CO_2 encountered by the beam of light. The rate of exponential decay is the absorption factor. When the product of the absorption factor times the amount of CO_2 encountered equals one, then the amount of light is reduced by a factor of $1/e$, that is, $1/2.71282\ldots$. For this, or larger amounts of CO_2, the atmosphere is optically thick at the corresponding wavelength. If we double the amount of CO_2, the proportion of surviving light is squared, or about a tenth; if we halve the amount of CO_2 instead, the amount surviving is $1/\sqrt{e}$, or about 60%, and the atmosphere is optically thin. Precisely where we draw the line between "thick" and "thin" is somewhat arbitrary, given that the absorption shades smoothly from small values to large values as the product of absorption factor with amount of CO_2 increases.

The units of absorption factor depend on the units we use to measure the amount of CO_2 in the column of the atmosphere encountered by the beam of light. Let us measure our units relative to the amount of CO_2 in an atmospheric column of base one square meter, present when the concentration of CO_2 is 300 parts per million (about the preindustrial value). In such units, an atmosphere with the present amount of CO_2 is optically thick where the absorption coefficient is one or greater, and optically thin where the absorption coefficient is less than one. If we double the amount of CO_2 in the atmosphere, then the absorption coefficient only needs to be 1/2 or greater in order to make the atmosphere optically thick. The absorption factor, so defined, is given in Fig. 3.6, based on the thousands of measurements in the HITRAN spectroscopic archive. The "fuzz" on this graph is because the absorption actually takes the form of thousands of closely spaced partially overlapping spikes. If one were to zoom in on a very small portion of the wavelength axis, one would see the fuzz resolve into discrete spikes, like the pickets on a fence. At the coarse resolution of the graph, one only sees a dark band marking out the maximum and minimum values swept out by the spike. These absorption results were computed for typical laboratory conditions, at sea level pressure and a temperature of 20°C. At lower pressures, the peaks of the spikes get higher and the valleys between them get deeper, leading to a broader "fuzzy band" on absorption curves like that shown below.

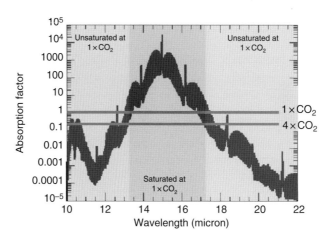

Fig. 3.6 **Infrared absorption spectrum of CO_2 with regions of band saturation indicated at 1 and 4 times atmospheric CO_2 concentration.**

We see that for the preindustrial CO_2 concentration, it is only the wavelength range between about 13.5 and 17 micron (millionths of a meter) that can be considered to be saturated. Within this range, it is indeed true that adding more CO_2 would not significantly increase the amount of absorption. All the red M&Ms are already eaten. But waiting in the wings, outside this wavelength region, there are more goodies to be had. In fact, noting that the graph is on a logarithmic axis, the atmosphere still would not be saturated even if we increased the CO_2 to 10 000 times the present level. What happens to the absorption if we quadruple the amount of CO_2? That story is told in Fig. 3.7.

The horizontal thick grey lines give the threshold CO_2 needed to make the atmosphere optically thick at 1× the preindustrial CO_2 level and 4× that level. Quadrupling the CO_2 makes the portions of the spectrum in the yellow bands optically thick, essentially adding new absorption there and reducing the transmission of infrared through the layer. One can relate this increase in the width of the optically thick region to the "thinning and cooling" argument determining infrared loss to space as follows. Roughly speaking, in the part of the spectrum where the atmosphere is optically thick,

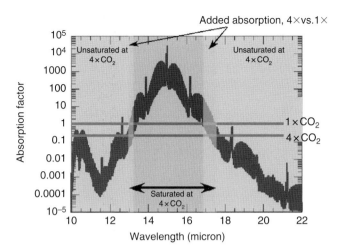

Fig. 3.7   **With an increase in atmospheric CO_2 concentration, the saturated band for CO_2 expands.**

the radiation to space occurs at the temperature of the high, cold parts of the atmosphere. That is practically zero compared to the radiation flux at temperatures comparable to the surface temperature; in the part of the spectrum that is optically thin, the planet radiates at near the surface temperature. Increasing CO_2 then increases the width of the spectral region where the atmosphere is optically thick, which replaces more of the high-intensity surface radiation with low-intensity upper-atmosphere radiation, and thus reduces the rate of radiation loss to space.

This box is based on material that originally appeared on RealClimate.org. © 2007 R.T. Pierrehumbert, used with permission of the author.

Angstrom, K. (1900). Ueber die Bedeutung des Wasserdampfes und der Kohlensäure bei der Absorption der Erdatmosphäre. *Annalen der Physik Bd 3*, 720–732. 13 pages.

On the Influence of Carbonic Acid in the Air upon the Temperature of the Ground*

SVANTE ARRHENIUS

I. INTRODUCTION: OBSERVATIONS OF *LANGLEY* ON ATMOSPHERICAL ABSORPTION

A GREAT deal has been written on the influence of the absorption of the atmosphere upon the climate. Tyndall[1] in particular has pointed out the enormous importance of this question. To him it was chiefly the diurnal and annual variations of the temperature that were lessened by this circumstance. Another side of the question, that has long attracted the attention of physicists, is this: Is the mean temperature of the ground in any way influenced by the presence of heat-absorbing gases in the atmosphere? Fourier[2] maintained that the atmosphere acts like the glass of a hot-house, because it lets through the light rays of the sun but retains the dark rays from the ground. This idea was elaborated by Pouillet;[3] and Langley was by some of his researches led to the view, that "the temperature of the earth under direct sunshine, even though our atmosphere were present as now, would probably fall to −200° C., if that atmosphere did not possess the quality of selective absorption".[4] This view, which was founded on wide a use of Newton's law of cooling, must be abandon as Langley himself in a later memoir showed that the full moon, which certainly does not possess any sensible heat-absorbing atmosphere, has a "mean effective temperature" of about 45° C.[5]

The air retains heat (light or dark) in two different ways. On the one hand, the heat suffers a selective diffusion on its passage through the air; on the other hand, some of the atmospheric gases absorb considerable quantities of heat. These two actions are very different. The selective diffusion is extraordinarily great for the ultra-violet rays, and diminishes continuously with increasing wave-length of the light, so that it is insensible for the rays that form the chief part of the radiation from a body of the mean temperature of the earth.[6]

The selective absorption of the atmosphere is, according to the researches of Tyndall, Lecher and Pernter, Röntgen, Heine, Langley, Ångström, Paschen, and others,[7] of a wholly different kind. It is not exerted by the chief mass of the air, but in a high degree by aqueous vapour and carbonic acid, which are present in the air in small quantities. Further, this absorption is not continuous over the whole spectrum, but nearly insensible in the light part of it, and chiefly limited to the long-waved part, where it manifests itself in very well-defined absorption-bands, which fall off rapidly on both sides.[8] The influence of this absorption is comparatively small on the heat from the sun, but must be of great importance in the transmission of rays from the earth. Tyndall held the opinion that the water-vapour has the greatest influence, whilst other authors, for instance Lecher and Pernter, are inclined to think that the carbonic acid plays the more important part. The researches of Paschen show that these gases are both very effective, so that probably sometimes the one, sometimes the other, may have the greater effect according to the circumstances.

In order to get an idea of how strongly the radiation of the earth (or any other body of the temperature +15° C.) is absorbed by quantities of water-vapour or carbonic acid in the proportions in which these gases are present in our atmosphere, one should, strictly speaking, arrange experiments on the absorption of heat from a body at 15° by means of appropriate quantities of both gases. But such experiments have not been made as yet, and, as they would require very expensive apparatus beyond that at my disposal, I have not been in a position to execute them. Fortunately there are other researches by Langley in his work on 'The Temperature of the Moon,' with the aid of which it seems not impossible to determine the absorption of heat by aqueous vapour and by carbonic acid in precisely the

* Arrhenius, S. (April 1896) On the influence of carbonic acid in the air upon the temperature of the ground. *The London, Edinburgh, and Dublin Philosophical Magazine and Journal of Science* 5th Series, Vol.41, no. 251.

conditions which occur in our atmosphere. He has measured the radiation of the full moon (if the moon was not full, the necessary correction relative to this point was applied) at different heights and seasons of the year. This radiation was moreover dispersed in a spectrum, so that in his memoir we find the figures for the radiant heat from the moon for 21 different groups of rays, which are defined by the angle of deviation with a rocksalt prism having a refracting angle of 60 degrees. The groups lie between the angles 40° and 35°, and each group is separated from its neighbours by an interval of 15 minutes. Now the temperature of the moon is nearly the same as that of the earth, and the moon-rays have, as they arrive at the measuring-instruments, passed through layers of carbonic acid and of aqueous vapour of different thickness according to the height of the moon and the humidity of the air. If, then, these observations were wholly comparable with one another, three of them would suffice for calculating the absorption coefficient relatively to aqueous vapour and carbonic acid for any one of the 21 different groups of rays. But, as an inspection of the 24 different series of observations will readily show, this is not the case. The intensity of radiation for any group of rays should always diminish with increasing quantity of aqueous vapour or carbonic acid traversed. Now the quantity of carbonic acid is proportional to the path of the ray through the atmosphere, that is, to the quantity called "Air-mass" in Langley's figures. As unit for the carbonic acid we therefore take air-mass = 1, *i. e.* the quantity of carbonic acid that is traversed in the air by a vertical ray. The quantity of aqueous vapour traversed is proportional partly to the "air-mass," partly to the humidity, expressed in grammes of water per cubic metre. As unit for the aqueous vapour I have taken the quantity of aqueous vapour that is traversed by a vertical ray, if the air contains 10 grammes per cubic metre at the earth's surface.[9] If we tabulate the 24 series of observations published by Langley in the work cited with respect to the quantities of carbonic acid and aqueous vapour, we immediately detect that his figures run very irregularly, so that very many exceptions are found to the rule that the transmitted heat should continuously decrease when both these quantities increase.

And it seems as if periodic alterations with the time of observation occurred in his series. On what circumstance these alterations with the

time depend one can only make vague conjectures: probably the clearness of the sky may have altered within a long period of observation, although this could not be detected by the eye. In order to eliminate this irregular variation, I have divided the observations into four groups, for which the mean quantities of carbonic acid (K) and of water-vapour (W) were 1·21 and 0·36, 2·21 and 0·86, 1·33 and 1·18, and 2·22 and 2·34 respectively. With the help of the mean values of the heat-radiation for every group of rays in these four groups of observations, I have roughly calculated the absorption coefficients (*x* and *y*) for both gases, and by means of these reduced the value for each observation to the value that it would have possessed if K and W had been 1·5 and 0·88 respectively. The 21 values for the different rays were then summed up, so that I obtained the total heat-radiation for every series of observations, reduced to K = 1·5 and W = 0·88. If the materials of observation were very regular, the figures for this total radiation should not differ very much from one another. In fact, one sees that observations that are made at nearly the same time give also nearly equal values, but if the observations were made at very different times, the values differ also generally very much. For the following periods I have found the corresponding mean values of the total radiation:–

Period			Mean value	Reduction factor
1885.	Feb.	21–June 24	4850	1·3
1885.	July.	29–1886. Feb. 16.	6344	1·00
1886.	Sept.	13–Sept. 18	2748	2·31
1886.	Oct.	11–Nov. 8	5535	1·15
1887.	Jan.	8–Feb. 9	3725	1·70

In order to reduce the figures of Langley to comparability with one another, I have applied the reduction factors tabulated above to the observations made in the respective periods. I have convinced myself that by this mode of working no systematic error is introduced into the following calculations.

After this had been done, I rearranged the figures of Langley's groups according to the values of K and W in the following table. (For further details see my original memoir.)

In this table the angle of deviation is taken as head-title. After K and W stand the quantities of carbonic acid and water-vapour traversed by the ray in the above-mentioned units. Under this comes after *i* obs. the intensity of radiation (reduced) observed by Langley on the bolometer, and after this the corresponding value *i* calc., calculated by means of the absorption-coefficients given in Table II. below. G is the "weight" given to the corresponding *i* obs. in the calculation, using the method of least squares.

For the absorption-coefficients, calculated in this manner, I give the following table. (The common logarithms of the absorption-coefficients are tabulated.)

The signification of these figures may be illustrated by an example. If a ray of heat, corresponding to the angle of deviation 39°·45, passes through the unit of carbonic acid, it decreases in intensity in the proportion 1 : 0·934 (log = −0·0296), the corresponding value for the unit of water-vapour is 1:0·775 (log =−0·1105). These figures are of course only valid for the circumstances in which the observations were made, viz., that the ray should have traversed a quantity of carbonic acid K=1·1 and a quantity of water-vapour W=0·3 before the absorption in the next quantities of carbonic acid and water-vapour was observed. And these second quantities should not exceed K=1·1 and W=1·8, for the observations are not extended over a greater interval than between K=1·1 and K=2·2, and W=0·3 and W=2·1 (the numbers for K and W are a little different for rays of different kind). Below A is written the relative value of the intensity of radiation for a given kind of ray in the moonlight after it has traversed K=1 and W=0·3. In some cases the calculation gives positive values for log *x* or log *y*. As this is a physical absurdity (it would signify that the ray should be strengthened by its passage through the absorbing gas), I have in these cases, which must depend on errors of observation, assumed the absorption equal to zero for the corresponding gas, and by means of this value calculated the absorption-coefficient of the other gas, and thereafter also A.

As will be seen from an inspection of Table I., the values of *i* obs. agree in most cases pretty well with the calculated values *i* calc. But in some cases the agreement is not so good as one could wish. These cases are mostly characterized by a small "weight" G, that is in other words, the material of observation is in these cases relatively insufficient. These cases occur also chiefly for such rays as are strongly absorbed by water-vapour. This effect is probably owing to the circumstance that the aqueous vapour in the atmosphere, which is assumed to have varied proportionally to the humidity at the earth's surface, has not always had the assumed ideal and uniform distribution with the height. From observations made during balloon voyages, we know also that the distribution of the aqueous vapour may be very irregular, and different from the mean ideal distribution. It is also a marked feature that in some groups, for instance the third, nearly all the observed numbers are less than the calculated ones, while in other groups, for instance the fourth, the contrary is the case. This circumstance shows that the division of the statistic material is carried a little too far; and a combination of these two groups would have shown a close agreement between the calculated and the observed figures. As, however, such a combination is without influence on the correctness of the calculated absorption-coefficients, I have omitted a rearrangement of the figures in greater groups, with consequent recalculation.

A circumstance that argues very greatly in favour of the opinion that the absorption-coefficient given in Table II. cannot contain great errors, is that so very few logarithms have a positive value. If the observations of Langley had been wholly insufficient, one would have expected to find nearly as many positive as negative logarithms. Now there are only three such cases, viz., for carbonic acid at an angle of 40°, and for water-vapour at the angles 36°·45 and 36°·15. The observations for 40° are not very accurate, because they were of little interest to Langley, the corresponding rays not belonging to the moon's spectrum but only to the diffused sunlight from the moon. As these rays also do not occur to any sensible degree in the heat from a body of 15° C., this non-agreement is without importance for our problem. The two positive values for the logarithms belonging to aqueous vapour are quite insignificant. They correspond only to errors of 0·2 and 1·5 per cent. for the absorption of the quantity W = 1, and fall wholly in the range of experimental errors.

It is certainly not devoid of interest to compare these absorption-coefficients with the results of the direct observations by Paschen and Ångström.[10] In making this comparison, we must bear in mind that an exact agreement cannot be expected, for the signification of the above coefficients is rather unlike that of the coefficients that are or may be calculated from the observations of these two authors. The above coefficients give the rate of absorption of a ray that has traversed quantities of carbonic acid (K = 1·1) and water-vapour (W = 0·3); whilst the coefficients of Paschen and Ångström represent the absorption experienced by a ray on the passage through the first layers of these gases. In some cases we may expect a great difference between these two quantities, so that only a general agreement can be looked for.

According to Paschen's figures there seems to exist no sensible emission or absorption by the aqueous vapour at wave-lengths between 0·9 μ and 1·2 μ (corresponding to the angle of deviation 40°). On the other hand, the representation of the sun's spectrum by Langley shows a great many strong absorption-bands in this interval, among which those marked ρ, σ, τ, and ϕ are the most prominent[11], and these absorption-bands belong most probably to the aqueous vapour, That Paschen has not observed any emission by water-vapour in this interval may very well be accounted for by the fact that his heat-spectrum had a very small intensity for these short-waved rays. But it may be conceded that the absorption-coefficient for aqueous vapour at this angle in Table II. is not very accurate (probably too great), in consequence of the little importance that Langley attached to the corresponding observations. After this occurs in Langley's spectrum the great absorption-band ψ at the angle 39·45 ($\lambda = 1\cdot4\ \mu$), where in Paschen's curve the emission first becomes sensible (log y = −0·1105 in Table II.). At wave-lengths of greater value we find according to Paschen strong absorption-bands at $\lambda = 1\cdot83\ \mu$ (Ω in Langley's spectrum), i. e. in the neighbourhood of 39°·30 and at $\lambda = 2\cdot64\ \mu$ (Langley's X) a little above the angle 39°·15. In accordance with this I have found rather large absorption-coefficients for aqueous vapour at these angles (log y = −0·0952 and −0·0862 resp.). From $\lambda = 3\cdot0\ \mu$ to $\lambda = 4\cdot7\ \mu$ thereafter, according to Paschen the absorption is very small, in agreement with my calculation (log y = −0·0068 at 39°, corresponding to $\lambda = 4\cdot3\ \mu$). From this point the absorption increases again and presents new maxima at $\lambda = 5\cdot5\ \mu$, $\lambda = 6\cdot6\ \mu$, and $\lambda = 7\cdot7\ \mu$, i. e. in the vicinity of the angles 38°·45 ($\lambda = 5\cdot6\ \mu$) and 38°·30 ($\lambda = 7\cdot1\ \mu$). In this region the absorption of the water-vapour is continuous over the whole interval, in consequence of which the great absorption-coefficient in this part (log y = −0·3114 and −0·2362) becomes intelligible. In consequence of the decreasing intensity of the emission-spectrum of aqueous vapour in Paschen's curve we cannot pursue the details of it closely, but it seems as if the emission of the water-vapour would also be considerable at $\lambda = 8\cdot7\ \mu$ (39°·15), which corresponds with the great absorption-coefficient (log y = −0·1933) at this place. The observations of Paschen are not exended further, ending at $\lambda = 9\cdot5\ \mu$, which corresponds to an angle of 39°·08.

For carbonic acid we find at first the value zero at 40°, in agreement with the figures of Paschen and Ångström[12]. The absorption of carbonic acid first assumes a sensible value at $\lambda = 1\cdot5\ \mu$, after which it increases rapidly to a maximum at $\lambda = 2\cdot6\ \mu$, and attains a new extraordinarily strong maximum at $\lambda = 4\cdot6$ (Langley's Y). According to Ångström the absorption of carbonic acid is zero at $\lambda = 0\cdot9\ \mu$, and very weak at $\lambda = 1\cdot69\ \mu$, after which it increases continuously to $\lambda = 4\cdot6\ \mu$ and decreases again to $\lambda = 6\cdot0\ \mu$. This behaviour is entirely in agreement with the values of log x in Table II. From the value zero at 40° ($\lambda = 1\cdot0\ \mu$) it attains a sensible value (−0·0296) at 39°·45 ($\lambda = 1\cdot4\ \mu$), and thereafter greater and greater values (−0·0559 at 39°·30, and −0·1070 at 39°·15) till it reaches a considerable maximum (−0·3412 at 39°, $\lambda = 4\cdot3\ \mu$). After this point the absorption decreases (at 38°·45 = 5·6 μ, log x = −0·2035). According to Table II. the absorption of carbonic acid at 38°·30 and 38°·15 ($\lambda = 7\cdot1\ \mu$ and 8·7 μ) has very great values (log x = −0·2438 and −0·3730), whilst according to Ångström it should be insensible. This behaviour may be connected with the fact that Ångström's spectrum had a very small intensity for the larger wavelengths. In Paschen's curve there are traces of a continuous absorption by the carbonic acid in this whole region with weak maxima at $\lambda = 5\cdot2\ \mu$, $\lambda = 5\cdot9\ \mu$, $\lambda = 6\cdot6\ \mu$ (possibly due to traces of water-vapour), $\lambda = 8\cdot4\ \mu$, and $\lambda = 8\cdot9\ \mu$. In consequence of the strong absorption of water-vapour in this region of the spectrum, the intensity of radiation was very small in Langley's observations, so that the calculated absorption-coefficients are there not very exact (cf. above, pp. 242–243). Possibly the calculated absorption of the carbonic acid may have come out too great, and that of the water-vapour too small in this part (between 38°·30 and 38°·0). This can happen the more easily, as in Table I. K and W in general increase together because they are both proportional to the "air-mass." It may be pointed out that this also occurs in the problems that are treated below, so that the error from this cause is not of so great importance as one might think at the first view.

For angles greater than 38° ($\lambda > 9\cdot5\ \mu$) we possess no direct observations of the emission or absorption of the two gases. The sun's spectrum, according to Langley, exhibits very great absorption-bands at about 37°·50, 37°·25, 37°, and 36°·40. According to my calculations the aqueous vapour has its greatest absorbing power in the spectrum from 38° to 35° at angles between 37°·15 and 37°·45 (the figures for 35°·45, 35°·30, and 35°·15 are very uncertain, as they depend upon very few measurements), and the carbonic acid between 36°·30 and 37°·0. This seems to indicate

Table I. Radiation (i) of the Full Moon for different Values of K and W.

	40°.	39°·45.	39°·30.	39°·15.	39°.	38°·45.	38°·30.	38°·15.	38°.	37°·45.	37°·30.
K	1·16	1·12	1·16	1·13	1·16	1·13	1·16	1·13	1·16	1·13	1·16
W	0.32	0·269	0.32	0·271	0.32	0·271	0.32	0·271	0.32	0·271	0.32
i obs.	28·7	26·6	27·0	26·4	24·8	24·8	12·6	20·1	43·8	65·9	74·4
i calc. ...	27·0	34·5	29·0	25·7	24·4	23·5	12·5	19·4	40·8	58·0	68·8
G	79	27	75	56	69	53	35	43	121	140	206
K	1·28	1·27	1·29	1·29	1·29	1·29	1·27	1·26	1·29	1·27	1·27
W	0.81	1·07	0.86	1·04	0.86	1·04	0.90	0.96	0.86	1·07	1·00
i obs.	22·9	31·2	26·7	21·3	18·2	11·0	5·8	3·7	14·0	32·0	52·3
i calc. ...	23·1	27·9	25·4	21·2	21·8	12·5	8·6	12·8	26·1	42·1	52·7
G	76	135	109	73	74	38	24	13	57	139	261
K	1·46	1·40	1·39	1·49	1·49	1·49	1·50	1·49	1·50	1·49	1·50
W	0.75	0.823	0.78	0.87	0.89	0.89	0.82	0.89	0.82	0.87	0.84
i obs.	11·9	28·2	23·0	18·9	18·0	9·2	9·9	14·4	24·6	34·8	46·6
i calc. ...	23·6	29·4	25·4	20·9	18·6	12·7	7·8	10·8	24·4	43·2	55·2
G	28	28	25	38	37	17	33	28	81	70	151
K	1·48	1·52	1·48	1·51	1·48	1·51	1·48	1·51	1·48	1·52	1·48
W	1·80	2·03	1·78	1·64	1·78	1·95	1·80	1·95	1·80	2·03	1·67
i obs.	25·2	27·6	24·6	18·3	27·6	4·8	3·7	3·6	17·6	45·5	43·9
i calc. ...	16·9	21·4	20·2	17·9	18·5	5·9	4·7	6·6	12·0	28·2	40·2
G	30	22	51	31	37	5	4	3	21	37	119
K	2·26	2·26	2·26	2·26	2·26	2·26	2·26	2·26	2·27	2·26	2·27
W	1·08	1·08	1·08	1·08	1·08	1·08	1·08	1·08	1·06	1·08	1·06
i obs.	21·3	23·4	20·8	16·4	11·1	8·2	4·5	3·5	17·3	36·1	47·1
i calc. ...	21·2	25·9	21·3	16·6	10·1	7·7	4·5	5·1	14·7	33·9	48·3
G	44	49	43	34	23	17	9	7	37	75	112
K	2·05	1·92	1·92	1·93	1·92	1·92	1·92	2·45	2·37	1·92	2·05
W	1·93	2·30	2·24	2·16	2·24	2·30	2·24	2·25	2·20	2·30	1·93
i obs.	13·4	12·8	14·8	15·1	10·3	6·6	3·4	3·4	7·9	20·8	31·5
i calc. ...	16·2	19·4	17·3	14·5	13·0	3·8	2·9	2·6	6·1	23·4	35·1
G	55	29	35	47	25	15	8	10	26	47	129

that the first two absorption-bands are due to the action of watervapour, the last two to that of carbonic acid. It should be emphasized that Langley has applied the greatest diligence in the measurement of the intensity of the moon's radiation at angles between 36° and 38°, where this radiation possesses its maximum intensity. It may, therefore, be assumed that the calculated absorption-coefficients for this part of the spectrum are the most exact. This is of great importance for the following calculations, for the radiation from the earth[13] has by far the greatest intensity (about two thirds, cf. p. 250) in this portion of the spectrum.

II.   THE TOTAL ABSORPTION BY ATMOSPHERES OF VARYING COMPOSITION.

As we have now determined, in the manner described, the values of the absorption-coefficients for all kinds of rays, it will with the help of Langley's figures[14] be possible to calculate the fraction of the heat from a body at 15° C., (the earth) which is absorbed by an atmosphere that contains specified quantities of carbonic acid and water-vapour. To begin with, we will execute this calculation with the values K = 1 and W = 0·3. We

	37°·15.	37°.	36°·45.	36°·30.	36°·15.	36°.	35°·45.	35°·30.	35°·15.	35°.
K	1·16	1·16	1·18	1·18	1·27	1·16	1·27	1·27	1·27	1·16
W	0·32	0·32	0·34	0·34	0·48	0·32	0·48	0·48	0·48	0·32
i obs.	68·6	59	56·2	48·3	43·4	40·7	39·0	32·6	31·5	19·7
i calc. ...	73·7	57·1	50·9	46·0	34·9	36·4	31·3	27·7	27·3	19·3
G	190	163	118	102	28	112	25	21	20	54
K	1·27	1·27	1·31	1·32	1·32	1·28	1·33	1·33	1·33	1·25
W	1·00	1·00	1·05	1·00	1·00	0·81	0·51	0·51	1·07	0·60
i obs.	58·9	50·3	47·9	41·2	31·7	29·7	25·7	18·8	27·5	16·6
i calc. ...	53·0	51·2	47·1	39·2	34·2	31·1	30·3	26·8	21·3	17·2
G	294	251	205	140	108	98	16	12	39	22
K	1·49	1·48	1·48	1·48	1·41	1·45	1·41	1·41	1·41	1·41
W	0·87	0·85	0·85	0·85	0·97	0·89	0·97	0·98	0·98	0·98
i obs.	43·1	36·4	35·4	31·2	28·3	24·9	16·6	15·4	10·3	9·2
i calc. ...	55·2	47·1	42·5	36·3	33·0	29·3	27·3	22·3	22·0	14·7
G	87	149	146	127	54	78	32	29	19	17
K	1·48	1·48	1·48	1·48	1·48	1·48	1·48	1·48	1·48	1·48
W	1·66	1·58	1·66	1·66	1·83	1·66	1·83	1·58	1·83	1·66
i obs.	47·5	48·7	45·8	34·5	35·0	27·5	28·7	21·4	17·4	15·4
i calc. ...	38·2	43·4	42·5	33·0	32·0	23·6	23·4	17·8	15·4	11·6
G	136	176	131	99	82	79	67	81	41	43
K	2·26	2·12	1·91	1·90	1·91	2·09	1·91	1·90	1·90	2·12
W	1·08	1·15	1·10	1·11	1·10	1·18	1·10	1·11	1·11	1·15
i obs.	44·6	32·0	27·8	24·7	26·6	24·5	19·0	16·0	13·9	10·1
i calc. ...	47·1	33·5	32·8	27·4	26·8	23·6	21·3	17·5	20·4	12·2
G	93	98	66	58	63	72	45	37	32	31
K	1·92	2·05	2·45	2·37	2·45	2·37	1·97	1·97	1·97	1·97
W	2·30	1·93	2·25	2·20	2·25	2·20	2·33	2·33	2·33	2·33
i obs.	24·7	33·2	26·7	19·4	22·6	18·8	16·4	10·9	12·1	7·9
i calc. ...	27·1	31·8	23·7	18·4	21·4	16·8	17·4	11·5	12·2	8·4
G	56	137	77	63	65	61	32	22	24	16

take that kind of ray for which the best determinations have been made by Langley, and this lies in the midst of the most important part of the radiation (37°). For this pencil of rays we find the intensity of radiation at K = 1 and W = 0·3 equal to 62·9; and with the help of the absorption-coefficients we calculate the intensity for K = 0 and W = 0, and find it equal to 105. Then we use Langley's experiments on the spectral distribution of the radiation from a body of 15° C., and calculate the intensity for all other angles of deviation. These intensities are given under the heading M. After this we have to calculate the values for K = 1 and W = 0·3. For

the angle 37° we know it to be 62·9. For any other angle we could take the values A from Table II. if the moon were a body of 15° C. But a calculation of the figures of Very[15] shows that the full moon has a higher temperature, about 100° C. Now the spectral distribution is nearly, but not quite, the same for the heat from a body of 15° C., and for that from one of 100° C. With the help of Langley's figures it is, however, easy to reduce the intensities for the hot body at 100° (the moon) to be valid for a body at 15° (the earth). The values of A reduced in this manner are tabulated below under the heading N.

Angle	40°	39·45	39·30	39·15	39·0	38·45	38·30	38·15	38·0	37·45	37·30
M	3·4	11·6	24·8	45·9	84·0	121·7	161	189	210	210	188
N	3·1	10·1	11·3	13·7	18·0	18·1	11·2	19·6	44·4	59	70

Angle	37°·15	37·0	36·45	36·30	36·15	36·0	35·45	35·30	35·15	35·0	Sum	P.c
M	147	105	103	99	60	51	65	62	43	39	2023	100
N	75·5	62·9	56·4	51·4	39·1	37·9	39·2	37·6	36·0	28·7	743·2	37·2

Table II. Absorption-Coefficients of Carbonic Acid (x) and Aqueous Vapour (y).

Angle of Deviation	log x	log y	A
40°	{ +0.0286 ⟶ 0.0000 }	{ −0.1506 ⟶ −0.1455 }	27.2
39.45	−0.0296	−0.1105	34.5
39.30	−0.0559	−0.0952	29.6
39.15	−0.1070	−0.0862	26.4
39.0	−0.3412	−0.0068	27.5
38.45	−0.2035	−0.3114	24.5
38.30	−0.2438	−0.2362	13.5
38.15	−0.3760	−0.1933	21.4
38.0	−0.1877	−0.3198	44.4
37.45	−0.0931	−0.1576	59.0
37.30	−0.0280	−0.1661	70.0
37.15	−0.0416	−0.2036	75.5
37.0	−0.2067	−0.0484	62.9
36.45	{ −0.2465 ⟶ −0.2466 }	{ +0.0008 ⟶ −0.0000 }	56.4
36.30	−0.2571	−0.0507	51.4
36.15	{ −0.1708 ⟶ −0.1652 }	{ +0.0065 ⟶ −0.0000 }	39.1
36.0	−0.0940	−0.1184	37.9
35.45	−0.1992	−0.0628	36.3
35.30	−0.1742	−0.1408	32.7
35.15	−0.0188	−0.1817	29.8
35.0	−0.0891	−0.1444	21.9

For angles less than 37° one finds, in the manner above described, numbers that are a little inferior to the tabulated ones, which are found by means of the absorption-coefficients of Table II. and the values of N. In this way the sum of the M's is a little greater (6·8 per cent.) than it would be according to the calculation given above. This non-agreement results probably from the circumstance that the spectrum in the observations was not quite pure.

The value 37·2 may possibly be affected with a relatively great error in consequence of the uncertainty of the M-values. In the following calculations it is not so much the value 37·2 that plays the important part, but rather the diminution of the value caused by increasing the quantities K and W. For comparison, it may be mentioned that Langley has estimated the quantity of heat from the moon that passed through the atmosphere (of mean composition) in his researches to be 38 per cent.[16] As the mean atmosphere in Langley's observations corresponded with higher values of K and W than K = 1 and W=0·3, it will be seen that he attributed to the atmosphere a greater transparence for opaque rays than I have done. In accordance with Langley's estimation, we should expect for K = 1 and W = 0·3 a value of about 44 instead of 37·2. How great an influence this difference may exert will be investigated in what follows.

The absorption-coefficients quoted in Table II. are valid for an interval of K between about 1·1 and 2·25, and for W between 0·3 and 2·22. In this interval one may, with the help of those coefficients and the values of N given above, calculate the value of N for another value of K and W, and so in this way obtain by means of summation the total heat that passes through an atmosphere of given condition. For further calculations I have also computed values of N for atmospheres that contain greater quantities of carbonic acid and aqueous vapour. These values must be considered as extrapolated. In the following table (Table III.) I have given these values of N. The numbers printed in italics are found directly in the manner described, these in ordinary type are interpolated from them with the help of Pouillet's exponential formula. The table has two headings, one which runs horizontally and represents the quantity of aqueous vapour (W), and another that runs vertically and represents the quantity of carbonic acid (K) in the atmosphere.

Quite different from this dark heat is the behaviour of the heat from the sun on passing through new parts of the earth's atmosphere. The first parts of the atmosphere exert without doubt a selective absorption of some ultra-red rays, but as soon as these are extinguished the heat seems not to diminish as it traverses new quantities of the gases under discussion. This can easily be shown for aqueous vapour with the help of Langley's actinometric observations from Mountain Camp and Lone Pine in Colorado.[17] These observations were executed at Lone Pine from the 18th of August to the 6th of September 1882 at 7^h 15^m and 7^h 45^m A.M., at 11^h 45^m A.M. and 12^h 15^m P.M., and at 4^h 15^m

Table III. The Transparency of a given Atmosphere for Heat from a body of 15° C.

$\overrightarrow{H_2O}$ $\downarrow CO_2$	0·3	0·5	1·0	1·5	2·0	3·0	4·0	6·0	10·0
1	37·2	35·0	30·7	26·9	23·9	19·3	16·0	10·7	8·9
1·2	34·7	32·7	28·6	25·1	22·2	17·8	14·7	9·7	8·0
1·5	31·5	29·6	25·9	22·6	19·9	15·9	13·0	8·4	6·9
2	27·0	25·3	21·9	19·1	16·7	13·1	10·5	6·6	5·3
2·5	23·5	22·0	19·0	16·6	14·4	11·0	8·7	5·3	4·2
3	20·1	18·8	16·3	14·2	12·3	9·3	7·4	4·2	3·3
4	15·8	14·7	12·7	10·8	9·3	7·1	5·6	3·1	2·0
6	10·9	10·2	8·7	7·3	6·3	4·8	3·7	1·9	0·93
10	6·6	6·1	5·2	4·3	3·5	2·4	1·8	1·0	0·26
20	2·9	2·5	2·2	1·8	1·5	1·0	0·75	0·39	0·07
40	0·88	0·81	0·67	0·56	0·46	0·32	0·24	0·12	0·02

	Morning				Noon				Evening			
	D	W	I	I_1	D	W	I	I_1	D	W	I	I_1
Lone Pine	29·3	0·61	1·424	1·554	23·3	0·46	1·692	1·715	26·6	0·51	1·417	1·351
	21·1	0·84	1·458	1·583	26·9	0·59	1·699	1·721	23·2	0·74	1·428	1·359
Mountain Camp	23·5	0·888	1·790		22·5	0·182	1·904	1·873	24·5	0·205	1·701	1·641
	23·5	0·153	1·749		24·5	0·245	1·890	1·917	22·5	0·32	1·601	1·527

and 4ʰ 45ᵐ P.M. At Mountain Camp the observations were carried out from the 22nd to the 25th of August at the same times of the day, except that only one observation was performed in the morning (at 8ʰ 0ᵐ). I have divided these observations into two groups for each station according to the humidity of the air. In the following little table are quoted, first the place of observation, and after this under D the mean date of the observations (August 1882), under W the quantity of water, under I the radiation observed by means of the actinometer, under I_1 the second observation of the same quantity.

At a very low humidity (Mountain Camp) it is evident that the absorbing power of the aqueous vapour has an influence, for the figures for greater humidity are (with an insignificant exception) inferior to those for less humidity. But for the observations from Lone Pine the contrary seems to be true. It is not permissible to assume that the radiation can be strengthened by its passage through aqueous vapour, but the observed effect must be caused by some secondary circumstance. Probably the air is in general more pure if there is more water-vapour in it than if there is less. The selective diffusion diminishes in consequence of this greater purity, and this secondary effect more than counterbalances the insignificant absorption

that the radiation suffers from the increase of the water-vapour. It is noteworthy that Elster and Geitel have proved that invisible actinic rays of very high refrangibility traverse the air much more easily if it is humid than if it is dry. Langley's figures demonstrate meanwhile that the influence of aqueous vapour on the radiation from the sun is insensible as soon as it has exceeded a value of about 0·4.

Probably the same reasoning will hold good for carbonic acid, for the absorption spectrum of both gases is of the same general character. Moreover, the absorption by carbonic acid occurs at considerably greater wave-lengths, and consequently for much less important parts of the sun's spectrum than the absorption by water-vapour[18]. It is, therefore, justifiable to assume that the radiation from the sun suffers no appreciable diminution if K and W increase from a rather insignificant value (K = 1, W = 0·4) to higher ones.

Before we proceed further we need to examine another question. Let the carbonic acid in the air be, for instance, the same as now (K = 1 for vertical rays), and the quantity of water-vapour be 10 grammes per cubic metre (W = 1 for vertical rays). Then the vertical rays from the earth traverse the quantities K = 1 and W = 1; rays that escape under an angle of 30° with the horizon (air-mass = 2)

Table IV. Mean path of the Earth's rays.

$\overrightarrow{H_2O}$ ↓CO_2	0·3	0·5	1	2	3
0·67	1·69	1·68	1·64	1·57	1·53
1	1·66	1·65	1·61	1·55	1·51
1·5	1·62	1·61	1·57	1·51	1·47
2	1·58	1·57	1·52	1·46	1·43
2·5	1·56	1·54	1·50	1·45	1·41
3	1·52	1·51	1·47	1·44	1·40
3·5	1·48	1·48	1·45	1·42	

traverse the quantities K = 2, W = 2; and so forth. The different rays that emanate from a point of the earth's surface suffer, therefore, a different absorption—the greater, the more the path of the ray declines from the vertical line. It may then be asked how long a path must the total radiation make, that the absorbed fraction of it is the same as the absorbed fraction of the total mass of rays that emanate to space in different directions. For the emitted rays we will suppose that the cosine law of Lambert holds good. With the aid of Table III. we may calculate the absorbed fraction of any ray, and then sum up the total absorbed heat and determine how great a fraction it is of the total radiation. In this way we find for our example the path (air-mass) 1·61. In other words, the total absorbed part of the whole radiation is just as great as if the total radiation traversed the quantities 1·61 of aqueous vapour and of carbonic acid. This number depends upon the composition of the atmosphere, so that it becomes less the greater the quantity of aqueous vapour and carbonic acid in the air. In the following table (IV.) we find this number for different quantities of both gases.

If the absorption of the atmosphere approaches zero, this number approaches the value 2.

III. THERMAL EQUILIBRIUM ON THE SURFACE AND IN THE ATMOSPHERE OF THE EARTH.

As we now have a sufficient knowledge of the absorption of heat by the atmosphere, it remains to examine how the temperature of the ground depends on the absorptive power of the air. Such an investigation has been already performed by Pouillet[19], but it must be made anew, for Pouillet used hypotheses that are not in agreement with our present knowledge.

In our deductions we will assume that the heat that is conducted from the interior of the earth to its surface may be wholly neglected. If a change occurs in the temperature of the earth's surface, the upper layers of the earth's crust will evidently also change their temperature ; but this later process will pass away in a very short time in comparison with the time that is necessary for the alteration of the surface temperature, so that at any time the heat that is transported from the interior to the surface (positive in the winter, negative in the summer) must remain independent of the small secular variations of the surface temperature, and in the course of a year be very nearly equal to zero.

Likewise we will suppose that the heat that is conducted to a given place on the earth's surface or in the atmosphere in consequence of atmospheric or oceanic currents, horizontal or vertical, remains the same in the course of the time considered, and we will also suppose that the clouded part of the sky remains unchanged. It is only the variation of the temperature with the transparency of the air that we shall examine.

All authors agree in the view that there prevails an equilibrium in the temperature of the earth and of its atmosphere. The atmosphere must, therefore, radiate as much heat to space as it gains partly through the absorption of the sun's rays, partly through the radiation from the hotter surface of the earth and by means of ascending currents of air heated by contact with the ground. On the other hand, the earth loses just as much heat by radiation to space and to the atmosphere as it gains by absorption of the sun's rays. If we consider a given place in the atmosphere or on the ground, we must also take into consideration the quantities of heat that are carried to this place by means of oceanic or atmospheric currents. For the radiation we will suppose that Stefan's law of radiation, which is now generally accepted, holds good, or in other words that the quantity of heat (W) that radiates from a body of the albedo $(1-\nu)$ and temperature T (absolute) to another body of the absorption-coefficient β and absolute temperature θ is

$$W = \nu\beta\gamma(T^4 - \theta^4),$$

where γ is the so-called radiation constant (1·21. 10^{-13} per sec. and cm²). Empty space may be regarded as having the absolute temperature 0^{20}.

Provisionally we regard the air as a uniform envelope of the temperature θ and the absorption-coefficient α for solar heat; so that if A calories arrive from the sun in a column of 1·cm² cross-section, αA are absorbed by the atmosphere and $(1 - \alpha)$ A reach the earth's surface. In the A calories there is, therefore, not included that part of the sun's heat which by means of selective reflexion in the atmosphere is thrown out towards space. Further, let β designate the absorption- coefficient of the air for the heat that radiates from the earth's surface; β is also the emission-coefficient of the air for radiation of low temperature—strictly 15°; but as the spectral distribution of the heat varies rather slowly with the temperature, β may be looked on as the emission-coefficient also at the temperature of the air. Let the albedo of the earth's crust be designated by $(1 - v)$, and the quantities of heat that are conveyed to the air and to the earth's surface at the point considered be M and N respectively. As unit of time we may take any period: the best choice in the following calculation is perhaps to take three months for this purpose. As unit of surface we may take 1 cm², and for the heat in the air that contained in a column of 1 cm² cross-section and the height of the atmosphere. The heat that is reflected from the ground is not appreciably absorbed by the air (see p. 252), for it has previously traversed great quantities of water-vapour and carbonic acid, but a part of it may be returned to the ground by means of diffuse reflexion. Let this part not be included in the albedo $(1 - v)$. γ, A, v, M, N, and α are to be considered as constants, β as the independent, and θ and T as the dependent variables.

Then we find for the column of air

$$\beta\gamma\theta^4 = \beta\gamma v(T^4 - \theta^4) + \alpha A + M. \qquad (1)$$

The first member of this equation represents the heat radiated from the air (emission-coefficient β, temperature θ) to space (temperature 0). The second one gives the heat radiated from the soil (1 cm², temperature T, albedo $1-v$) to the air; the third and fourth give the amount of the sun's radiation absorbed by the air, and the quantity of heat obtained by conduction (air-currents) from other parts of the air or from the ground. In the same manner we find for the earth's surface

$$\beta\gamma v(T^4 - \theta^4) + (1 - \beta)\gamma v\, T^4 = (1 - \alpha)vA + N. \qquad (2)$$

The first and second members represent the radiated quantities of heat that go to the air and to space respectively, $(1-\alpha)vA$ is the part of the sun's radiation absorbed, and N the heat conducted to the point considered from other parts of the soil or from the air by means of water- or air-currents.

Combining both these equations for the elimination of θ, which has no considerable interest, we find for T^4

$$T^4 = \frac{\alpha A + M + (1 - \alpha)A(1+v) + N(1+1/v)}{\gamma(1 + v - \beta v)} = \frac{K}{1 + v(1 - \beta)}.$$

$$(3)$$

For the earth's solid crust we may, without sensible error, put v equal to 1, if we except the snowfields, for which we assume $v=0.5$. For the water-covered parts of the earth I have calculated the mean value of v to be 0·925 by aid of the figures of Zenker[21]. We have, also, in the following to make use of the albedo of the clouds. I do not know if this has ever been measured, but it probably does not differ very much from that of fresh fallen snow, which Zöllner has determined to be 0·78, i. e. $v = 0.22$. For old snow the albedo is much less or v much greater; therefore we have assumed 0·5 as a mean value.

The last formula shows that the temperature of the earth augments with β, and the more rapidly the greater v is. For an increase of 1° if $v = 1$ we find the following increases for the values of $v=0.925$, 0·5, and 0·22 respectively:—

β	$v=0.925$	$v=0.5$	$v=0.22$
0·65	0·944	0·575	0·275
0·75	0·940	0·556	0·261
0·85	0·934	0·535	0·245
0·95	0·928	0·512	0·228
1·00	0·925	0·500	0·220

This reasoning holds good if the part of the earth's surface considered does not alter its albedo as a consequence of the altered temperature. In that case entirely different circumstances enter. If, for instance, an element of the surface which is not now snow-covered, in consequence of falling temperature becomes clothed with snow, we must in the last formula not only alter β but also v. In this case we must remember that α is very small compared to β. For α we will choose the value 0·40 in accordance with Langley's[22] estimate. Certainly a great part of this value depends upon the diffusely reflected part of the sun's heat, which is absorbed by the earth's atmosphere, and therefore should not be included in α, as we have defined it above. On the other hand, the sun may in general stand a little lower than in Langley's measurements, which were executed with a relatively high sun, and in consequence of this α may be a little greater, so that these circumstances may compensate each other. For β we will choose the value 0·70, which corresponds when K=1 and W=0·3 (a little below the freezing-point) with the factor 1·66 (see p. 253). In this case we find the relation between T (uncovered) and T_1 (snow-covered surface) to be

$$T^4 : T_1^4 = \frac{A(1+1-0\cdot40)+M}{\gamma(1+1-0\cdot70)} : \frac{A(1+0\cdot50-0\cdot20)+M}{\gamma(1+0\cdot50-0\cdot35)}$$
$$= \frac{1\cdot60+\phi}{1\cdot30} : \frac{1\cdot30+\phi}{1\cdot15},$$

if $M = \phi A$. We have to bear in mind that the mean M for the whole earth is zero, for the equatorial regions negative and for the polar regions positive. For a mean latitude M=0, and in this case T_1 becomes 267·3 if T=273, that is the temperature decreases in consequence of the snow-covering by 5°·7 C.[23] The decrease of temperature from this cause will be valid until $\phi = 1$, i.e. till the heat delivered by convection to the air exceeds the whole radiation of the sun. This can only occur in the winter and in polar regions.

But this is a secondary phenomenon. The chief effect that we examine is the direct influence of an alteration of β upon the temperature T of the earth's surface. If we start from a value T=273 and b=0·70, we find the alteration (t) in the temperature which is caused by the variation of β to the following values to be

$\beta = 0\cdot60$	$t = -5°$ C.
0·80	+5·6
0·90	+11·7
1·00	+18.6.

These values are calculated for $v = 1$, i. e. for the solid crust of the earth's surface, except the snowfields. For surfaces with another value of v, as for instance the ocean or the snowfields, we have to multiply this value t by a fraction given above.

We have now shortly to consider the influence of the clouds. A great part of the earth's surface receives no heat directly from the sun, because the sun's rays are stopped by clouds. How great a part of the earth's surface is covered by clouds we may find from Teisserene de Bort's work[24] on Nebulosity. From tab. 17 of this publication I have determined the mean nebulosity for different latitudes, and found:—

Latitude								
60	45	30	15	0	−15	−30	−45	−60
Nebulosity								
0·603	0·48	0·402	0·511	0·581	0·463	0·53	0·701	

For the part of the earth between 60° S. and 60° N. we find the mean value 0·525, i. e. 52·5 per cent of the sky is clouded. The heat-effect of these clouds may be estimated in the following manner. Suppose a cloud lies over a part of the earth's surface and that no connexion exists between this shadowed part and the neighbouring parts, then a thermal equilibrium will exist between the temperature of the cloud and of the underlying ground. They will radiate to each other and the cloud also to the upper air and to space, and the radiation between cloud and earth may, on account of the slight difference of temperature, be taken as proportional to this difference. Other exchanges of heat by means of air-currents are also, as a first approximation, proportional to this difference. If we therefore suppose the temperature of the cloud to alter (other circumstances, as its height and composition, remaining unchanged), the temperature of the ground under it must also alter in the same manner if the same supply of heat to both subsists—if there were no supply to the ground from neighbouring parts, the cloud and the ground would finally assume the same mean temperature. If, therefore, the temperature of the clouds varies in a determined manner (without alteration of their other properties, as height, compactness, & c.), the ground will undergo the same variations of temperature. Now it will be shown in the sequel that a variation of the carbonic acid of the atmosphere in the same proportion produces nearly the same thermal effect independently of its absolute magnitude (see p. 265). Therefore we may calculate the temperature-variation in this case as if the clouds covered the ground with a thin film of the albedo 0·78 ($v = 0.22$, see p. 256). As now on the average K = 1 and W = 1 nearly, and in this case β is about 0·79, the effect on the clouded part will be only 0·25 of the effect on parts that have $v = 1$. If a like correction is introduced for the ocean ($v = 0.925$) on the supposition that the unclouded part of the earth consists of as much water as of solid ground (which is approximately true, for the clouds are by preference stored up over the ocean), we find a mean effect of, in round numbers, 60 p. c. of that which would exist if the whole earth's surface had $v = 1$. The snow-covered parts are not considered, for, on the one hand, these parts are mostly clouded to about 65 p. c.; further, they constitute only a very small part of the earth (for the whole year on the average only about 4 p. c.), so that the correction for this case would not exceed 0·5 p. c. in the last number 60. And further, on the border countries between snowfields and free soil secondary effects come into play (see p. 257) which compensate, and perhaps overcome, the moderating effect of the snow.

In the foregoing we have supposed that the air is to be regarded as an envelope of perfectly uniform temperature. This is of course not true, and we now proceed to an examination of the probable corrections that must be introduced for eliminating the errors caused by this inexactness. It is evident that the parts of the air which radiate to space are chiefly the external ones, and on the other hand the layers of air which absorb the greatest part of the earth's radiation do not lie very high. From this cause both the radiation from air to space ($\beta\gamma\theta^4$ in eq. 1) and also the radiation of the earth to the air ($\beta\gamma\nu(T^4 - \theta^4)$ in eq. 2), are greatly reduced, and the air has a much greater effect as protecting against the loss of heat to space than is assumed in these equations, and consequently also in eq. (3). If we knew the difference of temperature between the two layers of the air that radiate to space and absorb the earth's radiation, it would be easy to introduce the necessary correction in formulæ (1), (2), and 3). For this purpose I have adduced the following consideration.

As at the mean composition of the atmosphere (K = 1, W = 1) about 80 p. c. of the earth's radiation is absorbed in the air, we may as mean temperature of the absorbing layer choose the temperature at the height where 40 p. c. of the heat is absorbed. Since emission and absorption follow the same quantitative laws, we may as mean temperature of the emitting layer choose the temperature at the height where radiation entering from space in the opposite direction to the actual emission is absorbed to the extent of 40 p. c.

Langley has made four measurements of the absorptive power of water-vapour for radiation from a hot Leslie cube of 100° C.[25] These give nearly the same absorption-coefficient if Pouillet's formula is used for the calculation. From these numbers we calculate that for the absorption of 40 p. c. of the radiation it would be necessary to intercalate so much water-vapour between radiator and bolometer that, when condensed, it would form a layer of water 3·05 millimetres thick. If we now suppose as mean for the whole earth K = 1 and W = 1 (see Table VI.), we find that vertical

rays from the earth, if it were at 100°, must traverse 305 metres of air to lose 40 p. c. Now the earth is only at 15° C., but this cannot make any great difference. Since the radiation emanates in all directions, we have to divide 305 by 1·61 and get in this way 209 metres. In consequence of the lowering of the quantity of water-vapour with the height[26] we must apply a slight correction, so that the final result is 233 metres. Of course this number is a mean value, and higher values will hold good for colder, lower for warmer parts of the earth. In so small a distance from the earth, then, 40 p. c. of the earth's radiation should be stopped. Now it is net wholly correct to calculate with Pouillet's formula (it is rather strange that Langley's figures agree so well with it), which gives necessarily too low values. But, on the other hand, we have not at all considered the absorption by the carbonic acid in this part, and this may compensate for the error mentioned. In the highest layers of the atmosphere there is very little watervapour, so that we must calculate with carbonic acid as the chief absorbent. From a measurement by Ångström,[27] we learn that the absorption-coefficients of water-vapour and of carbonic acid in equal quantities (equal number of molecules) are in the proportion 81 : 62. This ratio is valid for the least hot radiator that Ångström used, and there is no doubt that the radiation of the earth is much less refrangible. But in the absence of a more appropriate determination we may use this for our purpose ; it is probable that for a less hot radiator the absorptive power of the carbonic acid would come out a little greater compared with that of water-vapour, for the absorption-bands of CO_2 are, on the whole, less refrangible than those of H_2O (see pp. 246–248). Using the number 0·03 vol. p. c. for the quantity of carbonic acid in the atmosphere, we find that rays which emanate from the upper part of the air are derived to the extent of 40 p. c. from a layer that constitutes 0·145 part of the atmosphere. This corresponds to a height of about 15,000 metres. Concerning this value we may make the same remark as on the foregoing value. In this case we have neglected the absorption by the small quantities of water-vapour in the higher atmosphere. The temperature-difference of these two layers—the one absorbing, the other radiating—is, according to Glaisher's measurements[28] (with a little extrapolation), about 42° C.

For the clouds we get naturally slightly modified numbers. We ought to take the mean height of the clouds that are illuminated by the sun. As such clouds I have chosen the summits of the cumuli

Table V. Correction Factors for the Radiation.

$\beta =$	Solid ground, $v = 1$	Water, $v = 0.925$	Snow, $v = 0.5$	Clouds ($v = 0.22$) at a height of		
				0 m	2000 m	4000 m
0·65	1·53	1·46	0·95	0·49	0·42	0·37
0·75	1·60	1·52	0·95	0·47	0·40	0·35
0·85	1·69	1·59	0·95	0·46	0·38	0·33
0·95	1·81	1·68	0·94	0·43	0·36	0·31
1·00	1·88	1·74	0·94	0·41	0·35	0·30

that lie at an average height of 1855 metres, with a maximum height of 3611 metres and a minimum of 900 metres.[29] I have made calculations for mean values of 2000 and 4000 metres (corresponding to differences of temperature of 30° C., and 20° C., instead of 42° C., for the earth's surface).

If we now wish to adjust our formulæ (1) to (3), we have in (1) and (2) to introduce θ as the mean temperature of the radiating layer and $(\theta + 42)$, $(\theta + 30)$, or $(\theta + 20)$ respectively for the mean temperature of the absorbing layer. In the first case we should use $v = 1$ and $v = 0.925$ respectively, in the second and the third case $v = 0.22$.

We then find instead of the formula (3)

$$T^4 = \frac{K}{1 + v(1 - \beta)},$$

another very similar formula

$$T^4 = \frac{K}{1 + v(1 - \beta)}, \qquad (4)$$

where c is a constant with the values 1·88, 1·58, and 1·37 respectively for the three cases.[30] In this way we find the following corrected values which represent the variation of temperature, if the solid ground changes its temperature 1° C., in consequence of a variation of β as calculated by means of formula (3).

If we now assume as a mean for the whole earth K = 1 and W = 1, we get $\beta = 0.785$, and taking the clouded part to be 52·5 p. c. and the clouds to have a height of 2000 metres, further assuming the unclouded remainder of the earth's surface to consist equally of land and water, we find as average variation of temperature

$1.63 \times 0.2385 + 1.54 \times 0.2385 + 0.39 \times 0.525 = 0.979,$

or very nearly the same effect as we may calculate directly from the formula (3). On this ground I have used the simpler formula.

In the foregoing I have remarked that according to my estimation the air is less transparent for dark heat than on Langley's estimate and nearly in the proportion 37·2 : 44. How great an influence this difference may exercise is very easily calculated with the help of formula (3) or (4). According to Langley's valuation, the effect should be nearly 15 p. c. greater than according to mine. Now I think that my estimate agrees better with the great absorption that Langley has found for heat from terrestrial radiating bodies (see p. 260), and in all circumstances I have preferred to slightly under-estimate than to overrate the effect in question.

IV. CALCULATION OF THE VARIATION OF TEMPERATURE THAT WOULD ENSUE IN CONSEQUENCE OF A GIVEN VARIATION OF THE CARBONIC ACID IN THE AIR

We now possess all the necessary data for an estimation of the effect on the earth's temperature which would be the result of a given variation of the aërial carbonic acid. We only need to determine the absorption-coefficient for a certain place with the help of Table III. if we know the quantity of carbonic acid (K = 1 now) and water-vapour (W) of this place. By the aid of Table IV. we at first determine the factor ρ that gives the mean path of the radiation from the earth through the air and multiply the given K- and W-values by this factor. Then we determine the value of β which corresponds to ρK and ρW. Suppose now that the carbonic acid had another concentration K_1 (*e.g.* $K_1 = 1.5$). Then we at first suppose W unaltered and seek the new value of ρ, say ρ_1, that is valid on this supposition. Next we have to seek β, which corresponds to $\rho_1 K_1$ ($1.5\rho_1$) and ρ_1W. From formula (3) we can then easily calcu-

Table VI. Mean Temperature, Relative and Absolute Humidity.*

Latitude	Mean Temperature					Mean Relative Humidity					Mean Absolute Humidity				
	Dec.–Feb.	March–May	June–Aug.	Sept.–Nov.	Mean of the year	Dec.–Feb.	March–May	June–Aug.	Sept.–Nov.	Mean of the year	Dec.–Feb.	March–May	June–Aug.	Sept.–Nov.	Mean of the year
70	−21·1	−8·3	+7·5	−6·0	−7·0	86	81	77	84	82	1·15	2·14	6·22	2·84	3·09
60	−11·2	+0·2	+13·5	+2·2	+1·2	83	74	76	80	78·2	2·22	3·82	8·82	4·7	4·9
50	−1·4	+7·8	+18·7	+9·7	+8·7	78	73	69	76	74	3·86	5·98	10·8	7·16	6·95
40	+8·4	+14·5	+21·8	+16·6	+15·3	73	68	67	71	69·7	6·53	8·63	13·4	10·13	9·7
30	+17·0	+21·5	+26·0	+23·0	+21·9	71	68	70	73	70·5	10·36	12·63	17·1	15·0	13·8
20	+23·2	+25·5	+26·8	+25·9	+25·4	74	73	78	77	75·5	15·3	17·0	19·6	16·8	17·2
10	+25·5	+25·8	+25·4	+25·5	+25·5	77	78	82	81	79·5	17·7	18·9	19·9	19·3	18·9
0	+25·7	+25·5	+24·0	+25·0	+25·1	81	81	82	80	81	19·4	19·0	17·9	18·3	18·7
−10	+24·9	+24·0	+20·8	+23·1	+23·2	79	78	80	77	78·5	18·0	17·1	14·6	16·0	16·4
−20	+22·4	+20·5	+16·4	+19·3	+19·7	75	79	80	75	77·2	14·8	14·0	11·1	13·0	13·2
−30	+17·5	+15·2	+11·3	+14·2	+14·5	75	80	80	79	78·5	11·1	10·4	8·1	9·6	9·8
−40	+11·6	+9·5	+5·9	+8·2	+8·7	81	81	83	79	81	8·34	7·08	5·94	6·63	6·99
−50	+5·3	+2·0	−0·4	+1·6	+2·1	83	79	—	—	—	5·74	4·46	—	—	—
−60															

* From the figures for temperature and relative humidity I have calculated the absolute humidity in grams per cubic metre.

late the alteration (*t*) (here increase) in the temperature at the given place which will accompany the variation of β from β to β_1. In consequence of the variation (*t*) in the temperature, W must also undergo a variation. As the relative humidity does not vary much, unless the distribution of land and water changes (see table 8 of my original memoir), I have supposed that this quantity remains constant, and thereby determined the new value W_1 of W. A fresh approximation gives in most cases values of W_1 and β_1 which may be regarded as definitive. In this way, therefore, we get the variation of temperature as soon as we know the actual temperature and humidity at the given place.

In order to obtain values for the temperature for the whole earth, I have calculated from Dr. Buchan's charts of the mean temperature at different places in every month the mean temperature in every district that is contained between two parallels differing by 10 and two meridians differing by 20 degrees, (*e.g.*, between 0° and 10° N. and 160° and 180° W.). The humidity has not as yet been sufficiently examined for the whole earth; and I have therefore collected a great many measurements of the relative humidity at different places (about 780) on the earth and marked them down in maps of the world, and thereafter estimated the mean values for every district. These quantities I have tabulated for the four seasons, Dec.–Feb., March–May, June–Aug., and Sept.–Nov. The detailed table and the observations used are to be found in my original memoir : here I reproduce only the mean values for every tenth parallel (Table VI.).

By means of these values, I have calculated the mean alteration of temperature that would follow if the quantity of carbonic acid varied from its present mean value (K = 1) to another, viz. to K = 0·67, 1·5, 2, 2·5, and 3 respectively. This calculation is made for every tenth parallel, and separately for the four seasons of the year. The variation is given in Table VII.

A glance at this Table shows that the influence is nearly the same over the whole earth. The influence has a minimum near the equator, and increases from this to a flat maximum that lies the further from the equator the higher the quantity of carbonic acid in the air. For K = 0·67 the maximum effect lies about the 40th parallel, for K = 1·5 on the 50th, for K = 2 on the 60th, and for higher K-values above the 70th parallel. The influence is in general greater in the winter than in the summer, except in the case of the parts that lie between the maximum and the pole. The influence will

also be greater the higher the value of *v*, that is in general somewhat greater for land than for ocean. On account of the nebulosity of the Southern hemisphere, the effect will be less there than in the Northern hemisphere. An increase in the quantity of carbonic acid will of course diminish the difference in temperature between day and night. A very important secondary elevation of the effect will be produced in those places that alter their albedo by the extension or regression of the snow-covering (see p. 257), and this secondary effect will probably remove the maximum effect from lower parallels to the neighbourhood of the poles.[31]

It must be remembered that the above calculations are found by interpolation from Langley's numbers for the values K = 0·67 and K = 1·5, and that the other numbers must be regarded as extrapolated. The use of Pouillet's formula makes the values for K = 0·67 probably a little too small, those for K = 1·5 a little too great. This is also without doubt the case for the extrapolated values, which correspond to higher values of K.

We may now inquire how great must the variation of the carbonic acid in the atmosphere be to cause a given change of the temperature. The answer may be found by interpolation in Table VII. To facilitate such an inquiry, we may make a simple observation. If the quantity of carbonic acid decreases from 1 to 0·67, the fall of temperature is nearly the same as the increase of temperature if this quantity augments to 1·5. And to get a new increase of this order of magnitude (3°·4), it will be necessary to alter the quantity of carbonic acid till it reaches a value nearly midway between 2 and 2·5. thus if the quantity of carbonic acid increases in geometric inogression, the augmentation of the temperature will increase nearly in arithmetic progression. This rule–which naturally holds good only in the part investigated– will be useful for the following summary estimations.

GEOLOGICAL CONSEQUENCES

I should certainly not have undertaken these tedious calculations if an extraordinary interest had not been connected with them. In the Physical Society of Stockholm there have been occasionally very lively discussions on the probable causes of the Ice Age ; and these discussions have, in my opinion, led to the conclusion that there exists as yet no satisfactory hypothesis that could explain

TABLE VII. Variation of Temperature caused by a given Variation of Carbonic Acid.

Latitude	Carbonic Acid = 0.67					Carbonic Acid = 1.5					Carbonic Acid = 2.0					Carbonic Acid = 2.5					Carbonic Acid = 3.0				
	Dec.–Feb.	March–May	June–Aug.	Sept.–Nov.	Mean of the year	Dec.–Feb.	March–May	June–Aug.	Sept.–Nov.	Mean of the year	Dec.–Feb.	March–May	June–Aug.	Sept.–Nov.	Mean of the year	Dec.–Feb.	March–May	June–Aug.	Sept.–Nov.	Mean of the year	Dec.–Feb.	March–May	June–Aug.	Sept.–Nov.	Mean of the year
70	−2.9	−3.0	−3.4	−3.1	−3.1	3.3	3.4	3.8	3.6	3.52	6.0	6.1	6.0	6.1	6.05	7.9	8.0	7.9	8.0	7.95	9.1	9.3	9.4	9.4	9.3
60	−3.0	−3.2	−3.4	−3.3	−3.22	3.4	3.7	3.6	3.8	3.62	6.1	6.1	5.8	6.1	6.02	8.0	8.0	7.6	7.9	7.87	9.3	9.5	8.9	9.5	9.3
50	−3.2	−3.3	−3.3	−3.4	−3.3	3.7	3.8	3.4	3.7	3.65	6.1	6.1	5.5	6.0	5.92	8.0	7.9	7.0	7.9	7.7	9.5	9.4	8.6	9.2	9.17
40	−3.4	−3.4	−3.2	−3.3	−3.32	3.7	3.6	3.3	3.5	3.52	6.0	5.8	5.4	5.6	5.7	7.9	7.6	6.9	7.3	7.42	9.3	9.0	8.2	8.8	8.82
30	−3.3	−3.2	−3.1	−3.1	−3.17	3.5	3.3	3.2	3.5	3.47	5.6	5.4	5.0	5.2	5.3	7.2	7.0	6.6	6.7	6.87	8.7	8.3	7.5	7.9	8.1
20	−3.1	−3.1	−3.0	−3.1	−3.07	3.5	3.2	3.1	3.2	3.25	5.2	5.0	4.9	5.0	5.02	6.7	6.6	6.3	6.6	6.52	7.9	7.5	7.2	7.5	7.52
10	−3.1	−3.0	−3.0	−3.0	−3.02	3.2	3.2	3.1	3.1	3.15	5.0	5.0	4.9	4.9	4.95	6.6	6.4	6.3	6.4	6.42	7.4	7.3	7.2	7.3	7.3
0	−3.0	−3.0	−3.1	−3.0	−3.02	3.1	3.1	3.2	3.2	3.15	4.9	4.9	5.0	5.0	4.95	6.4	6.4	6.6	6.6	6.5	7.3	7.3	7.4	7.4	7.35
−10	−3.1	−3.1	−3.2	−3.1	−3.12	3.2	3.2	3.2	3.2	3.2	5.0	5.0	5.2	5.1	5.07	6.6	6.6	6.7	6.7	6.65	7.4	7.5	8.0	7.6	7.62
−20	−3.1	−3.2	−3.3	−3.2	−3.2	3.2	3.2	3.4	3.3	3.27	5.2	5.3	5.5	5.4	5.35	6.7	6.8	7.0	7.0	6.87	7.9	8.1	8.6	8.3	8.22
−30	−3.3	−3.3	−3.4	−3.4	−3.35	3.4	3.5	3.7	3.5	3.52	5.5	5.6	5.8	5.6	5.62	7.0	7.2	7.7	7.4	7.32	8.6	8.7	9.1	8.8	8.8
−40	−3.4	−3.4	−3.3	−3.4	−3.37	3.6	3.7	3.8	3.7	3.7	5.8	6.0	6.0	6.0	5.95	7.7	7.9	7.9	7.9	7.85	9.1	9.2	9.4	9.3	9.25
−50	−3.2	−3.3	—	—	—	3.8	3.7	—	—	—	6.0	6.1	—	—	—	7.9	8.0	—	—	—	9.4	9.5	—	—	—
−60																									

how the climatic conditions for an ice age could be realized in so short a time as that which has elapsed from the days of the glacial epoch. The common view hitherto has been that the earth has cooled in the lapse of time; and if one did not know that the reverse has been the case, one would certainly assert that this cooling must go on continuously. Conversations with my friend and colleague Professor Högbom, together with the discussions above referred to, led me to make a preliminary estimate of the probable effect of a variation of the atmospheric carbonic acid on the temperature of the earth. As this estimation led to the belief that one might in this way probably find an explanation for temperature variations of 5°–10° C., I worked out the calculation more in detail, and lay it now before the public and the critics.

From geological researches the fact is well established that in Tertiary times there existed a vegetation and an animal life in the temperate and arctic zones that must have been conditioned by a much higher temperature than the present in the same regions.[32] The temperature in the arctic zones appears to have exceeded the present temperature by about 8 or 9 degrees. To this genial time the ice age succeeded, and this was one or more times interrupted by interglacial periods with a climate of about the same character as the present, sometimes even milder. When the ice age had its greatest extent, the countries that now enjoy the highest civilization were covered with ice. This was the case with Ireland, Britain (except a small part in the south), Holland, Denmark, Sweden and Norway, Russia (to Kiev, Orel, and Nijni Novgorod), Germany and Austria (to Harz, Erz-Gebirge, Dresden, and Cracow). At the same time an ice-cap from the Alps covered Switzerland, parts of France, Bavaria south of the Danube, the Tyrol, Styria, and other Austrian countries, and descended into the northern part of Italy. Simultaneously, too, North America was covered with ice on the west coast to the 47th parallel, on the east coast to the 40th, and in the central part to the 37th (confluence of the Mississippi and Ohio rivers). In the most different parts of the world, too, we have found traces of a great ice age, as in the Caucasus, Asia Minor, Syria, the Himalayas, India, Thian Shan, Altai, Atlas, on Mount Kenia and Kilimandjaro (both very near to the equator), in South Africa, Australia, New Zealand, Kerguelen, Falkland Islands, Patagonia and other parts of South America. The geologists in general are inclined to think that these glaciations were simultaneous on the whole earth[33]; and this most natural view would probably have been generally accepted, if the theory of Croll, which demands a genial age on the Southern hemisphere at the same time as an ice age on the Northern and *vice versâ*, had not influenced opinion. By measurements of the displacement of the snow-line we arrive at the result,—and this is very concordant for different places–that the temperature at that time must have been 4°–5° C., lower than at present. The last glaciation must have taken place in rather recent times, geologically speaking, so that the human race certainly had appeared at that period. Certain American geologists hold the opinion that since the close of the ice age only some 7000 to 10,000 years have elapsed, but this most probably is greatly underestimated.

One may now ask, How much must the carbonic acid vary according to our figures, in order that the temperature should attain the same values as in the Tertiary and Ice ages respectively? A simple calculation shows that the temperature in the arctic regions would rise about 8° to 9° C., if the carbonic acid increased to 2·5 or 3 times its present value. In order to get the temperature of the ice age between the 40th and 50th parallels, the carbonic acid in the air should sink to 0·62–0·55 of its present value (lowering of temperature 4°–5° C.). The demands of the geologists, that at the genial epochs the climate should be more uniform than now, accords very well with our theory. The geographical annual and diurnal ranges of temperature would be partly smoothed away, if the quantity of carbonic acid was augmented. The reverse would be the case (at least to a latitude of 50° from the equator), if the carbonic acid diminished in amount. But in both these cases I incline to think that the secondary action (see p. 257) due to the regress or the progress of the snow-covering would play the most important rôle. The theory demands also that, roughly speaking, the whole earth should have undergone about the same variations of temperature, so that according to it genial or glacial epochs must have occurred simultaneously on the whole earth. Because of the greater nebulosity of the Southern hemisphere, the variations must there have been a little less (about 15 per cent.) than in the Northern hemisphere. The ocean currents, too, must there, as at the present time, have effaced the differences in temperature at different latitudes to a greater extent than in the Northern hemisphere. This effect also results from the greater nebulosity in the arctic zones than in the neighbourhood of the equator.

There is now an important question which should be answered, namely:—Is it probable that such great variations in the quantity of carbonic acid as our theory requires have occurred in relatively short geological times? The answer to this question is given by Prof. Högbom. As his memoir on this question may not be accessible to most readers of these pages, I have summed up and translated his utterances which are of most importance to our subject[34]:—

"Although it is not possible to obtain exact quantitative expressions for the reactions in nature by which carbonic acid is developed or consumed, nevertheless there are some factors, of which one may get an approximately true estimate, and from which certain conclusions that throw light on the question may be drawn. In the first place, it seems to be of importance to compare the quantity of carbonic acid now present in the air with the quantities that are being transformed. If the former is insignificant in comparison with the latter, then the probability for variations is wholly other than in the opposite case.

"On the supposition that the mean quantity of carbonic acid in the air reaches 0·03 vol. per cent., this number represents 0·045 per cent. by weight, or 0·342 millim. partial pressure, or 0·466 gramme of carbonic acid for every cm^2 of the earth's surface. Reduced to carbon this quantity would give a layer of about 1 millim. thickness over the earth's surface. The quantity of carbon that is fixed in the living organic world can certainly not be estimated with the same degree of exactness; but it is evident that the numbers that might express this quantity ought to be of the same order of magnitude, so that the carbon in the air can neither be conceived of as very great nor as very little, in comparison with the quantity of carbon occurring in organisms. With regard to the great rapidity with which the transformation in organic nature proceeds, the disposable quantity of carbonic acid is not so excessive that changes caused by climatological or other reasons in the velocity and value of that transformation might be not able to cause displacements of the equilibrium.

"The following calculation is also very instructive for the appreciation of the relation between the quantity of carbonic acid in the air and the quantities that are transformed. The world's present production of coal reaches in round numbers 500 millions of tons per annum, or 1 ton per km^2 of the earth's surface. Transformed into carbonic acid, this quantity would correspond to about a thousandth part of the carbonic acid in the atmosphere. It represents a layer of limestone of 0·003 millim. thickness over the whole globe, or 1·5 km.^{3} in cubic measure. This quantity of carbonic acid, which is supplied to the atmosphere chiefly by modern industry, may be regarded as completely compensating the quantity of carbonic acid that is consumed in the formation of limestone (or other mineral carbonates) by the weathering or decomposition of silicates. From the determination of the amounts of dissolved substances, especially carbonates, in a number of rivers in different countries and climates, and of the quantity of water flowing in these rivers and of their drainage-surface compared with the land-surface of the globe, it is estimated that the quantities of dissolved carbonates that are supplied to the ocean in the course of a year reach at most the bulk of 3 km.^{3} As it is also proved that the rivers the drainage regions of which consist of silicates convey very unimportant quantities of carbonates compared with those that flow through limestone regions, it is permissible to draw the conclusion, which is also strengthened by other reasons, that only an insignificant part of these 3 km^3 of carbonates is formed directly by decomposition of silicates. In other words, only an unimportant part of this quantity of carbonate of lime can be derived from the process of weathering in a year. Even though the number given were on account of inexact or uncertain assumptions erroneous to the extent of 50 per cent. or more, the comparison instituted is of very great interest, as it proves that the most important of all the processes by means of which carbonic acid has been removed from the atmosphere in all times, namely the chemical weathering of siliceous minerals, is of the same order of magnitude as a process of contrary effect, which is caused by the industrial development of our time, and which must be conceived of as being of a temporary nature.

"In comparison with the quantity of carbonic acid which is fixed in limestone (and other carbonates), the carbonic acid of the air vanishes. With regard to the thickness of sedimentary formations and the great part of them that is formed by limestone and other carbonates, it seems not improbable that the total quantity of carbonates would cover the whole earth's surface to a height of hundreds of metres. If we assume 100 metres,—a number that may be inexact in a high degree, but probably is underestimated,—we find that about 25,000 times as much carbonic acid is fixed to lime in the sedimentary formations as exists free in the air. Every molecule of carbonic acid in this mass of limestone has, however, existed in and passed through the atmosphere in the course of time. Although we neglect all other factors which may have influenced the quantity of carbonic acid in the air, this number lends but very slight probability to the hypothesis, that this quantity should in former geological epochs have changed within limits which do not differ much from the present amount. As the process of weathering has consumed quantities of carbonic acid many thousand times greater than the amount now disposable in the air, and as this process from different geographical, climatological and other causes has in all likelihood proceeded with very different intensity at different epochs, the probability of important variations in the quantity of carbonic acid seems to be very great, even if we take into account the compensating processes which, as we shall see in what follows, are called forth as soon as, for one reason or another, the production or consumption of carbonic acid tends to displace the equilibrium to any considerable degree. One often hears the opinion expressed, that the quantity of carbonic acid in the air ought to have been very much greater formerly than now, and that the diminution should arise from the circumstance that carbonic acid has been taken from the air and stored in the earth's crust in the form of coal and carbonates. In many cases this hypothetical diminution is ascribed only to the formation of coal, whilst the much more important formation of carbonates is wholly overlooked. This whole method of reasoning on a continuous diminution of the carbonic acid in the air loses all foundation in fact, notwithstanding that enormous quantities of carbonic acid in the course of time have been fixed in carbonates, if we consider more closely the processes by means of which carbonic acid has in all times been supplied to the atmosphere. From these we may well conclude that enormous variations have occurred, but not that the variation has always proceeded in the same direction.

"Carbonic acid is supplied to the atmosphere by the following processes:—(1) volcanic exhalations and geological phenomena connected therewith; (2) combustion of carbonaceous meteorites in the higher regions of the atmosphere; (3) combustion and decay of organic bodies; (4) decomposition of carbonates; (5) liberation of carbonic acid mechanically inclosed in minerals on their fracture or decomposition. The carbonic acid of the air is consumed chiefly by the following processes:—(6) formation of carbonates from silicates on weathering; and (7) the consumption of carbonic acid by vegetative processes. The ocean, too, plays an important rôle as a regulator of the quantity of carbonic acid in the air by means of the absorptive power of its water, which gives off carbonic acid as its temperature rises and absorbs it as it cools. The processes named under (4) and (5) are of little significance, so that they may be omitted. So too the processes (3) and (7), for the circulation of matter in the organic world goes on so rapidly that their variations cannot have any sensible influence. From this we must except periods in which great quantities of organisms were stored up in sedimentary formations and thus subtracted from the circulation, or in which such stored-up products were, as now, introduced anew into the circulation. The source of carbonic acid named in (2) is wholly incalculable.

"Thus the processes (1), (2), and (6) chiefly remain as balancing each other. As the enormous quantities of carbonic acid (representing a pressure of many atmospheres) that are now fixed in the limestone of the earth's crust cannot be conceived to have existed in the air but as an insignificant fraction of the whole at any one time since organic life appeared on the globe, and since therefore the consumption through weathering and formation of carbonates must have been compensated by means of continuous supply, we must regard volcanic exhalations as the chief source of carbonic acid for the atmosphere.

"But this source has not flowed regularly and uniformly. Just as single volcanoes have their periods of variation with alternating relative rest and intense activity, in the same manner the globe as a whole seems in certain geological epochs to have exhibited a more violent and general volcanic activity, whilst other epochs have been marked by a comparative quiescence of the volcanic forces. It seems therefore probable that the quantity of carbonic acid in the air has undergone nearly simultaneous variations, or at least that this factor has had an important influence.

"If we pass the above-mentioned processes for consuming and producing carbonic acid under review, we find that they evidently do not stand in such a relation to or dependence on one another that any probability exists for the permanence of an equilibrium of the carbonic acid in the atmosphere. An increase or decrease of the supply continued during geological periods must, although it may not be important, conduce to remarkable alterations of the quantity of carbonic acid in the air, and there is no conceivable hindrance to imagining that this might in a certain geological period have been several times greater, or on the other hand considerably less, than now."

As the question of the probability of quantitative variation of the carbonic acid in the atmosphere is in the most decided manner answered by Prof. Högbom, there remains only one other point to which I wish to draw attention in a few words, namely: Has no one hitherto proposed any acceptable explanation for the occurrence of genial and glacial periods? Fortunately, during the progress of the foregoing calculations, a memoir was published by the distinguished Italian meteorologist L. De Marchi which relieves me from answering the last question.[35] He examined in detail the different theories hitherto proposed—astronomical, physical, or geographical, and of these I here give a short *résumé*. These theories assert that the occurrence of genial or glacial epochs should depend on one or other change in the following circumstances:

(1) The temperature of the earth's place in space.
(2) The sun's radiation to the earth (solar constant).
(3) The obliquity of the earth's axis to the ecliptic.
(4) The position of the poles on the earth's surface.
(5) The form of the earth's orbit, especially its eccentricity (Croll).
(6) The shape and extension of continents and oceans.
(7) The covering of the earth's surface (vegetation).
(8) The direction of the oceanic and aërial currents.
(9) The position of the equinoxes.

De Marchi arrives at the conclusion that all these hypotheses must be rejected (p. 207). On the other hand, he is of the opinion that a change in the transparency of the atmosphere would possibly give the desired effect. According to his calculations, "a lowering of this transparency would effect a lowering of the temperature on the whole earth, slight in the equatorial regions, and increasing with the latitude into the 70th parallel, nearer the poles again a little less. Further, this lowering would, in non-tropical regions, be less on the continents than on the ocean and would diminish the annual variations of the temperature. This diminution of the air's transparency ought chiefly to be attributed to a greater quantity of aqueous vapour in the air, which would cause not only a direct cooling but also copious precipitation of water and snow on the continents. The origin of this greater quantity of water-vapour is not easy to explain." De Marchi has arrived at wholly other results than myself, because he has not sufficiently considered the important quality of selective absorption which is possessed by aqueous vapour. And, further, he has forgotten that if aqueous vapour is supplied to the atmosphere, it will be condensed till the former condition is reached, if no other change has taken place. As we have seen, the mean relative humidity between the 40th and 60th parallels on the northern hemisphere is 76 per cent. If, then, the mean temperature sank from its actual value + 5·3 by 4°–5° C., *i. e.* to + 1·3 or + 0·3, and the aqueous vapour remained in the air, the relative humidity would increase to 101 or 105 per cent. This is of course impossible, for the relative humidity cannot exceed 100 per cent. in the free air. *A fortiori* it is impossible to assume that the absolute humidity could have been greater than now in the glacial epoch.

As the hypothesis of Croll still seems to enjoy a certain favour with English geologists, it may not be without interest to cite the utterance of De Marchi on this theory, which he, in accordance with its importance, has examined more in detail than the others. He says, and I entirely agree with him on this point:—"Now I think I may conclude that from the point of view of climatology or meteorology, in the present state of these sciences, the hypothesis of Croll seems to be wholly untenable as well in its principles as in its consequences".[36]

It seems that the great advantage which Croll's hypothesis promised to geologists, viz. of giving them a natural chronology, predisposed them in favour of its acceptance. But this circumstance, which at first appeared advantageous, seems with the advance of investigation rather to militate against the theory, because it becomes more and more impossible to reconcile the chronology demanded by Croll's hypothesis with the facts of observation.

I trust that after what has been said the theory preposed in the foregoing pages will prove useful in explaining some points in geological climatology which have hitherto proved most difficult to interpret.

ADDENDUM[37]

As the nebulosity is very different in different latitudes, and also different over the sea and over the continents, it is evident that the influence of a variation in the carbonic acid of the air will be

Latitude	Nebulosity		Continent per cent	Reduction factor			K = 0·67		K = 1·5	
	Continent	Ocean		Continent	Ocean.	Mean	Continent	Ocean	Continent	Ocean
70										
	58·1	66·7	72·1	0·899	0·775	0·864	−2·8	−2·4	3·1	2·7
60										
	56·3	67·6	55·8	0·924	0·763	0·853	−3·0	−2·4	3·3	2·7
50										
	45·7	63·3	52·9	1·057	0·813	0·942	−3·5	−2·7	3·8	2·9
40										
	36·5	52·5	42·9	1·177	0·939	1·041	−3·9	−3·1	4·1	3·3
30										
	28·5	47·2	38·8	1·296	1·009	1·120	−4·1	−3·2	4·5	3·5
20										
	28·5	47·0	24·2	1·308	1·017	1·087	−4·1	−3·2	4·3	3·4
10										
	50·1	56·7	23·3	1·031	0·903	0·933	−3·1	−2·7	3·3	2·9
0										
	54·8	59·7	24·2	0·97	0·867	0·892	−2·9	−2·6	3·1	2·8
−10										
	47·8	54·0	22·5	1·056	0·932	0·96	−3·3	−2·9	3·4	3·0
−20										
	29·6	49·6	23·3	1·279	0·979	0·972	−4·1	−3·1	4·2	3·2
−30										
	38·9	51·0	12·5	1·152	0·958	0·982	−3·8	−3·2	4·0	3·4
−40										
	62·0	61·1	2·5	0·86	0·837	0·838	−2·9	−2·8	3·2	3·1
−50										
	71·0	71·5	0·9	0·749	0·719	0·719				
−60										

somewhat different from that calculated above, where it is assumed that the nebulosity is the same over the whole globe. I have therefore estimated the nebulosity at different latitudes with the help of the chart published by Teisserenc de Bort, and calculated the following table for the value of the variation of temperature, if the carbonic acid decreases to 0·67 or increases to 1·5 times the present quantity. In the first column is printed the latitude; in the second and third the nebulosity over the continent and over the ocean; in the fourth the extension of the continent in hundredths of the whole area. After this comes, in the fifth and sixth columns, the reduction factor with which the figures in the table are to be multiplied for getting the true variation of temperature over continents and over oceans, and, in the seventh column, the mean of both these correction factors. In the eighth and ninth columns the temperature variations for K = 0·67, and in the tenth and eleventh the corresponding values for K = 1·5 are tabulated.

The mean value of the reduction factor N. of equator is for the continent (to 70° N. lat.) 1·098 and for the ocean 0·927, in mean 0·996. For the southern hemisphere (to 60° S. lat.) it is found to be for the continent 1·095, for the ocean 0·871, in mean 0·907. The influence in the southern hemisphere will, therefore, be about 9 per cent. less than in the northern. In consequence of the minimum of nebulosity between 20° and 30° latitude in both hemispheres, the maximum effect of the variation of carbonic acid is displaced towards the equator, so that it falls at about 25° latitude in the two cases of K = 0·67 and K = 1·5.

NOTES

1 'Heat a Mode of Motion,' 2nd ed. p. 405 (Lond., 1865).
2 *Mém. de l'Ac. R. d. Sci. de l'Inst. de France*, t. vii. 1827.
3 *Comptes rendus*, t. vii. p. 41 (1838).
4 Langley, 'Professional Papers of the Signal Service,' No. 15. "Researches on Solar Heat," p. 123 (Washington, 1884).

5 Langley, "The Temperature of the Moon." Mem. of the National Academy of Sciences, vol. iv. 9th mem. p. 193 (1890).

6 Langley, 'Prof. Papers,' No. 15, p. 151. I have tried to calculate a formula for the value of the absorption due to the selective reflexion, as determined by Langley. Among the different formulae examined, the following agrees best with the experimental results:—

$$\log a = b\,(1/\lambda) + c\,(1/\lambda)3.$$

I have determined the coefficients of this formula by aid of the method of least squares, and have found—

$$b = -0{\cdot}0463, \qquad c = -0{\cdot}008204.$$

a represents the strength of a ray of the wave-length λ (expressed in μ) after it has entered with the strength 1 and passed through the air-mass 1. The close agreement with experiment will be seen from the following table:–

λ	$a^{1/7 \cdot 6}$ (obs.)	$a^{1/7 \cdot 6}$ (calc.)	Prob. error
$0{\cdot}358\ \mu$	0·904	0·911	
0·383	0·920	0·923	0·0047
0·416	0·935	0·934	
0·440	0·942	0·941	
0·468	0·950	0·947	0·0028
0·550	0·960	0·960	
0·615	0·968	0·967	
0·781	0·978	0·977	
0·870	0·982	0·980	0·0017
1·01	0·985	0·984	
1·20	0·987	0·987	
1·50	0·989	0·990	0·0011
2·59	0·990	0·993	0·0018

For ultra-violet rays the absorption becomes extremely great in accordance with facts.

As one may see from the probable errors which I have placed alongside for the least concordant values and also for one value (1·50 μ), where the probable error is extremely small, the differences are just of the magnitude that one might expect in an exactly fitting formula. The curves for the formula and for the experimental values cut each other at four points ($1/\lambda = 2{\cdot}43$, 1·88, 1·28, and 0·82 respectively). From the formula we may estimate the value of the selective reflexion for those parts of the spectrum that prevail in the heat from the moon and the earth (angle of deviation $= 38$ –$36°$, $\lambda = 10{\cdot}4$–$24{\cdot}4$ μ). We find that the absorption from this cause varies between 0·5 and 1 p. c. for air-mass 1. This insensible action, which is wholly covered by the experimental errors, I have neglected in the following calculations.

7 Vide Winkelmann, *Handbuch der Physik.*

8 Cf., e.g., Trabert, *Meteorologische Zeitschrift*, Bd. ii. p. 238 (1894).

9 This unit nearly corresponds to the mean humidity of the air (see Table VI. p. 264).

10 Paschen, Wied. *Ann.* 1. p. 409, 1893; li. p. 1, lii. p. 2C9, and liii. p. 334, 1894, especially vol. l. tab. ix. fig. 5, curve 1 for carbonic acid, curve 2 for aqueous vapour.

Ångström, *Bihang till K. Vet.-Ak. Handlingar*, Bd. xv. Afd. 1, No. 9, p. 15, 1889; *Öfversigt af K. Vet.-Ak, Förhandl.* 1889, No. 9, p. 553.

11 Langley, *Ann. Ch. et Phys.* sér. 6, t. xvii. pp. 323 and 326, 1889, Prof. Papers, No. 15, plate 12. Lamansky attributed his absorption-bands, which probably had this place, to the absorbing power of aqueous vapour (Pogg. *Ann.* cxlvi. p. 200, 1872).

12 It must be remembered that at this point the spectrum of Paschen was very weak, so that the coincidence with his figure may be accidental,

13 After having been sifted through an atmosphere of K = 1·1 and W = 0·3.

14 'Temperature of the Moon,' plate 5.

15 "The Distribution of the Moon's Heat," Utrecht Society of Arts and Sc. The Hague, 1891.

16 Langley, 'Temperature of the Moon,' p. 197.

17 Langley, 'Researches on Solar Heat,' pp. 94, 98, and 177.

18 *Cf.* above, pages 246–248, and Langley's curve for the solar spectrum, *Ann. d. Ch. et d. Phys.* sér. 6, t. xvii. pp. 323 and 326 (1889); 'Prof. Papers,' No. 15, plate 12.

19 Pouillet, *Comptes rendus*, t. vii. p. 41 (1838).

20 Langley, 'Prof. Papers,' No. 15, p. 122. "The Temperature of the Moon," p. 206.

21 Zenker, *Die Vertheilung der Wärme auf der Erdoberfläche*, p. 54 (Berlin, 1888).

22 Langley, "Temperature of the Moon," p. 189. On p. 197 he estimates a to be only 0·33.

23 According to the correction introduced in the sequel for the different heights of the absorbing and radiating layers of the atmosphere, the number $5°{\cdot}7$ is reduced to $4°{\cdot}0$. But as about half the sky is cloud-covered, the effect will be only half as great as for cloudless sky, *i.e.* the mean effect will be a lowering of about $2°$ C.

24 Teisserenc de Bort, "Distribution moyenne de la nébulosité," *Ann. du bureau central météorologique de France*, Année 1884, t. iv. 2de partie, p. 27.

25 Langley, "Temperature of the Moon," p. 186.

26 Hann, *Meteorologische Zeitschrift*, xi. p. 196 (1894).

27 Angström, *Bihang till K. Vet.-Ak. Handl.* Bd. xv. Afd. 1, No. 9, pp. 11 and 18 (1889).

28 Joh. Müller's *Lehrbuch d. kosmischen Physik*, 5te–Aufl. p. 539 (Braunschweig, 1894).

29 According to the measurements of Ekholm and Hagström, *Bihang till K. Sv, Vet-Ak. Handlingar*, Bd. xii. Afd. 1, No. 10, p. 11 (1886).

30 $1{\cdot}88 = \left(\dfrac{288}{246}\right)^{4}, 1{\cdot}58 = \left(\dfrac{276}{246}\right)^{4},$ and $1{\cdot}37 = \left(\dfrac{266}{246}\right)^{4}.$

$246°$ is the mean absolute temperature of the higher radiating layer of the air.

31 See Addendum, p. 275.

32 For details *cf.* Neumayr, *Erdgeschichte*, Bd. 2, Leipzig, 1887; and Geikie, 'The Great Ice-Age,' 3rd ed. London, 1894; Nathorst, *Jordens historia*, p. 989, Stockholm, 1894.

33 Neumayr, *Erdgeschichte*, p. 648; Nathorst, *l. c.* p. 992.

34 Högbom, *Svensk kemisk Tidskrift*, Bd. vi. p. 169 (1894).

35 Luigi De Marchi: *Le cause dell' era glaciale*, premiato dal R. Istituto Lombardo, Pavia, 1895.

36 De Marchi, *l. c.* p. 166.

37 *Cf.* p. 265.

4

Radiative Transfer

Plass, G.N. (1956). The influence of the $15\,\mu$ carbon-dioxide band on the atmospheric infra-red cooling rate. *Quarterly Journal of the Royal Meteorological Society*, **82**, 310–324.

The first half of the twentieth century was anything but a fallow period for radiative transfer. Astronomers and astrophysicists in particular showed an intense interest in this subject, and some of them dabbled in applications to the Earth's atmosphere as well. Observational astronomers were motivated to study absorption and emission spectra by the need to correct for the effects of the Earth's atmosphere. They also needed a good understanding of such things in order to interpret the infrared spectra of solar system planets, which during the early part of the century could be increasingly well observed from telescopes on the Earth.

In 1908, Frank Very published some wide-ranging thoughts on the operation of the greenhouse effect on the Earth and other planets, with a particular emphasis on multilayer models of radiative transfer (The greenhouse theory and planetary temperatures. *Philosophical Magazine*, **6**, 16, 478). Astrophysicists were primarily motivated by the problem of stellar structure, which required an understanding of how energy was transported radially by radiation and convection within the star – a problem that has very close affinities with the problem of determining the temperature of a planet. In 1906, Karl Schwartzchild (of black hole fame) presented his paper "On the equilibrium of the solar atmosphere," in which he set forth the basic equations of radiative transfer that bear his name today, and which are extensively used in the study of the Earth's energy balance. The Swiss physicist Jacob Robert Emden took this work further during the subsequent decade, and developed solutions bearing on the vertical structure of the Earth's atmosphere. (Radiative transfer seems to have been something of a family affair – Emden was the uncle of Martin Schwartzschild, Karl's son, and a noted astrophysicist in his own right.) The British atmospheric physicist E. Gold was also active in the early twentieth century, and made substantial contributions to the understanding of the vertical temperature structure of the atmosphere. In 1950, the great Chandrasekhar published his definitive tome on radiative transfer, building on the work of the previous half century.

Radiative transfer was a very active field of inquiry in Earth atmospheric science as well, both because it was necessary to the understanding of the vertical temperature structure of the atmosphere, and because it was recognized that accurate radiative heating and cooling rates were a necessary prerequisite for weather prediction beyond a day or so. Hugh Elsasser, Richard Goody, and Lou Kaplan were active in this area in the 1940s and beyond, and their papers provide a window into the thinking of the day. With very few exceptions, the effect of increasing CO_2 on climate did not seem to be particularly near the top of the list of reasons for studying radiative transfer during this time. A notable exception is George Callendar, who dabbled in radiative transfer in an effort to bring better spectroscopy and vertical resolution into Arrhenius' calculation.

By the time Gilbert Plass came along, much more accurate data on the infrared spectroscopy of CO_2 had become available. Equally importantly, computer power had advanced to the point that this data could be made use of in multilevel-multiband radiative transfer calculations. Plass was the first to put all that together, in our next paper of this collection. Because of the use of advanced computers, Plass was able to dispense with many of the approximations that compromised the accuracy of earlier work on the subject. He also fully took into account the pressure-broadening effect, which makes CO_2 infrared absorption less "saturated" at high altitudes than it is at the higher surface pressures. Plass only dealt with the effects of CO_2, however. The state of understanding of water vapor spectroscopy had not yet caught up to the point where a similarly detailed treatment of water vapor could be accomplished. From the standpoint of global warming, Plass' breakthrough accomplishment was the determination of the infrared cooling rate and the top-of-atmosphere emission, and the way these quantities are affected by changes in atmospheric CO_2. The key results are found in his Fig. 7, which can be said to be the first accurate calculation of the CO_2 radiative forcing.

Plass did a superb state-of-the-art job with CO_2 radiative transfer, but when it came to applying his results to the warming or cooling expected from changes in CO_2 he became ensnared in the Surface Budget Fallacy (see Box, The Surface Budget Fallacy), just as did George Callendar before him. Namely, he computed the surface warming by holding the atmospheric temperature fixed, and looking at the warming that would result from the increase in downward infrared flux into the surface that would occur when the atmosphere is made more emissive through doubling CO_2. This approach ignores the fact that doubling CO_2 throws the top-of-atmosphere budget out of balance, and would ultimately cause the atmospheric temperature to increase. It is the increase in atmospheric temperature that gives rise to most of the heating of the surface, and that heating is communicated to the surface by turbulent as well as radiative heat transfer. Interestingly, the *Quarterly Journal* published a discussion of Plass' paper by Goody, Kaplan, and Callendar, none of whom flagged the error in the temperature calculation. In fact, there is no indication that these luminaries considered that to be a particularly important part of the paper.

In some historical accounts, Plass is seen has having improved on Arrhenius' estimate through the use of a state-of-the-art multilayer radiative calculation incorporating realistic CO_2 spectroscopy and the pressure-broadening effect. Plass deserves credit for advancing the state of the art in computation of CO_2 effects on radiative transfer, and also deserves credit for helping to revive interest in the effects of doubling CO_2. However, he never used the top-of-atmosphere radiative forcing computed in his paper, nor did he appear to be aware of its significance. Thus, his update of Arrhenius' estimate of surface temperature change is a stumble down a garden path, not a waypoint on the way to the truth. Plass had better spectroscopy and radiative transfer calculations than Arrhenius, but Arrhenius had a much more correct concept of how to translate all that into changes in surface temperature.

Box: The Surface Budget Fallacy

A common fallacy in thinking about the effect of doubled CO_2 on climate is to assume that the additional greenhouse gas warms the surface by leaving the atmospheric temperature unchanged, but increasing the downward radiation into the surface by making the atmosphere a better infrared emitter. A corollary of this fallacy would be that increasing CO_2 would not increase temperature if the lower atmosphere is already essentially opaque in the infrared, as is nearly the case in the Tropics today, owing to the high water vapor content of the lower atmosphere. This reasoning is faulty because increasing the CO_2 concentration while holding the atmospheric temperature fixed reduces the OLR. This throws the top-of-atmosphere budget out of balance, and the atmosphere must warm up in order to restore balance. The increased temperature of the whole troposphere increases all the energy fluxes into the surface, not just the radiative fluxes. Further, if one is in a regime where the surface fluxes tightly couple the surface temperature to the overlying air temperature, there is no need to explicitly consider the surface balance in determining how much the surface warms. Surface and overlying atmosphere simply warm in concert, and the top-of-atmosphere balance rules the roost.

Arrhenius properly took both the top-of-atmosphere and surface balances into account in his estimate of the effect of doubling CO_2, though he did so using a crude one-layer model of the atmosphere. Guy Stewart Callendar (1938) and Gilbert Plass (1959) employed more sophisticated multi-level models, but when it came to translating their radiation results into surface temperature change both got mired in the surface budget fallacy. The prime importance of the top-of-atmosphere balance was emphasized with crystal clarity in Manabe's work of the early 1960s, but one still encounters the surface budget fallacy in discussions of global warming from time to time even today.

Figure 4.1 shows how the budgets change when CO_2 is doubled from 300 ppmv. The case shown is typical for the Earth's tropics, for which water vapor makes the lower atmosphere optically thick. The system starts off in balance, at a surface temperature of 300 K. If CO_2 is immediately doubled, the downward radiation into the surface increases by a mere 1.2 W/m². However, the OLR goes down by over 4 W/m². The atmosphere-ocean system is receiving more solar energy than it is losing, and so it warms up. The top-of-atmosphere balance is restored when the surface air temperature has warmed to 302 K. This increases the radiation into the ground by an additional 7.3 W/m². Part of this increase comes from the fact that the warmer boundary layer contains more water vapor, and therefore is closer to an ideal blackbody. Most of the increase, however, comes about simply because the low level air temperature T_{sa} increases, and hence σT_{sa}^4 increases along with it. This increase occurs even if the boundary layer is an ideal blackbody – that is, completely opaque to infrared. In addition, the increase of T_{sa} would increase the turbulent heat fluxes into the surface if the surface temperature were to stay fixed, and this increase also contributes to the warming of the surface.

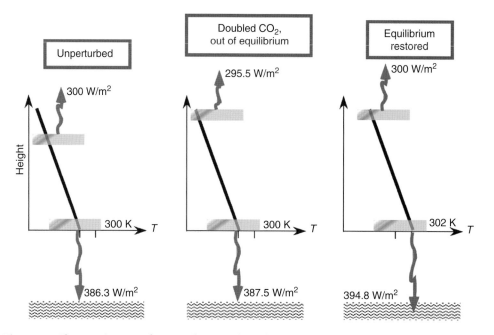

Fig. 4.1　**Changes in top-of-atmosphere and surface radiative fluxes upon doubling CO_2. Calculations were carried out with the radiation model used in the National Center for Atmospheric research climate model, employing an idealized representation of the vertical temperature and humidity variations in the atmosphere. The low level relative humidity is fixed at 80%, while the relative humidity in the free troposphere is 50%. (Figure and associated discussion in this box are reproduced from Pierrehumbert RT 2010: *Principles of Planetary Climate*, Cambridge: 780pp. © 2010 R.T. Pierrehumbert; used with permission.)**

The Influence of the $15\,\mu$ Carbon-dioxide Band on the Atmospheric Infra-red Cooling Rate

G. N. PLASS

*The Johns Hopkins University, Baltimore, U.S.A.**
(Manuscript received 12 October 1955, in revised form 23 March 1956)

SUMMARY

The upward and downward radiation flux and cooling rate are calculated for the $15\,\mu$ band of carbon dioxide. Results are obtained for three different carbon-dioxide concentrations from the surface of the earth to 75 km, and for six frequency intervals covering the band. The infra-red absorption measurements of Cloud (1952) are used for calculations, on a high-speed electronic computer, by a method which takes account of the pressure and Doppler broadening, the overlapping of the spectral lines, and the variation of the intensity and half-width of the spectral lines with temperature and pressure. The numerical integration is performed over intervals that are never larger than 1 km and average values over layers are not used. The cooling rate for the present atmospheric carbon-dioxide concentration is greater than 1°C/day from 24 km to 70 km and is greater than 4°C/day from 38 km to 55 km. The sum of the ozone and carbon-dioxide cooling rates is greater than 4°C/day from 33 km to 57 km and agrees reasonably well with the heating due to ozone absorption. The results for different carbon-dioxide concentrations indicate that the average temperature at the surface of the earth would rise by 3·6°C if the carbon-dioxide concentration were doubled and would fall by 3·8°C if the carbon-dioxide concentration were halved, on the assumption that nothing else changed to affect the radiation balance.

1. INTRODUCTION

The results of calculations of the infra-red radiation flux and the heating and cooling rates in the atmosphere for the spectral region from $12\,\mu$ to $18\,\mu$ are presented in this paper. The magnitude of the radiation flux in this region is determined largely by the properties of the $15\,\mu$ carbon-dioxide band. For atmospheric layers very close to the surface of the earth a correction must be introduced for absorption by water vapour. The radiation flux has been computed for three different carbon-dioxide concentrations when there are clear skies, and when clouds are present at one or the other of two different heights. All calculations were made from the surface of the earth to a height of 75 km. The assumed variation of temperature and pressure with height agrees closely with the values given by the Rocket Panel (1952).

The calculations are based on the carbon-dioxide absorption measurements of Cloud (1952). The radiation flux was obtained from these laboratory measurements by the same method as was used by Plass (1956a) for the spectral region near the $9·6\,\mu$ ozone band. Complete details of the method are given in that paper. However, for the carbon-dioxide calculations it was necessary to take into account the variation with temperature of the absorption in a given frequency interval; this was calculated from the partition function and the known dependence on temperature of the intensity and half-width of the spectral lines. The atmosphere is not divided into homogeneous layers for the purpose of this calculation, as this procedure may introduce large errors in the determination of the cooling rates. At each stage a careful estimate is made of the accuracy of the calculation.

* This paper was completed while the author was on leave of absence at Michigan State University. At present with Systems Research Corporation, Van Nuys, California.

An excellent survey of previous work on the carbon-dioxide absorption problem is given by Elsasser and King (1953). The pioneer work of Ladenburg and Reiche (1911) has been extended in recent years by Callendar (1941), Elsasser (1942), Kaplan (1950), and Yamamoto (1952). Kaplan (1952) and Elsasser and King (1953) have made the most extensive previous calculations of the absorption of the carbon-dioxide band.

2. METHOD FOR CALCULATION OF ATMOSPHERIC TRANSMISSION FROM LABORATORY DATA

The extensive and detailed study of the $15\,\mu$ carbon-dioxide band by Cloud (1952) was made on the large 100-foot absorption cell built by Professor John Strong at Johns Hopkins University. Path lengths up to 600 ft could be obtained by means of a six-fold optical path. The measurements were made with a wide slit so that the individual spectral lines were not resolved and the mean transmission could easily be obtained for various intervals. All the measurements made by Cloud are shown in Fig. 1 for the entire spectral interval from $12\,\mu$ to $18\,\mu$. The logarithm of the fractional absorption, A, is plotted against the logarithm of the product of the pressure, p, and path length w, used in the laboratory. All the measured points lie on a single universal absorption curve within the experimental accuracy of about two per cent.

However, the entire interval from $12\,\mu$ to $18\,\mu$ must be divided into smaller intervals in order to calculate the radiation flux in the atmosphere. The black-body intensity is replaced by an average value over the interval in question in order to perform the integration over frequency. Since the black-body intensity varies with frequency very slowly compared with the extremely rapid variations of the absorption coefficient, it can be shown numerically that the substitution of an appropriate average value for the black-body intensity introduces no appreciable error for frequency intervals as large as one micron. For this reason the interval from $12\,\mu$ to $18\,\mu$ was divided into six intervals, each one micron wide, and all quantities were calculated for each of these intervals. The absorption measurements of Cloud (1952) for each of these intervals are shown in Figs. 2, 3 and 4.

The atmospheric radiation flux and cooling rate were calculated from these measured absorption curves by the method described by Plass (1952a). Details of the procedure, a discussion of the influ-

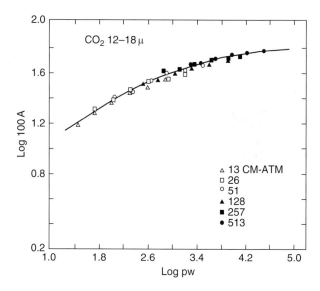

Fig. 1. The fractional absorption, A, of CO_2 as a function of the product of the pressure, p, and the optical path length used in the laboratory, w, for the spectral interval from $12\,\mu$–$18\,\mu$. The measurements were made by Cloud (1952).

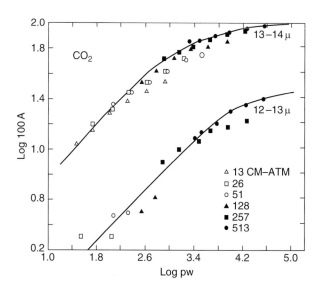

Fig. 2. The fractional absorption of CO_2 for the spectral intervals from $12\,\mu$–$13\,\mu$ and from $13\,\mu$–$14\,\mu$ as measured by Cloud (1952).

ence of the assumed temperature distribution on the final results, and detailed calculations of the probable error of the final results are described by Plass (1956a) for the similar calculations for ozone. In particular, the following corrections were made to the transmission function: (1) for Doppler broadening above $50\,\mathrm{km}$ (Plass and Fivel 1953); (2) extrapolation of experimental results to very small path lengths (Plass 1952a, 1954; Plass and

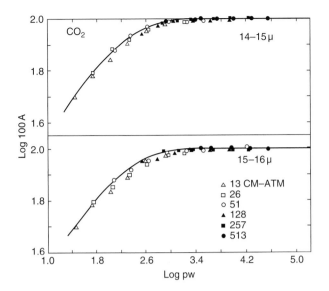

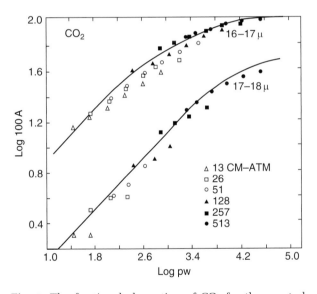

Fig. 3. The fractional absorption of CO_2 for the spectral intervals from 14μ–15μ and from 15μ–16μ as measured by Cloud (1952).

Fig. 4. The fractional absorption of CO_2 for the spectral intervals from 16μ–17μ and from 17μ–18μ as measured by Cloud (1952).

Fivel 1955b); (3) multiplication of the path length by a numerical factor varying with the transmission, in order to obtain the diffuse radiation (Plass 1952b); (4) for the variation of line strength and half-width with temperature (the procedure for making this correction is given in the Appendix).

It was assumed that the mixing ratio for carbon dioxide is a constant independent of height up to 75 km. This is a good approximation for the average carbon-dioxide amount, except possibly very

near the ground. The fluctuations of the carbon-dioxide mixing ratio are small compared to those of water vapour and ozone and should have very little influence on the radiation flux. The pressure and density curves given by the Rocket Panel (1952) were used with the necessary corrections for the small differences between their temperature and our assumed temperature.

The following temperature distribution was assumed for all the carbon-dioxide calculations:

$$T = 288 - 6z \qquad 0 \leqslant z \leqslant 13 \, \text{km}$$
$$T = 210 \qquad 13 \leqslant z \leqslant 22 \, \text{km}$$
$$T = 210 + 3(z - 22) \qquad 22 \leqslant z \leqslant 43 \, \text{km}$$
$$T = 273 \qquad 43 \leqslant z \leqslant 54 \, \text{km}$$
$$T = 273 - 3(z - 54) \qquad 54 \leqslant z \leqslant 75 \, \text{km},$$

where z is the height in kilometres. These values agree closely with those given by the Rocket Panel (1952) and should represent an average temperature distribution in middle latitudes. A more accurate result can be obtained from the same number of steps in the numerical calculation if the atmosphere can be divided into regions each of which has a linear variation of temperature with height, as here.

The necessary integrations over the transmission functions (Eqs. 1–3, Plass 1956a) were performed numerically for values of the integrand at intervals of 1 km. When the integrand was a rapidly varying function, even smaller intervals were chosen. The atmosphere was not divided into homogeneous layers for the purpose of this calculation, as this procedure may introduce large errors in the cooling rate. For each of the six spectral intervals, and for each of the three different carbon-dioxide concentrations, it was necessary to calculate the transmission function for approximately 3,000 pairs of values of z_0 and z_1 between the surface of the earth and 75 km, where z_0 and z_1 are the heights of the lower and upper level respectively. The entire numerical calculation, including the corrections to the transmission function, was coded for the MIDAC high-speed digital computer at the University of Michigan.

3. THE INFRA-RED RADIATION FLUX AND COOLING RATE

The results of the calculation of the infra-red radiation flux and cooling rate by the above method are shown in Figs. 5–8. The upward and downward radiation flux is given for the entire spectral

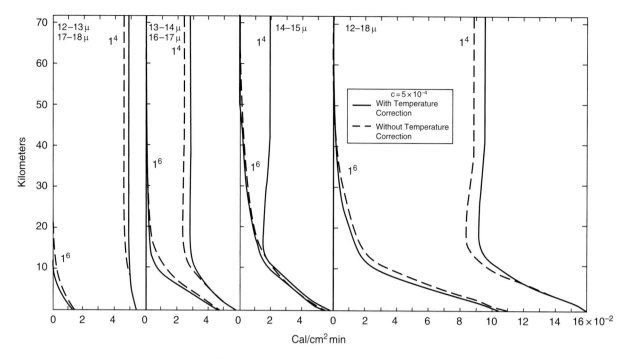

Fig. 5. The upward and downward radiation flux as a function of height for the combined frequency intervals from 12μ–13μ and 17μ–18μ ; from 13μ–14μ and 16μ–17μ ; from 14μ–16μ ; and for the entire interval from 12μ–18μ. The temperature correction described in the Appendix was made in the calculations for the solid curves and was omitted for the dashed curves. The present CO_2 concentration ($c = 5 \times 10^{-4}$; 0·033 per cent by volume) was assumed for this calculation. The difference between the two curves for the upward flux for the 14μ–16μ interval cannot be shown on the scale of this graph.

interval from 12μ–18μ and for three smaller intervals. In order to present as many of the results as possible in a reasonable space, the results for three pairs of intervals have been combined. Thus instead of showing the flux for the 12μ–13μ and 17μ–18μ intervals separately, as calculated, only the total for both intervals is given; the variation with height of the results for these two intervals is almost the same and no essential information is lost by combining the results in this manner. For the same reason the results for the 13μ–14μ and 16μ–17μ intervals were combined, as were those for the 14μ–15μ and 15μ–16μ intervals.

Figs. 5 and 6 give the results for the present concentration of CO_2 ($c = 5 \times 10^{-4}$; 0·033 per cent by volume) when the temperature correction in the Appendix is made (solid curve) and is not made (dashed curve).

The change in the upward flux with height depends on the transmission of the spectral region in question. If the transmission were unity in a given spectral interval, the higher atmospheric layers at lower temperatures would be unable to change the upward flux and it would be constant with height. If the transmission were zero, the upward flux would be equal to the black-body intensity at every height and, for our assumed temperature distribution, would have a minimum from 13 km to 22 km and maxima at the ground and from 43 km to 54 km.

The upward flux for various spectral intervals as shown in Fig. 5 is close to these limits. For the 12μ–13μ; 17μ–18μ interval the transmission is near unity; the upward flux is nearly constant with a value above 13 km only 8 per cent lower than the value at the ground. For the 13μ–14μ; 16μ–17μ interval the transmission has an intermediate value at the lower altitudes; between 13 km and 22 km the upward flux falls to almost one-half of its value at the ground. It increases slightly above 22 km, but this rise is very small, since the transmission is so near unity for paths at the higher altitudes that higher temperatures there can have little effect. For the 14μ–16μ interval the transmission is nearly zero up to 15 km; here the upward flux is very close to the black-body curve. Above this height the upward flux increases as the atmospheric temperature increases to a value 28 per cent above its minimum value. The upward flux for the entire range from 12μ–18μ exhibits the characteristics of a curve for a transmission function with a value intermediate between zero and unity.

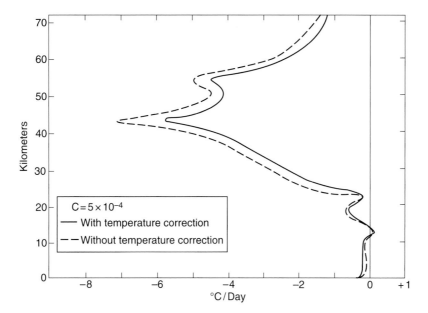

Fig. 6. The cooling rate in °C/day for the entire frequency interval from $12\,\mu$–$18\,\mu$.

There would be no downward flux if the transmission function were unity in a given frequency interval, since the atmosphere could not radiate at these frequencies. If the transmission function were zero, the downward flux would equal the black-body intensity for the appropriate temperature at a given height. For the interval from $12\,\mu$–$13\,\mu$; $17\,\mu$–$18\,\mu$ the downward flux is quite small and has an appreciable value only below 10 km. For the interval from $13\,\mu$–$14\,\mu$; $16\,\mu$–$17\,\mu$ the downward flux has a considerably larger value. For the interval from $14\,\mu$–$16\,\mu$, the downward flux is larger than for either of the previous intervals and approaches the black-body flux below 13 km. The upward and downward flux would have an identical value equal to the black-body intensity below 13 km for the $14\,\mu$–$16\,\mu$ interval, if the transmission function were actually zero for all possible path lengths below 13 km. Actually for sufficiently small path lengths, even near the surface of the earth, the transmission function approaches unity; this is the reason for the small difference between the upward and downward flux in this case.

The cooling rate for the entire CO_2 band from $12\,\mu$–$18\,\mu$ is shown in Fig. 6. From the ground up to 10 km the cooling rate is of the order of a few tenths of a degree Celsius per day. This rate rises rapidly from 22 km to 43 km and reaches a maximum of 5·8°C per day at 43 km. The cooling rate is above 4·0°C per day from 38 km to 55 km. The rate decreases rapidly at higher altitudes. The maxima

Table 1. Cooling rates in the stratosphere.

Height (km)	Ozone (°C/day)	Carbon Dioxide (°C/day)	Total (°C/day)
25	0·1	1·3	1·4
30	0·5	2·4	2·9
35	1·3	3·3	4·6
40	1·8	4·5	6·3
45	2·1	4·9	7·0
50	1·5	4·6	6·1
55	0·8	4·0	4·8
60	0·3	2·3	2·6
65	0·1	1·7	1·8
70	0·0	1·3	1·3

and minima in this curve at 13 km, 22 km, 43 km and 54 km arise from the discontinuities in the derivative of the assumed temperature curve at these altitudes (Plass 1956a). Such discontinuities seldom, if ever, actually exist in the atmosphere. Therefore the values in the neighbourhood of these maxima and minima should be smoothed out slightly when comparisons are made with the actual atmosphere. As discussed above, this temperature distribution was assumed, as it greatly increases the accuracy of numerical calculations carried to a given number of figures; it approximates closely the actual temperature distribution.

A comparison of the cooling rates of the ozone and carbon-dioxide bands is given Table 1. The ozone cooling rates are those calculated by Plass (1956a) for the ozone-concentration curve L and the temperature curve B as defined in that article.

This temperature curve is very close to the one used for the present carbon-dioxide calculations, except that it does not have discontinuous derivatives at certain heights. A small correction has been made to the carbon-dioxide results between 40 km and 60 km to bring them more closely into agreement with the results that would have been obtained had the temperature curve B been used. The largest relative contribution from the ozone is at 45 km where it contributes 30 per cent to the total energy radiated from both ozone and carbon dioxide. The carbon dioxide is much more important than the ozone in cooling the atmosphere above 55 km.

The absorption of solar energy by ozone at these altitudes has recently been calculated by Johnson (1953) and Pressman (1955). Since the heating rates vary greatly with latitude and season (Pressman 1955), it is difficult to make quantitative comparisons. However, it is clear that the heating and cooling rates agree qualitatively and have their maxima at the same altitude (45 km) with approximately the same value of 7·0°C/day. The cooling due to H_2O is not included in Table 1. Differences between the heating and cooling rates could be ascribed to the possible additional cooling effect of H_2O and to uncertainties in the calculations themselves.

The correction for the variation of the line intensities and half-widths with temperature, discussed in the Appendix, greatly complicates the calculation and introduces an additional factor of uncertainty in the final results. In order to study the effect of this correction, the calculations were made with and without the temperature correction, for each of the three CO_2 concentrations. The results are shown in Figs. 5 and 6 for the present CO_2 concentration ($c = 5 \times 10^{-4}$). The principal temperature correction is due to the increase in line intensity of the stronger lines of the band with increasing temperature over the temperature range that occurs in the atmosphere. Since the laboratory measurements of the transmission function are made at room temperature, the temperature-corrected transmission function is greater than the uncorrected value for the colder regions of the atmosphere. Thus the results are changed most when the temperature correction is made at altitudes from 10 to 30 km where the atmospheric temperature is relatively low.

The value of the upward flux at high altitudes is 5 per cent higher for the interval from 12μ–13μ; 17μ–18μ, and 14 per cent higher for the interval from 13μ–14μ; 16μ–17μ, when the temperature correction is made than when it is not made. The upward flux has a larger value since the transmission function is closer to unity after the temperature correction. Thus the upward flux cannot follow so well the dip in the black-body curve in the lower stratosphere. The change for the 14μ–16μ interval cannot be shown on the scale of Fig. 5, since the transmission function is virtually zero in the lower part of the atmosphere in both cases.

For the downward flux the difference between the calculated values with and without the temperature correction is largest in each case near the base of the stratosphere. In the intervals from 12μ–13μ; 17μ–18μ, and from 13μ–14μ; 16μ–17μ, most of the downward radiation reaching, for example, 8 km, originates in the colder region of the atmosphere from 10 km to 30 km. It is not surprising therefore that the increased values for the transmission after the temperature correction is made give values for the downward flux that are as little as half those obtained without the correction. This is the only region of the atmosphere where the temperature correction must be made in order to obtain even approximately correct values of the flux. At the surface of the earth so much of the downward radiation comes from the lower, warmer layers of the atmosphere that the difference between the two curves is less than 10 per cent for the interval from 12μ–13μ; 17μ–18μ, and is very small for the other intervals.

The cooling curve with and without the temperature correction is given in Fig. 6. Above 14 km the cooling rates are slightly smaller when the temperature correction is made. However, the shapes of the two curves are remarkably similar and the percentage deviation of the two curves is not large except below 14 km where in any case the cooling rate due to CO_2 is not of great importance. Thus it would appear that reasonably accurate calculations of the cooling rate above 14 km can be made if desired without introducing the complexities of the temperature correction.

The infra-red flux and cooling rates are shown in Figs. 7 and 8 for three different CO_2 concentrations, 0·025, 0·05 and 0·10 per cent by weight or 0·0165, 0·033 and 0·066 per cent by volume. The lowest and highest values correspond to halving or doubling the present atmospheric CO_2 amount. The temperature correction has been made to all the results given in these figures, since

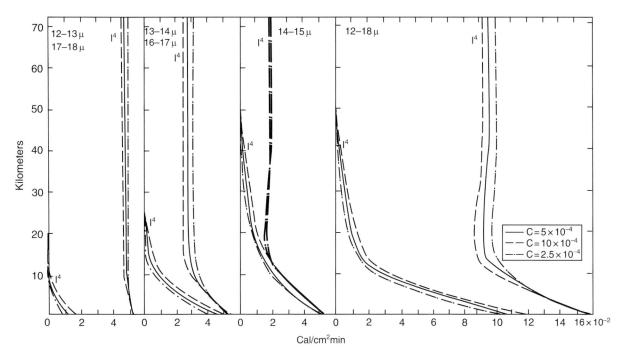

Fig. 7. The upward and downward radiation flux as a function of height for the combined frequency intervals from $12\,\mu$–$13\,\mu$ and $17\,\mu$–$18\,\mu$; from $13\,\mu$–$14\,\mu$ and $16\,\mu$–$17\,\mu$; from $14\,\mu$–$16\,\mu$; and for the entire interval from $12\,\mu$–$18\,\mu$. Curves are given for the following CO_2 concentrations: $c = 5 \times 10^{-4}$ (0·033 per cent by volume); $c = 10 \times 10^{-4}$ (0·066 per cent by volume); $c = 2·5 \times 10^{-4}$ (0·0165 per cent by volume). The temperature correction described in the Appendix was made in the calculation of all these curves.

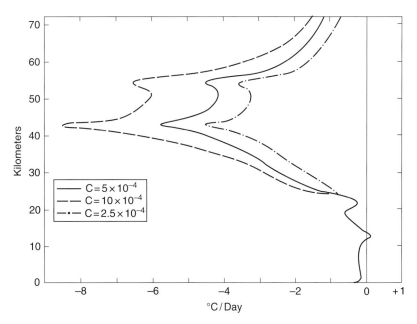

Fig. 8. The cooling rate in °C/day for the entire frequency interval from $12\,\mu$–$18\,\mu$. Curves are given for the following CO_2 concentrations : $c = 5 \times 10^{-4}$ (0·033 per cent by volume); $c = 10 \times 10^{-4}$ (0·066 per cent by volume); $c = 2·5 \times 10^{-4}$ (0·0165 per cent by volume). The temperature correction described in the Appendix was made in the calculation of all these curves. There is no difference in the cooling rates below 24 km within the accuracy of the calculation and only one curve is shown in this region.

the complete procedure had been coded for the MIDAC electronic computer.

For the intervals from $12\,\mu$–$13\,\mu$; $17\,\mu$–$18\,\mu$, and from $13\,\mu$–$14\,\mu$; $16\,\mu$–$17\,\mu$, the upward flux is smaller the larger the CO_2 concentration, since the smaller transmission function allows the flux to approach closer to the black-body flux at the temperature minimum from 13 km to 22 km. The same is true for the interval from $14\,\mu$–$16\,\mu$ below 32 km. Above this height the upward flux is slightly greater for larger CO_2 concentrations, since the transmission function is still sufficiently small in this region for the upward radiation to increase slightly owing to the higher temperature in the upper part of the stratosphere. Similar remarks apply to the downward flux. In this case the larger CO_2 concentrations give the larger values for the downward flux.

The cooling rates are given in Fig. 8 for these three different CO_2 concentrations. There was no definite difference within the accuracy of the calculation, in the cooling rates below 24 km, and only one curve is shown in this region. At higher altitudes the cooling rate increases appreciably if the CO_2 concentration changes. At the maximum of the cooling curve at 43 km, the cooling rate increases from 5·8°C/day to 8·5°C/day if the CO_2 concentration is doubled.

Numerous checks were made during the progress of these calculations in order to determine the accuracy of the final results. The procedure was similar to the one used by Plass (1956a) for ozone. The upward and downward flux is determined with a percentage error smaller than that of the original laboratory absorption measurements. Cloud's estimate of a 2 per cent error in the laboratory measurements of the transmission function may be somewhat optimistic. A conservative estimate of the error in the upward and downward flux, taking into account the various corrections made in the course of the calculations, is 3 per cent. However, the relative changes of the curves for different concentrations are determined with considerably greater accuracy than this. The probable error of the cooling rate is estimated by introducing arbitrary variations into the original transmission functions and calculating their influence on the final result. The probable error obtained in this manner is about 10 per cent below 20 km, increasing to 30 per cent at 50 km and becoming rather uncertain above 60 km. Again the relative differences between the various curves should be considerably more accurate than their magnitude.

4. TEMPERATURE VARIATIONS AT THE SURFACE OF THE EARTH

The change in the equilibrium temperature at the surface of the earth with CO_2 concentration can be estimated from the corresponding variations of the radiation flux, if it is assumed that nothing else changes to alter the radiation balance when the CO_2 amount varies. In order to obtain the temperature change it is also assumed that an additional amount of radiant energy equal to 0·0033 Cal cm^{-2} min^{-1} would be radiated to space, if the average temperature of the earth's surface were to increase by 1°C. This number cannot be calculated accurately until a detailed study has been made of the H_2O spectrum, but it represents the best value that can be given with our present knowledge of this spectrum. When a more accurate value for this number is obtained in the future, all the temperature changes given here should be multiplied by the ratio of the old to the new value.

When the CO_2 amount is doubled and halved the change in the downward flux at the surface of the earth is 0·0119 and 0·0125 Cal cm^{-2} sec^{-1} respectively. Thus, in order to restore equilibrium the surface temperature must rise by 3·6°C if the CO_2 concentration is doubled and the surface temperature must fall by 3·8°C if the CO_2 concentration is halved. It is interesting to note that the four intervals from $12\,\mu$–$13\,\mu$, $13\,\mu$–$14\,\mu$, $16\,\mu$–$17\,\mu$ and $17\,\mu$–$18\,\mu$ contribute about equally to this temperature change. The interval from $14\,\mu$–$16\,\mu$ does not contribute since it is opaque near the ground at all these concentrations. The argument has sometimes been advanced that the CO_2 cannot cause a temperature change at the surface of the earth because the CO_2 band is always black at any reasonable concentration. This argument is true for the lines near the centre of the band from $14\,\mu$–$16\,\mu$, but neglects completely the important contribution of the lines farther from the band centre.

This same method shows that the average temperature is lowered 2·2°C and 1·3°C at the upper surface of a cloud at 4 km and 9 km respectively, if the amount of CO_2 in the atmosphere is halved. This temperature change at the upper surface of a cloud could be the cause of the increased precipitation at the beginning of a glacial period (Plass 1956b).

If the CO_2 concentration is doubled, Arrhenius (1896) calculated a temperature increase of 6°C, Hulbert (1931) of 4°C, and Callendar (1938) of 2°C. The value of 3·6°C obtained from the present study is based on more accurate experimental data than were available previously and it was possible to perform the elaborate calculations only because an electronic computer was available. Such factors as the weak lines that are relatively far from the band centre and the change in intensity and half-width of the spectral lines with temperature and pressure were taken into account in the present calculations.

The radiation calculations predict a definite temperature change for every variation in CO_2 amount in the atmosphere. These temperature changes are sufficiently large to have an appreciable influence on the climate. Plass (1956b) has considered the various factors that control the CO_2 content of the atmosphere and oceans and has discussed the current status of the carbon dioxide theory of climatic change.

ACKNOWLEDGMENTS

This research was supported by the Office of Naval Research, Washington. I wish to thank Mr. L. Leopold for his extensive and careful work in calculating the radiative flux in the early stages of this study. His work made it possible to code the final calculation for the MIDAC high-speed electronic computer.

APPENDIX

VARIATION OF TRANSMISSION FUNCTION WITH TEMPERATURE

Laboratory measurements of the absorption of atmospheric gases should ideally be made over the range of temperatures that occur in the atmosphere. This would eliminate the need to make a theoretical estimate of the variation of the transmission function with temperature. However, to our knowledge no suitable measurements have yet been made at temperatures below room temperature.

First let us assume that the temperature decreases linearly with height between z_0 and z_1. Without any loss of generality z_0 may be set equal to zero. Then the temperature dependence is given by

$$T = T_0 - a z \qquad (1)$$

and the pressure dependence by

$$p = p_0 (T/T_0)^{Mg/aR} \qquad (2)$$

The temperature dependence of the line intensity, S, and, to a smaller extent, the temperature variation of the half-width of the spectral line, α, cause the transmission function to vary with temperature. The theoretical variation with temperature of both these factors is known. From the Maxwell-Boltzmann distribution law and the rotational partition function, it follows that the temperature dependence of S is given by the expression

$$\frac{S(T)}{S(T_0)} = \frac{T_0}{T} \exp\left[-\frac{E_J}{kT} + \frac{E_J}{kT_0}\right] \qquad (3)$$

where T_0 is a standard temperature, $S(T)$ and $S(T_0)$ are the line intensities at the temperatures T and T_0 respectively, E_J is the energy of the initial state of the transition, and k is the Boltzmann constant. Other terms in the general expression for S which change very slowly with T are omitted from Eq. (3), since numerical integration shows that their effect is very small for the range of temperatures in the atmosphere.

From kinetic theory it follows that

$$\alpha = \alpha_0 (p/p_0)(T_0/T)^{\frac{1}{2}} \qquad (4)$$

where α_0 is the half-width at a standard pressure and temperature, p_0 and T_0. Although the dependence of α on pressure has been verified, a square-root temperature dependence should not be expected for all lines. However, the variation of α with temperature introduces only a very small correction.

For a single spectral line, the result of the usual method for the calculation of the absorption in the atmosphere integrated over frequency, Λ, (Plass 1954, Plass and Fivel 1955a) is that

$$\Lambda(u_0, u_1) = \left\{4 \sec \theta \int_{u_0}^{u_1} S \alpha \, du\right\}^{\frac{1}{2}} \qquad (5)$$

for the strong-line approximation and that

$$\Lambda(u_0, u_1) = \sec \theta \int_{u_0}^{u_1} S \, du \qquad (6)$$

for the weak-line approximation, where

$$u = \int_z^\infty c \, p \, dz,$$
$$\Lambda(u_0, u_1) = \int_0^\infty [1 - \tau(u_0, u_1)] \, dv,$$

θ is the angle the radiation makes with the vertical, c is the ratio of the density of the radiating gas to the total density, ρ, and τ is the transmission function. Eqs. (5) and (6) are valid regardless of the functional dependence of S on height. The results for overlapping spectral lines are discussed later in this section.

In order to obtain the integrated absorption, Λ, for the strong-line approximation, substitute Eqs. (1), (2), (3) and (4) into Eq. (5). Define an effective temperature, T_e, between z_0 and z_1 so that the integrated absorption, $\Lambda(z_0, z_1)$, for a radiating gas in a hypothetical atmosphere with the same temperature, T_e, at all heights, is the same as for the actual radiating gas with variable temperature. The effective temperature, T_e, is a function of both z_0 and z_1. It can be calculated by setting the expression for Λ, obtained from Eq. (5) when the entire atmospheric layer is at the temperature T_e, equal to the expression for Λ, derived above, which is valid when the temperature decreases linearly with height.

In this manner it is found that T_e can be determined from the other known constants of the problem, from the equation

$$\int_1^\beta \exp\left(-\frac{E_J y}{kT_0}\right) y^{-\frac{2Mg}{aR}+\frac{1}{2}}$$

$$dy = \left(\frac{T_0}{T_e}\right)^{\frac{3}{2}} \frac{aR}{2Mg}\left[1 - \beta^{-\frac{2Mg}{aR}}\right]\exp\left(-\frac{E_J}{kT_e}\right). \quad (7)$$

where

$$\beta^{-1} = 1 - (az_1/T_0).$$

The value of the effective temperature, T_e, can be obtained as a function of z_0 and z_1 from the numerical solution of Eq. (7). Some typical results are shown in Fig. 9, where T_e is plotted against the temperature at the boundary of the layer. The upper set of three curves gives T_e for a layer between z_0, where the temperature is 293°K, and z_1, where the temperature has the value given by the abscissa. The lower set of three curves gives T_e for a layer extending from z_0, where the temperature has the value given by the abscissa, to z_1, where the temperature is 218°K. Each set of curves is drawn for three different values of the ratio $m = E_J/kT_{293}$. Since the value of m does not change by more than a factor of three when averaged over a frequency interval of $0\cdot1\,\mu$ or more in the range from $12\,\mu$–$18\,\mu$, Fig. 9 shows that the value of T_e is insensitive to the assumed value of m.

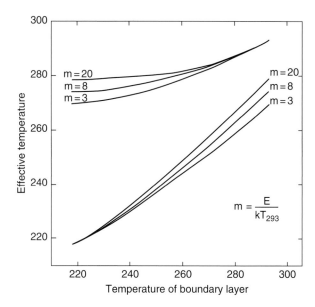

Fig. 9. The upper set of three curves gives the value of the effective temperature, T_e, for an atmospheric layer extending from a fixed height where the temperature is 293°K to a height where the temperature has the value given by the abscissa. The lower set of three curves gives the value of T_e for an atmospheric layer extending from a height where the temperature has the value given by the abscissa to a fixed height where the temperature is 218°K. Each set of curves is drawn for three different values of the ratio $m = E_J/kT_{293}$, where E_J is the energy of the initial state of the transition.

When the equation for the weak-line approximation, Eq. (6), is used instead of Eq. (5), a similar derivation to the one given above shows that

$$\int_1^\beta \exp\left(-\frac{E_J y}{kT_0}\right) y^{-\frac{Mg}{aR}-1}$$

$$dy = \left(\frac{T_0}{T_e}\right)\frac{aR}{Mg}\left[1 - \beta^{-\frac{Mg}{aR}}\right]\exp\left(-\frac{E_J}{kT_e}\right) \quad (8)$$

The results for T_e are very similar to those obtained from Eq. (7).

Eqs. (7) and (8) have been derived for a single spectral line. However, the identical equations are also valid for the Elsasser model of a band in the limits of the strong- and weak-line approximations. This result can be derived from the equations for the Elsasser model given by Plass and Fivel (1955b). Since T_e does not depend on the degree of overlapping of the spectral lines for the Elsasser model in the limits of the strong- and weak-line approximations, it appears that T_e cannot vary appreciably with the degree of overlapping of the spectral lines for any model of a band.

Equations similar to Eqs. (7) and (8) were derived for the cases when the temperature increases linearly with height or when there are several regions with a constant temperature in one region and a linear increase or decrease in another. In practice, for the temperature distribution assumed for this calculation, it is never necessary to use more than two regions with different lapse rates for the calculation of T_e. If more than two regions exist with different lapse rates between z_0 and z_1, the regions above the first two make a negligible contribution.

The values of the ratio E_j/kT, where T is the temperature at which the laboratory absorption measurements were made, were calculated from a tabulation of the carbon dioxide lines kindly made available to us by Dr. L. D. Kaplan in advance of publication. The region from 12μ to 18μ was divided into intervals 0.1μ wide. In each of these 60 intervals the average energy of the lines was computed. It was weighted either with the square-root of the line intensity or the first power of the intensity depending on whether the absorption for the entire interval followed the strong-line or the weak-line approximation respectively for the path length between z_0 and z_1. The average energy for the six intervals of one micron width from 12μ to 18μ was then calculated from the averages for the 0.1μ intervals. The deviations of the values for the 0.1μ intervals from the average for the 1μ intervals were relatively small. For example, for the interval from 12μ to 13μ the average deviation of the results for the smaller intervals was only 10 per cent of the average value for the entire interval.

The calculated value of E_j/kT was between 3 and 8 for each of the six intervals of 1μ width. Fig. 9 shows that T_e is insensitive to the value of E_j/kT. Thus it is possible to obtain the value of T_e for any combination of values of z_0 and z_1 with an estimated probable error of no more than a few degrees Kelvin.

Curves similar to those in Fig. 9 were plotted at intervals of 1 km from the surface of the earth to 75 km. They were calculated from Eqs. (7) and (8) or, when the temperature did not decrease linearly with height, from equivalent equations. The effective temperature, T_e, was determined in this manner for each pair of values of z_0 and z_1. The temperature correction to the integrated absorption was then calculated from Eqs. (3), (4) and the known equations for the dependence of the absorption on these variables (Plass 1952a, 1954; Plass and Fivel 1955b). Fortunately it was possible to code a programme of calculations for the MIDAC high-speed digital computer to calculate this correction entirely automatically for each pair of z_0 and z_1.

All the carbon-dioxide calculations were made both with and without this temperature correction. The correction was largest for the intervals, from 12μ–13μ and 17μ–18μ, that are far from the centre of the band, and was smallest for the region from 14μ–16μ at the centre of the band.

REFERENCES

Arrhenius, S. 1896 *Phil. Mag.*, **41**, p. 237.
Callendar, G. S. 1938 *Quart. J. R. Met. Soc.*, **64**, p. 223.
    1941 *Ibid.*, **67**, p. 263.
Cloud, W. H. 1952 'The 15μ band of CO_2 broadened by nitrogen and helium,' Johns Hopkins Univ.
Elsasser, W. M. 1942 'Heat transfer by infra-red radiation in the atmosphere,' *Harvard Met. Studies*, No. 6.
Elsasser, W. M. and King, J. I. 1953 'Stratospheric radiation,' *Tech. Rep. Univ. Utah*, No. 9.
Hulbert, E. O. 1931 *Phys. Rev.*, **38**, p. 1876.
Johnson, F. S. 1953 *Bull. Amer. Met. Soc.*, **34**, p. 106.
Kaplan, L. D. 1950 *J. Chem. Phys.*, **18**, p. 186.
Kaplan, L. D. 1952 *J. Met.*, **9**, p. 1.
Ladenburg, R. and Reiche, F. 1911 *Ann. Phys.*, **42**, p. 181.
Plass, G. N. 1952a *J. Opt. Soc. Amer.*, **42**, p. 677.
Plass, G. N. 1952b *J. Met.*, **9**, p. 429.
Plass, G. N. 1954 *Ibid.*, **11**, p. 163.
Plass, G. N. 1956a *Quart. J. R. Met. Soc.*, **82**, p. 30.
Plass, G. N. 1956b *Tellus*, **8**, p. 140.
Plass, G. N. and Fivel, D. I. 1953 *Astrophys. J.*, **117**, p. 225.
Plass, G. N. and Fivel, D. I. 1955a *Quart. J. R. Met. Soc.*, **81**, p. 48.
Plass, G. N. and Fivel, D. I. 1955b *J. Met.*, **12**, p. 191.
Pressman, J. 1955 *Ibid.*, **12**, p. 87.
Rocket Panel 1952 *Phys. Rev.*, **88**, p. 1027.
Yamamoto, G. 1952 *Sci. Rep. Tohoku Univ.*, Ser. 5, Geophys., **4**, p. 9.

5

The Balance of Energy

Manabe, S. & Wetherald, R.T. (1967). Thermal equilibrium of atmosphere with a given distribution of relative humidity. *Journal of the Atmospheric Sciences*, **24** (3), 241–258. Times Cited: 692. 18 pages.

Budyko, M.I. (1969). The effect of solar radiation variations on the climate of the Earth. *Tellus*, 611–619.

Sellers, W.D. (1969). A global climatic model based on the energy balance of the Earth-atmosphere system. *Journal of Applied Meteorology*, **8**, 392–400.

Three more building blocks had to fall into place before a proper, quantitative estimate of the effect of CO_2 changes on temperature could be carried out. First, the spectroscopy of water vapor had to be mapped out as comprehensively as had been done for CO_2 at the time of Plass. This was quite challenging, since the absorption of infrared by water vapor has a very complex dependence on wavelength, extending over a far greater range of wavelengths than is the case for CO_2. Second, a means had to be found to represent the effects of convection on the temperature structure of the atmosphere. For the most part, investigators had been content to compute the radiation fluxes for an observed or hypothetical temperature profile, without determining the effects of convection in a self-consistent way. Third – indeed the keystone building block of the edifice – the forgotten importance of the top-of-atmosphere energy balance had to be reintroduced into the calculation. All of this had to be knit together in a numerical model sufficiently efficient to permit solution on the computers of the day. This was achieved in the remarkable 1967 paper by Manabe and Weatherald, which we include in the following. Manabe and Weatherald (1967) can with confidence be described as the first fully sound estimate of the warming that would arise from a doubling of CO_2.

The required groundwork on water vapor had been laid in the preceding years. Manabe himself was involved in one of the very first computations of pure radiative equilibrium including accurate water vapor spectroscopy (Manabe and Möller (1961), cited in the paper we include below). This paper did not include a representation of convection, though. That came with Manabe and Strickler (1964), which introduced *convective adjustment* – the first parameterization of convection, and one which is still widely used today. At the same time Manabe was doing all that, he was developing the world's first general circulation model, the first results of which were published in Manabe, Smagorinsky, and Strickler (1965). In the 1967 paper below, the convective adjustment scheme and the numerical representation of both water vapor and CO_2 radiative effects were knit together in a numerical model, which was used to find solutions that satisfied the required top-of-atmosphere energy balance. The changes of water vapor in a changing climate are represented using the same assumption of fixed relative humidity as introduced by Arrhenius. The paper is as remarkable for its clarity of exposition as for its scientific content. Notably, it introduces the use of a curve of outgoing longwave radiation (OLR) vs. surface temperature as a means of graphically explaining how a change in CO_2 leads to surface warming, and also as a means of explaining how the water vapor feedback increases climate sensitivity. Graphs of this sort are the basis of most enlightened modern discussions of climate variations on the

Earth and other planets. With Manabe and Weatherald, the study of global warming can be said to have entered the modern era. It represents the culmination of the art of the radiative-convective model, in which the entire climate of a planet is represented by a single column subjected to vertical heat exchange by radiation and by convection.

We also include two short papers playing variations on the theme of energy balance. These papers, one by Budyko and the other by Sellers, introduce the very simplest kind of model in which the effects of the pole to equator temperature gradient can be represented. The models in these papers do not try to solve the fluid equations governing atmospheric heat transport in the horizontal, but instead use a simplified representation of heat transfer, based on heat diffusion. Models of this type are generally called *energy balance models*, though the term is a misnomer since all proper climate models are in some sense energy balance models. The key defining features of the ones introduced here are the simplified representation of horizontal heat transfer and the simplified representation of infrared cooling, which is parameterized as a function of surface temperature. The most important insight to come out of such models is the nature if ice-albedo feedback. These papers show quantitatively the destabilizing effect of ice-albedo feedback: if it gets colder, more of the planet is covered by ice, which reflects more sunlight, which in turn leads to further cooling. The papers are also important in introducing the notion of an ice-albedo bifurcation, and multiple equilibrium. For the same conditions of illumination by the Sun, the Earth can support at least three different states: a stable state with little or no polar ice, an unstable state with a large polar ice cap, and a stable globally glaciated state. The last of these has come to be referred to as Snowball Earth. At the time of Budyko and Sellers, the Snowball state was just thought of as a mathematical curiosity, but it has in recent years become a subject of intense inquiry, as strong geological evidence has emerged indicating that the Earth may have indeed passed through Snowball states during the Neoproterozoic (around 700 million years ago) and during the Paleoproterozoic (about 2.5 billion years ago). Energy balance models similar to those introduced by Budyko and Sellers are still important tools for exploratory work in climate science, though they have been supplanted by general circulation models for those cases in which quantitatively accurate predictions are needed.

Thermal Equilibrium of the Atmosphere with a Given Distribution of Relative Humidity

SYUKURO MANABE AND RICHARD T. WETHERALD

Geophysical Fluid Dynamics Laboratory, ESSA, Washington, D. C.
(Manuscript received 2 November 1966)

ABSTRACT

Radiative convective equilibrium of the atmosphere with a given distribution of relative humidity is computed as the asymptotic state of an initial value problem.

The results show that it takes almost twice as long to reach the state of radiative convective equilibrinm for the atmosphere with a given distribution of relative humidity than for the atmosphere with a given distribution of absolute humidity.

Also, the surface equilibrium temperature of the former is almost twice as sensitive to change of various factors such as solar constant, CO_2 content, O_3 content, and cloudiness, than that of the latter, due to the adjustment of water vapor content to the temperature variation of the atmosphere.

According to our estimate, a doubling of the CO_2 content in the atmosphere has the effect of raising the temperature of the atmosphere (whose relative humidity is fixed) by about 2C. Our model does not have the extreme sensitivity of atmospheric temperature to changes of CO_2 content which was adduced by Möller.

1. INTRODUCTION

This study is a continuation of the previous study of the thermal equilibrium of the atmosphere with a convective adjustment which was published in the JOURNAL OF THE ATMOSPHERIC SCIENCES (Manabe and Strickler, 1964). Hereafter, we shall identify this study by M.S. In M.S. the vertical distribution of absolute humidity was given for the computation of equilibrium temperature, and its dependence upon atmospheric temperature was not taken into consideration. However, the absolute humidity in the actual atmosphere strongly depends upon temperature. Fig. 1 shows the distribution of relative humidity as a function of latitude and height for summer and winter. According to this figure, the zonal mean distributions of relative humidity of two seasons closely resemble one another, whereas those of absolute humidity do not. These data suggest that, given sufficient time, the atmosphere tends to restore a certain climatological distribution of relative humidity responding to the change of temperature. If the moisture content of the atmosphere depends upon atmospheric temperature, the

effective height of the source of outgoing long-wave radiation also depends upon atmospheric temperature. Given a vertical distribution of relative humidity, the warmer the atmospheric temperature, the higher the effective source of outgoing radiation. Accordingly, the dependence of the outgoing long-wave radiation is less than that to be expected from the fourth-power law of Stefan-Boltzman. Therefore, the equilibrium temperature of the atmosphere with a fixed relative humidity depends more upon the solar constant or upon absorbers such as CO_2 and O_3, than does that with a fixed absolute humidity, in order to satisfy the condition of radiative convective equilibrium. In this study, we will repeat the computation of radiative convective equilibrium of the atmosphere, this time for an atmosphere with a given distribution of relative humidity instead of that for an atmosphere with a given distribution of absolute humidity as was carried out in M.S.

As we stated in M.S., and in the study by Manabe and Möller (1961), the primary objective of our study of radiative convective equilibrium is the incorporation of radiative transfer into the general circulation model of the atmosphere. Adopting

Manabe, M. (1967) Thermal equilibrium of atmosphere with a given distribution of relative humidity. *Journal of the Atmospheric Sciences* 24(3): 241–258. Reprinted with permission of American Meteorological Society.

Summer

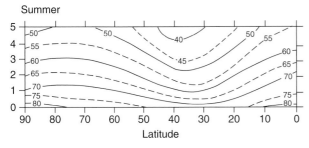

Winter

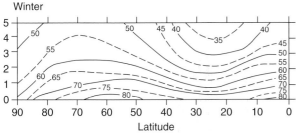

Fig. 1. Latitude-height distribution of relative humidity for both summer and winter (Telegadas and London, 1954).

the scheme of the computation of radiative transfer which was developed in M.S., Manabe *et al.* (1965) successfully performed the numerical integration of the general circulation of the atmosphere involving the hydrologic cycle. In order to avoid a substantial increase in the number of degrees of freedom, the distribution of water vapor, which emerged as the result of the hydrologic cycle of the model atmosphere, was not used for the computation of radiative transfer. Instead, the climatological distribution of absolute humidity was used. The next step is the numerical integration of the model with complete coupling between radiative transfer and the hydrologic cycle. Before undertaking this project, it is desirable to answer the following questions by performing a series of computations of radiative-convective equilibrium of the atmosphere with fixed relative humidity.

1) How long does it take to reach a state of thermal equilibrium when the atmosphere maintains a realistic distribution of relative humidity that is invariant with time?
2) What is the influence of various factors such as the solar constant, cloudiness, surface albedo, and the distributions of the various atmospheric absorbers on the equilibrium temperature of the atmosphere with a realistic distribution of relative humidity?
3) What is the equilibrium temperature of the earth's surface corresponding to realistic values of these factors?

There is no doubt that this information is indispensable for the successful integration of the fundamental model of the general circulation mentioned above.

Recently, Möller (1963) discussed the influence of the variation of CO_2 content in the atmosphere on the magnitude of long-wave radiation at the earth's surface, and on the equilibrium temperature of the earth's surface. Assuming that the absolute humidity is independent of the atmospheric temperature, he obtained an order-of-magnitude dependence of equilibrium temperature upon CO_2 content similar to those obtained by Plass (1956), Kondratiev and Niilisk (1960), and Kaplan (1960). However, he obtained an extremely large dependence for a certain range of temperature when he assumed that relative humidity (instead of absolute humidity) of the atmosphere was given. One shortcoming of this study is that the conclusion was drawn from the computation of the heat balance of earth's surface instead of that of the atmosphere as a whole. Therefore, it seems to be highly desirable to reevaluate this theory, using as a basis the computation of radiative convective equilibrium of the atmosphere with a fixed relative humidity. The results are presented in this study.

2.   RADIATIVE CONVECTIVE EQUILIBRIUM

a. Description of the model. As we explained in the previous paper and in the introduction, the radiative convective equilibrium of the atmosphere with a given distribution of relative humidity should satisfy the following requirements:

1) At the top of the atmosphere, the net incoming solar radiation should be equal to the net outgoing long-wave radiation.
2) No temperature discontinuity should exist.
3) Free and forced convection, and mixing by the large-scale eddies, prevent the lapse rate from exceeding a critical lapse rate equal to 6.5C km^{-1}.
4) Whenever the lapse rate is subcritical, the condition of local radiative equilibrium is satisfied.
5) The heat capacity of the earth's surface is zero.
6) The atmosphere maintains the given vertical distribution of relative humidity (new requirement).

In the actual computation, the state of radiative convective equilibrium is computed as an asymptotic state of an initial value problem. Details of

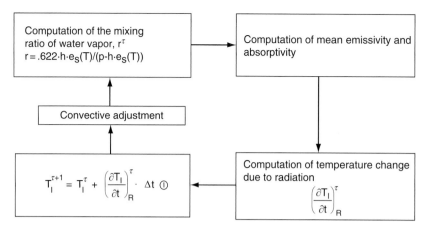

Fig. 2. Flow chart for the numerical time integration.

the procedure are described in Appendix 1. The flow chart of the marching computation is shown in Fig. 2. In this figure $e_s(T)$ denotes the saturation vapor pressure of water vapor as a function of temperature T, and h denotes the relative humidity. τ denotes the number of the time steps of numerical integration, and I is the indexing of the finite differences in the vertical direction. (Refer to Appendix 3 for the illustration of levels adopted for vertical differencing.) The exact definitions of mean emissivity and mean absorptivity are also given in M.S., pp. 365–366.

Since the changes of absolute humidity correspond to the change of air temperature, the equivalent heat capacity of moist air with relative humidity h may be defined as

$$C_p' = C_p\left[1+\frac{L}{C_p}\cdot\frac{\partial}{\partial T}\left(\frac{0.622he_s(T)}{p-he_s(T)}\right)\right], \quad (1)$$

where L and C_p are the latent heat of evaporation and the specific heat of air under constant pressure, respectively. The second term in the bracket appears due to the change of latent energy of the air.

The reader should refer to M.S. for the following information.

1) Computation of the flux of long-wave radiation.
2) Computation of the depletion of solar radiation.
3) Determination of mean absorptivity and emissivity.

Some additional explanation of how we determine the absorptivity is given in Appendix 2.

b. Standard distribution of atmospheric absorbers. In this subsection, the vertical distributions of

water vapor, carbon dioxide, ozone, and cloud, which are used for the computations of thermal equilibrium, and those of heat balance in the following section, are described. They are adopted unless we specify otherwise.

The typical vertical distribution of relative humidity can be approximated with the help of the data in Fig. 3. In this figure, the hemispheric mean of relative humidity obtained by Telegadas and London (1954) and that of relative humidity obtained by Murgatroyd (1960) are shown in the upper and lower troposphere, respectively. The stratospheric distributions of relative humidity obtained by Mastenbrook (1963) at Minneapolis and Washington, D. C., are also plotted after some smoothing of data. Referring to this figure, the following linear function is chosen to represent the vertical distribution of relative humidity, i.e.,

$$h = h_*\left(\frac{Q-0.02}{1-0.02}\right), \quad (2)$$

where h_* is the relative humidity at the earth's surface, equal to 0.77, $Q = p/p_*$, and p_* is surface pressure. When Q is smaller than 0.02, Eq. (2) gives negative value of h. Therefore, it is necessary to specify the humidity distribution for small Q values. According to the measurements by Mastenbrook (1963) and Houghton (1963), the stratosphere is very dry and its mixing ratio is approximately 3×10^{-6} gm gm^{-1} of air. We have therefore assumed that the minimum value of mixing ratio r_{min} to be 3×10^{-6} gm gm^{-1} of air, i.e., if

$$r(T,h)\left(=\frac{0.622he_s(T)}{p-he_s(T)}\right) < r_{min}, \quad (3)$$

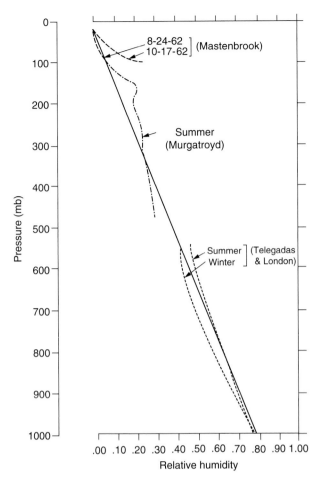

Fig. 3. Vertical distribution of relative humidity (Mastenbrook, 1963; Murgatroyd, 1960; Telegadas and London, 1954).

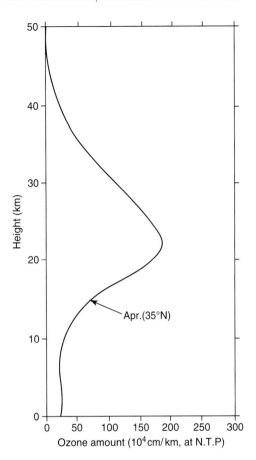

Fig. 4. Vertical distribution of ozone at 35N, April (Herring and Borden, 1965), normalized by the total amount from London (1962).

Table 1. Cloud characteristics employed in radiative convective equilibrium model.

Cloud	Height (km)	Amount	Albedo
High	10.0	0.228	0.20
Middle	4.1	0.090	0.48
Low			
top	2.7	0.313	0.69
bottom	1.7		

set

$$r = r_{min} \, (= 3 \times 10^{-6} \text{ gm gm}^{-1} \text{ of air}).$$

The mixing ratio of carbon dioxide in the atmosphere is assumed to be constant. The mixing ratio adopted for the present computation is 0.0456% by weight (300 ppm by volume).

The vertical distribution of ozone which is used for the computation is shown in Fig. 4. This data is obtained by Herring and Borden (1965) using chemiluminescent ozonesondes for the period September 1963 to August 1964. The vertical distribution at 35N, April, is taken from the figure of his paper, and is normalized to the total amount of ozone obtained by London (1962).

The heights, albedo, and the amounts of cloud adopted for the computation are tabulated in Table 1. The albedo of the earth's surface is assumed to be 0.102.

c. Hergesell's problem and radiative convective equilibrium. Before discussing the results of the study of radiative convective equilibrium in detail, we shall briefly discuss the problem of pure radiative equilibrium (no convection) of the atmosphere with a given distribution of relative humidity. This problem was first investigated by Hergesell (1919) who criticized Emden's solution of pure radiative equilibrium because it allows a layer of supersaturation. Using the assumption of grey body radiation, he obtained, numerically, the state of pure

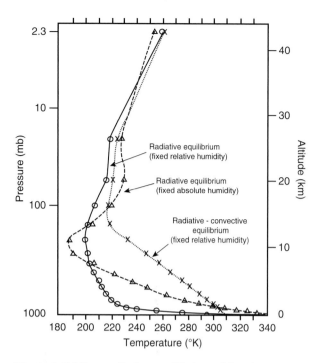

Fig. 5. Solid line, radiative equilibrium of the clear atmosphere with the given distribution of relative humidity; dashed line, radiative equilibrium of the clear atmosphere with the given distribution of absolute humidity; dotted line, radiative convective equilibrium of the atmosphere with the given distribution of relative humidity.

radiative equilibrium of the atmosphere with a realistic distribution of relative humidity. The atmosphere in pure radiative equilibrium thus obtained is almost isothermal, and its temperature is extremely low due to the self-amplification effect of water vapor on the equilibrium temperature of the atmosphere. (For example, since the water content of the atmosphere decreases with decreasing temperature, the greenhouse effect of the atmosphere decreases, and so on.) Fig. 5 shows the solution of this problem which is obtained by our method without using the assumption of grey body radiation (cloudiness = 0). Although the surface equilibrium temperature is much higher than that obtained by Hergesell due to the effect of line center absorption, a sharp decrease of temperature with increasing altitude appears near the ground, and the temperature of the troposphere is much lower than that which is obtained for the atmosphere with a realistic distribution of absolute humidity. This is shown in the same figure. For the sake of comparison, the distribution of equilibrium temperature with convective adjustment is also shown in the figure. The distribution of humidity adopted for this computation is given by Eqs. (2) and (3).

This comparison clearly demonstrates the role of convective adjustment in maintaining the existing distribution of atmospheric temperature. Without this effect, the temperature of the troposphere as well as the height of the tropopause would have been unrealistically low due to the positive feedback effect of water vapor on the air temperature, which we discussed in the introduction.

d. Approach towards the equilibrium state. It should take longer for the atmosphere with a given distribution of relative humidity than for the atmosphere with a given distribution of absolute humidity to reach the state of thermal equilibrium. Two of the reasons for this difference are as follows:

1) As we explained in the introduction, the dependence of outgoing radiation of the atmosphere with a given distribution of relative humidity depends less on the atmospheric temperature than does that of an atmosphere with a given distribution of absolute humidity. Accordingly, the speed of approach towards the equilibrium state is significantly less.
2) Since the vertical distribution of relative humidity is constant throughout the course of the time integration, absolute humidity depends upon the atmospheric temperature, and the variation of absolute humidity involves the variation of the latent energy of the air. Therefore, the effective heat capacity of the air with the given relative humidity is larger than the heat capacity of dry air. Accordingly, the speed of approach is slower.

In Fig. 6, the approaches of each of three idealized atmospheres towards the equilibrium are shown. These are as follows:

Atmosphere I: Vertical distribution of absolute humidity is constant with time.

Atmosphere II: Vertical distribution of relative humidity is constant with time. The heat capacity of the air is assumed to be 0.24 cal gm^{-1}, i.e., the heat capacity of dry air.

Atmosphere III: Vertical distribution of relative humidity is constant with time. The effective heat capacity of air which is given by Eq. (1) is adopted.

Two initial conditions which are chosen for the time integrations shown in Fig. 6 are obtained by adding 15K to the temperature distribution of radiative convective equilibrium.

Because of the first of the two reasons mentioned above, it takes about 1.5 times longer for Atmosphere II than for Atmosphere I to reach the

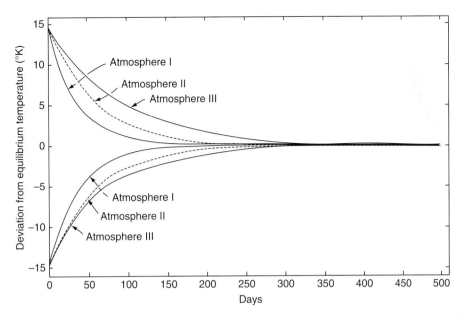

Fig. 6. Approach of vertical mean temperature toward the state of equilibrium. Atmosphere I (dotted line), Atmosphere II (dashed line), Atmosphere III (solid line).

state of equilibrium; and it takes even longer for Atmosphere III to reach equilibrium due to the second reason described above. In short, "the radiation-condensation relaxation" is much slower than pure radiation relaxation. Therefore, it is probable that the radiation-condensation relaxation is one of the important factors in determining the seasonal variation of atmospheric temperature. Also, Fig. 6 shows that it takes longer for the warm atmosphere to reach the state of equilibrium than for the cold atmosphere. This result suggests that it is not advisable to perform the numerical integration of the general circulation model by starting from the very warm initial condition.

3. SOLAR CONSTANT AND RADIATIVE CONVECTIVE EQUILIBRIUM

a. Thermal equilibrium for various solar constants. One of the most fundamental factors which determines the climate of the earth is the solar constant. In order to evaluate the effect of the solar constant upon the climate of the earth's surface, a series of computations of thermal equilibrium was performed. Fig. 7 shows the dependence of the surface equilibrium temperature upon the solar constant for both the atmosphere with a given distribution of relative humidity, and that with a given distribution of absolute humidity.

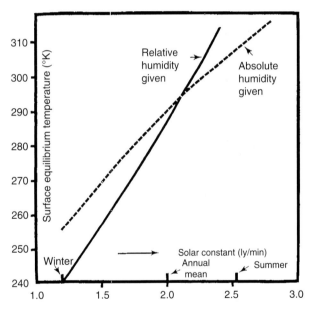

Fig. 7. Solar constant and surface temperature of radiative convective equilibrium. Solid line and dashed line show the case of fixed relative humidity and that of fixed absolute humidity, respectively. Insolation of summer and that of winter is obtained by taking the mean value for the period of June–July–August, and that of December–January–February.

According to this figure, the equilibrium temperature of the atmosphere with a given distribution of relative humidity is twice as sensitive to the change of the solar constant as that with a

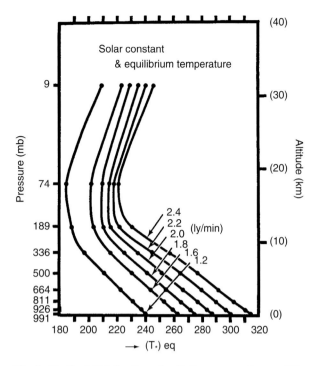

Fig. 8. Vertical distribution of radiative convective equilibrium temperature of the atmosphere with a given distribution of relative humidity for various values of the solar constant.

given distribution of absolute humidity in the range of temperature variation of middle latitudes. This difference in the sensitivity decreases with decreasing temperature. When the temperature is very low, say 240K, the difference is practically negligible, because the mixing ratio of water vapor is extremely small. On the other hand, the equilibrium temperature is very sensitive to the change of the solar constant when the temperature is much above 300K. This result clearly demonstrates the self-amplification effect of water vapor on the equilibrium temperature of the atmosphere with a given distribution of relative humidity. As a reference, the vertical distributions of equilibrium temperature corresponding to various values of the solar constant are shown in Fig. 8.

b. Outgoing radiation and atmospheric temperature. In order to satisfy the condition of thermal equilibrium, the variation of the solar constant must be compensated for by a corresponding change of the outgoing long-wave radiation at the top of the atmosphere. In this subsection, we shall investigate the dependence of outgoing radiation on atmospheric temperature in order that we may understand the results described just above.

In Fig. 9, various vertical distributions of temperature adopted for the present computations

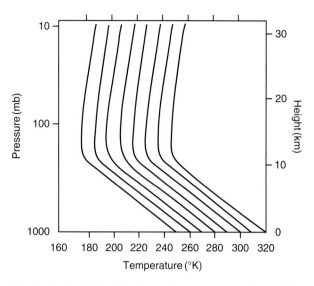

Fig. 9. Vertical distributions of temperature adopted for the computation of radiative flux shown in Fig. 10.

are shown, and in Fig. 10, the upward long-wave radiation at the top of the atmosphere is plotted versus the temperature of the earth's surface T_*. The distribution of relative humidity, cloudiness, and other atmospheric absorbers adopted for this computation are described in Section 2b. In Fig. 9 curves representing blackbody radiation temperatures T_*, (T_*-10), (T_*-20), (T_*-30), and (T_*-50) are also drawn for the sake of comparison. According to this comparison, the outgoing long-wave radiation at the top of the atmosphere with a given distribution of relative humidity depends less upon the atmospheric temperature than is expected from the fourth-power law of Stefan-Boltzmann.

As we explained in the introduction, the deviation from the fourth-power law is mainly due to the dependence of the effective source of outgoing radiation upon the temperature of the atmosphere. This result explains why the atmosphere with the fixed distribution of relative humidity is more sensitive to the variation of solar radiation than the atmosphere with the fixed distribution of absolute humidity.

4.   EQUILIBRIUM TEMPERATURE AND ATMOSPHERIC ABSORBERS

In this section, we shall discuss the dependence of equilibrium temperature upon the vertical distribution of atmospheric absorbers such as water vapor, carbon dioxide, ozone, and cloud. It is hoped that the results of this section will be useful

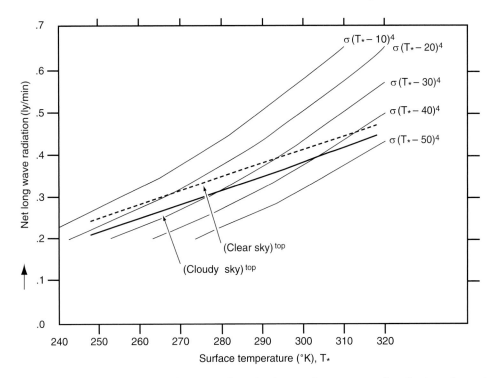

Fig. 10. Net long-wave radiation at the top of the atmosphere for the set of temperature distributions shown in Fig. 9. Thin lines in the background show the black body radiation at the temperatures $(T_* - 10)$, $(T_* - 20)$, $(T_* - 30)$, $(T_* - 40)$, and $(T_* - 50)$.

for evaluating the possibility of various climatic changes in the earth's atmosphere.

a. Tropospheric relative humidity. In order to evaluate the dependence of equilibrium temperature upon the distribution of relative humidity of the atmosphere, a series of computations of thermal equilibrium was performed for various distributions of relative humidity. The vertical distribution of relative humidity adopted for this series of computations is described by Eqs. (2) and (3), except that we assigned various values to h_*. For the distribution of other gaseous absorbers and clouds, see Section 2b.

Fig. 11 shows the vertical distributions of equilibrium temperature corresponding to h_* values of 0.2, 0.6, and 1.0. The following features are noteworthy.

1) The higher the tropospheric relative humidity, the warmer is the equilibrium temperature of the troposphere.
2) The equilibrium temperature of the stratosphere depends little upon the tropospheric relative humidity.

Table 2 shows the dependence of the net upward radiation at the top of the atmosphere R_L, and that of the surface equilibrium temperature T_*^e upon

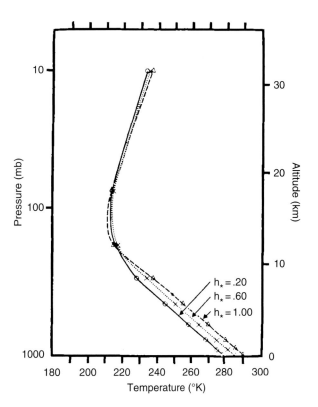

Fig. 11. Vertical distributions of radiative convective equilibrium temperature for various distributions of relative humidity.

Table 2. Variation of surface equilibrium temperature T_*^e and net upward radiation at the top of the atmosphere R_L for various values of the relative humidity at the earth's surface h_*.

h_*	T_*^e	R_L (1 y min^{-1})
0.2	278.1	0.3214
0.6	285.0	0.3274
1.0	289.9	0.3313

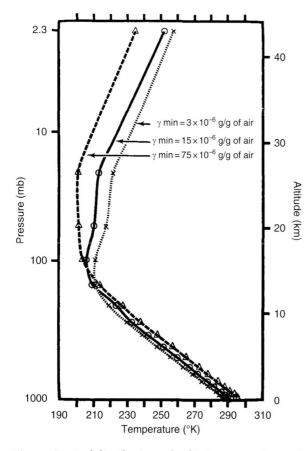

Fig. 12. Vertical distributions of radiative convective equilibrium temperature for various values of water vapor mixing ratio in the stratosphere.

the relative humidity of the earth's surface. Based upon this table, it is possible to obtain the following approximate relationship between T_*^e and h_*:

$$\frac{\partial T_*^e}{\partial (100 h_*)} \simeq 0.133,$$

the units being degrees Celsius per unit percentage change of relative humidity, and where $1.0 \geq h^* \geq 0.2$. This table also indicates that the net outgoing radiation depends very little upon the tropospheric relative humidity. This is why the dependence of

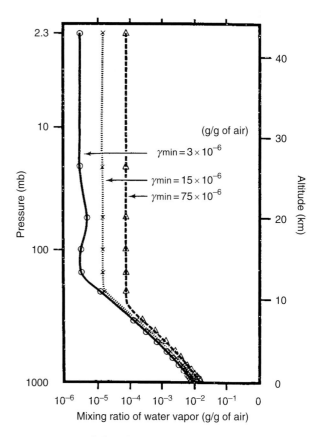

Fig. 13. Vertical distribution of water vapor mixing ratio corresponding to the equilibrium status shown in Fig. 12.

stratospheric temperature upon tropospheric relative humidity is so small.

b. Water vapor in the stratosphere. Recently, a panel on weather and climate modification appointed by the National Academy of Science (1966) suggested that the temperature of the earth's atmosphere may be altered significantly by an increase of stratospheric water vapor anticipated with an increasing number of supersonic transport aircraft flights. It should be useful to evaluate the effect of the variation of stratospheric water vapor upon the thermal equilibrium of the atmosphere, with a given distribution of relative humidity. The distribution of humidity adopted for this series of computations is given by Eqs. (2) and (3), except that the absolute humidity of the stratosphere r_{min} is different for each experiment.

The values of r_{min} chosen for this series of computations are 3×10^{-6}, 15×10^{-6}, and 75×10^{-6} gm gm^{-1} of air. Figs. 12 and 13 show the states of thermal equilibrium thus computed, and the corresponding vertical distributions of water vapor mixing ratio. Examination of Fig. 12 reveals the following features.[1]

1) The larger the stratospheric mixing ratio r_{min}, the warmer is the tropospheric temperature.
2) The larger the water vapor mixing ratio in the stratosphere, the colder is the stratospheric temperature.
3) The dependence of the equilibrium temperature in the stratosphere upon the stratospheric water vapor mixing ratio is much larger than that in the troposphere.

Table 3 shows the equilibrium temperature of the earth's surface corresponding to various water vapor mixing ratios in the stratosphere. Recent measurements by Mastenbrook (1963) and others suggest that the mixing ratio in the atmosphere is about 3×10^{-6} gm gm^{-1} of air. According to this table, a 5-fold increase of stratospheric water vapor over its present value would increase the temperature of the earth's surface by about 2.0C. It is highly questionable that such a drastic increase of stratospheric water vapor would actually take place due to the release of water vapor from the supersonic transport. Recently, Manabe *et al.* (1965) performed a numerical experiment of the general circulation model of the atmosphere with

Table 3. The variation of the equilibrium temperature of the earth's surface T_*^e with stratosphere water vapor mixing ratio r_{min}.

r_{min} (gm gm^{-1})	T_*^e (°K)
3×10^{-6}	288.4
15×10^{-6}	290.4
75×10^{-6}	296.0

the hydrologic cycle. They concluded that, in the model atmosphere, the large-scale quasi-horizontal eddies are very effective in removing moisture from the high- and middle-latitude stratosphere by freezing out near the cold equatorial tropopause. Their results must be viewed with caution, however, because their computation involves a large truncation error in evaluating vertical advection of water vapor in the upper troposphere and the lower stratosphere. This study, nonetheless, suggests the possible importance of dynamical process in the water balance of the stratosphere.

c. Carbon dioxide. As we mentioned in the introduction, Möller (1963) discussed the influence of the change of CO_2 content in the atmosphere with a given value of relative humidity on the temperature of the earth's surface. He computed the magnitudes of downward long-wave radiation corresponding to various CO_2 contents, and estimated the change of surface temperature required to compensate for the change of net downward radiation due to the change of CO_2 content. His results suggest that the increase in the water content of the atmosphere with increasing temperature causes a self-amplification effect, which results in an almost arbitrary change of temperature at the earth's surface. In order to re-examine Möller's computation by use of the present scheme of computing radiative transfer, the net upward long-wave radiation into the atmosphere with given distributions of relative humidity was computed for various distributions of surface temperature as shown in Fig. 9. The vertical distribution of cloudiness and that of relative humidity have already been specified in Section 2b. In Fig. 14,

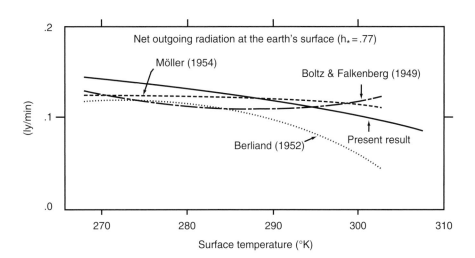

Fig. 14. Values of net upward long-wave radiation at the earth's surface which are computed from the empirical formulas obtained by various authors as well as the values from the present contribution.

the magnitude of net radiation thus computed is plotted versus the temperature of the earth's surface. For the sake of comparison, the magnitudes of net flux obtained by using the formulas proposed by Möller (1963), Berliand and Berliand (1952), and Boltz and Falkenberg (1950) are added to the same figure. The relative humidity at the earth's surface needed for these computations is assumed to be 77%. Möller (1963) and Berliand and Berliand (1952) obtained their empirical formulas from radiation chart computations, whereas Boltz and Falkenberg (1950) obtained their empirical formula using measurements with a carefully calibrated vibrational pyranometer. Our comparison shows the results obtained by the various methods to be fairly consistent. Generally speaking, the dependence of net radiation upon temperature is small. It increases or decreases with increasing temperature depending upon the method of computation. For example, the results of the present computation and those of Berliand indicate that the net upward radiation decreases monotonically with increasing surface temperature for the ordinary range of temperature. If one discusses the effect of carbon dioxide upon the climate of the earth's surface based upon these results, one could conclude that the greater the amount of carbon dioxide, the colder would be the temperature of the earth's surface, i.e., to compensate for the increase of downward radiation due to the increase of carbon dioxide, it is necessary to have a lower temperature. On the other hand, the result of Boltz and Falkenberg (1950) may lead us to the opposite conclusion for temperatures above 290K. As Möller (1963) suspected, these results do not always indicate the extreme sensitivity of the actual earth's climate. The basic shortcoming of this line of argument may be that it is based upon the heat balance only of the earth's surface, instead of that of the atmosphere as a whole. In Fig. 15, the net upward long-wave radiation at the top of the atmosphere, together with that at the earth's surface, are plotted against the temperature of the earth's surface. As we have already discussed in Section 3b, the former increases significantly with increasing temperature in contrast to the latter. In order to compensate for the decrease of net outgoing radiation at the top of the atmosphere due to the increase of CO_2 content, it is necessary to increase the atmospheric temperature. Therefore, one may expect that the larger the CO_2 content in the atmosphere, the warmer would be the temperature of the earth for the ordinary range of atmospheric temperature.

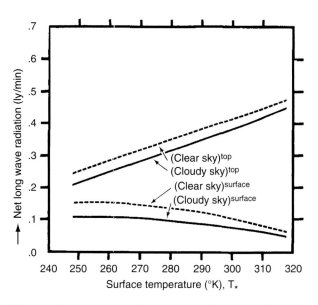

Fig. 15. The net upward long-wave radiation both at the top and bottom of the atmosphere.

This result is not in agreement with the conclusion which we reached based upon the earth's surface.

In order to obtain the complete picture, it is also necessary to consider the effect of convection. Therefore, a series of radiative convective equilibrium computations were performed. Fig. 16 shows the vertical distributions of equilibrium temperature corresponding to the three different CO_2, i.e., 150, 300, and 600 ppm contents by volume. In this figure, the following features are noteworthy:

1) The larger the mixing ratio of carbon dioxide, the warmer is the equilibrium temperature of the earth's surface and troposphere.
2) The larger the mixing ratio of carbon dioxide, the colder is the equilibrium temperature of the stratosphere.
3) Relatively speaking, the dependence of the equilibrium temperature of the stratosphere on CO_2 content is much larger than that of tropospheric temperature.

Table 4 shows the equilibrium temperature of the earth's surface corresponding to various CO_2 contents of the atmosphere, and Table 5 shows the change of surface equilibrium temperature corresponding to the change of CO_2 content. In these tables, values for both the atmosphere with given distribution of absolute humidity, and that with the given distribution of relative humidity are shown together. According to this comparison,

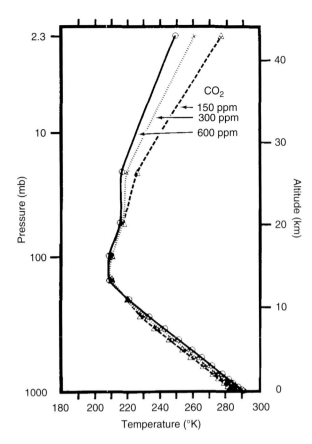

Fig. 16. Vertical distributions of temperature in radiative convective equilibrium for various values of CO_2 content.

Table 4. Equilibrium temperature of the earth's surface (°K) and the CO_2 content of the atmosphere.

CO_2 content (ppm)	Average cloudiness		Clear	
	Fixed absolute humidity	Fixed relative humidity	Fixed absolute humidity	Fixed relative humidity
150	289.80	286.11	298.75	304.40
300	291.05	288.39	300.05	307.20
600	292.38	290.75	301.41	310.12

the equilibrium temperature of the former is almost twice as sensitive to the change of CO_2 content as that of the latter, but not as sensitive as the results of Möller suggest. These results indicate that the extreme sensitivity he obtained was mainly for the reason already stated.

Although our method of estimating the effect of overlapping between the 15-μ band of CO_2 and the rotation band of water vapor is rather crude, we believe that the general conclusions which have been obtained here on the atmosphere with a fixed

Table 5. Change of equilibrium temperature of the earth's surface corresponding to various changes of CO_2 content of the atmosphere.

Change of CO_2 content (ppm)	Fixed absolute humidity		Fixed relative humidity	
	Average cloudiness	Clear	Average cloudiness	Clear
$300 \rightarrow 150$	−1.25	−1.30	−2.28	−2.80
$300 \rightarrow 600$	+1.33	+1.36	+2.36	2.92

Table 6. Change of equilibrium temperature of the earth's surface corresponding to various changes of CO_2 content of the atmosphere [computed by Möller using the absorption value of Yamamoto and Sasamori (1958)].

Variation of CO_2 content (ppm)	Fixed absolute humidity	
	Average cloudiness	Clear
$300 \rightarrow 150$	−1.0	−1.5
$300 \rightarrow 600$	+1.0	+1.5

relative humidity should not be altered by this inaccuracy. It is interesting to note that the dependencies of surface temperature on the CO_2 content, which were obtained by Möller (1963) and present authors for the atmosphere with a fixed relative humidity, agree reasonably well with each other (see Table 6 for Möller's results).

d. Ozone. States of thermal equilibrium were computed for three different distributions of ozone as shown in Fig. 17. These distributions were read off from the results which were obtained by Herring and Borden (1965), using the chemiluminescent ozonesonde. The total amounts of ozone are adjusted such that they coincide with those obtained by London (1962). Among the three distributions shown in the figure, the total amount for 0N, April, is a minimum, and that for 80N, April, a maximum. Fig. 18 shows the vertical distribution of equilibrium temperature corresponding to each ozone distribution. The following features are noteworthy:

1) The larger the amount of ozone, the warmer is the temperature of troposphere and the lower stratosphere, and the colder is the temperature of the upper stratosphere.
2) The influence of ozone distribution upon equilibrium temperature is significant in the stratosphere, but is small in the troposphere.
3) As we pointed out in M.S., the ozone distribution of 0N, April, tends to make the tropopause

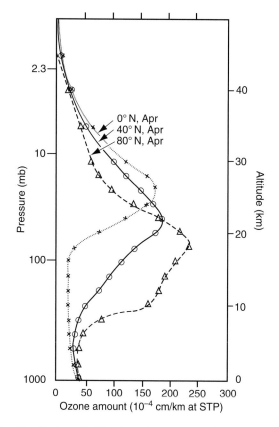

Fig. 17. Vertical distribution of O_3 adopted for the computation of radiative equilibrium shown in Fig. 18. Vertical distribution (Herring and Borden, 1965); total amount (London, 1962).

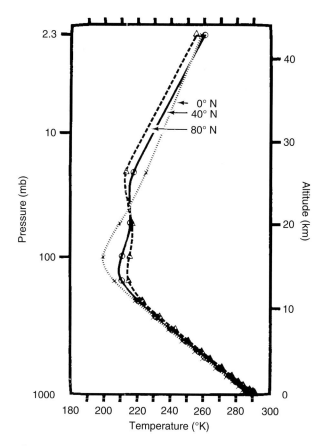

Fig. 18. Vertical distributions of the temperature of radiative convective equilibrium, which correspond to the ozone distributions shown in Fig. 17.

height higher and the tropopause profile sharper than those of 80N, April. As reference, equilibrium temperatures of the earth's surface as well as the total amounts for the three distributions adopted here are tabulated in Table 7.

e. Surface albedo. A series of thermal equilibrium states of the atmosphere with the given distribution of relative humidity was computed for various albedos of the earth's surface. Fig. 19 shows the results. Examination of this figure reveals the following features:

1) The larger the value of albedo of the earth's surface, the colder the temperature of the atmosphere.
2) The influence of the surface albedo decreases with increasing altitude. It is a maximum at the earth's surface, and is almost negligible at the 9-mb level.

Table 8 shows the surface equilibrium temperature T_*^e for various values of surface albedo α_*.

Table 7. Equilibrium temperatures of the earth's surface for three ozone distributions.

Latitude, month	Total amount O_3 (cm, STP)	T_*^e (°K)
0N, April	0.260	287.9
40N, April	0.351	288.8
80N, April	0.435	290.3

According to this table, the sensitivity of the equilibrium temperature of the earth's surface on the surface albedo may be approximately expressed by

$$\partial T_*^e/\partial(100\alpha_*) = -1,$$

where the units are degrees Celcius change per unit percentage change of surface albedo.

Again, this sensitivity is almost twice as large as that of the atmosphere with a fixed absolute humidity for the ordinary range of solar constant.

f. Cloudiness. A series of thermal equilibrium computations was performed for various distribu-

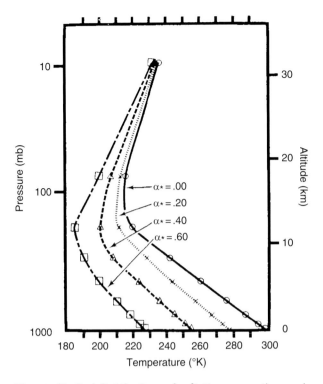

Fig. 19. Vertical distributions of radiative convection equilibrium for various values of surface albedo.

Table 8. Surface equilibrium temperatures T_*^e for various values of surface albedo α_*.

Albedo	T_*^e
0.00	297.2
0.20	276.4
0.40	253.2
0.60	227.0

tions of cloudiness. The equilibrium temperatures of the earth's surface for a variety of cloud distributions are tabulated in Table 9 and shown in Fig. 20.

Generally speaking, the larger the cloud amount, the colder is the equilibrium temperature of the earth's surface, though this tendency decreases with increasing cloud height and does not always hold for cirrus. The equilibrium temperature of the atmosphere with average cloudiness specified in the table is about 20.7C colder than that for a clear atmosphere. This difference is significantly larger than the difference of about 13C, which was obtained for the atmosphere with a fixed absolute humidity (see M.S.). The dependence of equilibrium temperature of the earth's surface on the amount of low (C_L) middle (C_M), and high (C_H) clouds may be expressed by the following equations:

Table 9. Effect of cloudiness on surface equilibrium temperature T_*^e. FB and HB refer to full black and half black, respectively.

Experiment no.	Cloudiness (amount) High	Middle	Low	T_*^e (°K)
C1	0.000(HB)	0.072(FB)	0.306(FB)	280.1
C2	0.500(HB)	0.072(FB)	0.306(FB)	281.6
C3	1.000(HB)	0.072(FB)	0.306(FB)	284.2
C1	0.000(FB)	0.072(FB)	0.306(FB)	280.1
C4	0.500(FB)	0.072(FB)	0.306(FB)	298.4
C5	1.000(FB)	0.072(FB)	0.306(FB)	318.0
C6	0.218(FB)	0.000(FB)	0.306(FB)	290.5
C7	0.218(FB)	0.500(FB)	0.306(FB)	271.5
C8	0.218(FB)	1.000(FB)	0.306(FB)	251.8
C9	0.218(FB)	0.072(FB)	0.000(FB)	311.3
C10	0.218(FB)	0.072(FB)	0.500(FB)	272.0
C11	0.218(FB)	0.072(FB)	1.000(FB)	229.3
C12	0.000	0.000	0.000	307.8
C13	0.218(FB)	0.072(FB)	0.306(FB)	287.1

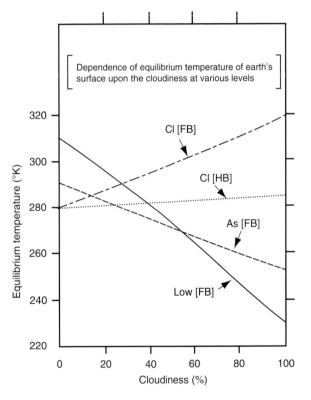

Fig. 20. Radiative convective equilibrium temperature at the earth's surface as a function of cloudiness (cirrus, altostratus, low cloud). FB and HB refer to full black and half black, respectively.

$$\partial T_*^e / \partial (100C_L) = -8.2$$
$$\partial T_*^e / \partial (100C_M) = -3.9$$
$$\partial T_*^e / \partial [100C_H(FB)] = +0.17$$
$$\partial T_*^e / \partial [100C_H(HB)] = +0.04$$

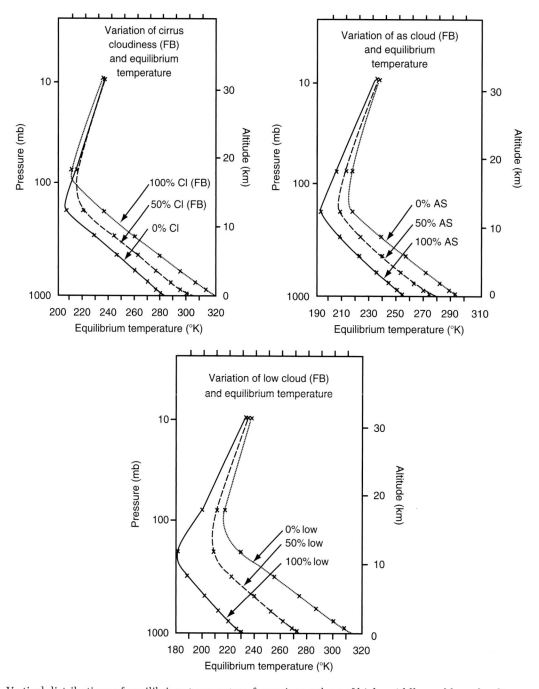

Fig. 21. Vertical distributions of equilibrium temperature for various values of high, middle, and low cloudiness.

All units are in degrees Celcius change per unit percentage increase in cloudiness. FB and HB refer to full black and half black, respectively.

The reader should refer to Section 2b for the albedo of each cloud type and the distributions of the gaseous absorbers adopted for this computation. Whether cirrus clouds heat or cool the equilibrium temperature depends upon both the height and the blackness of cirrus cloud. This subject was previously discussed in M.S.

Relatively speaking, the influence of cloudiness upon the equilibrium temperature is more pronounced in the troposphere than in the stratosphere, where the influence decreases with increasing altitude. (Refer to Figs. 21 and 22, which show the vertical distribution of equilibrium temperatures for various distributions of cloudiness.) Accordingly, middle and low clouds have a tendency to lower the height of the tropopause.

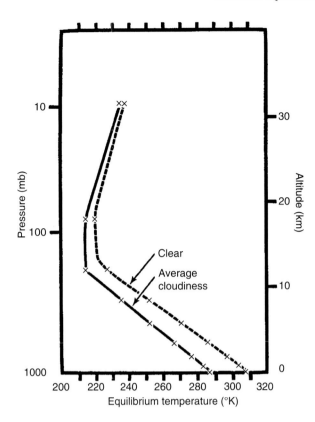

Fig. 22. Vertical distributions of equilibrium temperature for a clear atmosphere and that for an atmosphere with average cloudiness.

5. SUMMARY AND CONCLUSIONS

1) A series of radiative convective equilibrium computations of the atmosphere with a given distribution of relative humidity were performed successfully.

2) Generally speaking, the sensitivity of the surface equilibrium temperature upon the change of various factors such as solar constant, cloudiness, surface albedo, and CO_2 content are almost twice as much for the atmosphere with a given distribution of relative humidity as for that with a given distribution of absolute humidity.

3) The speed of approach towards the state of equilibrium is half as much for the atmosphere with a given distribution of relative humidity as for that with the given distribution of absolute humidity. In other words, the time required for radiation-condensation relaxation is much longer than that required for radiation relaxation of the mean atmospheric temperature.

4) Doubling the existing CO_2 content of the atmosphere has the effect of increasing the surface temperature by about 2.3C for the atmosphere with the realistic distribution of relative humidity and by about 1.3C for that with the realistic distribution of absolute humidity. The present model atmosphere does not have the extreme sensitivity of atmospheric temperature to the CO_2 content which Möller (1963) encountered in his study when the atmosphere has a given distribution of relative humidity.

5) A five-fold increase of stratospheric water vapor from the present value of 3×10^{-6} gm gm^{-1} of air causes an increase of surface equilibrium temperature of about 2.0C, when the vertical distribution of relative humidity is fixed. Its effect on the equilibrium temperature of the stratosphere is larger than that of troposphere.

6) The effects of cloudiness, surface albedo, and ozone distribution on the equilibrium temperature were also presented.

Acknowledgments. The authors wish to thank Dr. Smagorinsky and Dr. Möller whose encouragement started this study, Dr. J. Murray Mitchell, Jr., who read the manuscript carefully and who gave us many useful comments, and Mrs. Marylin Varnadore and Mrs. Clara Bunce who assisted in the preparation of the manuscript. Finally, the constant encouragement of Dr. Bryan is sincerely appreciated.

APPENDIX 1

DETAIL OF THE METHOD OF CONVECTIVE ADJUSTMENT

Since we did not describe the detail of the method of convective adjustment in M.S., we shall explain the method in this appendix. The following procedures are executed at each timestep (see Fig. 23).

1) Compute $T_K^{(0)}$ ($K = 1, 2, \cdots, N$) by use of the equation

$$T_K^{(0)} = T_K^{\tau} + \left(\frac{\partial T_K^{\tau}}{\partial \tau} \right)_{\text{RAD}} \Delta t,$$

where T_K is the temperature of the Kth level at the τth time step, $(\partial T_K^{\tau}/\partial t)_{\text{RAD}}$ is the rate of change of T_K at the τth time step, and Δt is the time interval of forward time integration.

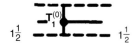

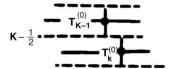

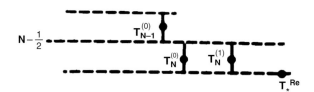

Fig. 23. Notations used for the explanations of convective adjustment.

2) Compute the radiative equilibrium temperature of the earth's surface T_*^{Re} such that it satisfies the relationship

$$(SR)_*^{\tau} + (DLR)_*^{\tau} = \sigma (T_*^{\mathrm{Re}})^4,$$

where $(SR)_*^{\tau}$ and $(DLR)_*^{\tau}$ are net solar radiation and downward long-wave radiation at the τth time step, respectively.

3) Compute $T_N^{(1)}$ such that it satisfies the relationship

$$C_p \frac{\Delta p N}{g}(T_N^{(1)} - T_N^{(0)}) = \sigma \{(T_*^{\mathrm{Re}})^4 - (T_N^{(1)})^4\}\Delta t,$$

where Δp_N is the pressure thickness of the Nth layer and σ is the Stefan-Boltzmann constant.

4) If $T_N^{(1)} - T_{N-1}^{(0)} > (LRC)_{N-\frac{1}{2}}$ (unstable), compute $T_N^{(2)}$ and $T_{N-1}^{(1)}$ such that they satisfy the relationships

$$T_N^{(2)} - T_{N-1}^{(1)} = (LRC)_{N-\frac{1}{2}},$$

$$C_p \left\{ \frac{\Delta p N}{g}(T_N^{(2)} - T_N^{(1)}) + \frac{\Delta p_{N-1}}{g}(T_{N-1}^{(1)} - T_{N-1}^{(0)}) \right\}$$
$$= \sigma \{(T_N^{(1)})^4 - (T_N^{(2)})^4\}\Delta t,$$

where $(LRC)_{N-\frac{1}{2}}$ is the critical (neutral) temperature difference between the Nth and $(N-1)$th level.

If $T_N^{(1)} - T_{N-1}^{(0)} < (LRC)_{N-\frac{1}{2}}$ (stable), set $T_N^{(2)} = T_N^{(1)}$ and $T_{N-1}^{(1)} = T_{N-1}^{(0)}$.

5) Repeat the following procedures for $K = N-1$, $N-2, \cdots, 1$.

If $T_K^{(1)} - T_{K-1}^{(0)} > (LRC)_{K-\frac{1}{2}}$ (unstable),

compute $T_K^{(2)}$ and $T_{K-1}^{(1)}$ such that they satisfy the relationships

$$T_K^{(2)} - T_{K-1}^{(1)} = (LRC)_{K-\frac{1}{2}},$$

$$C_p \left\{ \frac{\Delta p_K}{g}(T_K^{(2)} - T_K^{(1)}) + \frac{\Delta p_{K-1}}{g}(T_{K-1}^{(1)} - T_{K-1}^{(0)}) \right\} = 0.$$

6) Repeat processes 4) and 5) after making the following replacement in the equations:

$$T_K^{(n)} \leftarrow T_K^{(n+2)}(K = 1, 2, \cdots, N).$$

7) Repeat process 6) until the layer of supercritical lapse rate is completely eliminated.

Effectively, the temperature of the earth's surface at the $(\tau + 1)$th step $(T_*^{\tau+1})$ and that of the atmosphere at the τth step are used for the computation of net radiative flux at the earth's sun face. This method is adopted for the sake of computational stability. Since we reach the final equilibrium which satisfies the requirement described in Section 2a, this inconsistency should not cause any error in the final equilibrium.

APPENDIX 2

ABSORPTIVITIES

Absorptivity data, which are used for this study, were given in Figs. A1–A6 of M.S. In these figures, mean slab absorptivities, emissivity, or the rate of absorption of solar insolation were taken as the ordinate, and the logarithm of optical thickness[2] was taken as the abscissa. The curves which were obtained for various pressures using the experimental data are shown in each figure.

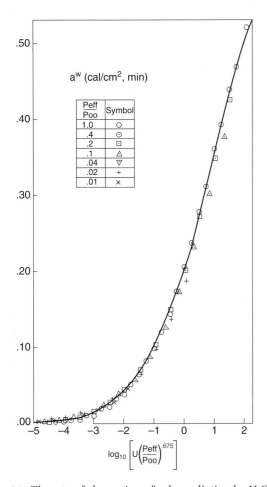

Fig. 24. The rate of absorption of solar radiation by H_2O.

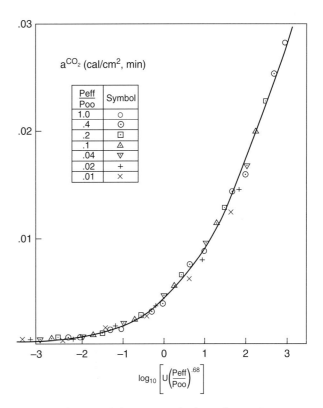

Fig. 25. The rate of absorption of solar radiation by CO_2.

As Howard *et al.* (1955) suggested, one can attempt to parameterize the dependence of the absorptivities upon pressure by taking the following effective optical thickness $\boldsymbol{u}_\tau$ as abscissa instead of the optical thickness u, i.e.,

$$u_r = (p/p_0)^k u,$$

where p is pressure, p_0 the standard atmospheric pressure, and k the constant to be determined from experimental values. It is not possible, however, to express the absorption curves as a function of u_r alone for the wide range of pressure and optical thickness covered by Figs. A1–A5 of M.S. In order to overcome this difficulty, we limited ourselves to the (p, u) range which is usually needed for the present computation of radiative convective equilibrium of the atmosphere. For example, the combination of a large u and small p or that of small u and large p is not encountered in our computation. This restriction enables us to construct the universal curves which are shown in Figs. 24 or 29. The

values of absorptivity which correspond to one-half integral values of the logarithm of optical thickness, are plotted in these figures. However, for the 9.6-μ band only, the values of absorptivity which are obtained from the experiments of Walshaw (1957) are plotted. That part of the (p,u) range which is covered by Figs. A1–A5 of M.S., but is not encountered in our computation, is omitted from this compilation. Needless to say, it is desirable to use a two parameter u and p. However, one parameter u_r is adopted here for simplicity of programming. On page 368 of M.S. the method of obtaining a^W (Fig. 24) and a^{CO_2} (Fig. 25) are given. Refer to Eq. (13) of M.S. for the definition of $\tilde{\varepsilon}_f^W$ (Fig. 26), to Eq. (12) of M.S. for the definition of ε_f^W (Fig. 27), and to Eqs. (16a), (16b), and (17) of M.S. for the definition of $\varepsilon_f^{CO_2}$ (Figs. 28 and 29), $\varepsilon_f^{O_3}$ (Fig. 30), and $\tilde{\varepsilon}_f^{WOV}$ (Fig. 26). For the curve of absorption of solar radiation by ozone, refer to Fig. A6 of M.S.

APPENDIX 3

Both 18 and 9 atmospheric levels are used in the present computations. As in our previous computations, the location of each level is based on a

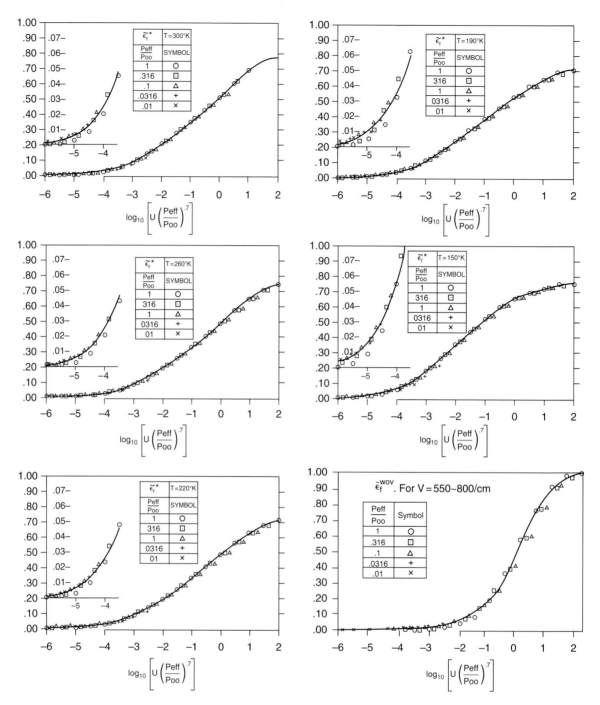

Fig. 26. The mean slab absorptivity of H_2O, from which the contribution of the range of wave number 550–800 cm⁻¹ is omitted. In the lower right corner is shown the slab absorptivity of H_2O for the omitted range of wave number.

suggestion by J. Smagorinsky. Let the quantity σ be defined as the following function of pressure:

$$Q = p/p_* = \sigma^2(3 - 2\sigma),$$

where p_* is the pressure at the earth's surface and assumed here to be 1000 mb. If we divide the

atmosphere into equal σ-intervals, the pressure thickness of the layer is thin both near the earth's surface and the top of the atmosphere. Tables 10 and 11 show the σ-level adopted for 18- and 9-level models, respectively. For our study, we used both 18- and 9-level models.

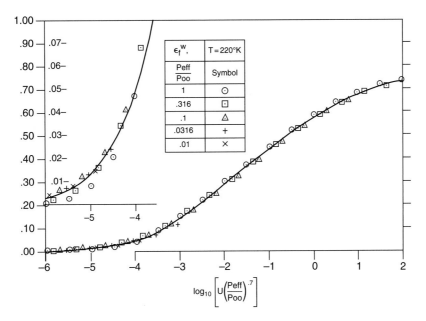

Fig. 27. Emissivity of H_2O from which the contribution of the wave number range 550–800 cm^{-1} is omitted.

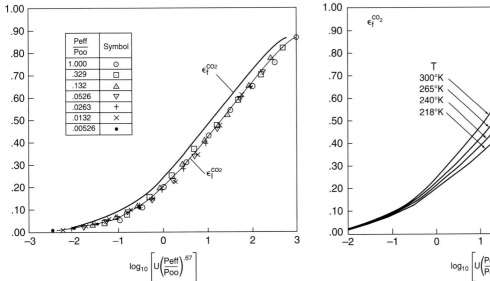

Fig. 28. Slab absorptivity ε_f and column absorptivity ε_l of CO_2 at 300K. Bandwidth is assumed to be 250 cm^{-1}.

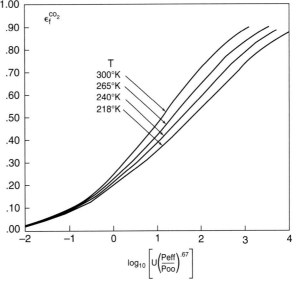

Fig. 29. Slab absorptivities of CO_2 at various temperatures.

In order to compare the equilibrium solutions obtained from these two coordinate systems, reference should be made to Fig. 31. The standard distribution of atmospheric absorbers, which is described in Section 2b is used for both of these computations. The coincidence between the two equilibrium solutions is reasonable.

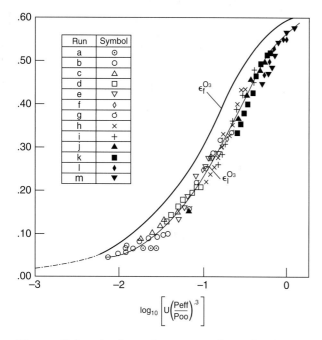

Fig. 30. Slab and column absorptivity of 9.6-μ band of O_3. Run means the run of experiments by Walshaw (1957). Bandwidth is assumed to be 138 cm^{-1}.

Table 10. Illustration of the 18-level σ-coordinate system based on $p_* = 1000$ mb. H denotes the approximate height of the level, and Δp is the pressure thickness of the layer.

Level	σ	p(mb)	Δp(mb)	H(km)
1	0.0277	2	9	42.9
2	0.0833	20	25	26.4
3	0.1388	53	40	20.1
4	0.1944	99	52	16.1
5	0.2500	156	62	13.3
6	0.3055	223	71	11.0
7	0.3611	297	77	9.0
8	0.4166	376	81	7.5
9	0.4722	458	83	6.1
10	0.5277	542	83	4.9
11	0.5833	624	81	3.7
12	0.6388	703	77	2.9
13	0.6944	777	71	2.1
14	0.7500	844	62	1.4
15	0.8055	901	52	0.86
16	0.8611	947	40	0.46
17	0.9166	980	25	0.18
18	0.9722	998	9	0.02

Table 11. Illustration of the 9-level σ-coordinate system based on $p_* = 1000$ mb.

Level	σ	p(mb)	Δp(mb)	H(km)
1	0.0555	9	34	31.6
2	0.1666	74	92	18.0
3	0.2777	189	133	12.0
4	0.3888	336	158	8.3
5	0.4999	500	166	5.5
6	0.6110	664	158	3.3
7	0.7221	811	133	1.7
8	0.8332	926	92	0.64
9	0.9443	991	34	0.07

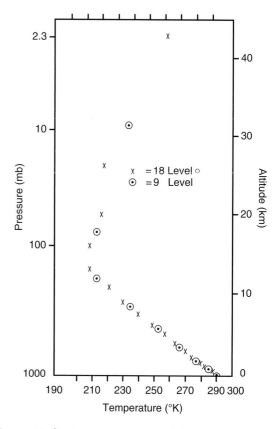

Fig. 31. Radiative convective equilibrium of the atmosphere from the 9- and 18-level models. See text for discussion.

NOTES

1 Qualitatively similar conclusions were obtained for the atmosphere with a given distribution of absolute humidity [see M. S. and Manabe and Möller (1961)].

2 In Figs. A1, A2, A3, A4, and A5 of M. S., the abscissas show the logarithm of optical thickness instead of effective optical thickness, which was implied by captions.

REFERENCES

Berliand, M. E., and T. G. Berliand, 1952: Determination of the effective outgoing radiation of the earth, taking into account the effect of cloudiness. *Izv. Akad. Nauk SSSR, Ser. Geofiz.*, No. 1, 64–78.

Boltz, H., and G. Falkenberg, 1950: Neubestimmung der Konstanten der Angströmschen Strahlungsformel. *Z. Meteor.*, **7**, 65–66.

Hergesell, M., 1919: Die Strahlung der Atmosphäre unter Zungrundlegung bon Lindeberger Temperatur: und Feuchtigkeits Messungen. *Die Arbeiten des Preusslichen Aeronautischen Observatoriums bei Lindenberg*, Vol. 13, Braunschweig, Germany, Firedr., Vieweg and Sohn, 1–24.

Herring, W. S., and T. R. Borden, Jr., 1965: Mean distributions of ozone density over North America, 1963–1964. Environmental Research Papers, No. 162., AFCRL-65-913, Air Force Cambridge Research Laboratories, Bedford, Mass., 19 pp.

Houghton, J. T., 1963: Absorption in the stratosphere by some water vapor lines in the v_2 band. *Quart. J. Roy. Meteor. Soc.*, **89**, 332–338.

Howard, J. N., D. L. Burch and D. Williams, 1955: Near-infrared transmission through synthetic atmosphere. Geophysics Research Papers, No. 40, Air Force Cambridge Research Center, AFCRC-TR-55-213, 214 pp.

Kaplan, L. D., 1960: The influence of carbon dioxide variations on the atmospheric heat balance. *Tellus*, **12**, 204–208.

Kondratiev, K. Y., and H. I. Niilisk, 1960: On the question of carbon dioxide heat radiation in the atmosphere. *Geofis. Pura Appl.*, **46**, 216–230.

London, J., 1962: Mesospheric dynamics, Part III. Final Report, Contract No. AF19(604)-5492, Department of Meteorology and Oceanography, New York University, 99 pp.

Manabe, S., and F. Möller, 1961: On the radiative equilibrium and heat balance of the atmosphere. *Mon. Wea. Rev.*, **89**, 503–532.

———, J. Smagorinsky and R. F. Strickler, 1965: Simulated climatology of general circulation with a hydrologic cycle. *Mon. Wea. Rev.*, **93**, 769–798.

———, and R. F. Strickler, 1964: Thermal equilibrium of the atmosphere with a convective adjustment. *J. Atmos. Sci.*, **21**, 361–385.

Mastenbrook, H. J., 1963: Frost-point hygrometer measurement in the stratosphere and the problem of moisture contamination. *Humidity and Moisture*, Vol. 2, New York, Reinhold Publishing Co., 480–485.

Möller, F., 1963: On the influence of changes in the CO_2 concentration in air on the radiation balance of the earth's surface and on the climate. *J. Geophys. Res.*, **68**, 3877–3886.

Murgatroyd, R. J., 1960: Some recent measurements by aircraft of humidity up to 50,000 ft in the tropics and their relationship to meridional circulation: *Proc. Symp. Atmos. Ozone*, Oxford, 20–25 July 1959, IUGG Monogr. No. 3, Paris, p. 30.

National Academy of Science, Panel on Weather and Climate Modification, 1966: Weather and climate modification, problem and prospects. Vol II (Research and Development). Publication No. 1350, National Academy of Science—National Research Council, Washington, D. C., 198 pp.

Plass, G. N., 1956: The influence of the 15-micron carbon dioxide band on the atmospheric infrared cooling rate. *Quart. J. Roy. Meteor. Soc.*, **82**, 310–324.

Telegadas, K., and J. London, 1954: A physical model of Northern Hemisphere troposphere for winter and summer. Scientific Report No. 1, Contract AF19(122)-165, Research Div. College of Engineering, New York University, 55 pp.

Walshaw, C. D., 1957: Integrated absorption by 9.6 μ band of ozone. *Quart. J. Roy. Meteor. Soc.*, **83**, 315–321.

Yamamoto, G., and T. Sasamori, 1958: Calculation of the absorption of the 15 μ carbon dioxide band. *Sci. Rept. Tohoku Univ. Fifth Ser.*, **10**, No. 2, 37–57.

The Effect of Solar Radiation Variations on the Climate of the Earth

M. I. BUDYKO

Main Geophysical Observatory, Leningrad, M. Spasskaja 7
(Manuscript received September 25, 1968, revised version December 18, 1968)

ABSTRACT

It follows from the analysis of observation data that the secular variation of the mean temperature of the Earth can be explained by the variation of short-wave radiation, arriving at the surface of the Earth. In connection with this, the influence of long-term changes of radiation, caused by variations of atmospheric transparency on the thermal regime is being studied. Taking into account the influence of changes of planetary albedo of the Earth under the development of glaciations on the thermal regime, it is found that comparatively small variations of atmospheric transparency could be sufficient for the development of quaternary glaciations.

As paleogeographical research including materials on paleotemperature analyses has shown (Bowen, 1988, *et al*), the Earth's climate has long differed from the present one. During the last two hundred million years the temperature difference between the poles and equator has been comparatively small and there were no zones of cold climate on the Earth. By the end of the Tertiary period the temperature at temperate and high latitudes had decreased appreciably, and in the Quaternary time subsequent increase in the thermal contrast between the poles and equator took place, that was followed by the development of ice cover on the land and oceans at temperate and high latitudes.

The size of Quaternary glaciations changed several times, the present epoch corresponding to the moment of a decrease in the area of glaciations that still occupy a considerable part of the Earth's surface.

To answer the question of in what way the climate will change in future, it is necessary to establish the causes of Quaternary glaciations initiation and to determine the direction of their development. Numerous studies on this problem contain various and often contradictory hypotheses on the causes of glaciations. The absence of the generally accepted viewpoint as regards this seems to be explained by the fact that the existing hypotheses were based mainly on qualitative considerations allowing different interpretation.

Taking into account this consideration, we shall examine in the present paper the possibility of using quantitative methods of physical climatology to study the problem in question.

Firstly we shall dwell upon the problem of climate change regularities during the last century. Fig. 1 represents the secular variation of annual temperature in the northern hemisphere that was calculated from the maps of temperature anomalies for each month for the period of 1881 to 1960 which were compiled at the Main Geophysical Observatory. Line 1 in this figure characterizes the values of anomalies that are not smoothed, line 2 the anomalies averaged by ten-year periods.

As is seen from this figure, a rise in temperature that began at the end of last century stopped in about 1940, and a fall in temperature started. The temperature in the northern hemisphere that increased in the warming period by 0.6°C then decreased by the middle of the fifties by 0.2°C. A comparatively short-period rise in temperature with smaller amplitude was also observed in the last years of the XIXth century.

The curve of secular temperature variation can be compared with the curve of secular variation of direct solar radiation with cloudless sky that was

Budyko, M. I. (1969) The effect of solar radiation variations on the climate of the Earth. *Tellus* 611–619. Reprinted with permission of John Wiley & Sons.

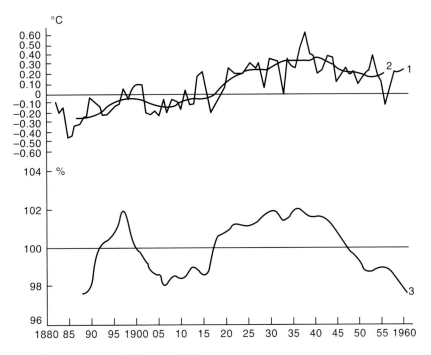

Fig. 1. Secular variation of temperature and direct radiation.

drawn by the data from a group of stations in Europe and America with the longest-period series of observations. This curve presenting the values of solar radiation smoothed for ten-year periods corresponds to line 3 in Fig. I. As is seen from the above figure, the direct radiation had two maxima—a short-period one at the end of the XIXth century and a longer-period one with the maximum values of radiation in the thirties.

The problem of the causes of secular variation of direct radiation was already discussed by Humphreys (1929, and others) who considered that it was determined by the change in the atmospheric transparency due to the propagation of volcanic eruption dust in it. Having agreed to this point of view that is confirmed by many now data, it should be suggested that a decrease in radiation after 1940 could also depend on the increase of dust in the atmosphere due to man's activity.

As can be seen from Fig. 1, the curves of secular variations of temperature and radiation are more or less similar.

To find out the dependence between the radiation change and that of temperature, let us compare the radiation and thermal regimes of the northern hemisphere for two thirty-year periods: 1888–1917 and 1918–1947. It follows from the data given in Fig. 1 that the temperature in the latter of these periods was by 0.33°C higher than that in the former, and the direct radiation by 2.0% higher.

To estimate the corresponding change in total radiation, it should be taken into account that the atmospheric transparency changes after volcanic eruptions as a result of propagation of dust with particles of the order of 1 μ in the lower stratosphere. This dust considerably increases the short-wave radiation diffusion, as a result of which the planetary albedo of the Earth becomes higher. Because of the radiation diffusion by dust mainly in the direction of an incident ray (Mie effect) the direct radiation decreases with diffusion to a greater extent than the total radiation does. Using the calculation method developed by K. S. Shifrin and his collaborators (K. S. Shifrin, I. N. Minin, 1957; K. S. Shifrin, N. P. Pyatovskaya, 1959), one can estimate the ratio of decrease in total radiation to that in direct radiation.

Such a calculation shows that this ratio computed for the average annual conditions changes slightly with the change of latitude, and on an average for the Earth equals 0.15.

Thus, the difference in total radiation for the periods under consideration amounts to 0.30%. In this case the ratio of temperature change to the change of radiation turns out to be equal to 1.1°C per 1% of radiation change.

This value should be compared with the values of similar ratio obtained as a result of calculating the radiation influence on the thermal regime of the Earth.

To determine the dependence of temperature on solar radiation with the average relationship between temperature, air humidity and other factors influencing the long-wave radiation, we used the results of calculations of monthly mean values of radiation at the outer boundary of the atmosphere that were made when preparing *Atlas of the heat balance of the Earth* (1963).

On the basis of these data relating to each month for 260 stations an empirical formula was derived

$$I = a + BT - (a_1 + B_1 T)n \qquad (1)$$

where I = outgoing radiation in kcal/cm^2 month,

   T = temperature at the level of Earth's surface in °C,

   n = cloudiness in fractions of unit,

the values of dimensional coefficients of which equal: $a = 14.0$; $B = 0.14$; $a_1 = 3.0$; $B_1 = 0.10$.

The root-mean-square deviation of the results of calculation by this formula from the initial data accounts for less than 5% of the radiation values.

Comparing formula (1) with similar dependence that can be obtained from the work by Manabe and Wetherald (1967), it is possible to conclude that they practically coincide for the conditions of cloudless sky and differ in considering the cloudiness effect on radiation.

For mean annual conditions, the equation of the heat balance of the Earth-atmosphere system has the following form:

$$Q(1 - \alpha) - I = A \qquad (2)$$

where Q = solar radiation coming to the outer boundary of the atmosphere;

   α = albedo;

   A = gain or loss of heat as a result of the atmosphere and hydrosphere circulation, including heat redistribution of phase water transformations.

Taking into account that for the Earth as a whole $A = 0$, we shall find from formulae (1) and (2) the dependence of the Earth's mean temperature on the value of solar radiation. In this case it turns out that the change of solar radiation by 1% with the average for the Earth value of cloudiness equal to 0.50 and constant albedo equal to 0.33, causes the temperature change by 1.5°.

This result can be compared with similar estimate obtained from the work by Manabe and Wetherald from which it follows that with constant relative air humidity the mean temperature at the Earth's surface changes by 1.2° solar radiation changes by 1%.

It is clear that both of these values agree satisfactorily with the relation between changes in temperature and radiation that was obtained from observational data. One can believe that some excess of the computed temperature changes as compared to observational data reflects the thermal inertia effect of oceans the heating or cooling of which smoothes the Earth's temperature variations in comparison with the computed values for stationary conditions.

Thus, it seems probable that present changes of the Earth's temperature are determined mainly by the atmosphere transparency variations that depend on the level of volcanic activity.

If the present changes in voleanic activity cause radiation variations by several tenths of per cent and the planetary temperature variations by several tenths of a degree, one can believe that in the past respective variations of radiation and temperature reached appreciably larger values.

It is evident that the number of volcanic eruptions for the given interval of time is different with constant mean level of volcanic activity for statistic reasons, these differences being the greater, the longer general period of time being considered. The standard of volcanic activity in different geological epochs is also known to change noticeably in connection with the change of toctonic processes intensity.

Since the volcanic activity variations caused by tectonic factors are characterized by long periods of time to calculate the influence of radiation variations associated with them on the thermal regime, changes in the Earth's albedo should be taken into account that are due to expansion or reduction of the area covered with ice on the land and oceans.

As observations from meteorological satellites have shown (see Raschke, Möller, Bandeen, 1968), the albedo of the Earth-atmosphere system over areas with ice cover is greater than that over ice-free areas, due to which fact the change in area covered with ice increases the radiation variation effect on thermal regime.

To estimate the radiation variation influence on the temperature of latitudinal zones, taking into account the indicated effect, one of numerical

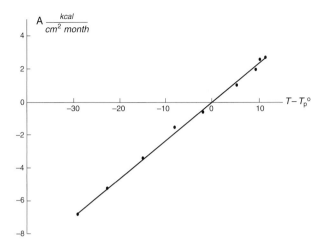

Fig. 2. The dependence of horizontal heat transfer upon temperature difference.

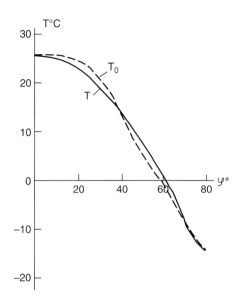

Fig. 3. The average latitudinal temperature distribution.

models of the average latitudinal temperature distribution should be used. Since in this case we are only interested in temperature distribution near the Earth's surface it is possible to use, instead of existing comparatively complicated models, a simple scheme based on the solution of equations (1) and (2) to which the relation should be added that characterizes the relationship between temperature distribution and horizontal heat transfer in the atmosphere and hydrosphere.

Such a relation can be obtained by comparing the mean latitudinal values of term A calculated from formula (1) with quantities $T - T_p$, where T is annual mean temperature at a given latitude, T_p is the planetary mean temperature.

The result of the above comparison is shown in Fig. 2 from which it follows that the corresponding dependence can be expressed in the form of equation

$$A = \beta(T - T_p) \qquad (3)$$

where $\beta = 0.235$ kcal/cm^2 month degree

From formulae (1), (2) and (3), taking into consideration that for the Earth as a whole $A = 0$, we obtain equations

$$T = \frac{Q(1-\alpha) - a + a_1 n + \beta T_p}{\beta + B - B_1 n} \qquad (4)$$

$$T_p = \frac{Q_p(1-\alpha_p) - a + a_1 n}{B - B_1 n} \qquad (5)$$

(where Q_p and α_p are planetary values of radiation and albedo) by which the average latitudinal annual mean temperatures were computed for present climatic conditions of the northern hemisphere. The values of Q, Q_p accepted in this calculation correspond to the value of solar constant 1.92 cal/cm^2 min, the albedo, according to observational data available, at the latitudes of 0° to 60° is considered to be equal to 0.32, at the latitude of 70° to 0.50, at the latitude of 80° to 0.62. In the calculation, the influence of deviations of cloudiness values from its mean planetary value equal to 0.50 on temperature is neglected.

The possibility of such an assumption results from the conclusion established in the calculations made using the above formulae concerning a comparatively weak effect of cloudiness on the mean indices of thermal regimewithin a rather wide range of conditions. Such a conclusion drawn, taking into account the dependence of albedo on cloudiness, implies that the effect of cloudiness on the change in absorbed radiation in a number of cases is compensated for by its influence on the outgoing long-wave radiation. The results of calculating the contemporary average latitudinal distribution of temperature are presented in Fig. 3 where they correspond to line T_0. As is seen, these results are in good agreement with the observed temperature at different latitudes that is represented in Fig. 3 by line T. Such an agreement allows us to use the scheme described for evaluation of the radiation variation effect on the Earth's thermal regime and glaciations.

The southern boundary of the existing ice cover on the seas and land in the Arctic corresponds to the mean latitude of 72° N. Let us consider that with a decrease in solar radiation the surface of ice cover expands in accordance with the extension of the surface area with temperature equal to or lower than the temperature observed now at 72° N. In this case let us assume that albedo on the ice-covered area is equal to 0.62 and at the southern boundary of this ice cover to 0.50.

It follows from the above values of albedo that with the change of ice cover area the mean albedo of the Earth changes by value 0.30 S where coefficient 0.30 corresponds to the difference of albedo values with the presence and absence of ice cover, and quantity $S = lq$, (l = the ratio of ice area change to the whole area of the Earth, q = the ratio of mean radiation in the same zone of ice area change to the mean value of radiation for the Earth as a whole).

To take into account the influence of the glaciation area change on the annual mean temperature of the Earth, we shall use formula

$$\Delta T_p = \frac{Q_p}{B - B_1 n}\left[\frac{\Delta Q_p}{Q_p}(1 - \alpha_p - 0.30S) - 0.30S\right] \quad (6)$$

which is obtained from formulae (1) and (2), where ΔT_p is the Earth's temperature change with the change of mean radiation Q_p by value ΔQ_p.

From (1), (2), (3), (6) we shall deduce a formula for temperature at some latitude

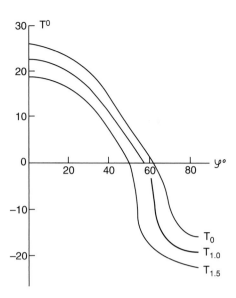

Fig. 4. The dependence of temperature distribution on radiation amount.

where T_p' is the existing mean temperature of the Earth.

Using this formula and considering the dependence of values Q and S on latitude, one can compute the position of glaciation boundary for different values of $\Delta Q_p / Q_p$. By this formula it is also possible to calculate the distributions of temperature at different latitudes that correspond to these values. The results of such a calculation are shown in Fig. 4, where lines $T_{1.0}$ and $T_{1.5}$ correspond to temperature distributions with the decrease in radiation income by 1.0% and 1.5% respectively. In the above-mentioned calculation the interrelationship

$$T = \frac{Q(1 - \alpha)\left(1 + \dfrac{\Delta Q_p}{Q_p}\right) - a + a_1 n + \beta T_p' + \dfrac{\beta Q_p}{B - B_1 n}\left[\dfrac{\Delta Q_p}{Q_p}(1 - \alpha_p - 0.30S) - 0.30S\right]}{\beta + B - B_1 n} \quad (7)$$

between the thermal regimes of the northern and southern hemispheres is neglected (which assumption is reasonable with the similar change of thermal regime in both hemispheres). It is assumed in calculation that the relative decrease in radiation at different latitudes is the same.

Fig. 5 represents the values obtained from this calculation for the mean planetary temperature T_p and mean latitude to which glaciation extends φ_0 depending on relative radiation changes. As is seen from this figure, the radiation variation effect on thermal regime considerably increases as a

result of glaciation development, the corresponding dependence becoming nonlinear.

If with the decrease in radiation by 1% the mean temperature of the Earth drops by 5°, then with the decrease in radiation by 1.5% such a drop reaches 9°. Simultaneously with the above temperature drop the glaciation displacement 10 − 18° to the south takes place, i.e. the distances approximately corresponding to the expansion of quaternary glaciations. When radiation decreases by 1.6% the ice cover reaches the mean latitude of about 50°, after that it starts shifting towards

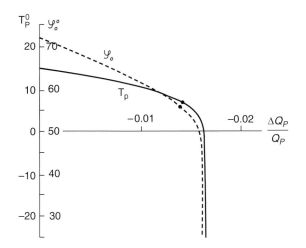

Fig. 5. The dependence of the Earth's temperature and ice cover boundary on radiation variations.

lower latitudes up to the equator as a result of self-development. At the same time the planetary temperature drops sharply and reaches the value of several tens of degrees below zero.

A conclusion on the possibility of complete glaciation of the Earth after ice cover reaches some critical latitude follows from the calculation, using the above formulae, of the values of decrease in radiation necessary for further movement of ice to the equator. Such a calculation shows that to the south of critical latitude ice will move to the equator with the decrease in radiation by less than 1.6%, and at lower latitudes ice will move in the indicated direction with the existing values of radiation and even with its values exceeding those in the present epoch.

It should be noted that similar conclusion from other considerations was drawn previously by Öpik (1953, *et al.*) who considered, however, that for glaciating the Earth a considerable decrease in solar constant is necessary. The possibility of existence of complete glaciation of the Earth with the present value of solar constant was mentioned in the author's works (Budyko, 1961, 1966).

Thus, the present thermal regime and glaciations of the Earth prove to be characterized by high instability. Comparatively small changes of radiation—only by 1.0–1.5%—are sufficient for the development of ice cover on the land and oceans that reaches temperate latitudes.

It should be noted that such changes in radiation are only several times as great as its variations observed due to the changeability of volcanic activity in the last century.

Taking into consideration that according to the data of geological investigations the level of volcanic activity for long periods of time in the past changed by a factor of several times (see Ronov, 1959), one can believe that the influence of long-period variations of volcanic activity is a probable factor of glaciation development.

This conclusion is confirmed by the fact, established by Fuchs and Patterson, of correspondence between the main epochs of quaternary glaciations and the periods of considerable increase in volcanic activity in a number of regions of low latitudes (1947).

Though in this paper the author has no possibility to discuss numerous other hypotheses as to be causes of quaternary glaciations, nevertheless it is necessary to dwell upon popular idea concerning the influence of changes of the Earth's orbit elements on glaciations.

Such a conception substantiated by Milankovich (1930 and others) and other authors is shared by many specialists studying quaternary glaciations.

As is known, the effect of changes in the Earth's orbit elements leads to appreciable redistribution of radiation amount coming to different latitudes. Considering these changes and using the model of latitudinal temperature distribution suggested by him, Milankovich concluded that with the changes of orbit elements at temperate and high latitudes considerable changes in temperature occur that can result in glaciation.

It should be mentioned that the model of temperature distribution suggested by Milankovich did not take into account horizontal heat transfer in the atmosphere and hydrosphere due to which it had to overestimate considerably the influence of changes in radiation in a given latitudinal zone on the thermal regime of the same zone.

To verify the hypothesis of Milankovich, there were calculated, using the above-mentioned scheme, changes in thermal regime and glaciations for the case of considerable change of the Earth's orbit elements 22 thousand years ago which is usually associated with the last glaciation. The calculations made have shown that though the variations of orbit elements influence in a definite way the thermal regime and glaciation, this influence is comparatively small and corresponds to possible displacement of the glaciation boundary by a little less than 1° of latitude. It should be borne in mind that such a calculation allows for the change in annual radiation totals. According to Milankovich, the main influence on the glaciation is exerted by the variations of the summer radiation values that at latitudes 65–75° are 2 to 3 times as large as the variations of annual values. Emphasizing the necessity of further study of the problem on the effect of annual radiation variation on the glaciation, it should be noted that the above-obtained result casts some doubt on the hypothesis that the effect of the Earth's orbit changes is sufficient for the explanation of the quaternary glaciations.

Now we shall proceed to the question of why the volcanic activity variations, that occured during the whole history of the Earth, did not result in the development of glaciations during hundreds of millions of years previous to the quaternary period.

It has been established in geological investigations that in the pre-quaternary time the gradual rise of continents level took place. This caused weakening of water circulation in the oceans between low and high latitudes.

It was ascertained long ago (Budyko, 1948, *et al.*) that the heat transfer between the equator and the poles in the hydrosphere is a considerable portion of the corresponding transfer in the atmosphere, in connection with which the changes in water circulation in the oceans should influence essentially the thermal regime at high and temperature latitudes.

To clear up this question, temperature distribution was calculated using the above-mentioned scheme for the case of absence of ice at high latitudes.

The results of such calculations are shown in Fig. 6 where line T_0 represents the present-day temperature distribution, and line T_α temperature distribution with the absence of polar glaciations. In these calculations the albedo at high latitudes is accepted to be equal to the albedo of ice-free areas and the coefficient β is considered to be equal to its value accepted above.

As is seen from Fig. 6, the polar ice changing little the temperature at low latitudes considerably decreases the temperature at high latitudes. As a result, the mean difference in temperature between the pole and the equator decreases and the annual mean temperature in polar zone turns out to be equal to several degrees below zero.

One can believe that with ice-free regime the meridional heat transfer in the polar ocean will increase as compared to present conditions since this ocean, that is now isolated from the atmosphere by ice, will give off a considerable amount of heat to the atmosphere through turbulent heat exchange.

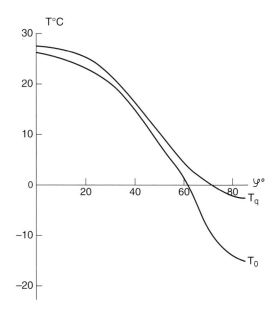

Fig. 6. Ice cover effect on temperature distribution.

If to consider that with the absence of ice the Arctic Ocean receives additionally an amount of heat equal to the mean value coming now to the ice-free areas of the oceans at high latitudes, the mean air temperature in the Arctic must be somewhat higher than the above value, i.e. close to zero.

This result is in agreement with the conclusions drawn using other methods in previous works by the author (Budyko, 1961, 1962, 1966), L. R. Rakipova (1962, 1966), Donn & Shaw (1966), and others. It confirms once more the possibility of existence of ice-free regime in the polar basin in the present epoch and at the same time indicates high instability of such a regime.

It is evident that with the annual mean temperature in the Central Arctic close to water freezing point comparatively small anomalies of radiation income may lead to ice restoration.

Thus, with the present distribution of continents and oceans the existence of two climatic regimes is possible one of which is characterized by the presence of polar ice and large thermal contrast between the pole and the equator, and the other by the absence of glaciation and small meridional mean gradient of temperature.

Both of these regimes are unstable since even small variations of solar radiation income could be sufficient either for freezing of the ice-free polar ocean or melting of the existing ice. Such a peculiarity of climatic regime seems to determine the main features of climate variations in the Quaternary period.

In the periods of decreased volcanic activity the temperature distribution corresponded to ice-free regime which characterizes the climate of comparatively warm inter-ice epochs. When volcanic activity increased, ice formed firstly in the arctic seas, and then greater or smaller glaciations were developed on the land.

As it was mentioned in the author's work (Budyko, 1968), in the mesozoic era and in the paleogene the northern polar basin was connected with the oceans of low latitudes with much wider straits as compared to the Quaternary period. In this case the heat income to the polar basin as a result of activity of sea currents seemed to exceed these values that are observed at high latitudes under present conditions. If this income was 1.5 to 2 times as great as its present mean value for the ice-free areas, then according to the calculations by the above formulae, the annual mean temperature in the Arctic reached 10°, which fact excluded the possibility of glaciation even with appreciable anomalies of radiation.

During the Tertiary period the isolation of polar basin from the tropic regions of ocean gradually developed, which caused the temperature decrease near the pole and approaching of temperature distribution to the values characteristic of inter-ice epochs.

It follows from the above considerations that the present epoch is a part of glacial period since any noticeable increase in volcanic activity should lead to new glaciation development.

Moreover, it seems probable that one of the following glaciers expansion could reach the critical latitude after which the complete glaciation of the Earth would set in. Such a possibility was on the point of being realized in the period of maximum quaternary glaciation when the temperature of the Earth and the position of ice cover corresponded to dots plotted on lines T_p and φ_0 in Fig. 5.

As is seen from this figure, the ice cover under these conditions has moved about 0.8 of the way from the present ice boundary to the critical latitude.

From such a viewpoint the Quaternary History of the Earth seems to be the period of coming climatic catastrophe due to which the existence of higher forms of organic life on our planet may be exterminated.

When estimating the probability of such a catastrophe being realized in future, the character of man's activity should be taken into account which influences to some extent the climate at present. Without touching upon the possibility of implementing in future some projects of active influence on the climate which could affect the glaciation development, ever increasing influence of man's activity on the energy budget of the Earth should be mentioned.

All the energy used by man is transformed into heat, the main portion of this energy being an additional source of heat as compared to the present radiation gain. Simple calculations show (Budyko, 1961) that with the present rate of growth of using energy the heat produced by man in loss than two hundred years will be comparable with the energy coming from the sun. Since glaciations are greatly influenced by the changes in energy budget which are a small part of solar radiation income then it is probable that in the comparatively near future the possibility of glaciation expansion will be excluded and there will appear the reverse one of polar ice melting on the land and oceans with all the changes in the Earth's climate that are associated with it.

It should be mentioned that the conclusions stated in this paper on the effect of changes in solar radiation on climate have been drawn as a result of using a strongly schematized model of the Earth's thermal regime. It is considered desirable to make similar calculations with the use of more general models.

REFERENCES

Atlas of the heat balance of the Earth. Ed. M.I. Budyko, (In Russian), 1963, Moscow.

Bowen, R. 1966. *Paleotemperature analysis.* Amsterdam.

Budyko, M. I. 1948. The heat balance at the northern hemisphere. (In Russian), *Proceedings of the Main Geophysical Observatory,* 18.

Budyko, M. I. 1961. On the thermal zones of the Earth. (In Russian), *Meteorology and Hydrology,* No. 11.

Budyko, M. I. 1962. Polar ice and climate. (In Russian), *Proceedings Ac. Sci. USSR,* ser. geograph., No. 6.

Budyko, M. I. 1962a. Some ways of climate modification. (In Russian), *Meteorology and Hydrology,* No. 2.

Budyko, M. I. 1966. Polar ice and climate. Proceedings of the Symposium on the Arctic heat budget and atmospheric circulation.

Budyko, M. I. 1968. On the causes of climate variations. *Sveriges Meteorologiska och Hydrologiska Institut Meddelanden,* ser. B, No. 28.

Donn, W. L. & Shaw, M. 1966. The heat budgets of ice-free and ice-covered Arctic Ocean. *Journ. of Geoph. Res.,* 71, No. 4.

Fuchs, V. S. & Pattorson, T. T. 1947. The relation of volcanicity and eregeny to climate change. *Geological Magazine,* v. LXXXIV, No. 6.

Humpbreys, N. J. 1920. Physics of the air. Second edition, N.Y.

Manabe, S. & Wetherald, R. 1967. Thermal equilibrium of the atmosphere with a given distribution of relative humidity. *Journ. of Atmosph. Sciences, 24*, No. 3.

Milankovich, M. 1930. Matematische Klimalehro and Astronomische Theorie der Klimaschwanungen.

Rakipova, L. R. 1962. Climate change when influenceing the Arctic basin ice. (In Russian), *Meteorology and Hydrology*, No. 9.

Rakipova, L. R. 1966. The influence of the Arctic ice cover on the zonal distribution of atmospheric temperature. *Proceedings of the Symposium on the Arctic heat budget and atmospheric circulation.*

Raschke, E., Möller, F. & Bandeen, W. R. 1968. The radiation balance of Earth-atmosphere system.... *Sveriges Meteorologiska och Hydrologiska Institut Meddelanden*, ser. B, No. 28.

Ronov, A. B. 1959. To the post-Cambrian geochemical history of the atmosphere and hydrosphere. (In Russian), *Geochemistry*, No. 5.

Shifrin, K. S. & Minin, I. N. 1957. To the theory of non-horizontal visibility. (In Russian), *Proceedings of the Main Geophysical Observatory, 68.*

Shifrin, K. S. & Pyatovskaya, N. V. 1959. Tables of slant visibility range and day sky brightness. (In Russian), Loningrad.

Öpik, E. Y. 1953. On the causes of palaeoclimatic variations and of ice ages in particular. *Journ. of Glaciology, 2*, No. 13.

A Global Climatic Model Based on the Energy Balance of the Earth–Atmosphere System

WILLIAM D. SELLERS

Institute of Atmospheric Physics, The University of Arizona, Tucson
(Manuscript received 23 December 1968, in revised form 4 March 1969)

ABSTRACT

A relatively simple numerical model of the energy balance of the earth-atmosphere is set up and applied. The dependent variable is the average annual sea level temperature in 10° latitude belts. This is expressed basically as a function of the solar constant, the planetary albedo, the transparency of the atmosphere to infrared radiation, and the turbulent exchange coefficients for the atmosphere and the oceans.

The major conclusions of the analysis are that removing the arctic ice cap would increase annual average polar temperatures by no more than 7C, that a decrease of the solar constant by 2–5% might be sufficient to initiate another ice age, and that man's increasing industrial activities may eventually lead to a global climate much warmer than today.

1. INTRODUCTION

One of the most intriguing and interesting problems in the atmospheric sciences today is the estimation of the effect of changes in the planetary albedo, the solar constant, or atmospheric turbidity on the large-scale surface temperature field. Most of the work accomplished so far in this area has been done in Russia and is concerned mainly with the removal of the arctic ice sheet. Budyko (1966), for example, concludes that the polar ice decreases the surface temperature in the central arctic by 30–35C in winter and 5C in summer. For his model he used the surface energy balance equation and an empirical relationship between the vertical turbulent heat flux in the arctic and the temperature difference between the equator and 80N. Rakipova (1966a), using a much more elaborate approach, reached essentially the same conclusion, obtaining a surface temperature decrease of 19–26C in winter and 3–4C in summer. Her model utilized the energy balance of an atmospheric column and yielded the vertical temperature distribution up to a height of 20 km between the equator and 90N. Both Budyko and Rakipova apparently neglected heat transfer by ocean currents.

Rakipova (1966b) also considered the zonal distribution of temperature that would result from contamination of the upper atmosphere by dust. Decreasing the solar constant by 10% and assuming no change in the flux of infrared radiation, she obtained a surface temperature drop ranging from 8C at the equator to 14C at 90N. She also cites the conclusion of Veksler (1958) that decreasing the planetary albedo from 0.43 to 0.35 will raise the average temperature of the earth-atmosphere system by 8C.

Eriksson (1968) describes a generalized model of the dynamics of glaciations and presents functional relationships between the amount of ice coverage and the mean surface temperature. He concludes that a 4% decrease in the solar constant might be enough to reduce the global mean temperature by 9C and produce typical ice age conditions. He also mentions the possibility that, once the ice coverage reaches a certain low latitude, any further growth would be explosive and eventually ice might cover the entire earth.

The purpose of this paper is to investigate a different approach than used by either Rakipova or Eriksson. It is based on the concept or conviction that the steady-state average latitudinal distribution of surface temperature, to a first approximation,

Sellers, W. D. (1969) A global climatic model based on the energy balance of the earth-atmosphere system. *Journal of Applied Meteorology* 8: 392–400. Reprinted with permission of American Meteorological Society.

should depend entirely on the incoming solar energy, the transparency of the atmosphere to terrestrial radiation, the planetary albedo, the ability of the atmosphere to carry heat and water vapor from source to sink, and the heat storage potential of the oceans and land. The latter factor drops out when average annual temperatures are considered. Bernard (1967) has essentially stated the problem in this form.

The model outlined below is a preliminary model with the dependent variable being the average annual sea level temperature in each 10° latitude belt. It is based on the energy balance equation for the earth-atmosphere system, with the boundary conditions that there can be no meridional energy transport across the poles. One of the main purposes of the model is to show that attempts to modify the climate of a small section of the world must ultimately affect the whole globe before a new steady-state regime can be attained. Several empirical and relatively crude relationships are involved. Unfortunately, this is true of all present models and probably will continue to be true in the future. The best one can hope for is that the empirical relationships are realistic and will continue to give valid results when the basic parameters are changed slightly from their current values.

2. THE MODEL

Neglecting heat storage in the oceans, land or atmosphere, and hence assuming no long-term climatic trend, the energy balance equation for the earth-atmosphere system may be written

$$R_s = L\Delta c + \Delta C + \Delta F, \tag{1}$$

where R_s is the radiation balance of a given latitude belt, L the latent heat of condensation, assumed to equal 590 cal gm^{-1}, and Δc, ΔC and ΔF are the net fluxes out of the belt of, respectively, water vapor by atmospheric currents, sensible heat by atmospheric currents, and sensible heat by ocean currents.

Eq. (1) may be rewritten in the expanded form

$$-R_s \frac{A_0}{l_1} = Lc_1 + C_1 + F_1 - P_0 \frac{l_0}{l_1}, \tag{2}$$

where

$$P_0 + Lc_0 + C_0 + F_0.$$

A_0 is the horizontal area of the latitude belt, and l_0 and l_1 are, respectively, the lengths of its northern and southern latitude boundaries. P represents the total heat transport across the given latitude circle. The units of each term in Eq. (2) are cal cm^{-1} sec^{-1}. With the boundary conditions that c, C and F are all equal to zero at the poles, Eq. (2) reduces to

$$-R_s \frac{A_0}{l_1} = Lc_1 + C_1 + F_1$$

between 80 and 90N and to

$$P_0 = R_s \frac{A_0}{l_0} \tag{3}$$

between 80 and 90S.

Eq. (2) must be expressed in terms of the single unknown ΔT, the sea level temperature difference $T_0 - T_1$ between successive latitude belts. T_0 between 80 and 90N is then specified and adjusted until Eq. (3) is satisfied to the required accuracy. This is an iterative process and ultimately relates the temperature in each latitude belt to the temperature in all other belts.

The components of Eq. (2) can be expressed in the following forms:

$$R_s = Q_s(1-\alpha_s) - I_s, \tag{4}$$

$$\alpha_s = b - 0.009T_g, \quad T_g < 283.16, \tag{5a}$$

$$\alpha_s = \alpha_d = b - 2.548, \quad T_g > 283.16, \tag{5b}$$

$$I_s = \sigma T_0^4[1 - m\tanh(19T_0^6 \times 10^{-16})], \tag{6}$$

$$c = \left(\upsilon q - K_w \frac{\Delta q}{\Delta y}\right)\frac{\Delta p}{g}, \tag{7}$$

$$\upsilon = -a(\Delta T + |\overline{\Delta T}|), \quad \text{north of 5N}, \tag{8a}$$

$$\upsilon = -a(\Delta T - |\overline{\Delta T}|), \quad \text{south of 5N}, \tag{8b}$$

$$q = \frac{\varepsilon e}{p}, \tag{9}$$

$$\Delta q = \frac{\varepsilon^2 Le\Delta T}{pR_d T_0^2}, \tag{10}$$

Table 1. Average values of b, Z and α_d for each latitude belt and of Δp and Δz for each latitude circle.

Latitude belt	b	Z (m)	α_d	Latitude circle	Δp (mb)	Δz (km)
80–90N	2.924	137	0.376	80N	709	2
70–80N	2.927	220	0.379	70N	710	1
60–70N	2.878	202	0.330	60N	713	2
50–60N	2.891	296	0.343	50N	750	3
40–50N	2.908	382	0.360	40N	800	4
30–40N	2.870	496	0.322	30N	833	4
20–30N	2.826	366	0.278	20N	880	4
10–20N	2.809	146	0.261	10N	904	4
0–10N	2.808	158	0.260	0	906	4
0–10S	2.801	154	0.253	10S	904	4
10–20S	2.798	121	0.250	20S	880	4
20–30S	2.815	156	0.267	30S	833	4
30–40S	2.865	106	0.317	40S	800	4
40–50S	2.922	5	0.374	50S	750	4
50–60S	2.937	5	0.389	60S	713	4
60–70S	2.989	388	0.441	70S	710	3
70–80S	2.992	1420	0.444	80S	709	0
80–90S	2.900	2272	0.352			

$$e = e_0 - 0.5\frac{\varepsilon L e_0 \Delta T}{R_d T_0^2}, \tag{11}$$

$$C = \left(\upsilon T_0 - K_h \frac{\Delta T}{\Delta y}\right)\frac{c_p}{g}\Delta p, \tag{12}$$

$$F = -K_0 \Delta z \frac{l'}{l_1}\frac{\Delta T}{\Delta y}. \tag{13}$$

In these equations the notation is as follows:

a. For each latitude belt

Q_s incident solar radiation
α_s planetary albedo
I_s infrared emission to space
b empirical coefficient (see Table 1)
T_g average surface temperature (= $T_0 - 0.0065Z$)
Z average surface elevation in meters (see Table 1)
σ Stefan-Boltzmann constant [1.356×10^{-12} ly sec^{-1} ($^\circ$K)$^{-4}$]
m atmospheric attenuation coefficient (0.5 for present conditions)
e_0 mean sea level saturation vapor pressure

b. For each latitude circle

υ mean meriodional wind speed (positive northward)
q mean saturation specific humidity at sea level
K_w eddy diffusivity for water vapor in air

Δy 1.11×10^8 cm
Δp pressure depth of the troposphere (see Table 1)
g gravity (10^3 cm sec^{-2})
a meridional exchange coefficient
$|\overline{\Delta T}|$ average absolute value of ΔT (weighted with respect to l_1)
ε 0.622
e mean sea level saturation vapor pressure
p average sea level pressure (1000 mb)
R_d gas constant [6.8579×10^{-2} cal gm^{-1} ($^\circ$K)$^{-1}$]
K_h eddy thermal diffusivity for air
c_p specific heat at constant pressure [0.24 cal gm^{-1} ($^\circ$K)$^{-1}$]
K_0 eddy thermal diffusivity for ocean currents
Δz ocean depth (see Table 1)
l' length of the ocean-covered portion of l_1

There are a number of approximations involved in Eqs. (4)–(13). Probably the most important is, in Eqs. (7), (12) and (13), that average atmospheric temperatures and specific humidities and ocean temperatures are assumed to be proportional to their respective sea level values. Further, the actual specific humidity is replaced by the saturation specific humidity, thus permitting the use of Eq. (10), the finite difference form of the Clausius-Clapeyron equation, and Eq. (11) to relate temperature and vapor pressure. The constants of proportionality are absorbed into the diffusivities and the meridional exchange coefficient.

In Eq. (5a) the planetary albedo is given as a function of the surface temperature T_g. It is assumed that variations in the albedo are associated mainly with variations in surface snow cover and that the relationship is valid only as long as the surface temperature <10C and the albedo <0.85. For higher temperatures, Eq. (5b) is used to give the planetary albedo α_d in the absence of a snow cover. These latter values are listed in Table 1. If the computed albedo >0.85, a value of 0.85 is used. Actually, Eq. (5a) plays a critical role in the analysis. It indicates that the albedo will increase by 0.009 for every 1C drop in surface temperature. This figure was obtained by comparing present-day albedos and temperatures in the same latitude belts of the two hemispheres. Eriksson (1968), using a maximum snow surface albedo of 0.65, implies a value laying between 0.004 and 0.008. The coefficient b was selected so that the equation would fit the observed data. The effects of variations in cloud cover on the albedo were ignored, mainly because there is no easy way to include them. Eriksson points out that large changes in the mean cloudiness of the planet as a whole should not be expected. However, the meridional distribution of cloud cover can vary significantly.

Eq. (6), relating the infrared radiation loss to space to the sea level temperature, was obtained from present data (see Fig. 1)), where the bracketed quantity can be viewed as an atmospheric transmission factor or, as Bryson (1968) puts it, an effective emissivity, which ranges from about 0.8 at −30C to 0.5 at 30C when $m = 0.5$. It represents the effect of water vapor, carbon dioxide, dust and clouds on terrestrial radiation. The attenuation coefficient m should go up or down as the atmospheric infrared turbidity increases or decreases, respectively. Very roughly, an increase or decrease in the atmospheric water vapor content by a factor of 1.4 or in the CO_2 content by a factor of 10 would be needed to change m by 10%.

In Eqs. (7) and (12), the poleward transport of water vapor and sensible heat in the atmosphere is assumed to consist of two parts, one associated with a mean meridional motion and the other with large-scale eddies or cyclones and anticyclones. The inclusion of the mean meridional motion is necessary in order to avoid having to deal with negative diffusivities. The dependence of the north-south velocity component v on the meridional temperature gradient was determined empirically. Eqs. (8a) and (8b) are based on the observations, first, that both v and the zonal velocity component in the lower troposphere average

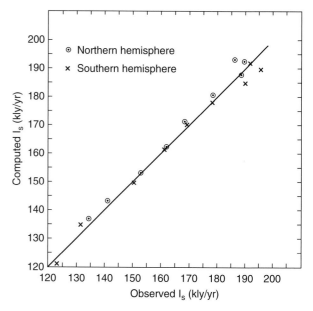

Fig. 1. A plot of the computed and observed values of the annual infrared emission to space in 10° latitude belts. The computed values were obtained from Eq. (6). The observed values are those given by Sellers (1965), modified slightly by recent satellite measurements.

close to zero for the earth as a whole, and, second, that the strongest westerlies have a poleward component and the strongest easterlies an equatorward component. The resulting correlation, plus that between the zonal speed and the temperature gradient, yields Eqs. (8a) and (8b).

Eqs. (7) and (12) involve the three unknowns u, K_w and K_h, which, therefore, cannot be determined directly from presently available data. However, since the diffusivities must be greater than zero, upper or lower limits on u can be determined [by setting K_w and K_h equal to zero in Eqs. (7) and (12)]. These, then, can be used as a guide for specifying the coefficient a. When this is done, the values of a, K_w and K_h given in Table 2 result. The magnitude 10^{10}–10^{11} cm² sec⁻¹ for K_h in middle latitudes is in agreement with values used or suggested by Rakipova (1966b), Adem (1964), Panchev (1968) and Eriksson (1968), among others. The relative magnitudes of K_w, K_h and K_0 seem to be quite reasonable. Because they all apply to horizontal transfer, the large values near the equator are difficult to explain. However, since the temperature and moisture gradients here are small, the resulting poleward transfer is no greater, and often smaller, than that associated with the mean meridional circulation.

In Eq. (13) average ocean depths given by Dietrich (1963) were used for Δz. If it is believed

Table 2. Average values of Q_s for each latitude belt and of K_h, K_w, K_0 and a for each latitude circle.

Latitude belt	Q_s (kly yr^{-1})	Latitude circle	K_h (10^{10} cm^2 sec^{-1})	K_w (10^9 cm^2 sec^{-1})	K_0 (10^6 cm^2 sec^{-1})	a [cm sec^{-1} ($°$K)$^{-1}$]
80–90N	135.7	80N	1.9	4.6	0.7	0.5
70–80N	145.1	70N	1.2	2.0	6.6	1.0
60–70N	167.3	60N	1.7	2.5	6.9	1.0
50–60N	202.2	50N	1.5	5.9	7.6	1.0
40–50N	237.7	40N	0.9	5.1	5.3	1.0
30–40N	269.0	30N	1.3	2.9	9.6	2.0
20–30N	293.9	20N	10.3	2.2	13.3	3.0
10–20N	311.1	10N	133.2	34.4	59.9	3.0
0–10N	319.8	0	68.2	16.3	7.0	3.0
0–10S	319.8	10S	30.8	0.6	19.0	3.0
10–20S	311.1	20S	8.5	0.8	9.5	3.0
20–30S	293.9	30S	2.3	4.3	5.1	2.0
30–40S	269.0	40S	2.3	11.9	2.9	1.0
40–50S	237.7	50S	1.9	12.8	1.8	1.0
50–60S	202.2	60S	1.4	8.2	0.5	1.0
60–70S	167.3	70S	1.0	7.4	0.2	0.5
70–80S	145.1	80S	0.5	6.3	0.0	0.5
80–90S	135.7					

more satisfactory to use instead the depth of the mixed layer, which averages about 60 m, the values of K_0 should be multiplied by approximately 10^2.

Substituting Eqs. (4)–(13) into Eq. (2) results in a second-order equation in ΔT, with only the negative root (giving ΔT negative in the Northern Hemisphere and positive in the Southern Hemisphere) yielding realistic temperatures. It is interesting to speculate that a more complex model might give a higher order equation with several realistic roots and, hence, several plausible equilibrium temperature distributions for a given set of input variables. Lorenz (1968) touches on this subject.

In this model, the primary variables that determine the equilibrium latitudinal distribution of sea level temperature are the incoming solar radiation, the eddy diffusivities, the meridional exchange coefficient, and the atmospheric attenuation coefficient. Current average values of Q_s, a, K_h, K_w and K_0 are given in Table 2 for each latitude zone or circle. As mentioned earlier, the attenuation coefficient $m = 0.5$. With these data, the model was first used to reconstruct the present state of the earth-atmosphere system. The results are shown in Figs. 2, 3 and 4. The observed data are essentially those given by Sellers (1965), with minor changes instigated by recent satellite measurements of both the short- and longwave radiative fluxes at the top of the atmosphere (Vonder

Haar, 1968; Raschke et al., 1967; Arking and Levine, 1967).

Since the eddy diffusivities were determined from the data, the observed and computed values should coincide, which, in general, they do. The latitudinal distributions of ΔC, $L\Delta c$, ΔF and R_s are shown in Fig. 2. Superimposed on the ΔC curve is that for Lr, the energy released by condensation. There are obvious similarities between the two, suggesting that the condensation process provides most of the energy transferred poleward in the atmosphere as sensible heat. It is felt, however, that the relationship is not close enough to incorporate into the model and, thus, to allow estimates of global precipitation. The major source region for both atmospheric and oceanic sensible heat lies in the tropics between 20N and 20S. Latent heat, on the other hand, originates primarily in the subtropics (15–35N, 15–35S) where annual evaporation far exceeds annual precipitation.

In Fig. 3 are shown the latitudinal distributions of the meridional transport terms, C, Lc, F and P. In general, the transport is poleward in both hemispheres. The one exception is the latent heat or water vapor flux, which, as a result of a strong Hadley circulation, is directed equatorward between 20N and 20S. The bulk of the required transport occurs in the atmosphere in the form of sensible heat.

The latitudinal distributions of the sea level temperature, the meridional wind speed, and the

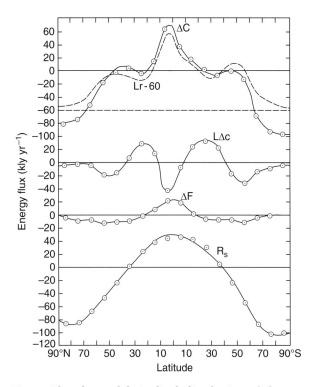

Fig. 2. The observed latitudinal distribution of the net fluxes out of each 10° latitude belt of sensible heat by atmospheric currents, ΔC, latent heat by atmospheric currents, $L\Delta c$, and sensible heat by ocean currents, ΔF. Also shown are the distribution of the energy released by condensation, Lr, and the radiation balance R_s of the earth-atmosphere system. The circled points represent the values computed using the model.

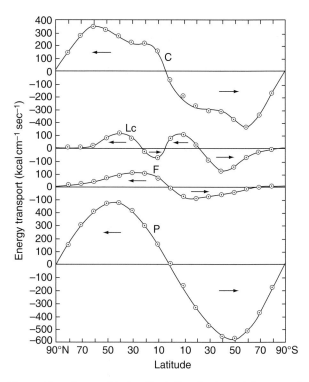

Fig. 3. The observed latitudinal distribution of the annual meridional sensible heat (atmosphere) transport C, latent heat transport Lc, sensible heat (ocean) transport F, and the total transport P. The circled points represent the values computed using the model. The arrows indicate the direction of the transport.

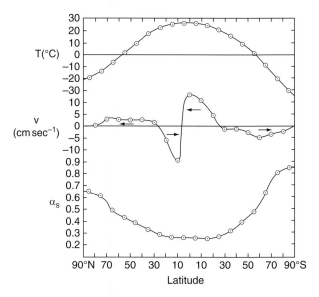

Fig. 4. The observed latitudinal distribution of the mean annual sea level temperature T, meridional wind speed u, and planetary albedo α_s. The circled points represent the values computed using the model.

planetary albedo are shown in Fig. 4. Because of the presence of the Antarctic continent, sea level temperatures are lower and albedos higher at the south pole than at the north pole. This asymmetry should assure some degree of interaction between the two hemispheres. The meridional wind component is small everywhere except between 20N and 20S where the trade wind systems of the two hemispheres produce considerable equatorward mass transfer in the lower troposphere. Mintz (1968) gives similar results. Because of the assumptions involved in the model, the values given for v should be considered only proportional to those which actually exist. They appear to be low by a factor of about 10.

The model obviously specifies present conditions very well. This, however, does not assure its reliability when applied to situations where one or more of the variables in Table 2 are different than currently observed. Therefore, the following applications should be taken only as an indication of what might happen. As further improvements are made in the functional relationships

among the variables involved and the model is extended to the individual months, the picture might change considerably. Also, recall that steady-state conditions are specified, and these may take many years to attain, considering the great heat storage capacity of the world's oceans. Eriksson (1968), for example, gives a lag time of the order of 1000 years.

3.   APPLICATIONS

a.   Modification of the polar ice caps

As mentioned earlier, Budyko (1966) and Rakipova (1966a) estimate that the surface temperature in the arctic would rise 15–20C if the ice sheet were either melted or covered with a black powder. However, both authors dealt mainly with the local energy balance and neglected possible interactions with other regions at lower latitudes.

The basic parameter, and the only one that need be changed in the model, is the planetary albedo at high latitudes. Rakipova (1966b) uses an average value of 0.619 with the ice in, 0.484 with the ice out. The corresponding values used by Fletcher (1966) are 0.51 and 0.34, both of which seem to be low, judging from available satellite data. In the present analysis, a number of different combinations were used, with albedos at one or both poles (70–90°) being set equal to and held fixed at either 0.50 or 0.40 (with the exception of 70–80S, where, in line with Table 1, a minimum value of 0.44 was used). Before proceeding to a discussion of the results, it should be emphasized again that the present state of the earth-atmosphere system is assumed in the model to be the only possible equilibrium state as long as the variables in Table 2 are not changed.

The results, shown in Fig. 5, bring out a number of interesting points. First, removing the north polar ice would increase the temperature poleward of 70N by, at most, ~7C. At the same time, temperatures would rise by about 1C in the tropics and 1–3C near the south pole. A similar treatment in the antarctic would produce a temperature increase there of 12–15C, with a concomitant rise of about 4C in the arctic. Treating both polar regions would yield a temperature increase of 7–10C in the arctic and 13–17C in the antarctic. Warming in the tropics would never exceed 2C.

Thus, albedo manipulation at either pole would have worldwide repercussions. As an extreme illustration, reducing the albedo south of 70S to 0.50 would apparently produce a larger temperature rise in the arctic than would a reduction of

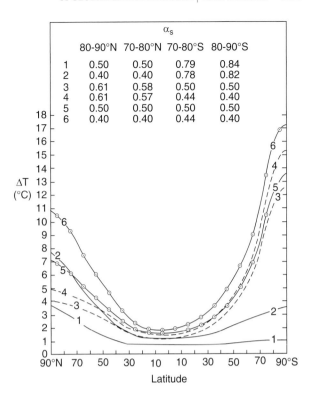

Fig. 5. Predicted latitudinal distribution of the mean annual temperature rise associated with albedo manipulation at one or both poles.

the albedo north of 70N to 0.50 (compare curves 1 and 3 in Fig. 5). The greater effects of treatment in the antarctic than in the arctic are simply the result of its higher initial albedo (see Fig. 4). Along this line, Fletcher (1968) has recently concluded that the extent of sea ice in the Southern Hemisphere has a much more potent effect on the global circulation than the extent of sea ice in the Northern Hemisphere.

b.   Variations in the solar constant

One of the favorite theories of climatic change during the last million years attributes the ice ages to variations in the intensity of solar radiation or, equivalently, to variations in the contamination of the upper atmosphere by dust of volcanic or cosmic origin. The general feeling seems to be that a slight reduction in the solar constant (or increase in global dust contamination) would be sufficient to initiate another ice age. This appears to be borne out by the calculations made in this study and summarized in Figs. 6 and 7. In Fig. 6 the global mean sea level temperature $\bar{T}$ is plotted as a function of the solar constant S, currently assumed to equal 2.00 ly min^{-1}. The present value of $\bar{T}$ is about 14.0C. Three

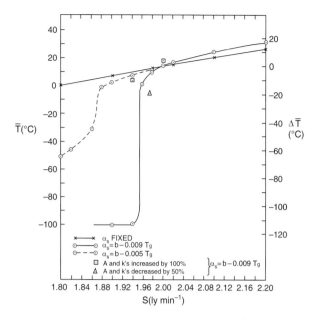

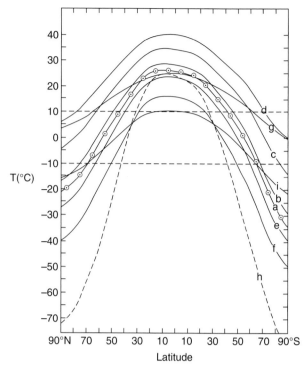

FIG. 6. The mean global sea level temperature $\overline{T}$ as a function of the solar constant S for different responses of the planetary albedo α_s, and different values of the meridional exchange coefficient and the eddy diffusivities, K_h, K_w and K_0.

FIG. 7. The predicted latitudinal distribution of the mean annual sea level temperature when the solar constant is increased by 0%, a, (present conditions), 1%, b, 5%, c, and 10%, d, or reduced by 1%, e, 2%, f. Curve i shows the distribution resulting from a 3% decrease of the solar constant and a 100% increase in the exchange coefficient and eddy diffusivities. In curves g and h the solar constant is kept fixed at its present value of 2.00 ly min⁻¹ and the exchange coefficients and eddy diffusivities are increased by 100% and decreased by 50%, respectively. In all cases the planetary albedo is given by Eqs. (5a) and (5b).

analyses were carried out, two with the planetary albedo allowed to vary, according to $\alpha_s = b - 0.009\,T_g$ and $\alpha_s = b - 0.005\,T_g$, respectively, and the other with the albedo held fixed at its current values (Fig. 4). The latter case corresponds to that reported by Rakipova (1966b), who, as mentioned earlier, obtained temperature decreases ranging from 8C at the equator to 14C at 90N when the solar constant was decreased by 10%. In the present study the magnitudes of the computed drops were the same, but the range was from 14.3C between latitudes of 0° and 20° and 50–60° to 13.0C between 80 and 90°. With a 10% increase in the solar constant, the global mean temperature should rise about 12C if the albedos are held fixed. These results are in good agreement with those of Manabe and Wetherald (1967) and Eriksson (1968).

When the albedos are more realistically permitted to vary with temperature according to Eq. (5), the picture changes considerably. If all other variables are held constant, a decrease in the solar constant by about 2% would be sufficient to create another ice age, with the ice caps extending equatorward to 50° ($T = -10C$) and mountain glaciers and heavy winter snow to 30° ($T = 10C$) as shown in Fig. 7. Any further drop in the solar constant is, as a result of Eq. (5), accompanied in the model by a rapid transition to an ice-covered earth with an albedo of 0.85 and an equilibrium

temperature of −100C. Since the model is based more on functional data fitting techniques than on physical laws, extrapolation to such extreme conditions is hazardous. Generally speaking, valid results should be expected only as long as the response to small changes in the basic parameters is itself small. However, Eriksson (1968) mentions that an explosive development of the ice caps might take place once the ice coverage extended beyond a certain latitude, which he indicates to be about 50°.

Although not in line with observed data, the preceding analysis was repeated using a constant of 0.005 instead of 0.009 in Eq. (5), the values of b being adjusted accordingly. With this weaker dependence of the planetary albedo on surface temperature and ice cover, a severe ice age would not occur until the solar constant decreased by 5–6%.

On the positive side, an increase in the solar constant of ~3% would probably be sufficient to melt the ice sheets. An increase of something more than 10% would be needed, however, to eliminate completely snowfall anywhere on the earth.

Up to this point, the only variable in Table 2 allowed to change was the solar intensity Q_s. Suppose now that this is held fixed at its current values and the eddy diffusivities and meridional exchange coefficients are arbitrarily either increased by 100% or decreased by 50%. As shown in Figs. 6 and 7, the higher values would raise the global mean temperature by about 5C, greatly reduce the equator to pole temperature difference (from 48C to 28C in the Northern Hemisphere), and melt the ice sheets. The lower values would initiate an ice age probably worse than any that has ever been experienced; average temperatures would drop below −70C at the poles and below −10C at 45N and 45S. However, there would be little change in temperatures near the equator.

If the values of Q_s are reduced by 3% and those of a, K_h, K_w and K_0 are increased by 100%, there would still be a worldwide temperature drop, but not by nearly as much as when only Q_s was reduced. The enhanced meridional heat exchange would decrease the pole to equator temperature difference to 26C in the Northern Hemisphere. The ice caps would remain near their present positions, but seasonal snowfall would occur almost to the equator, where the temperature would be about 16C lower than it is today.

Thus, a relatively small change in the ability of the atmosphere and oceans to transfer heat poleward could conceivably offset any increase or decrease in the intensity of solar radiation. Hence, it might be quite risky to change one variable without accounting for possible changes in the others. Nevertheless, the model seems to indicate quite conclusively that a decrease in the solar constant of less than 5% would be sufficient to start another ice age.

c. Variations in the infrared transmissivity of the atmosphere

The global mean temperature responds to changes in the atmospheric attenuation coefficient m in Eq. (6) in much the same way as it does to changes in the solar constant. A 3% decrease in m should be sufficient to put the globe on the brink of an ice age. Fortunately, because of the increasing carbon dioxide content of the atmosphere, m is more likely to increase than decrease. Hence, the global mean temperature should slowly rise due to this factor.

d. Increased use of stored energy by man

Budyko et al. (1966) point out that the quantity of energy used by man and converted into heat, presently about 0.02 kly yr^{-1} averaged over all continental areas, is increasing by 4% each year. Should this rate of increase continue, in less than 200 years the heat produced will reach 50 kly yr^{-1}, or approximately the value of the radiation balance of the continents. Hence, from that time on, "solar energy will no longer be the main climate forming factor, and the climate will become primarily a result of the activity of man."

The latter statement is probably a slight exaggeration, since 50 kly yr^{-1} averaged over all continental areas is equivalent to about 15 kly yr^{-1} averaged over the entire earth surface or little more than 5% of the solar radiation intercepted by the globe annually. Nevertheless, the effects could be considerable, especially in middle and high latitudes of the Northern Hemisphere where the energy used would presumably be greatest.

The latitudinal sea level temperature distribution that might result from the added heating was obtained from the model by spreading the heating H, rather arbitrarily, among the various latitude belts, as shown in Table 3. The only guide used here was the latitudinal distribution of large cities, the main source of combustion. It was assumed that this distribution will not change significantly during the next 200 years.

The results given in Table 3 show that the temperature rise ΔT should average about 15C, ranging from 11C near the equator to 27C at the North pole. This should be enough to eliminate all permanent ice fields, leaving only a few high mountain glaciers on Antarctica and perhaps Greenland. These conclusions ignore any possible decrease, ΔQ_s, in the intensity of solar radiation because of the increased air pollution. When this is taken into account, the temperature rise decreases only slightly (by 1–2C). The values of ΔQ_s in Table 3 are based partly on information given by Bornstein (1968). His data imply a value of about 90 kly yr^{-1} for H at New York City. Radiation data, on the other hand, indicate that the incoming solar radiation at the surface is decreased there by 9.4 kly yr^{-1} or about 3.7% of the extraterrestrial radiation. In general, ΔQ_s appears to equal about 10% of H.

Table 3. Annual sea level temperature rise ΔT_1 due solely to an increase in the energy consumption by man from 0.02–50 kly yr^{-1} averaged over all continental areas and the temperature rise ΔT_2 taking into account the reduction in incoming solar radiation ΔQ_s due to increased air pollution.

Latitude belt	H^* (kly yr^{-1})	ΔT_1 (°C)	ΔQ_s^* (kly yr^{-1})	ΔT_2 (°C)
80–90N	0.0	27.0	0.0	25.2
70–80N	1.4	25.9	0.1	24.2
60–70N	23.0	24.2	1.6	22.6
50–60N	50.5	22.0	4.2	20.5
40–50N	114.8	19.0	11.3	17.6
30–40N	96.4	15.1	10.7	13.9
20–30N	76.2	12.6	9.2	11.5
10–20N	41.3	11.4	5.3	10.5
0–10N	23.0	11.3	3.0	10.3
0–10S	23.9	11.3	3.2	10.4
10–20S	34.4	11.4	4.4	10.5
20–30S	46.9	11.9	5.7	10.9
30–40S	24.8	12.8	2.8	11.7
40–50S	6.9	14.2	0.7	13.7
50–60S	0.5	16.5	0.0	15.3
60–70S	2.3	19.1	0.2	17.7
70–80S	3.7	22.2	0.2	20.5
80–90S	0.0	24.4	0.0	22.5

* Values are expressed per cm² of land.

Thus, man's activity, if it continues unabated, should eventually lead to the elimination of the ice caps and to a climate much warmer than today. Annual mean temperatures of 26C, now characteristic of the tropics, would exist as far poleward as 40°. Considering the thermal inertia of the world's oceans, it is impossible to state how long it will take for this warming to occur—possibly as little as 100 years or as long as 1000 years. During this time, it is not inconceivable that the solar constant will change. A decrease of slightly more than 7% in its value would yield a global mean temperature equal to that existing today. Since such a large drop in the solar constant over any extended period is on the fringe of being highly unlikely, if one believes the earlier results of this paper, it follows that Budyko *et al.* (1966) may be correct after all in stating that eventually man may inadvertently generate his own climate.

4. CONCLUSIONS

In this paper an attempt has been made to develop a consistent, yet simple, climatic model based on the global conservation of energy. The major conclusions—that removing the arctic ice cap would have less effect on climate than previously suggested, that a decrease of the solar constant by 2–5% would

be sufficient to initiate another ice age, and that man's increasing industrial activities may eventually lead to the elimination of the ice caps and to a climate about 14C warmer than today—should all be viewed in the light of the assumptions made in the model. Although these were specified so as to be physically realistic, the possibility still remains that neglected higher order or nonlinear effects could alter the picture considerably.

While the assumptions made must limit the accuracy of the model, it is believed that the use of only mean annual values of the parameters involved is even more restrictive. The logical solution here is to go to mean monthly values. This would, of course, entail a great deal more work, not only because of the much greater volume of data involved, but also because energy storage in the atmosphere and oceans would have to be taken into account. The distribution of land and water would become an important factor. The end result, however, would be a much more flexible model than presented here, and one that could be used to study a broad spectrum of possible climatic effects.

REFERENCES

Adem, J., 1964: On the physical basis for the numerical prediction of monthly and seasonal temperatures in the troposphere-ocean-continent system. *Mon. Wea. Rev.*, **92**, 91–103.

Arking, A., and J. S. Levine, 1967: Earth albedo measurements: July 1963 to June 1964. *J. Atmos. Sci.*, **24**, 721–724.

Bernard, E. T., 1967: Climatic zonation theory. *Encyclopedia of Atmospheric Science and Astrogeology*. New York, Reinhold, 213–217.

Bornstein, R. D., 1968: Observations of the urban heat island effect in New York City. *J. Appl. Meteor.*, **7**, 575–582.

Bryson, R. A., 1968: "All other factors being constant ...", *Weatherwise*, **21**, 56–61, 94.

Budyko, M. I., 1966: Polar ice and climate. *Proc. Symp. Arctic Heat Budget and Atmospheric Circulation*, Memo. RM-5233-NSF, The RAND Corporation, Santa Monica, Calif., 3–21.

———, O. A. Drosdov and M. I. Yudin, 1966: Influence of economic activity on climate. *Modern Problems of Climatology* (*Collection of Articles*), FTD-HT-23-1338-67, Foreign Tech. Div., Wright-Patterson Air Force Base, Ohio, 484–500.

Dietrich, G., 1963: *General Oceanography*. New York, Wiley, 588 pp.

Eriksson, E., 1968: Air-ocean-ice cap interactions in relation to climatic fluctuations and glaciation cycles. *Meteor. Monogr.*, **8**, No. 30, 68–92.

Fletcher, J. O., 1966: The arctic heat budget and atmospheric circulation. *Proc. Symp. Arctic Heat Budget and Atmospheric Circulation*, Memo. RM-5233-NSF, The RAND Corporation, Santa Monica, Calif., 23–43.

———, 1968: The polar oceans and world climate. Paper presented at the Symposium on Beneficial Modifications of the Marine Environment, 60 pp. (Available from the Clearinghouse for Federal Scientific & Technical Information Springfield, Va.).

Lorenz, E. N., 1968: Climatic determinism. *Meteor. Monogr.*, **8**, No. 30, 1–3.

Manabe, S., and R. T. Wetherald, 1967: Thermal equilibrium of the atmosphere with a given distribution of relative humidity. *J. Atmos. Sci.*, **24**, 241–259.

Mintz, Y., 1968: Very long-term global integration of the primitive equations of atmospheric motion: An experiment in climate simulation. *Meteor. Monogr.*, **8**, No. 30, 20–36.

Panchev, S., 1968: Coefficient of horizontal macroturbulent exchange in the atmosphere. *J. Atmos. Sci.*, **25**, 933–935.

Rakipova, L. R., 1966a: The influence of the arctic ice cover on the zonal distribution of atmospheric temperature. *Proc. Symp. Arctic Heat Budget and Atmospheric Circulation*, Memo. RM-5233-NSF, The RAND Corporation, Santa Monica, Calif., 411–441.

———, 1966b: Changes in the zonal distribution of the atmospheric temperature as a result of active influence on the climate. *Modern Problems of Climatology* (*Collection of Articles*), FTD-HT-23-1338-67, Foreign Tech. Div., Wright-Patterson Air Force Base, Ohio, 388–421.

Raschke, E., F. Möller and W. R. Bandeen, 1967: The radiation balance of the earth-atmosphere system over both polar regions obtained from radiation measurements of the Nimbus II meteorological satellite. Rept. No. X-622-67-460, Goddard Space Flight Center, Greenbelt, Md, 27 pp.

Sellers, W. D., 1965: *Physical Climatology*. University of Chicago Press, 272 pp.

Veksler, Kh., 1958: The radiation balance of the earth as a factor in changing the climate. *Izmeneniye Klimata* 49, Moscow.

Vonder Haar, T. H., 1968: Variations of the earth's radiation budget. Ph.D. thesis, University of Wisconsin, Madison, 118 pp.

6

The Birth of the General Circulation Climate Model

Manabe, S. and Wetherald, R.T. (1975). The effects of the doubling CO_2 concentration on the climate of a general circulation model. *Journal of the Atmospheric Sciences*, **32** (1), 3–15. Times Cited: 404. 12 pages.

Hansen, J., Lacis, A., Rind, D., *et al.* (1984). Climate sensitivity: Analysis of feedback mechanisms. In *Climate Processes and Climate Sensitivity*, AGU Geophysical Monograph 29, Maurice Ewing, Vol. 5 (eds. J.E. Hansen & T. Takahashi), pp. 130–163. American Geophysical Union.

Manabe and Wetherald's estimate of the warming expected from a doubling of CO_2 was the first such estimate to be based on completely sound and quantitatively accurate implementations of radiative and convective physics, and it is an estimate that has stood the test of time very well. Even that remarkable paper had its limitations, though, stemming from the representation of the entire atmosphere by a single column. The need to go further is very well articulated in the abstract to our next landmark paper, Manabe and Wetherald (1975). The temperature of the Earth's surface is far from uniform, and to obtain a more complete picture of climate change one needs to be able to predict the geographical distribution of the warming. Further, many climate impacts descend from rainfall changes (drought, flood) rather than just temperature, so one needs to characterize changes in the hydrological cycle. The factors governing atmospheric humidity and hence the all-important water vapor feedback are also intimately connected with the hydrological cycle; it would be highly desirable to replace the assumption of fixed relative humidity with a calculation more closely tied to the basic physics leading to moistening and drying of the atmosphere. Melting of sea ice and snow provides an important feedback on climate, one that was recognized already by Arrhenius but that is virtually impossible to treat accurately in a one-column model. Ultimately, one needs to be able to predict changes in cloud patterns as well, though this challenge was not taken up in Manabe's 1975 paper.

The next step, required to treat these processes, was a very big step indeed. It involved nothing less than solving the full three-dimensional fluid dynamical and thermodynamical equations governing transfer of heat, moisture, and momentum around the globe. The fluid equations needed to be coupled to equations governing the physics of radiation, convection, and thermal exchange with the surface. The result is known as a *General Circulation Model* (GCM). Now, Manabe had to solve his radiation problem not for just a single column, but for thousands of columns, and a grid of these columns had to be linked through the air currents that transport heat, momentum, and moisture from one column to another. This was a problem to tax the very biggest computers of the time, which were in fact physically behemoths but mere mice by the standards of today's computer power. The Univac 1108 on which Manabe's original GCM was developed had all of half a megabyte of RAM – not even enough to store a single MP3 track. It took 20 min to simulate a single day of the atmospheric circulation (Fig. 6.1).

Manabe's prose speaks for itself, but it is worth highlighting a few of the accomplishments of the paper. The work rediscovered *polar amplification* first predicted by Arrhenius – the fact that the high latitudes (especially the Arctic) warm more than low latitudes. In addition to the obvious explanation of this as being due to feedback from melting ice and snow, Manabe invokes a clever and little-

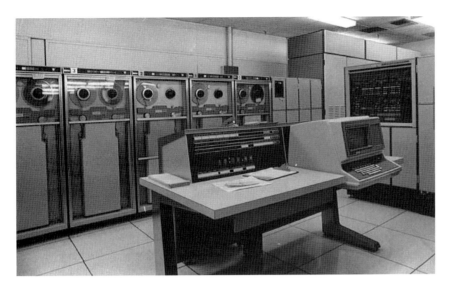

Fig. 6.1   **A Univac 1108 computer which ran Manabe's original GCM climate model.**

remembered hypothesis involving the bottom-heavy vertical structure of polar warming. Manabe and Wetherald (1975) also discovered that land warms more than ocean; this is quite important, given that land is where most people live and where all agriculture is carried out. The paper also for the first time demonstrated that global mean precipitation increases as the world warms.

Manabe's GCM was the first to be able to handle a doubling of CO_2, but others gradually came on the scene. NASA's Goddard Institute for Space Studies (GISS) was an important early entrant to the field. GISS, whose modeling effort was led by James Hansen (from whom we will hear more anon), was an offshoot of NASA's main climate lab left behind above Tom's Diner in New York when diehard Manhattanites balked at making the move to the wilds of suburban Greenbelt, Maryland The paper we include by Hansen *et al.* represents a landmark in analysis of GCMs, particularly with regard to quantitative analysis of climate feedbacks. This paper is particularly notable in that it provided one of the early indications that cloud feedbacks had the potential to affect the climate sensitivity greatly. Another important result of the paper was the analysis of the nature of the delay in warming caused by ocean heat storage, and the consolidation of the notion of "committed warming," which is basically warming that is in the pipeline and will be realized even if CO_2 concentrations are frozen. The very clever discussion of the intimate relation between climate sensitivity and the time required for climate to reach equilibrium repays careful study even today.

The Effects of Doubling the CO_2 Concentration on the Climate of a General Circulation Model[1]

SYUKURO MANABE AND RICHARD T. WETHERALD

Geophysical Fluid Dynamics Laboratory/NOAA, Princeton University, Princeton, N.J. 08540
(Manuscript received 6 June 1974, in revised form 8 August 1974)

ABSTRACT

An attempt is made to estimate the temperature changes resulting from doubling the present CO_2 concentration by the use of a simplified three-dimensional general circulation model. This model contains the following simplifications: a limited computational domain, an idealized topography, no heat transport by ocean currents, and fixed cloudiness. Despite these limitations, the results from this computation yield some indication of how the increase of CO_2 concentration may affect the distribution of temperature in the atmosphere. It is shown that the CO_2 increase raises the temperature of the model troposphere, whereas it lowers that of the model stratosphere. The tropospheric warming is somewhat larger than that expected from a radiative-convective equilibrium model. In particular, the increase of surface temperature in higher latitudes is magnified due to the recession of the snow boundary and the thermal stability of the lower troposphere which limits convective heating to the lowest layer. It is also shown that the doubling of carbon dioxide significantly increases the intensity of the hydrologic cycle of the model.

1. INTRODUCTION

According to the estimate by Machta (1971), the concentration of carbon dioxide in the atmosphere may increase as much as 20% during the latter half of this century as a result of fossil fuel combustion. It has been speculated that an increase in the CO_2 content in the atmosphere may result in the gradual rise of the atmospheric temperature (Callendar, 1949). Plass (1956), Kaplan (1960) and Kondratiev and Niilisk (1960) estimated the increase of the temperature of the earth's surface required to compensate for the increase in the downward terrestrial radiation due to the increase in the CO_2 content in the atmosphere. According to Plass, the doubling or halving of the CO_2 results in a temperature change of +3.8°C or −3.6°C, respectively. Kaplan (1960) attempted to improve Plass' estimate by taking into consideration the effect of cloudiness, as well as by improving the computation scheme of radiative transfer. Kondratiev and Niilisk (1960) considered the effect of overlapping

between the 15-μm band of CO_2 and the rotation band of water vapor. The magnitudes of the temperature changes estimated by both Kaplan, and Kondratiev and Niilisk are significantly less than those estimated by Plass, indicating the importance of such factors as cloudiness and overlapping between bands in this problem.

Möller (1963) reviewed these studies critically and tried to improve these estimates by making a more realistic assumption. According to Möller, the atmosphere tends to restore a certain climatological distribution of relative humidity in response to the change of temperature. The change in surface temperature is accomplished by a change in air temperature, which in turn causes the changes in absolute humidity of air and in the downward terrestrial radiation. The change in absolute humidity also affects absorption of solar radiation and, accordingly, the amount of solar radiation reaching the earth's surface. Taking into consideration all these factors, Möller reevaluated the effect of the increase of carbon dioxide content upon the tempera-

Manabe, S. & Wetherald, R. T. (1975) The effects of doubling the CO_2 concentration on the climate of a general circulation model. *Journal of the Atmospheric Sciences* 32(1): 3–15. Reprinted with permission of American Meteorological Society.

ture of the earth's surface. To his surprise, he obtained results which are quite different from those of the earlier studies mentioned above. According to his estimate, an increase in the water vapor content of the atmosphere with rising temperature causes a self-amplifying effect which results in an almost arbitrary temperature change. For example, when the air temperature is around 15°C, the doubling of CO$_2$ content results in an increase of temperature by as much as 10°C. For other air temperatures, the result may be completely different.

Examining Möller's method, Manabe and Wetherald (1967) felt that it was necessary to take into consideration another physical factor in order to obtain reasonable results. They maintained that the change in CO$_2$ content not only affects the flux of net radiation at the earth's surface but also the fluxes of sensible and latent heat from the earth's surface to the atmosphere. In order to find out how the flux of total heat energy depends upon the CO$_2$ content, it is necessary to take into consideration the heat balance of the atmosphere as well as that of the earth's surface. Therefore, Manabe and Wetherald proposed to use a one-dimensional model of radiative-convective equilibrium to investigate this problem. By comparing the state of radiative-convective equilibrium which was obtained for various CO$_2$ concentrations, they estimated the dependence of atmospheric temperature upon CO$_2$ concentration. Following the suggestion of Möller, they incorporated into their model the mechanism of the water vapor greenhouse feedback by assuming a fixed distribution of relative (rather than absolute) humidity in the model atmosphere. Although their results show that the water vapor greenhouse feedback is responsible for increasing the sensitivity of the temperature of the model atmosphere, it does not yield an indefinite result such as that obtained by Möller. They found that the doubling of CO$_2$ in the air increases the equilibrium temperature of the earth's surface by about 2.3°C provided that no change in cloudiness takes place. Manabe (1971) attempted to improve this result further by replacing his scheme of computing radiative transfer with the method of Rodgers and Walshaw (1966) as modified by Stone and Manabe (1968). In this case, he obtained a somewhat smaller increase of about 1.9°C resulting from the doubling of carbon dioxide content.

Rasool and Schneider (1971) also estimated the thermal effects of a change in the concentration of CO$_2$ based upon the consideration of the heat balance of the atmosphere. It turned out that the change in surface temperature which they esti-

mated is much smaller than the change obtained by Manabe and Wetherald. According to our more recent studies, a significant part of the difference stems from the following effects:

1) Rasool and Schneider did not take into consideration the fact that the temperature change in the stratosphere has an opposite sign to that in the troposphere.
2) The absorption of solar radiation is altered if the atmospheric temperature and accordingly also the water vapor content changes. This factor was not considered by Rasool and Schneider.

The models of Rasool and Schneider (1971) and Manabe and Wetherald (1967) are globally averaged models. However, the climate of the planet Earth is maintained by the nonlinear coupling between various processes, such as the poleward heat transfer by the atmospheric and oceanic circulations, as well as the vertical heat transfer by radiative transfer and convection. Thus, it is clear that one cannot obtain a definitive conclusion, using a globally averaged model, concerning the effect of an increase in CO$_2$ upon climate. Recently, Budyko (1969) and Sellers (1969) constructed a very simple model of the atmosphere in which the effects of a poleward heat transport as well as radiative transfer were included although in highly parameterized forms. In addition, they incorporated into their model the following positive feedback mechanism that would tend to enhance the sensitivity of the model climate:

decrease (increase) of atmospheric temperature
→ wider (narrower) area of snow cover or ice pack
→ larger (smaller) albedo of earth's surface
→ decrease (increase) of atmospheric temperature.

By using their model, Budyko and Sellers suggested that the thermal regime of the model atmosphere is highly sensitive to small changes in various parameters, such as the solar constant, when the process of the snow cover feedback is incorporated. Their study suggests that it is important to consider this feedback mechanism for our problem. Since the heat transport by large-scale eddies is parameterized in an extremely simple manner and many idealizations are made in their model, it is clear that we need a better model for a quantitative study of the climatic change resulting from an increase of the amount of CO$_2$ in the atmosphere.

The present study represents a preliminary attempt to do this by using a three-dimensional general circulation model in which the heat transport by large-scale eddies is computed explicitly rather than by parameterization. The long-term integrations of the model are carried out for two different concentrations of CO_2, i.e., the present and twice the present concentration. By comparing the two quasi-equilibrium states which emerge from these experiments, we will evaluate how the concentration affects the state of the model atmosphere.

It should be noted here that a sensitivity study such as described in this paper is meaningful provided that the following conditions are satisfied.

(i) The model has a stable equilibrium climate.
(ii) An external forcing (such as the doubling of CO_2 concentration) is not large enough to force the model climate out of the stable equilibrium into a markedly different state.

Obviously, it is useful to inquire how large an external forcing is required in order to push the model climate out of stable equilibrium or how the model climate changes thereafter. However, it is not the subject of the present study.

In view of the many simplifications adopted for the construction of the model, the result from this study should not be regarded as a definitive study of this problem. Instead, we are satisfied if this study yields some valuable insight into the physical factors which control the response of the atmosphere to the change in the carbon dioxide content in the atmosphere.

2. DESCRIPTION OF THE MODEL

The general circulation model used for this study is essentially the same as that described by Manabe (1969). Therefore, only a brief description of the model is given here.

The model solves the primitive equations on a Mercator projection using an energy conserving form of the finite-difference formulation. The vertical coordinate is defined by pressure normalized by surface pressure. To simulate the effects of sub-grid mixing, a nonlinear viscosity is included in the model. The topography adopted is the same as that described by Manabe (1969). Free-slip insulated walls are placed at the equator and at 81.7N, whereas cyclic continuity is assumed for the two meridional boundaries 120° of longitude apart. Grid-point spacing is such that the resolution is approximately 500 km. Nine vertical levels are

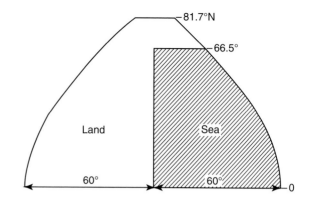

Fig. 1. Diagram illustrating the distribution of continent and "ocean." Cyclic continuity is assumed at the eastern and western ends of the domain.

chosen so that the model can simulate the structure of both the stratosphere and the planetary boundary layer. The computational domain is divided into two equal areas up to 66.5N latitude, continent and "ocean," respectively. From 66.5N to the polar boundary at 81.7N, continent is assumed throughout. A diagram of the computational region is shown in Fig. 1. It should be stressed here that this model does not contain a separate ocean computation. The "ocean" portion is simply considered to be an area of wet land or an area possessing an infinite source of soil moisture for evaporation. The model ocean resembles the actual ocean in the sense that it is wet, but it lacks the effects of heat transport by ocean currents.

The scheme for computing radiative heating and cooling is described by Manabe and Strickler (1964) and Manabe and Wetherald (1967), and computes both solar and longwave radiation fluxes.[2] The distribution of cloudiness is specified from annual mean observations and is a function of latitude and height only. Three atmospheric gases are taken into account, i.e., water vapor, ozone and carbon dioxide. The distribution of water vapor is determined by the prognostic equations of the model or, in other words, the radiative computation is "coupled" with the hydrologic cycle. The spatial distribution of ozone is specified in a manner analogous to clouds. The mixing ratio of carbon dioxide is taken to be constant everywhere. The surface temperature over the continent and the hypothetical ocean is determined from the boundary condition that no heat is stored at the earth's surface, i.e., net fluxes of solar and terrestrial radiation and the turbulent fluxes of sensible and latent heat locally sum to zero.

The prognostic system of water vapor involves the three-dimensional advection of water vapor, evaporation, vertical mixing, nonconvective condensation, and an idealized moist convective adjustment. Over continental surfaces, the depth of snow cover and the amount of soil moisture are based upon detailed balance computations of snow and soil moisture, respectively. In particular, the snow depth is increased by snowfall and depleted by evaporation and snowmelt. The latter quantity is computed from the requirement of the heat balance when conditions for the snowmelt are satisfied. [See Manabe (1969) for further description of the prognostic system of soil moisture and snow.] Differentiation between rain or snow is determined by the temperature at a height of approximately 350 m. If this temperature is equal to less than 0°C, precipitation is in the form of snow; otherwise, it occurs in the form of rain. Over the oceanic region, the area of sea ice is identified as the area where the surface temperature over the ocean is less than −2°C.

For the computation of the heat balance at the earth's surface, it is necessary to know the distribution of surface albedo. It is assumed that the albedo of the soil surface is a function of latitude and that its distribution with latitude is the same as that used by Manabe (1969). The albedos of snow cover and sea ice are assumed to be much larger than the albedo of bare soil. As pointed out in the Introduction, this difference in albedo accounts for the snow cover feedback.

Both snow cover and sea ice are classified into two categories, i.e., permanent and temporary snow cover (sea ice). Different values of albedo are assigned to each category. Referring to the study of Budyko (1956), unstable snow cover and unstable sea ice are assigned an albedo of 0.45 and 0.35, respectively, whereas both permanent snow cover and permanent sea ice are assigned a common albedo of 0.70. The discrimination between the permanent and temporary snow cover (sea ice) is made according to the surface temperature of snow cover (sea ice). Originally, we intended to assume this critical surface temperature to be −10°C. Because of a mistake in the programming, the value of −25°C was actually used instead of −10°C. This error will have the effect of displacing the permanent ice cap with high albedo further northward (by about 10° latitude) than it would have been otherwise. Therefore, the results presented in this study must be interpreted with this in mind.[3]

In concluding this section, a block diagram, which illustrates the structure of the model, is shown in Fig. 2.

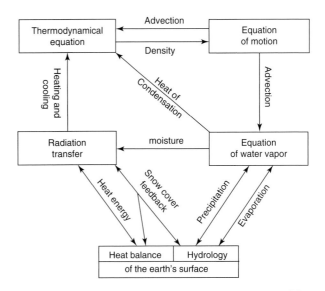

Fig. 2. Box diagram indicating the major components of the model. Arrows represent the links between components.

3. APPROACH TOWARD THE EQUILIBRIUM

As mentioned in the Introduction, the numerical experiments are performed for two different concentrations of CO$_2$, i.e., the present and twice the present amount. The specific values of mixing ratio chosen for these two cases are 0.456×10^{-3} g kg^{-1} and 0.912×10^{-3} g kg^{-1}, respectively. The climatic effects of this doubling of the CO$_2$ concentration will be identified next by comparing the two quasi-equilibrium states of the model atmospheres which are obtained by the method described below.

Starting from the initial condition of an isothermal atmosphere at rest, the model is integrated with respect to time over a period of approximately 800 days. One version of the equilibrium climate is thus obtained by averaging over the last 100-day period. In order to reduce the bias in the equilibrium climate resulting from the specific choice of initial conditions, another integration is carried out starting from quite different initial conditions, which are described in Appendix B. The final state of equilibrium is now obtained by averaging the two 100-day mean states.

In order to attain a satisfactory convergence toward statistical equilibrium, it is necessary to carry out the time integration for sufficiently long periods. The degree of convergence obtained by the procedure described above is evident in Fig. 3. This figure shows how the (mass-weighted) mean temperature of the model atmosphere changes with time starting from the two initial conditions. Note

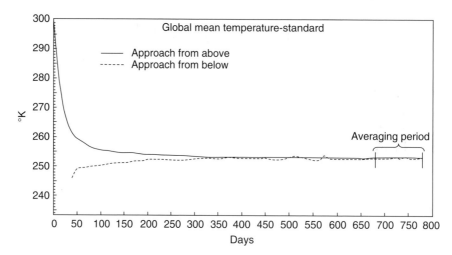

Fig. 3. Time variation of (mass-weighted) mean temperature (K) for the entire period of integration of the two standard runs.

that the initial values for the two runs are considerably different from one another, but that they are practically indistinguishable toward the end of the runs. The difference between the two mean temperatures averaged over the last 100 days of each integration is about 0.10°C. For a further discussion of the convergence of the solution, see Appendix B.

4. DOUBLING OF CO_2 CONCENTRATION

This section deals with the comparison between the two statistical equilibrium states corresponding to the present and twice the present concentration of CO_2. For ease of identification, the case of the present concentration is identified as the "standard" case and that of double concentration as the "$2 \times CO_2$" case. Unless we specify otherwise, the results presented here are obtained by taking the average of two time-mean states, which are computed from the last 100-day period of each integration as described in the preceding section.

a. Temperature

The latitude-height distribution of the zonal mean temperature in the model atmosphere with a normal concentration of carbon dioxide is shown in Fig. 4a. According to the comparison between this result and the distribution of observed annual mean temperature,[4] the thermal structure of the model atmosphere generally resembles that of the actual atmosphere. However, the surface temperature in the model tropics is significantly higher than the

observed value due to the lack of the poleward transport of heat by ocean currents (see Fig. 5 also). In the lower stratosphere of the model, the zonal mean temperature in middle latitudes is significantly colder than observed. The failure of the model to produce the relatively warm belt in the lower stratosphere of the middle latitudes seems to be a characteristic of the limited domain model which cannot accommodate ultra-long waves (see Manabe, 1969).

Fig. 4b shows the difference in zonal mean temperature between the $2 \times CO_2$ and the standard case. Owing to the increase in the greenhouse effect resulting from the increase in the concentration of CO_2, there is a general warming in the model troposphere. On the other hand, large cooling occurs in the model stratosphere. Qualitatively similar results were obtained by Manabe and Wetherald (1967) in their study of radiative-convective equilibrium of the atmosphere. According to Fig. 4b, the tropospheric warming is most pronounced in the lower troposphere in high latitudes. This large warming is associated with the decrease in the area of snow (or ice) cover, which has a much larger albedo than the soil surface. The increase in downward terrestrial radiation due to the increase in the amount of CO_2 contributes to the decrease in the area of snow cover and, as a result, to the increase in the amount of solar radiation absorbed by the earth's surface. As mentioned above, the warming in high latitudes is confined within a relatively shallow layer next to the earth's surface because the vertical mixing by turbulence is suppressed in the stable layer of the troposphere in polar regions. Therefore, most of the thermal energy involved goes into raising the

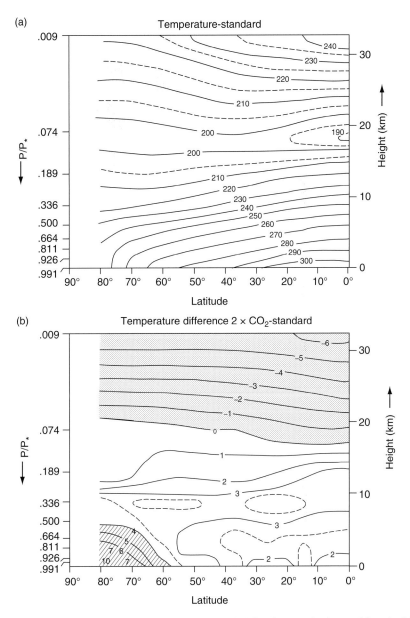

Fig. 4. Latitude-height distribution of the zonal mean temperature (K) for the standard case (a) and of the increase in zonal mean temperature (K) resulting from the doubling of CO_2 concentration (b). Stippling indicates a decrease in temperature.

temperature of this shallow surface layer rather than being spread throughout the entire depth of the troposphere. In short, the effects of suppression of vertical mixing together with those of snowmelt are responsible for the large warming in the polar region. It should be noted here that the maximum increase of temperature right next to the polar boundary is partly due to the recession of the permanent ice cover, which has a larger albedo than the variable snow cover. See Section 4d for further discussion of this subject. In the model tropics, the warming spreads throughout the entire troposphere due to intense moist convection. Accordingly, its

magnitude is relatively small as compared with the warming in the polar region.

It should be noted that in low and middle latitudes, the warming is greater in the upper troposphere (~336 mb) than near the surface. This is due to the fact that the moist convective processes in the model tend to adjust temperatures in a column toward the moist adiabatic lapse rate. Since this lapse rate is more stable in a warmer atmosphere than in a colder atmosphere, the greatest difference in temperature will be found near the top of the moist convective layer. Therefore, the area of maximum tropospheric

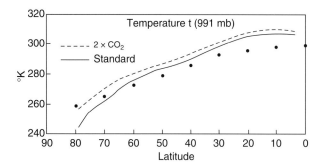

Fig. 5. Zonal mean temperature at the lowest prognostic level (i.e., ~991 mb). Dots indicate the observed distribution of zonal mean surface air temperature (Oort and Rasmusson, 1971).

Table 1. Increase in the equilibrium temperature (K) of the earth's surface resulting from the doubling of concentration CO_2. The figure for the G-C (general circulation) model represents the average value over the entire domain.

Change of CO_2 content (ppm)	R-W model	M-W model	G-C model
$300 \rightarrow 600$	+1.95	+2.36	+2.93

warming occurs in the upper troposphere instead of near the earth's surface.

As pointed out above, large cooling occurs in the model stratosphere. This is caused by the increase in the emission from the stratosphere to space resulting from the increase in the concentration of CO_2. Since the total amount of CO_2 above a given level decreases with increasing altitude, the absorption of the emission from above also decreases correspondingly. This is one reason why the magnitude of the cooling increases with increasing height in the model stratosphere.

The differences between the distributions of surface air temperature for the two cases are further illustrated by Fig. 5, which shows the latitudinal distributions of zonal mean temperature at the lowest prognostic level (i.e., ~991 mb). In particular, there is a considerable difference between the two temperature curves in higher latitudes, whereas there is a smaller change at equatorial latitudes. This is consistent with the results shown in Fig. 4b.

In order to evaluate the sensitivity of the mean surface temperature to changes in CO_2 concentration, Table 1 is presented. This table shows the difference of mean surface temperature as a function of CO_2 concentration for both the general circulation model and the associated Manabe and Wetherald

(M-W) radiative-convective equilibrium model (1967). (Note that the scheme of computing radiative transfer which is incorporated into the general circulation model is identical to that adopted by M-W.) Also shown, for the sake of comparison, are the corresponding values obtained using a version of the Rodgers-Walshaw (R-W) radiation model (1966) which is modified by Stone and Manabe (1968)[5] and is, in our opinion, a superior model. According to this table, the magnitude of the surface temperature difference is considerably greater for the general circulation model than for the corresponding radiative-convective model by itself. (Note that the general circulation model incorporates the M-W radiation model.) This suggests that the former is more sensitive to changes in CO_2 concentration than the latter. This difference in sensitivity is due, in part, to the snow cover feedback mechanism which is present in the general circulation model but is not accounted for in the radiative-convective equilibrium model. It is of interest to note that the R-W model is slightly less sensitive to CO_2 changes than the M-W model. The reason for this difference in sensitivity between the two models has not been pinpointed at present. Had we used the R-W radiation scheme instead of the M-W scheme, the sensitivity of the general circulation model to the change in CO_2 concentration would be slightly less than what is indicated in the table.

b. Relative humidity

The latitude-height distribution of relative humidity which emerged from the time integration of the model with the standard concentration of carbon dioxide is shown in Fig. 6a. According to the comparison between these results and the observed distribution of zonal mean relative humidity,[6] the relative humidity in the model atmosphere tends to be too high near the earth's surface. However, the model simulates some of the qualitative features of the annual mean distributions of relative humidity in the actual atmosphere. For further discussion of relative humidity of the model, see Manabe *et al.* (1965).

Fig. 6b shows the difference in zonal mean relative humidity between the $2 \times CO_2$ and the standard case. According to this figure, the pattern is rather irregular. In view of this irregularity, the details of this distribution may not be significant. However, it is probable that the large-scale features of this difference are meaningful. In general, relative humidity increases by a few percent in the lower troposphere (below 700 mb) and decreases

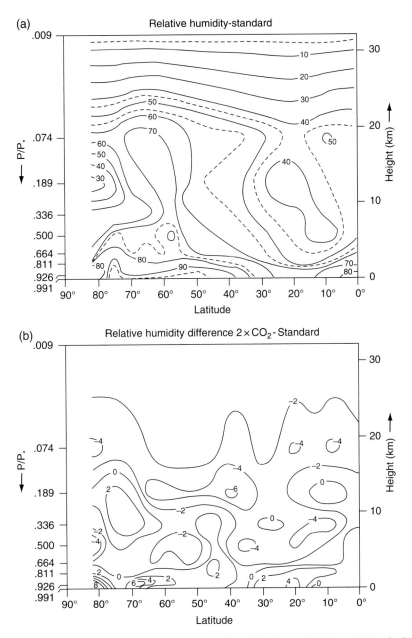

Fig. 6. Latitude-height distribution of zonal mean relative humidity (%) for standard case (a) and of the increase in zonal mean relative humidity resulting from the doubling of CO$_2$ concentration (b). Stippling indicates an increase of relative humidity.

in the upper troposphere (above 700 mb). The M-W results suggest that the change of this magnitude in relative humidity does not have very large effects on the state of the thermal equilibrium of the atmosphere. However, it could have changed the cloudiness and affected the temperature of the model atmosphere significantly had the model included the prognostic system of cloudiness. [See, for example, Smagorinsky (1960) for the relationship between cloud amount and relative humidity.] According to M-W and Schneider (1972),

cloudiness, particularly low cloud cover, has a very large effect upon the thermal equilibrium of the atmosphere. Therefore, it seems to be necessary to repeat this study later by use of a model with the capability of cloud prediction.

c. Hydrology

Fig. 7 shows the zonal mean precipitation and evaporation rates for the 2×CO$_2$ and the standard cases. For both quantities, the rates for the 2×CO$_2$

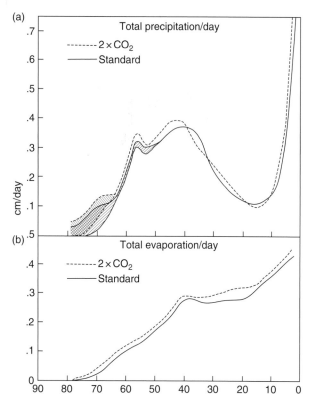

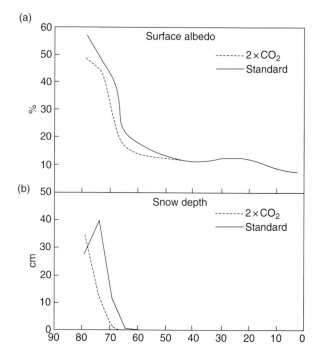

Fig. 8. Zonal mean surface albedo (a) and zonal mean snow depth in water equivalent (b).

Fig. 7. Zonal mean rates of total precipitation, where shaded areas denote the rates of snowfall (a), and zonal mean rates of evaporation (b).

case are significantly greater than the corresponding rates for the standard case over most of the region. Averaged over the entire computational domain, the time-mean rate of precipitation is practically equal to that of evaporation. Its value is 0.255 cm day^{-1} for the standard case and 0.275 cm day^{-1} for the $2 \times CO_2$ case. This suggests that, within the limits of this study, the general circulation model containing the higher CO_2 content has a significantly more active hydrologic cycle by about 7% than the one with the standard CO_2 content. The greater downward flux of terrestrial radiation resulting from the larger concentration of CO_2 increases the heat energy available for the evaporation from the earth's surface, and thus enhances the intensity of the hydrologic cycle in the model atmosphere. There is another important reason for the intensification of the hydrologic cycle, which is discussed in Section 4e.

Also shown in Fig. 7a are the relative snowfall rates (cross-hatched areas) associated with both cases. Here, it is evident that for the $2 \times CO_2$ case, the snowfall rate is less and extends over a considerably smaller latitude area than the snowfall rate for the standard case.

d. Snow cover and albedo

In the preceding discussion, it was stated that a change in the ground albedo (snow cover vs bare soil) was partly responsible for the large warming obtained in higher latitudes. Fig. 8 shows the correspondence between the ground albedo (8a) and snow depth (8b) for the standard and $2 \times CO_2$ cases. As expected, the $2 \times CO_2$ case is associated with a lower surface albedo and a lesser extent of the snow cover as compared with the standard case. These changes occur mainly in the zone of unstable new snow cover (assigned an albedo of 45%) and at the boundary of the permanent ice cap (assigned an albedo of 70%) which are displaced further northward as the CO_2 concentration is increased. According to Fig. 4b, the area of the largest temperature increase appears to be located near the polar boundary at the surface where the change of ground albedo from 45 to 70% mainly occurs (see the left-hand side of Fig. 8a). There is also a corresponding northward displacement of the zone of maximum snow depth for the $2 \times CO_2$ case.[7] It should be recalled here that due to the oversight mentioned in Section 2, the permanent ice cap boundary is displaced poleward by about 10° of latitude and, therefore, neither zonal mean surface albedo curve reaches the maximum value of 70%.

In order to evaluate the consequence of the code error in the snow albedo formulation, the latitudinal distribution of the planetary albedo for the standard case is compared with that of the actual atmosphere, which was determined from the data of the Nimbus 3 satellite by Raschke *et al.* (1973), in Fig. 9. (Here, "planetary albedo" is defined as the reflectivity of the earth-atmosphere system to solar radiation at the top of the atmosphere.) According to this figure, the planetary albedo of the model in middle and high latitudes is either larger than or approximately the same as that of the actual atmosphere, depending on whether the Northern or the Southern Hemisphere is selected for comparison. In other words, the planetary albedo in high latitudes of the model is not smaller than that of the actual atmosphere as one would expect from the nature of the code error. In fact, it is probable that the formulation of the albedos of snow and ice, which are erroneously adopted for this study, may be more realistic than the formulation originally intended. For a better formulation of snow cover albedo, further study is required to iron out the suspected inconsistency between the observed values of surface albedo and those of planetary albedo.

e. Heat balance

The heat balance of the model atmosphere may be summarized by the heat balance diagram shown in Fig. 10. The standard and $2\times CO_2$ cases are depicted by the left-hand and center diagrams, respectively.

According to this figure, the net downward radiation (i.e., the net downward solar radiation minus net upward longwave radiation) increases by about 3.4%, resulting from the doubling of CO_2 content. This is mainly due to the increase in greenhouse effects of carbon dioxide and water vapor. Responding to this increase in net downward radiation, the upward flux of latent heat (i.e., evaporation) increases by as much as 7%, whereas that of sensible heat actually *decreases* by about 8%. This interesting result can be partly explained by the change in the so-called Bowen ratio.[8] The increase in surface temperature due to the doubling

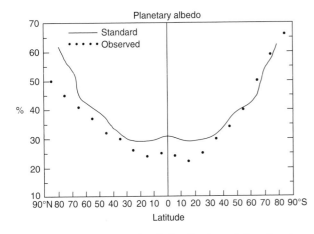

Fig. 9. Solid line: latitudinal distribution of the planetary albedo of the model with the standard concentration of CO_2. The distribution in the Southern Hemisphere represents the mirror image of that in the Northern Hemisphere. Dots: latitudinal distribution of annual mean planetary albedo based upon the data from Nimbus 3 satellite (Raschke *et al.* 1973).

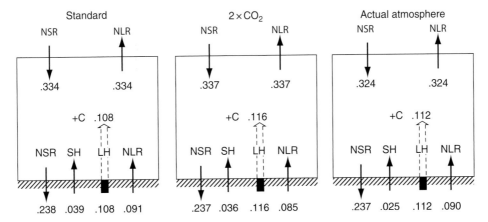

Fig. 10. Box diagrams illustrating area-mean heat balance components (units ly min⁻¹): Left, standard case; middle, $2\times CO_2$ case; right, the actual atmosphere. NLR, net longwave radiation; NSR, net solar radiation; SH, sensible heat flux; LH, latent heat flux. The observed net radiational fluxes are taken from London (1957), and the observed sensible and latent heat fluxes are taken from Budyko (1963).

of CO_2 raises the saturation vapor pressure at the earth's surface, and thus significantly reduces the Bowen ratio. In other words, evaporation becomes a more effective way of removing heat from the surface of the model earth than the upward flux of sensible heat. This accounts for why the percentage increase in the rate of evaporation is twice as large as that of net downward radiation. Since the model atmosphere reaches quasi-steady states toward the end of both experiments, this increase in evaporation is almost equal to that in precipitation rate. In short, the doubling of CO_2 increases significantly (by about 7%) the intensity of the hydrologic cycle in the model atmosphere as described in Section 4c.

Another important comparison between the two cases concerns the integrated net solar and long-wave radiation at the top of the model atmosphere. According to Fig. 10, net upward longwave radiation increases by ~1% due to the doubling of CO_2. This is because the stratospheric cooling, as well as the increase in the height of the effective source of upward emission caused by the increased CO_2, are overcome by the general warming of the model troposphere. At the same time, there is an increase of about 1% in the intensity of net downward solar radiation. This is due, not only to the lower surface albedo in the polar region, but also to the increased absorption of insolation caused by the increase in both water vapor and carbon dioxide in the model atmosphere.

For the sake of comparison, a similar representation of the actual atmosphere is shown as the right-hand diagram of Fig. 10. The data for this diagram were obtained from Budyko (1963) and London (1957). Despite the idealized land-sea configuration adopted for this study, the agreement between Budyko's results and those computed from the standard case is quite good with perhaps the exception of the sensible heat flux.

f.   Eddy kinetic energy

It is of interest to note the effect upon the dynamical processes taking place in the model atmosphere due to the increase of the CO_2 concentration. A measure of these processes is the distribution of eddy kinetic energy[9] which is shown in Fig. 11. Fig. 11a shows the zonal mean latitude-height cross section of eddy kinetic energy for the standard case, and Fig. 11b illustrates the difference of eddy kinetic energy between the 2×CO_2 and standard cases. In general, the eddy kinetic energy in the model atmosphere is much lower than that in the actual atmosphere. As pointed out by

Manabe *et al.* (1970), this is one of the characteristic features of the model atmosphere with coarse computational resolution of finite differences. According to Fig. 11b, a net decrease of eddy kinetic energy centers between 300 and 500mb, whereas there is an area of net increase in the layer above the 300-mb level for practically all latitudes. The lower area of maximum net decrease is situated in middle latitudes which approximately coincides with the region of a reduced tropospheric meridional temperature gradient (refer to Fig. 4b). Therefore, the change in meridional temperature gradient, and accordingly, the change in vertical wind shear may be partly responsible for this reduction of eddy kinetic energy. However, the area of net increase in the upper troposphere and lower stratosphere is more difficult to explain. If one compares the zonal mean temperature difference (Fig. 4b) with Fig. 11b, it may be seen that the region of maximum increase of eddy kinetic energy corresponds closely to the level where the static stability decreases the most. Therefore, one may expect a larger generation of eddy kinetic energy for the 2×CO_2 case as compared with the standard case. This may partly explain why the eddy kinetic energy in the upper troposphere and lower stratosphere of the model increases due to the increase in the concentration of CO_2.

g.   Poleward transport of energy

One of the important factors which indicate the general activity of the atmospheric circulation is the poleward transport of energy. Fig. 12 shows the poleward transport of total energy ($C_pT + \phi + K + Lr$), heat energy ($C_pT + \phi + K$), and latent energy (Lr) across each latitude circle. Here, T, ϕ, K and r represent temperature, geopotential height, kinetic energy and mixing ratio, respectively; and C_p and L denote the specific heat of air and latent heat of evaporation.

1) Heat energy

According to Fig. 12b, the doubling of the concentration of CO_2 in the model atmosphere reduces the poleward transport of heat energy in middle latitudes. This result is consistent with the decrease of the eddy kinetic energy in the lower troposphere where the major portion of heat transport by large-scale eddies is accomplished.

In the model tropics, the opposite situation holds. The poleward heat transport increases slightly due to the doubling of CO_2. Again, this

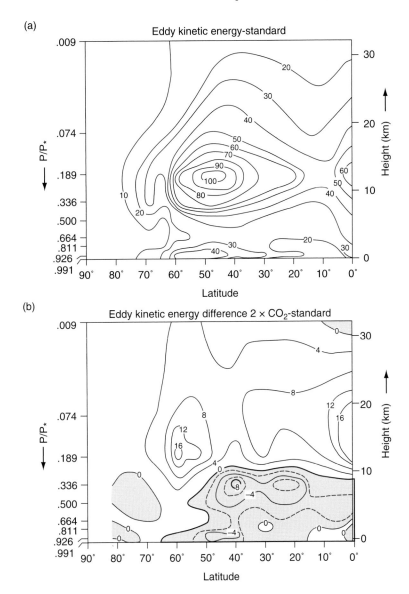

Fig. 11. Latitude-height distribution of zonal mean eddy kinetic energy for the standard case (a) and of the difference in zonal mean eddy kinetic energy resulting from the doubling of the concentration of CO$_2$ (b). Stippling indicates the region of decrease. Units are in 10^{-3} J cm^{-2} mb^{-1}.

result is consistent with the change in the distribution of precipitation described in Section 4c. The increase in rainfall results from intensification of the Hadley cell which increases the poleward transport of heat energy.

2) Latent energy

In middle latitudes of the model, the poleward transport of latent energy (Fig. 12c) increases slightly due to the doubling of CO$_2$. This increase takes place despite the reduction of lower tropospheric eddy kinetic energy mentioned above.

The general increase of the water vapor in the atmosphere associated with the rise of temperature may be responsible for this result.

In low latitudes, the equatorward transport of latent heat for the 2×CO$_2$ case is larger than that for the standard case. Again, the intensification of the Hadley cell mentioned above is mainly responsible for this difference.

3) Total energy

Fig. 12a indicates that the increase of the concentration of CO$_2$ results in reduction of the poleward

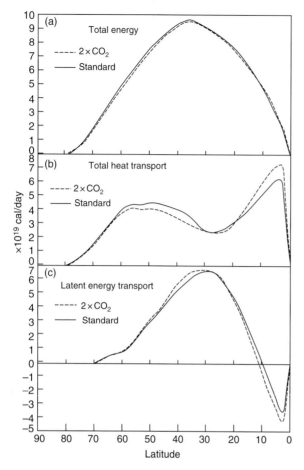

Fig. 12. Poleward transport of total energy ($C_pT+\phi+K+Lr$), a., poleward transport of heat energy ($C_pT+\phi+K$), b., and poleward transport of latent energy (Lr), c.

transport of total energy in the model atmosphere over most of the latitudes with the exception of the tropics. However, the magnitude of the change is very small because the change in heat transport is mostly offset by the change in the transport of latent heat.

5. SUMMARY AND CONCLUSIONS

In this study, an attempt is made to analyze the effect of doubling the CO_2 concentration in a highly simplified three-dimensional general circulation model. The main results of this study may be summarized as follows:

1) In general, the temperature of the model troposphere increases resulting from the doubling of the concentration of carbon dioxide. This warming in higher latitudes is magnified two to three times the overall amount due to the

effects of snow cover feedback and the suppression of the vertical mixing by a stable stratification. Because of the snow cover feedback mechanism, the overall sensitivity of the three- dimensional model is found to be significantly larger than that of the radiative-convective equilibrium model.

2) The temperature of the model stratosphere decreases because of the larger emission from the stratosphere into space caused by the greater concentration of CO_2. The magnitude of the cooling increases with increasing altitude. Hence, the static stability in the stratosphere tends to decrease.

3) A more active hydrologic cycle is obtained as indicated by the greater rates of total precipitation and evaporation computed in this study. This is due not only to the increase in the flux of downward radiation at the earth's surface, but also to the decrease in the Bowen ratio, resulting from the increase in the saturation vapor pressure for surface temperature. Also obtained is a poleward displacement of snow cover, which is consistent with the increase of surface temperature in high latitudes.

4) An analysis of the eddy kinetic energy indicates an overall reduction of this quantity in the lower troposphere, particularly in middle latitudes. This layer of decrease approximately coincides with the layer of reduced meridional temperature gradient. Conversely, there is an increase of eddy kinetic energy in the upper troposphere and lower stratosphere which corresponds approximately with the height at which the static stability decreases.

In evaluating these results, one should recall that the current study is based upon a model with a fixed cloudiness. As pointed out in Section 4, the results may be altered significantly if we used a model with the capability of predicting cloudiness. Other major characteristics of the model which can affect the sensitivities of the model climate are idealized geography, swamp ocean, and no seasonal variation.

Because of the various simplifications of the model described above, it is not advisable to take too seriously the quantitative aspect of the results obtained in this study. Nevertheless, it is hoped that this study not only emphasizes some of the important mechanisms which control the response of the climate to the change of carbon dioxide, but also identifies the various requirements that have to be satisfied for the study of climate sensitivity with a general circulation model.

Acknowledgments. It is a pleasure to acknowledge J. Smagorinsky, the Director of the Geophysical Fluid Dynamics Laboratory, who has given wholehearted support and encouragement during the course of this research. The very useful comments of I. Held, M. Suarez, A. Oort, C. Leith and M. MacCracken are appreciated. Finally, we would like to thank E. D'Amico, E. Groch, M. Stern and P. Tunison for assisting in the preparation of the manuscript.

APPENDIX A

Adjustment of the Radiation Model

Recently, we have suspected that the model of radiative transfer which is being used at the Geophysical Fluid Dynamics Laboratory (Manabe and Strickler, 1964; Manabe and Wetherald, 1967) has a systematic bias. Incorporated into a general circulation model, it tends to yield a model atmosphere which is significantly colder than the actual atmosphere. Such a bias does not cause serious difficulty in a model with a given distribution of sea-surface temperature as a lower boundary condition because the ocean is assumed to have an infinite heat capacity. However, it can yield unrealistic temperatures for a model in which the temperature of the earth's surface is computed explicitly. In order to eliminate this bias in the present study, the radiative model is adjusted such that the area-mean net flux of radiation at the top of the atmosphere is zero *given the observed distributions of temperature and atmospheric absorbers.* The data used for this adjustment test are taken from the following studies: zonal mean temperature, zonal mean humidity, and average cloudiness from London (1957); and surface albedo from Posey and Clapp (1964). Adopting these data, the net upward flux of radiation at the top of the atmosphere is computed for the four seasons. According to this computation, the model yields a net radiative cooling of the atmosphere as a whole, which confirms our earlier suspicions. In order to obtain a satisfactory balance, the following adjustments are performed even though these adjustments resulted in the choice of less realistic values for the relevant parameters.

(i) The contribution of molecular scattering to the planetary albedo is assumed to be 6% instead of 7%.

(ii) The reflectivity of low cloud to solar radiation is assumed to be 63% instead of 69%.

(iii) High cloud or cirrus is assumed to be fully black with regard to longwave radiation.

When these adjustments are applied, the net radiative flux[10] at the top of the atmosphere becomes -0.0003 ly min^{-1} which is negligible for most practical purposes. This adjusted model is used for all numerical experiments carried out in this study.

APPENDIX B

Reduction of Bias in a Quasi-Equilibrium State

It is stated in Section 2 that, in order to reduce the bias in the equilibrium climate resulting from a particular choice of initial conditions, the experiments are rerun starting from a considerably different initial condition. This second initial condition is illustrated schematically in Fig. B1.

Let $\mathbf{M}_1$ be the entire matrix of fields representing the state of the model atmosphere on the 40th day of the first run. A new matrix of fields $\mathbf{M}_1'$ is then generated such that $\mathbf{M}_1' = 2\mathbf{M}_2 - \mathbf{M}_1$, or $\mathbf{M}_1' - \mathbf{M}_2 = -(\mathbf{M}_1 - \mathbf{M}_2)$, where $\mathbf{M}_2$ represents the final state obtained from the first run. This $\mathbf{M}_1'$ matrix is then used to start the second run. To economize on computer time, we selected day 40 rather than day 1 of the first run as a point in model time for $\mathbf{M}_1$.

A measure of the degree of convergence discussed in Section 2 is the time variation of mean temperature for the two runs corresponding to the standard case (Fig. 3). This convergence can be further illustrated by the distribution of zonal mean temperature difference between the two standard runs averaged over the last 100 days of each experiment, shown in Fig. B2. This figure indicates that the absolute magnitude of the

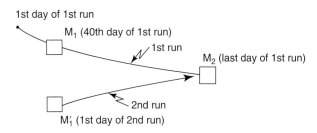

1st day of 1st run

M₁ (40th day of 1st run)

1st run

M₂ (last day of 1st run)

2nd run

M₁′ (1st day of 2nd run)

Fig. B1. Diagram of the procedure used to generate the initial condition for the second run in each case. See main text for further explanation.

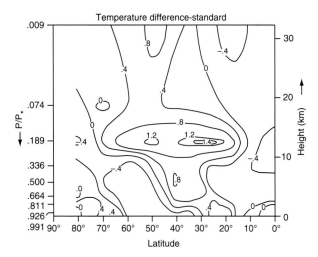

Fig. B2. Latitude-height distribution of the zonal mean temperature difference (K) between the two standard runs.

difference is generally less than 1 K except for a small region in the lower stratosphere. These differences were considered to be small enough to allow a sensitivity analysis to be performed. Similar graphs are obtained for the $2 \times CO_2$ case, but they differ little, qualitatively, from Figs. 3 and B2 and are, therefore, omitted.

It should be mentioned here that the time integration from the second initial condition may lead to a quasi-equilibrium state which is considerably different from the one obtained from the first integration. Such a situation did not occur during the course of the present effort. However, there are indications that the general circulation model used here has more than one stable equilibrium climate for a given set of external parameters, such as the solar constant (see, for example, Budyko, 1969). This subject will be discussed in more detail in a future study as it is beyond the scope of the present investigation.

NOTES

1   A brief description of this study appears on pp. 238–239 of the Study of Man's Impact on Climate (SMIC, 1971).
2   The model of radiative transfer is adjusted slightly such that it does not contain a systematic bias. For more details, see Appendix A.
3   According to Raschke *et al.* (1973), the annual zonal mean planetary albedo near the North Pole does not greatly exceed 50% as deduced from Nimbus 3 satellite data. A surface albedo of 70% according to our calculations at higher latitudes and average cloudiness yields a planetary albedo considerably greater than this or approximately 65%. On the other hand, a surface albedo of 45% under the same conditions yields a

planetary albedo of about 55% which is in better agreement with the observations cited above. See Section 4d for further discussion of this subject.
4   See, for example, Fig. 5.1 of Manabe *et al.* (1965).
5   The modified version of the Rodgers-Walshaw model has the following characteristics: 1) it uses the Curtis-Godson 2-parameter approximation instead of scaling approximation; and 2) it computes the contribution of water vapor bands by subdividing them into 19 subintervals and that of the CO_2 15-μm band by subdividing it into 4 subintervals.
6   See the bottom half of Fig. 7.3 in the paper by Manabe *et al.* (1965).
7   Note that, in both cases, the latitudinal range of the area of snowfall is wider than that of the area of significant snow accumulation. This difference is caused by the effects of snowmelt and sublimation.
8   Bowen ratio = $(C_p/L)(T_* - T_a)/(r_* - r_a)$, where the asterisk and a subscripts indicate earth's surface and lowest prognostic level, respectively, C_p the specific heat of air, L the latent heat of evaporation, T the temperature, and r the mixing ratio of water vapor.
9   Here, eddy kinetic energy is defined as the kinetic energy of the component of wind which represents the deviation from the zonal mean wind vector.
10 Net downward flux of solar radiation minus net upward flux of terrestrial radiation.

REFERENCES

Budyko, M. I., 1956: *The Heat Balance of the Earth's Surface.* 255 pp. (English transl., N. A. Stepanova, 1958, OTS.)

——, 1963: *Atlas of the Heat Balance of the Earth.* Moscow, Gl. Geofiz. Observ., 69 pp.

——, 1969: The effect of solar radiation variations on the climate of the Earth. *Tellus,* **21,** 611–619.

Callendar, G. S., 1949: Can carbon dixoide influence climate? *Weather,* **4,** 310–314.

Kaplan, L. D., 1960: The influence of carbon dioxide variation on the atmospheric heat balance. *Tellus,* **12,** 204–208.

Kondratiev, K. Ya., and H. I. Niilisk, 1960: On the question of carbon dioxide heat radiation in the atmosphere. *Geofis. Pura Appli.,* **46,** 216–230.

London, J., 1957: A study of the atmospheric heat balance. Final Report, Contract AF19(122)-165, New York University, 99 pp.

Machta, L., 1971: The role of the oceans and the biosphere in the carbon dioxide cycle. Nobel Symposium 20, Gothenburg, Sweden.

Manabe, S., 1969: Climate and the ocean circulation: I. The atmospheric circulation and the hydrology of the earth's surface. *Mon. Wea. Rev.,* **97,** 739–774.

——, 1971: Estimates of future change of climate due to the increase of carbon dioxide concentration in the air. *Man's Impact on the Climate,* W. H. Matthews, W. W. Kellogg, and G. D. Robinson, Eds., The MIT Press, 249–264.

——, J. Smagorinsky, J. L. Holloway, Jr., and H. M. Stone, 1970: Simulated climatology of a general circulation model with a hydrologic cycle: III. Effects of increased horizontal computational resolution. *Mon. Wea. Rev.,* **98,** 175–212.

——, ——and R. F. Strickler, 1965: Simulated climatology of a general circulation model with a hydrologic cycle. *Mon. Wea. Rev.*, **93**, 769–798.

——, and R. F. Strickler, 1964: Thermal equilibrium of the atmosphere with a convective adjustment. *J. Atmos. Sci.*, **21**, 361–385.

——, and R. T. Wetherald, 1967: Thermal equilibrium of the atmosphere with a given distribution of relative humidity. *J. Atmos. Sci.*, **24**, 241–259.

Möller, F., 1963: On the influence of changes in the CO_2 concentration in air on the radiation balance of the earth's surface and on the climate. *J. Geophys. Res.*, **68**, 3877–3886.

Oort, A. H., and E. M. Rasmusson, 1971: Atmospheric circulation statistics. NOAA Prof. Paper 5, 323 pp.

Plass, G. N., 1956: The carbon dioxide theory of climatic change. *Tellus*, **8**, 140–154.

Posey, J. W., and P. F. Clapp, 1964: Global distribution of normal surface albedo. *Geofis. Intern.*, **4**, 33–48.

Raschke, E., T. H. Vonder Haar, W. R. Bandeen and M. Pasternak, 1973: The annual radiation balance of the earth-atmosphere system during 1969–70 from Nimbus 3 measurements. *J. Atmos. Sci.*, **30**, 341–364.

Rooasl, S. I., and S. H. Schneider, 1971: Atmospheric carbon dixoide and aerosols: Effects of large increases on global climate. *Science*, **173**, 138–141.

Rodgers, C. D., and C. D. Walshaw, 1966: The computation of infra-red cooling rate in planetary atmospheres. *Quart. J. Roy. Meteor. Soc.*, **92**, 67–92.

Schneider, S. H., 1972: Cloudiness as a global climatic feedback mechanism: The effects on the radiation balance and surface temperature of variations in cloudiness. *J. Atmos. Sci.*, **29**, 1413–1422.

Sellers, W. D., 1969: A global climatic model based on the energy balance of the earth-atmosphere system. *J. Appl. Meteor.*, **8**, 392–400.

Smagorinsky, J., 1960: On the dynamical prediction of large-scale condensation by numerical methods. *Physics of Precipitation*, Monog. No. 5, Amer. Geophys. Union, 71–78.

SMIC, 1971: *Inadvertent Climate Modification.* Report of the Study of Man's Impact on Climate, The MIT Press, 308 pp.

Stone, H. M., and S. Manabe, 1968: Comparison among various numerical models designed for computing infrared cooling. *Mon. Wea. Rev.*, **96**, 735–741.

Climate Sensitivity: Analysis of Feedback Mechanisms

J. HANSEN, A. LACIS, D. RIND, G. RUSSELL

NASA/Goddard Space Flight Center, Institute for Space Studies 2880 Broadway, New York, NY 10025

P. STONE

Center for Meteorology and Physical Oceanography Massachusetts Institute of Technology, Cambridge, MA 02139

I. FUNG

Lamont-Doherty Geological Observatory of Columbia University Palisades, NY 10964

R. RUEDY, J. LERNER

M/A COM Sigma Data, Inc., 2880 Broadway, New York, NY 10025

ABSTRACT

We study climate sensitivity and feedback processes in three independent ways: (1) by using a three dimensional (3-D) global climate model for experiments in which solar irradiance S_o is increased 2 percent or CO_2 is doubled, (2) by using the CLIMAP climate boundary conditions to analyze the contributions of different physical processes to the cooling of the last ice age (18K years ago), and (3) by using estimated changes in global temperature and the abundance of atmospheric greenhouse gases to deduce an empirical climate sensitivity for the period 1850–1980.

Our 3-D global climate model yields a warming of ~4°C for either a 2 percent increase of S_o or doubled CO_2. This indicates a net feedback factor of f = 3–4, because either of these forcings would cause the earth's surface temperature to warm 1.2–1.3°C to restore radiative balance with space, if other factors remained unchanged. Principal positive feedback processes in the model are changes in atmospheric water vapor, clouds and snow/ice cover. Feedback factors calculated for these processes, with atmospheric dynamical feedbacks implicitly incorporated, are respectively $f_{water\ vapor}$ ~ 1.6, f_{clouds} ~ 1.3 and $f_{snow/ice}$ ~ 1.1, with the latter mainly caused by sea ice changes. A number of potential feedbacks, such as land ice cover, vegetation cover and ocean heat transport were held fixed in these experiments.

We calculate land ice, sea ice and vegetation feedbacks for the 18K climate to be $f_{land\ ice}$ ~ 1.2–1.3, $f_{sea\ ice}$ ~ 1.2, and $f_{vegetation}$ ~ 1.05–1.1 from their effect on the radiation budget at the top of the atmosphere. This sea ice feedback at 18K is consistent with the smaller $f_{snow/ice}$ ~ 1.1 in the S_o and CO_2 experiments, which applied to a warmer earth with less sea ice. We also obtain an empirical estimate of f = 2–4 for the fast feedback processes (water vapor, clouds, sea ice) operating on 10–100 year time scales by comparing the cooling due to slow or specified changes (land ice, CO_2, vegetation) to the total cooling at 18K.

The temperature increase believed to have occurred in the past 130 years (approximately 0.5°C) is also found to imply a climate sensitivity of 2.5–5°C for doubled CO_2 (f = 2–4), if (1) the temperature increase is due to the added greenhouse gases, (2) the 1850 CO_2 abundance was 270±10 ppm, and (3) the heat perturbation is mixed like a passive tracer in the ocean with vertical mixing coefficient k ~ 1 cm^2 s^{-1}.

These analyses indicate that f is substantially greater than unity on all time scales. Our best estimate for the current climate due to processes operating on the 10–100 year time scale is f = 2–4, corresponding to a climate sensitivity of 2.5–5°C for doubled CO_2. The physical process contributing the greatest uncertainty to f on this time scale appears to be the cloud feedback.

Hansen, J. *et al.* (1984) Climate sensitivity: analysis of feedback mechanisms. In: *Climate Processes and Climate Sensitivity*, AGU Geophysical Monograph 29, Maurice Ewing Volume 5, J. E. Hansen & T. Takahashi (eds). pp. 130–163. Copyright © 1984 American Geophysical Union. Reproduced by permission of American Geophysical Union.

We show that the ocean's thermal relaxation time depends strongly on f. The e-folding time constant for response of the isolated ocean mixed layer is about 15 years, for the estimated value of f. This time is sufficiently long to allow substantial heat exchange between the mixed layer and deeper layers. For f = 3–4 the response time of the surface temperature to a heating perturbation is of order 100 years, if the perturbation is sufficiently small that it does not alter the rate of heat exchange with the deeper ocean.

The climate sensitivity we have inferred is larger than that stated in the Carbon Dioxide Assessment Committee report (CDAC, 1983). Their result is based on the empirical temperature increase in the past 130 years, but their analysis did not account for *the dependence of the ocean response time on climate sensitivity*. Their choice of a fixed 15 year response time biased their result to low sensitivities.

We infer that, because of recent increases in atmospheric CO_2 and trace gases, there is a large, rapidly growing gap between current climate and the equilibrium climate for current atmospheric composition. Based on the climate sensitivity we have estimated, the amount of greenhouse gases *presently in the atmosphere* will cause an eventual global mean warming of about 1°C, making the global temperature at least comparable to that of the Altithermal, the warmest period in the past 100,000 years. Projection of future climate trends on the 10–100 year time scale depends crucially upon improved understanding of ocean dynamics, particularly upon how ocean mixing will respond to climate change at the ocean surface.

INTRODUCTION

Over a sufficient length of time, discussed below, thermal radiation from the earth must balance absorbed solar radiation. This energy balance requirement defines the effective radiating temperature of the earth, T_e, from

$$\pi R^2 (1 - A) S_o = 4\pi R^2 \sigma T_e^4 \qquad (1)$$

or

$$T_e = [S_o (1 - A) / 4\sigma]^{1/4} = (s / \sigma)^{1/4} \qquad (2)$$

where R is the earth radius, A the earth albedo, S_o the solar irradiance, s the mean flux of absorbed solar radiation per unit area and σ the Stefan-Boltzmann constant. Since A ~ 0.3 and S_o ~ 1367 W m^{-2}, s ~ 239 W m^{-2} and this requirement of energy balance yields T_e ~ 255K. The effective radiating temperature is also the physical temperature at an appropriately defined mean level of emission to space. In the earth's atmosphere this mean level of emission to space is at altitude H ~ 6 km. Since the mean tropospheric temperature gradient is ~5.5°C km^{-1}, the surface temperature is T ~ 288K, ~33K warmer than T_e.

It is apparent from (2) that for changes of solar irradiance

$$\frac{dT_e}{T_e} = \frac{1}{4} \frac{dS_o}{S_o} = \frac{1}{4} \frac{ds}{s}. \qquad (3)$$

Thus if S_o increases by a small percentage δ, T_e increases by $\delta/4$. For example, a 2 percent change in solar irradiance would change T_e by about 0.5 percent, or 1.2–1.3°C. If the atmospheric temperature structure and all other factors remained fixed, the surface temperature would increase by the same amount as T_e. Of course all factors are not fixed, and we therefore define the net feedback factor, f, by

$$\Delta T_{eq} = f \, \Delta T_o \qquad (4)$$

where ΔT_{eq} is the equilibrium change of global mean surface air temperature and ΔT_o is the change of surface temperature that would be required to restore radiative equilibrium if no feedbacks occurred.

We use procedures and terminology of feedback studies in electronics (Bode, 1945) to help analyze the contributions of different feedback processes. We define the system gain as the ratio of the net feedback portion of the temperature change to the total temperature change

$$g = \frac{\Delta T_{\text{feedbacks}}}{\Delta T_{eq}}. \qquad (5)$$

Since

$$\Delta T_{eq} = \Delta T_o + \Delta T_{\text{feedbacks}}, \qquad (6)$$

it follows that the relation between the feedback factor and gain is

$$f = \frac{1}{1-g}. \qquad (7)$$

In general a number of physical processes contribute to f, and it is common to associate a feedback factor f_i with a given process i, where f_i is the feedback factor which would exist if all other feedbacks were inoperative. If it is assumed that the feedbacks are independent, feedback contributions to the temperature change can be separated into portions identifiable with individual feedbacks,

$$\Delta T_{feedbacks} = \sum_i \Delta T_i, \qquad (8)$$

with

$$g = \sum_i \frac{\Delta T_i}{\Delta T_{eq}} = \sum_i g_i \qquad (9)$$

and

$$\Delta T_{eq} = \frac{1}{1 - \sum_i g_i} \Delta T_o. \qquad (10)$$

It follows that two feedback gains combine linearly as

$$g = g_1 + g_2, \qquad (11)$$

but the feedback *factors* combine as

$$f = \frac{f_1 f_2}{f_1 + f_2 - f_1 f_2}. \qquad (12)$$

Thus even when feedback processes are linear and independent the feedback factors are not multiplicative. For example, a feedback process with gain $g_i = 1/3$ operating by itself would cause a 50 percent increase in ΔT_{eq} compared to the no feedback radiative response, i.e., $f_i = 1.5$. If a second feedback process of the same strength is also operating, the net feedback is f = 3 (not 2.25). One implication is that, if strong positive feedback exists, a moderate additional positive feedback may cause a large increase in the net feedback factor and thus in climate sensitivity.

The feedback factor f provides an intuitive quantification of the strength of feedbacks and a convenient way to describe the effect of feedbacks on the transient climate response. The gain g allows clear comparison of the contributions of different mechanisms to total climate change. The above formalism relates f and g and provides a framework for analyzing feedback interactions and climate sensitivity.

A number of physical mechanisms have been identified as causing significant climate feedback (Kellogg and Schneider, 1974). As examples, we mention two of these mechanisms here. Water vapor feedback arises from the ability of the atmosphere to hold more water vapor as temperature increases. The added water vapor increases the infrared opacity of the atmosphere, raising the mean level of infrared emission to space to greater altitude, where it is colder. Because the planetary radiation to space temporarily does not balance absorbed solar energy, the planet must warm to restore energy balance; thus $f_W > 1$ and $g_W > 0$, a condition described as a positive feedback. Ice/snow feedback is also positive; it operates by increasing the amount of solar energy absorbed by the planet as ice melts.

Feedback analyses will be most useful if the feedback factors are independent to first order of the nature of the radiative forcing (at the top of the atmosphere). The similar model responses we obtain in our S_o and CO_2 experiments tend to corroborate this possibility, although there are some significant differences in the feedbacks for solar and CO_2 forcings. We expect the strength of feedbacks to have some dependence on the initial climate state and thus on the magnitude of the climate forcing; for example, the ice/snow albedo feedback is expected to change with climate as the cryospheric region grows or shrinks.

We examine feedback processes quantitatively in the following sections by means of 3-D climate model simulations and analysis of conditions during the last ice age (18K years ago). The 3-D experiments include doubling CO_2 and increasing S_o by 2 percent, forcings of roughly equal magnitude which have also been employed by Manabe and Wetherald (1975) and Wetherald and Manabe (1975). 18K simulations with a 3-D general circulation model have previously been performed by Williams *et al.* (1974), Gates (1976) and Manabe and Hahn (1977).

THREE-DIMENSIONAL CLIMATE MODEL

The global climate model we employ is described and its abilities and limitations for simulating today's climate are documented as model II (Hansen *et al.*, 1983b, hereafter referred to as paper 1). We note here only that the model solves the simultaneous equations for conservation of energy, momentum, mass and water and the equation of state on a coarse grid with horizontal resolution 8° latitude by 10° longitude and with 9 atmospheric layers. The radiation includes the radiatively significant atmospheric gases, aerosols and cloud particles. Cloud cover and height are computed. The diurnal and seasonal cycles are included. The ground hydrology and surface albedo depend upon the local vegetation. Snow depth is computed and snow albedo includes effects of snow age and masking by vegetation.

Ocean temperatures and ice cover are specified climatologically in the documented model II. In the experiments described here, ocean temperatures and ice cover are computed based on energy exchange with the atmosphere, ocean heat transport, and the ocean mixed layer heat capacity. The latter two are specified, but vary seasonally at each gridpoint. Monthly mixed layer depths are climatological, compiled from NODC mechanical bathythermograph data (NOAA, 1974) and from temperature and salinity profiles in the southern ocean (Gordon, 1982). The resulting global-mean seasonal-maximum mixed layer depth is 110 m. In our 3-D experiments a 65 m maximum is imposed on the mixed layer depth to minimize computer time; this yields a global-mean seasonal-maximum mixed layer depth of 63 m. The 65 m maximum depth is sufficient to make the mixed layer thermal response time much greater than one year and provide a realistic representation of seasonal temperature variations, so the mixed layer depth limitation should not significantly affect the modeled equilibrium climate.

The ocean heat transport was obtained from the divergence of heat implied by energy conservation at each ocean gridpoint in the documented model II (paper 1), using the specified mixed layer depths. The geographical distribution of the resulting annual mean heat flux into and out of the ocean surface is shown in Fig. 1a; averaged over the entire hemispheres, it yields 2.4 W m^{-2} into the Southern Hemisphere surface and an equal amount out of the Northern Hemisphere. The gross characteristics of the ocean surface heating and implied ocean heat transport appear to be realistic, with heat input at low latitudes, especially in regions of upwelling cold water, and release at high latitudes, especially in regions of poleward currents. Fig. 15 of paper 1 shows that the longitude-integrated heat transport is consistent with available knowledge of actual transports. A more comprehensive comparison with observations has been made by Miller et al. (1983), who show that the implied annual northward heat flux

at the equator is 6.2 × 10^{14} W. With the ocean heat transport specified in this manner, the control run with computed ocean temperature has a simulated climate nearly the same as the documented model II. It is not identical, as a result of changes in the sea ice coverage which arise when the sea ice is a computed quantity. There is 15 percent less sea ice in the standard control run with computed ocean temperature than in the documented model II, as discussed below. This has local effects, mainly around Antarctica, but otherwise simulated quantities are practically identical to the documented model II climatology.

In our experiments with changed solar irradiance and atmospheric CO_2, we keep the ocean heat transport identical to that in the control run. Thus no ocean transport feedback is permitted in these experiments. Our rationale for this approach as a first step is its simplicity for analysis, and the fact that it permits a realistic atmospheric simulation.

Ocean ice cover is also computed in the experiments described here on the basis of the local heat balance. When the ocean surface loses heat, the mixed layer temperature decreases as far as the freezing point of ocean water, −1.6°C. Further heat loss from the open ocean causes ice to grow horizontally with thickness 1 m until the gridbox is covered up to the limit set by the prescription for leads (open water). Still further heat loss causes the ice to thicken. Leads are crudely represented by requiring the fraction of open water in a gridbox to be greater than or equal to $0.1/z_{ice}$, where z_{ice} is the ice thickness in meters (paper 1).

Heat exchange between ocean ice and the mixed layer occurs by conduction in the climate model. A two-slab model is used for ice, with the temperature profile parabolic in each slab. This conduction is inefficient, and, if it were the only mechanism for heat exchange between the mixed layer and the ice, it would at times result in ocean ice coexisting with ocean water far above the freezing point; since this does not occur in nature, other mechanisms (such as lateral heat exchange) must contribute to the heat exchange. Therefore in our standard control run and S_o and CO_2 experiments we impose the condition that the mixed layer temperature, which represents a mean for an 8° × 10° gridbox, not be allowed to exceed 0°C until all the ice in the gridbox is melted; i.e., if the mixed layer temperature reaches 0°C additional heat input is used to meltice, decreasing its horizontal extent within the gridbox.

The annual mean sea ice cover in out standard control run is shown in Fig. 2b. Evidently there is

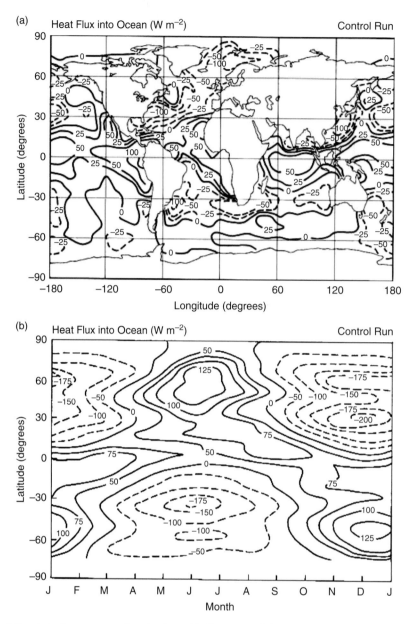

Fig. 1. Specified heat flux into the ocean surface in the 3–D climate model experiments, obtained from the model II run of paper 1 which had specified climatological seasonally-varying ocean surface temperature and ocean ice cover. (a) is the geographical distribution of the annual-mean flux. (b) is the latitude/season distribution of the zonal-mean flux.

too little sea ice in the model (15 percent less than the observations of Fig. 2a), especially at longitudes ~100°W and ~50°E in the Southern Hemisphere. Thus we also produced an alternate control run by removing the condition that all heat added to the mixed layer be used to melt ice if the mixed layer temperature reaches 0°C. This alternate control run has about 23 percent greater ocean ice cover (Fig. 2c) than observed, and thus the standard and alternate control runs bracket observations. We use the alternate control run for a second doubled CO_2 experiment, as one means

of assessing the role of ocean ice in climate sensitivity.

In the following we first describe our standard S_0 and CO_2 experiments.

S_0 AND CO_2 EXPERIMENTS

S_0 was increased 2 percent and CO_2 was doubled (from 315 ppm to 630 ppm) instantaneously on January 1 of year 1. Both experiments were run for 35 years. In this section we study the

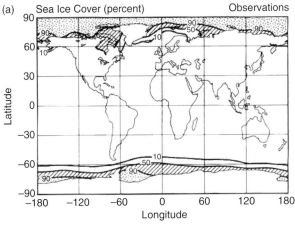

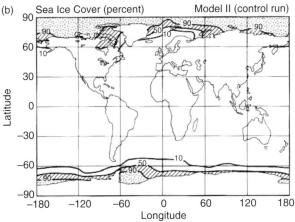

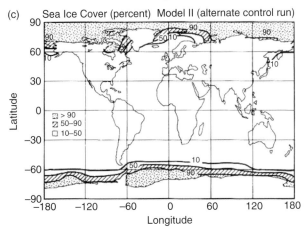

Fig. 2. Annual-mean sea ice cover. (a) observational climatology of Walsh and Johnson (1979) for the northern hemisphere and Alexander and Mobley (1976) for the southern hemisphere. (b) our standard control run of the 3-D climate model. (c) alternate control run, as described in the text.

equilibrium response of the climate model to the S_o and CO_2 forcings. The time dependence of the surface air temperature and the heat flux into the planetary surface are briefly noted, but only to verify that equilibrium has been achieved.

The time dependence of these experiments is discussed in greater detail in a subsequent section concerned with the transient response of the climate system.

Global mean heat balance and temperature

Model II (paper 1) has a global annual mean net heat flux into the top of the atmosphere of $7.5\,W\,m^{-2}$ (~2 percent of the insolation). $2.5\,W\,m^{-2}$ of this imbalance is due to conversion of potential energy to kinetic energy (which is not reconverted to heat in the model) and computer truncation. The other global $5\,W\,m^{-2}$ is absorbed by the ocean and ocean ice, at a rate of $7.1\,W\,m^{-2}$ for the ocean surface area. This portion of the imbalance must be due to inaccuracies such as in the cloud properties, surface albedo, thermal emission calculations, etc. In our control run and experiments with computed ocean temperature we multiply the solar radiation absorbed at the ocean surface by the factor 0.96, which cancels the entire energy imbalance. The radiation correction factor has no appreciable direct effect on model sensitivity since all results are differenced against a control run; however, it does enable physical processes, such as condensation and ice melting, to operate at temperatures as realistic as possible. Together with the specified ocean transports, this allows the control run with computed ocean temperature to have essentially the same ocean temperature and climate as the model II run with fixed climatological ocean temperatures (paper 1).

The global mean heat flux into the planetary surface and surface air temperature are shown in Fig. 3 for the S_o and CO_2 experiments. The heat flux peaks at $\sim 3\,W\,m^{-2}$ for both experiments; the radiative imbalance at the top of the atmosphere is essentially the same as this flux into the planetary surface, since the heat capacity of the atmosphere is small. Similar fluxes are expected in the two experiments because of the similar magnitudes of the radiative forcings. The 2 percent S_o change corresponds to a forcing of $4.8\,W\,m^{-2}$. The initial radiative imbalance at the top of the atmosphere due to doubling CO_2 is only $\sim 2.5\,W\,m^{-2}$, but after CO_2 cools the stratosphere (within a few months) the global mean radiative forcing is about $4\,W\,m^{-2}$ (Fig. 4, Hansen et al., 1981). Over the ocean fraction of the globe we find a peak flux into the surface of $4–5\,W\,m^{-2}$ in both experiments, of order 10 percent greater than the global mean forcing for an all-ocean planet. Thus heating of the air over land with subsequent mixing by the atmosphere increases the

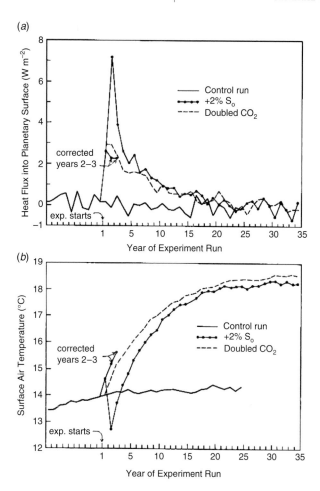

Fig. 3. Global net heat flux into planetary surface (a) and global surface air temperature (b). On April 1 of year 2 in the S_o experiment the computer was hit by a cosmic ray or some other disturbance which caused improper numbers to be stored in the ground temperature array. This affected the temporal development of that run, but should not influence its equilibrium results. In order to determine the maximum heat flux into the ocean, the S_o experiment was rerun for years 2 and 3 from March 31 year 2 thus eliminating the computer error for that period.

net heat flux into the ocean, but not by the ratio of global area to ocean area as assumed by Hansen *et al.* (1981). Apparently heating over continental areas is balanced substantially by increased cooling to space. A chief implication is that the time constant for the ocean to respond to global heating is longer than obtained from the common practice of averaging the ocean heat capacity over the entire globe (rather than over the ocean area).

The equilibrium global mean warming of the surface air is about 4°C in both the S_o and CO_2 experiments. This corresponds to a feedback factor $f = 3-4$, since the no-feedback temperature change required to restore radiative equilibrium with

space is $\Delta T_o = 1.2-1.3°C$. The heat flux and temperature approach their new equilibria with an e-folding time of almost a decade. We show in the section on transient climate response that the e-folding time is proportional to f, and that the value inferred from Fig. 3 is consistent with f = 3–4.

The mechanisms causing the global warmings in these experiments are investigated below, including presentation of the global distribution of key changes. These results are the means for years 26–35 of the control and experiment runs. Fig. 3 indicates that this should provide essentially the equilibrium response, since by that time the heat flux into the ocean is near zero and the temperature trend has flattened out.

Global temperature changes

The temperature changes in the S_o and CO_2 experiments are shown in Fig. 4 for the annual mean surface air temperature as a function of latitude and longitude, the zonal mean surface air temperature as a function of latitude and month, and the annual and zonal mean temperature as a function of altitude and latitude. We discuss the nature and causes of the temperature changes, and then make a more quantitative analysis below using 1-D calculations and the alternate CO_2 experiment with changed sea ice prescription.

The surface air warming is enhanced at high latitudes (Fig. 4, upper panel) partly due to the greater atmospheric stability there which tends to confine the warming to the lower troposphere, as shown by the radiation changes discussed below and the experiment with altered sea ice.

There is a very strong seasonal variation of the surface warming at high latitudes (Fig. 4, middle panel), due to the seasonal change of atmospheric stability and the influence of melting sea ice in the summer which limits the ocean temperature rise. At low latitudes the temperature increase is greatest in the upper troposphere (Fig. 4, lower panel), because the added heating at the surface primarily causes increased evaporation and moist convection, with deposition of latent heat and water vapor at high levels.

The statistical significance of these results can be verified from Fig. 5, which shows the standard deviation for the last 10 years of the control run for all the quantities in Fig. 4, and the ratio of the change of the quantity in the doubled CO_2 experiment to the standard deviation. The standard deviation is computed routinely for all of the quantities output from our 3-D model. We only

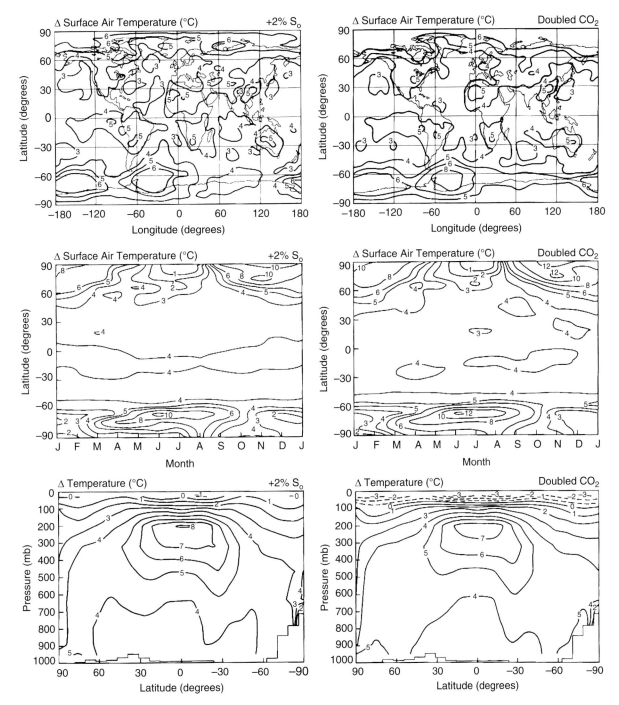

Fig. 4. Air temperature change in the climate model for a two percent increase of solar irradiance (left) and for doubled atmospheric CO_2 (right). The upper graphs show the geographical distribution of annual mean surface air warming, the middle graphs show the seasonal variation of the surface air warming averaged over longitude, and the lower graphs show the altitude distribution of the temperature change averaged over season and longitude.

discuss changes in the experiment runs which are far above the level of model fluctuations or 'noise' in the control run.

The patterns of temperature change are remarkably similar in the S_o and CO_2 experiments, sug-

gesting that the climate response is to first order a function of the magnitude of the radiative forcing. The only major difference is in the temperature change as a function of altitude; increased CO_2 causes substantial stratospheric cooling. This similarity

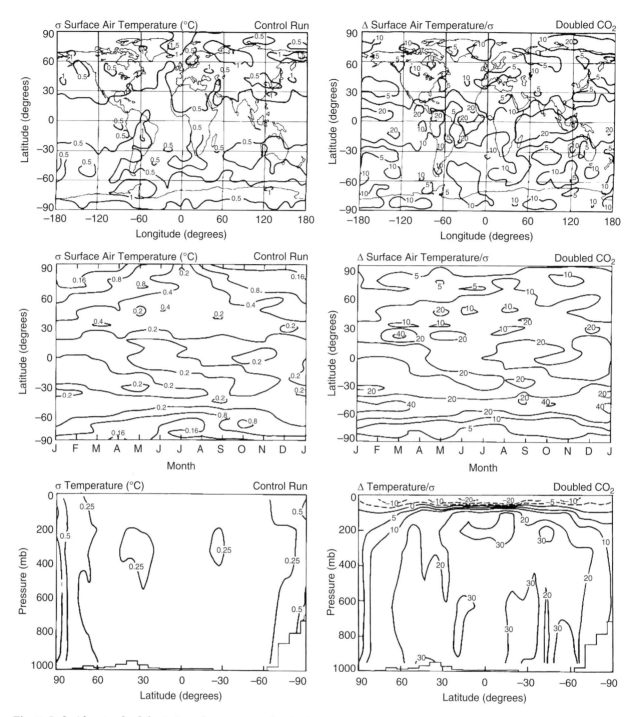

Fig. 5. Left side: standard deviation of temperature for the last 10 years in the control run. Right side: ratio of temperature change for years 26–35 of the doubled CO_2 experiment to the standard deviation of temperature in the control run.

suggests that, to first order, the climate effect due to several forcings including various tropospheric trace gases may be a simple function of the total forcing.

The global mean warming of surface air that we obtain for doubled CO_2 is similar to that obtained by Manabe and Stouffer (1980) for quadrupled

CO_2. This large difference in sensitivity of the two models appears to be associated mainly with the feedback mechanisms in the models, as discussed below. The patterns of the temperature changes in the two models show gross similarities, but also significant differences. We defer detailed comparison of the model results until after discussion of the feedback mechanisms.

1-D analysis of feedbacks in 3-D experiments

The processes chiefly responsible for the temperature rise in the 3-D model can be investigated with a 1-D radiative convective (RC) climate model. We use the 1-D model of Lacis *et al.* (1981) to evaluate the effect of changes in radiative forcing that take place in the 3-D model experiments. As part of the 3-D model diagnostics, we have available global average changes in surface and planetary albedo, and changes in amount and vertical distribution of clouds, water vapor and atmospheric lapse rate. We insert these changes one-by-one, or in combination, into the 1-D model and compute the change in global surface temperature. We employ the usual 'convective adjustment' procedure in our 1-D calculations, but with the global mean temperature profile of the 3-D model as the critical lapse rate in the troposphere. Contrary to usual practice, we allow no feedbacks to operate in the 1-D calculations, making it possible to associate surface temperature changes with individual feedback processes.

There is no a priori guarantee that the net effect of these changes will yield the same warming in the 1-D model as in the 3-D model, because simple global and annual averages of the changes do not account for the nonlinear nature of the physical processes and their 2-D and 3-D interactions. Also, changes in horizontal dynamical transports of heat and moisture are not entered explicitly into the 1-D model; the effects of dynamical feedbacks are included in the radiative factors which they influence, such as the cloud cover and moisture profile, but the dynamical contributions are not identified. Nevertheless, this exercise provides substantial information on climate feedbacks. Determination of how well the 1-D and 3-D results correspond also is a useful test for establishing the value of 1-D global climate models.

The procedure we use to quantify the feedbacks is as follows. The increase of total water vapor in the 3-D model (33 percent in the S_o experiment) is put in the 1-D model by multiplying the water vapor amount at all levels by the same factor (1.33); the resulting change in the equilibrium surface temperature of the 1-D model defines the second bars in Fig. 6. Next the water vapor at each level in the 1-D model is increased by the amount found in the 3-D experiment; the temperature change obtained in the first (total H_2O amount) test is subtracted from the

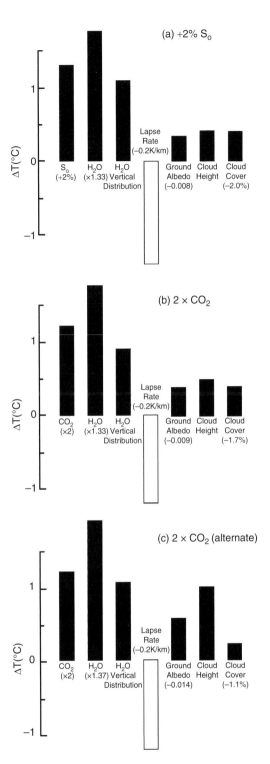

Fig. 6. Contributions to the global mean temperature rise in the S_o and CO_2 experiments as estimated by inserting changes obtained in the 3-D experiments into 1-D radiative convective model. (a) +2 percent S_o experiment, (b) doubled CO_2 experiment, and (c) doubled CO_2 experiment for alternate control run with greater sea ice.

temperature change obtained in this test to obtain the temperature change credited to the change in water vapor vertical distribution. The change of temperature gradient (lapse rate) between each pair of levels in the 3-D model is inserted in the control 1-D model to estimate the effect of lapse rate change on surface temperature, shown by the fourth bars in Fig. 6. Since the lapse rate changes are due mainly to changes of water vapor, we take the net of these three temperature changes in the 1-D model as our estimate of the water vapor contribution to the total temperature change. The global mean ground albedo change in the 3-D model (defined as the ratio of the global mean upward and downward solar radiation fluxes at the ground) is inserted into the 1-D control run to obtain our estimate of the ice/snow albedo contribution to the temperature change.

Cloud contributions are more difficult to analyze accurately because of the variety of cloud changes that occur in the 3-D model (see below), including changes in cloud overlap, and the fact that the changes do not combine linearly. We first estimate the total cloud impact by changing the cloud amounts at all levels in the 1-D model in proportion to changes obtained in the 3-D model. The total cloud effect on the temperature obtained in this way is subdivided by defining a portion to be due to the cloud cover change (by running the 1-D model with a uniform change of all clouds so as to match the total cloud cover change in the 3-D model) and by assigning the remainder of the total cloud effect to cloud height changes. These assumptions involve some arbitrariness. Nevertheless, the resulting total temperature changes in the 1-D model are found to be within $0.2°C$ of the global mean temperature changes in the 3-D experiments, providing circumstantial evidence that the procedure takes into account the essential radiative aspects of cloud cover change.

The temperature changes in the 1-D model are shown in Fig. 6 for the standard S_o and CO_2 experiments, and the CO_2 experiment with alternate sea ice computation. Resulting gains and feedback factors are given in Table 1.

Water vapor feedback

Water vapor provides the largest feedback, with most of it caused by the increase of water vapor amount. Additional positive feedback results from the water vapor distribution becoming weighted more to higher altitudes, but for the global and hemispheric means this is approximately cancelled by the negative feedback produced by the changes in lapse rate, also due mainly to the added H_2O. The near cancellation of these two effects is not surprising, since the amount of water the atmosphere holds is largely dependent on the mean temperature, and the temperature at which the infrared opacity occurs determines the infrared radiation. This tendency for cancellation suggests that the difficulty in modeling moist convection and the vertical distribution of water vapor may not have a great impact on estimates of global climate sensitivity (excluding the indirect effect on cloud distributions).

The net water vapor gain thus deduced from the 3-D model is $g_w \sim 0.4$, or a feedback factor $f_w \sim 1.6$. The same sensitivity for water vapor is obtained in 1-D models by using fixed relative humidity and fixed critical lapse rate (Manabe and Wetherald, 1967), thus providing some support for that set of assumptions in simple climate models. Relative humidity changed only slightly in our 3-D experiments; for example, in our standard doubled CO_2 experiment the average relative humidity increased 0.015 (1 = 100 percent humidity), with a 0.06 global increase at 200 mb being the largest change at any altitude. This compares with an increase of mean specific humidity of 33 percent. The global mean lapse rate change in the 3-D model ($-0.2°C\,km^{-1}$) is less than the change of the moist adiabatic lapse rate ($-0.5°C\,km^{-1}$), the decrease at low latitudes being partly offset by an increase at high latitudes. And, as explained above, the effect of the lapse rate change on temperature is largely balanced by the effect of the resulting displacement of water vapor to greater altitude.

Snow/ice feedback

Ground albedo decrease also provides a positive feedback. The ground albedo change (upper panel of Fig. 7) is largely due to reduced sea ice. Shielding of the ground by clouds and the atmosphere (middle panel of Fig. 7) makes this feedback several times smaller than it would be in the absence of the atmosphere. However, it is a significant positive feedback and is at least as large in the Southern Hemisphere as in the Northern Hemisphere. The geographic pattern of the temperature increase (Fig. 4) and the coincidence of warming maxima with reduced sea ice confirm that the sea ice effect is a substantial positive feedback.

Further insight into the sea ice feedback is provided by the experiment with alternate prescription

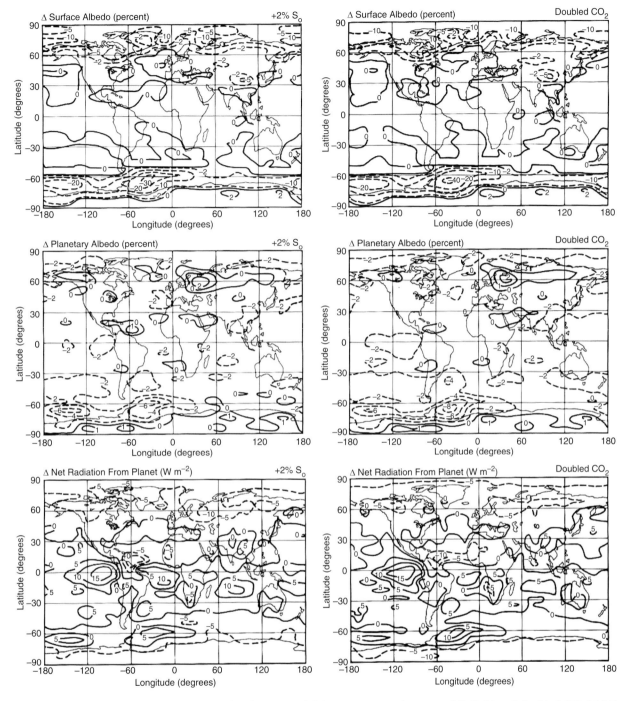

Fig. 7. Annual mean radiation changes in the climate model for two percent increase of S_o (left) and doubled CO_2 (right). In Figs. 7, 8, 10 and 11 "percent" change refers to the full range of the quantity, e.g., a change from 60 percent albedo to 50 percent albedo is defined to be a 10 percent change.

for computing sea ice cover. The greater sea ice cover in the control run for this experiment permits a greater surface albedo feedback, as indicated by the analysis with the 1-D model shown in Fig. 6c. These results illustrate the sensitivity of a system which already contains large positive feedbacks,

the gain due to increased surface albedo being augmented by in-water vapor and cloud gains.

Based on these experiments, we estimate the sea ice/snow feedback factor as ~1.1. However, this value refers to a climate change from today's climate to a climate which is warmer by about 4°C.

We expect $f_{snow/ice}$ to decrease as the area of sea ice and snow decreases, so its value is probably somewhat larger in the limit of a small increment about today's climate. Also, the prescription for computing sea ice in our standard experiments (which gives 15 percent too little sea ice for today's climate) probably causes an underestimate of $f_{snow/ice}$, as indicated by the value inferred for f in the experiment with altered sea ice prescription (which yielded 23 percent too much sea ice for today's climate).

The gain we obtain for ice/snow feedback in our 3-D model (~0.1) agrees well with the value (0.12) obtained by Wang and Stone (1980) from a 1-D radiative convective model. The feedback is much smaller than early estimates such as those of Budyko (1969) and Sellers (1969), who assigned a large albedo increment to ice/snow, but did not account for cloud shielding, vegetation masking of snow, and zenith angle variation of albedo (North, 1975; Lian and Cess, 1977).

Cloud feedback

Cloud changes (Fig. 8) also provide a significant positive feedback in this model, as a result of a small increase in mean cloud height and a small decrease in cloud cover. The gain we obtain for clouds is 0.22 in our standard doubled CO_2 experiment. This happens to be similar to the gain of 0.19 which is obtained in 1-D models if the cloud cover is kept fixed and the cloud height is determined by the assumption of fixed cloud temperature (Cess, 1974). However, a substantial part of the cloud gain in the 3-D model is due to the cloud cover change (Fig. 6). The portion of the cloud gain associated with cloud height change in the S_o experiment and the standard doubled CO_2 experiment is about midway between the two common assumptions used in simple climate models: fixed clouds altitude (gain = 0) and fixed cloud temperature (gain = 0.2).

The cloud height and cloud cover changes in the 3-D model seem qualitatively plausible. The reduced cloud cover primarily represents reduction of low and middle level clouds, due to increased vertical transport of moisture by convection and the large scale dynamics. The increase of high level cirrus clouds at low latitudes is consistent with the increase of penetrating moist convection at those latitudes. However, the cloud prescription scheme in the model (paper 1) is highly simplified; for example, it does not incorporate a liquid water budget for the cloud droplets or predict changes in cloud optical thickness at a given height. Thus the possibility of an increase in mean cloud optical thickness with the increased water vapor content of the atmosphere is excluded. Indeed, because the cloud optical thickness decreases with increasing altitude (paper 1), the increase of cloud height causes a decrease of optical thickness. This is a positive feedback for low and middle level clouds, but a negative feedback for cirrus clouds, which are a greenhouse material with suboptimal optical thickness. As a crude test of possible effects of changes in cloud optical thickness we let the cloud optical thickness in the 1-D model change in proportion to the absolute humidity: this practically eliminated the positive cloud feedback, i.e., it resulted in f_{clouds} ~ 1.0. Clearly, assessment of the cloud contribution to climate sensitivity depends crucially upon development of more realistic representation of cloud formation processes in climate models, as verified by an accurate global cloud climatology.

Summary of model feedbacks

Given the cancellation between the change in lapse rate and change in vertical distribution of water vapor, the processes providing the major radiative feedbacks in this climate model are total atmospheric water vapor, clouds and the surface albedo. Considering the earth from a planetary perspective, it seems likely that these are the principal feedbacks for the earth on a time scale of decades. The albedo of the planet for solar radiation is primarily determined by the clouds and surface, with the main variable component of the latter being the ice/snow cover. The thermal emission of the planet is primarily determined by the atmospheric water vapor and clouds. Thus the processes principally responsible for the earth's radiation balance and temperature are included in the 1-D model, and we have shown that these processes are the source of the primary radiative feedbacks in our 3-D model.

Table 1 summarizes the gains and feedback factors inferred from the changes which occurred in our 3-D model experiments, and the corresponding temperature changes for different combinations of these feedback processes. Note again that effects of dynamical feedbacks are implicitly included in these changes. The temperature changes illustrate the nonlinear way in which feedback processes combine [Eq. (9)]. For example, the ice/snow feedback adds about 1°C to the temperature response, but if the water vapor and cloud feedbacks did not exist the ice/snow feedback would add only a few tenths of a degree to the sensitivity. This nonlinear behavior is a result of the fact that when the ice/snow feedback occurs in the presence

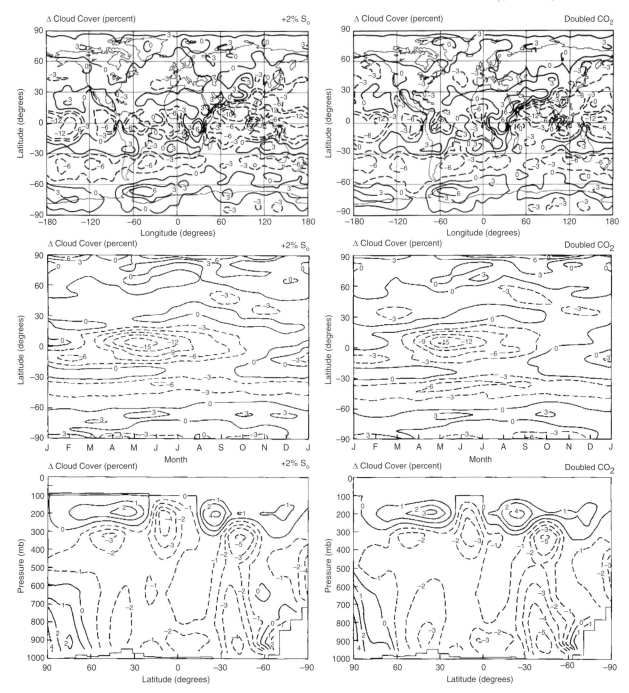

Fig. 8. Cloud changes in the climate model for two percent increase of S_o (left) and doubled CO_2 (right). The upper graphs show the geographical distribution of annual mean cloud cover change, the middle graphs show the seasonal variation of cloud cover change averaged over longitude, and the lower graphs show the altitude distribution of the cloud cover change averaged over season and longitude.

of the other (positive) feedbacks, it enhances the water vapor and cloud changes.

Comparison to Manabe and Stouffer

This analysis of the feedbacks in our model provides an indication of the causes of the difference

between our climate model sensitivity and that of Manabe and Stouffer (1980). They infer a warming of 2°C for doubled CO_2, based on an experiment with quadrupled CO_2 which yielded 4°C warming. Their model had fixed clouds (altitude and cloud cover), thus $f_{cloud} \equiv 1$. Also their control run

Table 1. Gain (g), feedback factor (f) and equilibrium temperature changes (ΔT) inferred from calculations with 1-D radiative-convective model for global mean changes in the 3-D GCM experiments. The subscripts w, c and s refer to water vapor, clouds and surface albedo. g is obtained as the ratio of the temperature change in the 1-D model (with only the indicated processes included) to the global mean temperature change in the 3-D experiment. f is from $f_i = 1/(1 - g_i)$. ΔT is the equilibrium surface air warming computed with the 1-D model for global mean changes of 3-D model constituents with only the indicated processes included; ΔT_o includes only the indicated radiative forcing, without feedbacks.

| | Experiment | | |
	+2% S_o	Doubled CO_2	Alternate Doubled CO_2
g_w	0.37	0.40	0.37
g_c	0.20	0.22	0.26
g_s	0.09	0.09	0.12
f_w	1.59	1.66	1.58
f_c	1.26	1.29	1.34
f_s	1.09	1.10	1.14
f_{wc}	2.36	2.62	2.67
f_{wcs}	2.96	3.45	3.95
ΔT_o	1.3	1.2	1.2
ΔT_{ow}	2.1	2.0	1.9
ΔT_{oc}	1.7	1.6	1.6
ΔT_{os}	1.5	1.3	1.4
ΔT_{owc}	3.2	3.2	3.2
ΔT_{owcs}	4.0	4.2	4.8

had less sea ice than our model, so their $f_{sea\ ice}$ should be between 1 and the value (~1.1) for our model. It is apparent from Table 1 that differences arising from the treatments of these two processes may account for most of the difference in global climate sensitivity.

Another major difference between our model and the model of Manabe and Stouffer is that we include a specified horizontal transport of heat by the ocean. This transport is identical in our control and experiment runs, i.e., the changed climate is not allowed to feed back on the ocean circulation. Of course Manabe and Stouffer do not allow feedback on ocean transport either, since the ocean transport is zero in both experiment and control. However, some other mechanisms must replace oceanic poleward transport of heat in their model, since their high latitude temperatures are at least as warm as in our model (and observations). Enhanced poleward transport of latent and sensible heat by the atmosphere must be the mechanism replacing ocean heat transport in their model. This atmospheric transport is expected to

provide a negative feedback (Stone, 1984), and indeed the total atmospheric energy transport did decrease poleward of 50°C latitude in the CO_2 experiments of Manabe and Wetherald (1975, 1980). It is not obvious whether the ocean transport feedback is positive or negative in the real world.

The contribution of ocean heat transport to climate sensitivity, like that of the atmospheric transports, does not appear as an identified component in an energy balance analysis such as in Fig. 6. This is irrelevant for our model, since it has no ocean transport feedback. However, in models which calculate the ocean heat transport or a surrogate energy transport, this feedback is included implicitly as a (positive or negative) portion of identified components of ΔT ($\Delta T_{water\ vapor}$, ΔT_{clouds}, $\Delta T_{ice/snow}$). The portion of these changes due to this feedback process could be identified by running those models with fixed (climatological) ocean heat transport.

Manabe (1983) suggests that our ice/snow feedback is unrealistically large and accounts for most of the difference between our climate model sensitivity and that of Manabe and Stouffer (1980). However, as summarized in Table 2, the amount of sea ice in the control run for our standard CO_2 experiment is actually somewhat less than observed sea ice cover. In our alternate CO_2 experiment, with sea ice cover greater than in observations, the ice/snow feedback increases significantly, suggesting that the ice/snow feedback in our standard experiment may be an underestimate. Also, we show in the next section that the sea ice feedback for the climate change from 18K to today, a warming of about 4°C, is about twice as large as in our doubled CO_2 experiments; this 18K sea ice feedback factor is based on measured changes of sea ice cover. The small ice/snow feedback in Manabe and Stouffer's model may be a result of their model being too warm at high latitudes; indeed, in the Southern Hemisphere (where the sea ice feedback is greatest in our model and in 18K measurements) their control run has almost no ice in the summer. Another likely reason for Manabe and Stouffer's albedo feedback being weaker is the stronger negative feedback in their meridional dynamical flux, as a result of that flux all being carried in the atmosphere. We conclude that our estimate for the sea ice feedback is conservative, i.e., it is more likely to be in error on the low side than on the high side.

We obtain a greater warming at low latitudes (~3–4°C for doubled CO_2) than that found by Manabe and Stouffer (~3°C for quadrupled CO_2). We analyzed the contributions to the warming in our 3-D model as a function of latitude by inserting

Table 2. Annual-mean sea ice cover as fraction of global or hemispheric area in several 3-D experiments. In run 1 the sea ice cover is specified to be today's climatology of Alexander and Mobley (1976) for the Southern Hemisphere and Walsh and Johnson (1979) for the Northern Hemisphere. Run 7 specifies the sea ice cover according to CLIMAP data for 18K (CLIMAP, 1981) and run 11 modifies the Southern Hemisphere CLIMAP data as discussed in the section on ice age experiments. In other runs the sea ice cover is computed.

Run	Experiment Description	Sea Ice Cover		
		Globe	Northern Hemisphere	Southern Hemisphere
1	Model II, Run 61 of paper 1 ; sea ice specified as today's climatology	0.048	0.042	0.054
2	Control run for standard CO_2 and S_0 experiments	0.041	0.039	0.043
3	Standard 2 × CO_2 experiment	0.023	0.028	0.017
4	Standard +2% S_0 experiment	0.025	0.030	0.020
5	Control run for alternate CO_2 experiment	0.060	0.046	0.073
6	Alternate 2 × CO_2 experiment	0.031	0.033	0.029
7	CLIMAP boundary conditions	0.089	0.048	0.131
11	CLIMAP boundary conditions with modified Southern Hemisphere sea ice	0.077	0.048	0.106

all zonal-mean radiative changes into the 1-D radiative-convective model. At low latitudes (0–30°) the clouds contribute a positive feedback of about 1–1.5°C; the larger part of this, nearly 1°C, is due to reduction of low level cloud cover with doubled CO_2, with increase of cirrus clouds contributing a smaller positive feedback. At high latitudes (60–90°) the clouds contribute a smaller negative feedback (0–1°C), due to increased low level clouds; this cloud increase (Fig. 8) probably is due to increased evaporation resulting from the reduced sea ice cover. The computed distributions of water vapor may also contribute to the difference between our result and that of Manabe and Stouffer. For example, in our model low latitude relative humidity at 200 mb increased by 0.085 with doubled CO_2. The cloud and water vapor characteristics depend on the modeling of moist convection and cloud formation; Manabe and Stouffer use the moist adiabatic adjustment of Manabe *et al.* (1965) and fixed clouds; we use a moist convection formulation which allows more penetrative convection (paper 1) and cloud formation dependent on local saturation. Presently available cloud climatology data has not permitted detailed evaluation of these moist convection and cloud formation schemes.

The high latitude enhancement of the warming is less in our model than in observed temperature trends for the past 100 years (Hansen *et al.*, 1983a). If this observed high latitude enhancement also occurs for large global temperature increases, the smaller high latitude enhancement in our 3-D model suggests the possibility that the 3-D model has either overestimated the low latitude climate

sensitivity (probably implicating the low latitude cloud feedback) or underestimated the high latitude sensitivity. If the former case is correct, the global climate sensitivity implied by the 3-D model may be only 2.5–3°C; but if the latter interpretation is correct, the global climate sensitivity may be greater than 4°C. A more precise statement requires the ability to analyze and verify the cloud feedback on a regional basis.

Conclusion

Atmospheric water vapor content provides a large positive feedback, and we find that in our model the effects of changes in lapse rate and water vapor vertical distribution largely cancel (for global or hemispheric means). The existence of the strong positive water vapor feedback implies that moderate additional positive feedback can greatly increase climate sensitivity, because of the nonlinear way in which feedbacks combine. In our model, sufficient ice/snow feedback occurs to increase the global sensitivity to ~2.5°C, and with cloud feedback to ~4°C for doubled CO_2. Although the cloud feedback is very uncertain, our 3-D study suggests that it is in the range from approximately neutral to strongly positive in global mean, and thus that global climate sensitivity is at least 2.5°C for doubled CO_2. The magnitudes of the global ice/snow and cloud feedbacks in our 3-D model are plausible, but confirmation requires improved ability to accurately model the physical processes as well as empirical tests of the climate model on a variety of time scales.

Records of past climate provide a valuable means to test our understanding of climate feedback mechanisms, even in the absence of a complete understanding of what caused the climate change. In this section we use the comprehensive reconstruction of the last ice age (18,000 years ago) compiled by the CLIMAP project (CLIMAP project members, McIntyre, project leader, 1981; Denton and Hughes, 1981). We first run our climate model with the 18K boundary conditions as specified by CLIMAP; this allows us to estimate the global mean temperature change between 18K and today. We then rerun the model changing feedback processes one-by-one and note their effect on the planetary radiation balance at the top of the atmosphere. This provides a measure of the gain or feedback factor for each process. We also examine the model for radiation balance when all of the known 18K feedbacks are included; this allows some inferences about the model sensitivity and the accuracy of the CLIMAP data. Finally, we compare different contributions to the 18K cooling; by considering the land ice and atmospheric CO_2 changes as slow or specified global climate forcings, we can infer empirically the net feedback factor for processes operative on 10–100 year time scales.

Global maps of the CLIMAP 18K boundary conditions, including the distributions of continental ice, sea ice and sea-surface temperature, are given by CLIMAP (1981) and Denton and Hughes (1981). These boundary conditions, obtained from evidence such as glacial scouring, ocean sediment cores containing detritus rafted by sea ice, and oxygen isotopic abundances in snowfall preserved in Greenland and Antarctic ice sheets, necessarily contain uncertainties. For example, Burckle et al. (1982) suggest that the CLIMAP Southern Hemisphere sea ice cover may be overestimated, and DiLabio and Klassen (1983) argue that the CLIMAP 'maximum extent' ice sheet model may be an overestimate. Questions have also been raised about the accuracy of the ocean surface temperatures, especially at low latitudes (Webster and Streten, 1978). We examine quantitatively the effect of each of these uncertainties on our feedback analyses.

Simulated 18K climate patterns

Our 18K simulation was obtained by running climate model II (paper 1) with the CLIMAP (1981) boundary conditions. The boundary conditions included the earth orbital parameters for that time (Berger, 1978). The run was extended for six years, with the results averaged over the last five years to define the 18K simulated climate. The control run was the five year run of model II with today's boundary conditions, which is documented in paper 1.

Temperature

Simulated 18K temperature patterns are shown in Fig. 9. The temperatures in the model, especially of surface air, are constrained strongly by the fixed boundary conditions, and thus their accuracy is dependent mainly on the reliability of the CLIMAP data.

Global surface air temperature in the 18K experiment is 3.6°C cooler than in the control run for today's boundary conditions. Much greater cooling, exceeding 20°C, occurs in southern Canada and northern Europe and cooling of more than 5°C is calculated for most of the Southern Hemisphere sea ice region. Some high latitude regions, including Alaska and parts of Antarctica, are at about the same temperature in the 18K simulation as today; thus there is not universal high latitude enhancement of the climate change.

Temperature changes over the tropical and subtropical oceans are only of the order of 1°C, and include substantial areas that are warmer in the 18K simulation than today. The latter aspect requires verification; diverse areas of the tropics and subtropics experienced mountain glaciation at 18K with snowline descent of about 1 km, and pollen data indicate substantial cooling of the order of 5°C at numerous low latitude areas. As indicated by our 3-D model experiment the CLIMAP sea surface temperatures are inconsistent with the observations of tropical cooling, since specification of relatively warm tropical and subtropical ocean temperatures effectively prohibits large cooling over land at these latitudes. We conclude that the low latitude ocean temperatures are probably overestimated in the CLIMAP data. A more quantitative analysis (Rind and Peteet, in preparation) suggests that large areas in the low latitude oceans may be too warm by 2–3°C in the CLIMAP data.

The middle parts of Fig. 9 show that the cooling at 18K occurred especially in the fall and winter. Although the surface air was substantially colder all year at latitude 60°N, this was largely a result of the change in mean surface altitude caused by the presence of ice sheets; the cooling at fixed altitude is considerably less. The zonal mean surface air in the tropics was cooler all year. The lower parts of Fig. 9 show substantial cooling throughout most of the troposphere. At high latitudes the greatest cooling occurs in the lower troposphere.

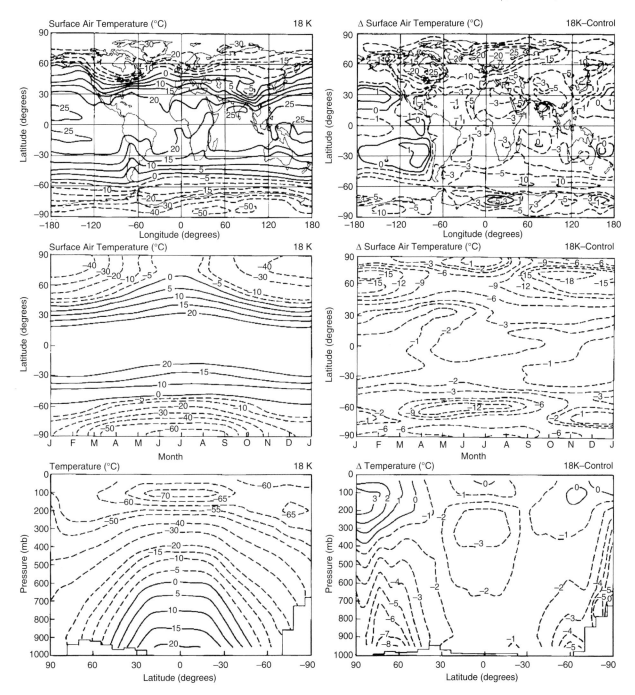

Fig. 9. Air temperature in the climate model experiment with boundary conditions for the ice age 18,000 years ago (left) and the temperature difference between the 18K simulation and the control run for today's climate boundary conditions (right). The control run is described in detail in paper 1.

Radiation

Changes in the planetary radiation budget in the 18K simulation are shown in Fig. 10. The surface albedo increases as much as 0.45 in the regions of ice sheets over northern Europe and southern Canada and about 0.30 in regions of large changes in sea ice coverage. Shielding by the atmosphere and the large zenith angles reduce the impact on planetary albedo to 0.15–0.20 over the ice sheets and 0.05–0.10 over sea ice. The effect of the planetary albedo change on the net radiation from the

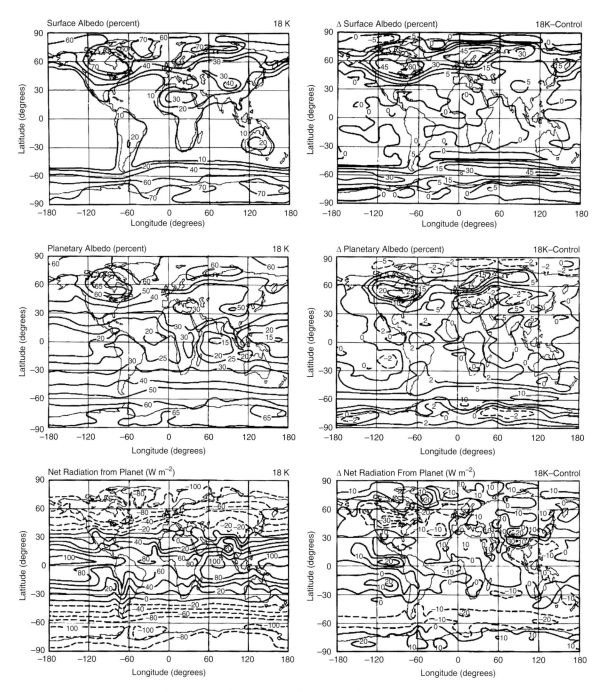

Fig. 10. Radiation quantities for the 18K simulation and differences with the control run. The control run is described in detail in paper 1.

planet is partially compensated over the ice sheets by reduced thermal emission, but nearly the full effect of the albedo change appears in the net radiation change over sea ice; these conclusions follow from comparison of the middle and lower parts of Fig. 10 and the fact that an albedo change of 0.10 is equivalent to $24\,\mathrm{W\,m^{-2}}$. Most of the more detailed changes in the geographical patterns of

the radiation budget are associated with changes of cloud cover or cloud top altitude.

Clouds

Cloud changes in the 18K simulation are illustrated in Fig. 11. There is a significant reduction of cloud cover in regions with increased sea ice,

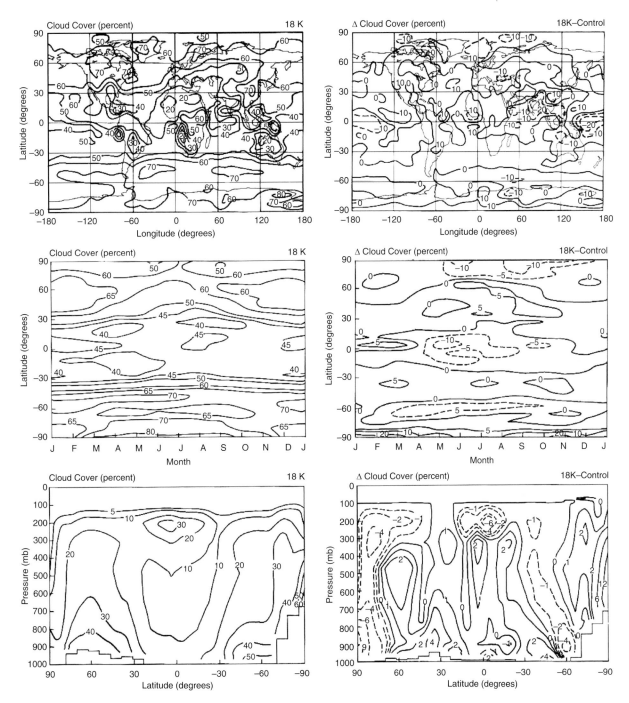

Fig. 11. Cloud cover for the 18K simulation and differences with the control run. The control run is described in detail in paper 1.

probably because of the reduced evaporation in those regions. The zonal mean cloud cover decreases slightly in the tropics during most of the year, increases slightly in the subtropics and increases at Northern Hemisphere midlatitudes in summer. The polar regions exhibit opposite behavior; at the north pole (a region of sea ice) the clouds decrease, while at the south pole (a continental region of high topography) the clouds increase in the 18K experiment. The lowest panel in Fig. 11 shows that the high level (cirrus) clouds are reduced substantially in the 18K simulation, presumably due to the reduction of penetrating moist convection and its associated transport of moisture. Most of these changes are consistent with those in the doubled CO_2 experiment, the cloud changes at 18K being the opposite of those which occur for the warmer doubled CO_2 climate.

Summary

The global mean surface air cooling of the Wisconsin ice age (compared to today) is computed from the CLIMAP boundary conditions to be ~4°C. Thus the mean temperature change between 18K and today is very similar to the projected warming for doubled CO_2. Below we analyze the contributions of different feedback processes to this global climate change.

18K feedback factors

We perform two types of experiments to study the feedback processes at 18K. In experiments of the first type, a given factor is modified (say the sea ice cover is changed) and the model is run for several years with the atmosphere free to adjust to the change, but with the ocean temperature and other boundary conditions fixed. Thus the only substantial feedback factors allowed to operate are water vapor and clouds (snow over land and ice can also change, but this represents only a small part of the ice/snow feedback). Experiments of this type enable us to relate surface temperature changes with flux imbalances at the top of the atmosphere under conditions of radiative/dynamic equilibrium in the atmosphere. Results of this type of experiment are summarized in the first part of Table 3 (experiments 8–14) along with the 18K control run (experiment 7). The method for converting the flux imbalance at the top of the atmosphere in these experiments to gain or feedback factors is described below in conjunction with experiment 8.

Experiments of the second type [labeled with a star (*) and tabulated in the lower part of Table 3] provide a faster, but more approximate, method for evaluating feedbacks which can be applied to certain types of radiative forcing. In the starred experiments we determine the radiative forcing by changing a factor in the control run (say sea ice cover) and calculating the instantaneous change in the planetary radiation balance at the top of the atmosphere. The atmospheric temperature and other radiative constituents and boundary conditions are not allowed to change; thus no feedbacks operate in these experiments. The flux change at the top of the atmosphere, ΔF, defines a change of planetary effective temperature

$$\Delta T_e(°C) = (\sigma T_e^3)^{-1} \Delta F(W\,m^{-2})$$
$$\sim 0.27 \Delta F(W\,m^{-2}) \qquad (13)$$

for $T_e = 255K$. This relation provides a good estimate of the no-feedback contribution to the equilibrium surface temperature change, if the radiative perturbation does not appreciably alter the vertical temperature structure. This procedure is applicable to solar flux, surface albedo and certain tropospheric gas perturbations (Hansen *et al.*, 1982), but does not work as simply for CO_2 perturbations, because CO_2 cools the stratosphere (Fig. 4 of Hansen *et al.*, 1981).

Although (13) provides a useful estimate of the (no feedback) surface temperature change resulting from a given radiative imbalance at the top of the atmosphere, it is a rough estimate because the radiation to space comes from a broad range of wavelengths and altitudes. In order to account for this spectral dependence, we used the 1-D radiative convective model for the following experiment. A flux of $1\,W\,m^{-2}$ was arbitrarily added to the ocean surface, and the lapse rate, water vapor and other radiative constituents were kept fixed. The surface temperature increase at equilibrium was 0.29°C, implying

$$\Delta T_s(°C) \sim 0.29 \Delta F(W\,m^{-2}). \qquad (14)$$

The coefficient in (14) is preferable to that in (13), for radiative perturbations which uniformly affect surface and atmospheric temperatures.

Water vapor and cloud feedbacks

Although we do not have measurements of the water vapor and cloud distribution for 18K, we can use experiment 8 to determine the combined water vapor/cloud feedback factor in our 3-D model for the 18K simulation. In this experiment the ocean surface temperature was arbitrarily decreased by 2°C everywhere. As shown in Table 3, the global mean surface temperature decreased by 2°C and the net radiation flux to space decreased by $2.7\,W\,m^{-2}$. Thus, with the sea ice and land ice fixed, the model sensitivity $\Delta T/\Delta F$ for the combined water vapor and cloud feedbacks is 0.76°C $(W/m^2)^{-1}$. If no feedbacks were allowed to operate the sensitivity would be ~0.29°C $(W/m^2)^{-1}$, cf. equation (14). Thus, since the atmospheric feedbacks are the only ones allowed to operate in experiment 8, we infer that the combined water vapor and cloud feedback factor in our model for 18K is $f_{WC} \sim 2.6$ and $g_{WC} \sim 0.6$. This is practically

Table 3. Changes in planetary radiation balance in climate model experiments. The control run for experiments 7, 11, and 13 is climate model II documented in paper 1. σ is the standard deviation of the annual mean about the 5 year mean for this control run; standard deviations for the other control runs were of similar magnitude. Experiment 7 was run for 6 years, with the results averaged over the final 5 years. The other unstarred experiments were run for 4 years and averaged over the final 3 years. The starred experiments were run for 3 years and averaged over 3 years. T_S is surface air temperature, F net radiation at the top of the atmosphere, A_g ground albedo and A_p planetary albedo. Global values are shown, with the numbers in parenthesis being the results for the Northern Hemisphere and Southern Hemisphere respectively. Experiment 7', the control run for experiment 12, was identical to experiment 7 except that the Koppen vegetation of Fig. 12(a) was substituted for Matthews (1983) 1° × 1° vegetation used in model II.

Experiment	Description	Control Run	ΔT_S (°C)	$100 \times \Delta A_g$	$100 \times \Delta A_p$	ΔF (W m^{-2})
7	18K boundary conditions	model II (paper 1)	−3.6 (−4.6,−2.5)	4.1 (5.0,3.1)	1.9 (2.2,1.6)	−1.6 (−0.1,−3.2)
8	ocean temperature reduced by 2°C	experiment 7	−2.0 (−2.0,−2.0)	0.4 (0.4,0.4)	0.5 (0.5,0.5)	2.7 (2.7,2.6)
9	today's sea ice	experiment 7	0.6 (0.5,0.7)	−1.8 (−0.9,−2.7)	−0.6 (−0.3,−0.8)	1.7 (0.6,2.8)
10	today's land ice	experiment 7	0.9 (1.7,0.1)	−1.9 (−3.6,−0.2)	−0.9 (−1.3,−0.5)	1.9 (2.4,1.4)
11	reduced 18K sea ice	model II (paper 1)	−3.5 (−4.6,−2.4)	3.7 (5.0,2.4)	1.8 (2.2,1.4)	−1.4 (0.1,−2.9)
12	18K vegetation	experiment 7'	−0.1 (−0.3,0.0)	0.6 (1.2,0.0)	0.3 (0.5, 0.1)	−0.9 (−1.4,−0.4)
13	modified 18K boundary conditions	model II (paper 1)	−3.7 (−5.0,−2.4)	3.4 (4.9,1.9)	1.7 (2.2,1.3)	−2.1 (−0.2,−3.9)
14	CO$_2$ (315→200 ppm)	experiment 7	−0.2 (−0.3,−0.1)	0.1 (0.1,0.1)	0.2 (0.2,0.1)	−2.2 (−2.3,−2.1)
9*	today's sea ice	experiment 7	–	−1.6 (−0.8,−2.6)	−0.5 (−0.4,−0.7)	3.1 (2.0,4.3)
10*	today's land ice	experiment 7	–	−1.5 (−2.7,−0.4)	−0.6 (−1.1,0.0)	3.6 (6.4,0.7)
11*	today's sea ice	experiment 11	–	−1.3 (−0.9,−1.8)	−0.5 (−0.4,−0.6)	2.6 (1.8,3.3)
$\sigma_{control}$			0.04 (0.03,0.06)	0.05 (0.07,0.03)	0.09 (0.09,0.10)	0.3 (0.3,0.4)

the same as the combined water vapor and cloud feedbacks for the doubled CO_2 experiment [Table 1 and equation (12)].

Experiment 8 can be used to convert the flux imbalances at the top of the atmosphere in the other unstarred experiments in Table 3 to equilibrium surface temperature changes. Thus, if the ocean temperature were free to change and water vapor and clouds were the only operative feedbacks, a flux imbalance ΔF at the top of the atmosphere would vanish as the surface temperature changed by an amount $\Delta T = 0.76 \Delta F$.

Sea ice and land ice feedbacks

Experiments 9 and 10, in which the 18K distributions of sea ice and land ice were replaced with today's distributions, show that both the sea ice and land ice changes made major contributions to the ice age cooling (Table 3). The CLIMAP sea ice and land ice distributions each affect the global ground albedo by ~0.02. Atmospheric shielding and zenith angle effects reduce the impact on planetary albedo to 0.006 for the sea ice change and 0.009 for the land ice change. The impact on the net radiation balance with space is between 1.5 and 2.0 W m^{-2} in each case, for these experiments in which the atmosphere was allowed to adjust to the changed sea ice and land ice.

The radiation imbalances in these experiments of the first type can be used to estimate the gain factors for these two feedback processes. Based on the conversion factor 0.76°C/(W m^{-2}), the flux imbalances in experiments 9 and 10 yield equilibrium surface temperature changes of $\Delta T_{sea\,ice} = 1.9$K and $\Delta T_{land\,ice} = 2.3$K. Since the feedback factor in these experiments is $f_{wc} = 2.6$, the radiative forcings produced by the sea ice and land ice changes in the absence of feedbacks are $\Delta T_{sea\,ice} = 1.9$K/$f_{wc} = 0.7$K and $\Delta T_{land\,ice} = 0.9$K, respectively. Thus the gain factors for sea ice and land ice changes, for the climate change from 18K to today, can be estimated as

$$g_{sea\,ice} \sim 0.7/\Delta T \sim 0.14 - 0.20$$
$$g_{land\,ice} \sim 0.9/\Delta T \sim 0.18 - 0.25, \qquad (15)$$

where ΔT is the change of global mean surface air temperature in °C between 18K and today. Experiment 7 yields $\Delta T = 3.6$°C, but indications that CLIMAP low latitude ocean temperatures are too warm (see above) suggest $\Delta T \sim 5$°C; the range given for g refers to $\Delta T = 3.6 - 5$°C.

In experiments 9* and 10* the 18K distributions of sea ice and land ice in experiment 1 were replaced with today's distributions, but only for

diagnostic calculation of the planetary radiation balance; all other quantities in the diagnostic calculation were from experiment 7. Based on the radiative forcings computed at the top of the atmosphere and Eq. (13) we estimate the gain factors, $g_i = \Delta T_i/\Delta T$, for the sea ice and land ice changes to be

$$g_{sea\,ice} \sim \frac{0.27 \times 3.1}{\Delta T} \sim 0.17 - 0.23$$
$$g_{land\,ice} \sim \frac{0.27 \times 3.6}{\Delta T} \sim 0.19 - 0.27. \qquad (16)$$

These gain factors include the effect of ice on thermal emission and planetary albedo. The fact that the gains estimated from (16) exceed those from (15) indicates that the emission from the added snow and ice surfaces on the average is from a somewhat higher temperature than the effective temperature, 255K.

The accuracy of these feedback gains depends primarily on the accuracy of the CLIMAP boundary conditions. Indeed, it is possible that the CLIMAP sea ice distribution is too extensive. Burckle *et al.* (1982), on the basis of satellite measurements of sea ice coverage and present sediment distributions, suggest that the sediment boundaries which CLIMAP had assumed to be the summer sea ice limit in the Southern Hemisphere are in fact more representative of the spring sea ice limit. Experiments 11 and 11* test the effect of this reduced sea ice coverage. In experiment 1 the CLIMAP February and August sea ice coverage were used as the extremes and interpolated sinusoidally to other months. For experiments 11 and 11* the winter (August) coverage was left unchanged, but the CLIMAP Southern Hemisphere February coverage was used for the spring (November) with linear extrapolation to February, and linear interpolation between the February and August extremes.

Experiment 11* implies that the sea ice gain estimated in experiments 9 and 9* should be reduced by 15–20 percent, if the arguments of Burckle *et al.* (1982) are correct. Although there is uncertainty about the true 18K sea ice distribution, it seems likely that the original CLIMAP data is somewhat an overestimate. On the basis of experiments 9, 9* and 11* our best estimate of the sea ice gain for the climate change from 18K to today is $g_{sea\,ice} \sim 0.15$ and thus a feedback factor $f_{sea\,ice} \sim 1.2$. This is larger than the snow/ice feedback obtained in the S_o and CO_2 experiments. However, the area of the sea ice cover change is about twice as large in the 18K experiment (-18.4×10^6 km^2 for the annual mean with our revised

CLIMAP sea ice) than in these other experiments $(7.8 \times 10^6 \, \text{km}^2, 9.2 \times 10^6 \, \text{km}^2$ and $14.8 \times 10^6 \, \text{km}^2$ in the S_o, CO_2 and alternate CO_2 experiments, respectively). Thus, the gains obtained from the ice age and the warmer climate experiments are consistent.

It also has been argued (DiLabio and Klassen, 1983) that the CLIMAP land ice cover is an overestimate, because the ice sheet peripheries probably did not all achieve maximual extent simultaneously. This possibility was recognized by the CLIMAP investigators, who therefore also presented a minimal extent ice sheet model for 18K (Denton and Hughes, 1981; CLIMAP, 1981). In this minimal ice model the area by which the ice sheets of 18K exceeded those of today is reduced to five-sixths of the value in the standard CLIMAP model. We conclude that the land ice gain for the climate change from 18K to today is 0.15–0.25. The corresponding feedback factor is 1.2–1.3.

Vegetation feedback

We also investigated the vegetation feedback, which Cess (1978) has estimated to provide a large positive feedback. We used the Koppen (1936) scheme, which relates annual and monthly mean temperature and rainfall to vegetation type, to infer expected global vegetation distributions for the GCM runs representing today's climate (model II in paper 1) and the 18K climate. The resulting vegetation distribution from the run with today's boundary conditions (Fig. 12a) suggests that the model and Koppen scheme can do a fair job of 'predicting' vegetation, in the case of today's climate for which the scheme was derived. Discrepancies with observed vegetation (Matthews, 1983) exist, e.g., there is too much rainforest on the east coast of Africa and too little boreal forest in central Asia, but the overall patterns are realistic.

The vegetation distribution obtained for 18K (Fig. 12b) from the Koppen scheme and our 18K experiment has more desert than today, less rainforest and less boreal forest. These changes are qualitatively consistent with empirical evidence of tropical aridity during the last glacial maximum based on a variety of paleoclimate indicators, such as pollen (Flenley, 1979), fauna (Vuilleumier, 1971), geomorphology (Sarnthein, 1978) and lake levels (Street and Grove, 1979). However, the magnitude of the desert and rainforest changes is smaller than suggested by the paleoclimate evidence. The smaller changes may result from (a) the CLIMAP tropical ocean temperatures being too warm, as discussed above, which would tend to cause an overestimate

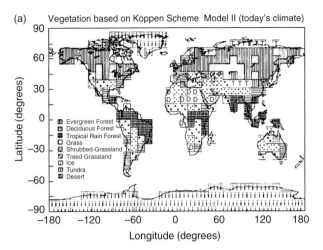

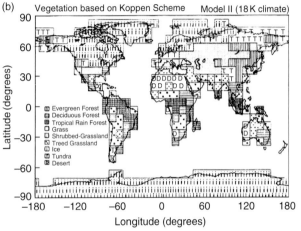

Fig. 12. Vegetation types (for gridboxes with more than 30 percent land) inferred from 3-D model simulations and the Koppen (1936) scheme, which relates annual and monthly mean temperature and precipitation to vegetation type. (a) is the control run for today's climate (paper 1), while (b) is the 18K simulation (experiment 7).

of rainforest and underestimate of desert area; (b) the lower atmospheric CO_2 abundance of 18K (Shackleton *et al.*, 1983), since CO_2 'fertilization' effects are not included in the Koppen scheme.

In experiment 12 today's vegetation was replaced with the Koppen 18K vegetation (Fig. 12b). The land, land ice and other boundary conditions were identical to those in the control run. In this experiment the modified vegetation directly affects the planetary albedo and also indirectly affects it through the masking depth for snow (paper 1). The 18K Kopppen vegetation of Fig. 12b increased the global ground albedo by 0.006 and the planetary albedo by 0.003 (Table 3) and left a flux imbalance of $-0.9 \, \text{W m}^{-2}$ at the top of the atmosphere. Based on the same analysis as for ice above, the no-feedback temperature change due to vegetation is 0.3°C, yielding $g_{vegetation}$ = 0.06–0.08. Because of the imprecisions in the Koppen 18K vegetation, we broaden the estimated gain to $g_{vegetation}$ = 0.05–0.09, and thus $f_{vegetation}$ = 1.05–1.1. Examination of global maps shows that the greatest impact of the changed vegetation was the replacement of European and Asian evergreen forests with tundra and grassland; the greatly reduced masking depths produced annual ground albedo increases of 0.1 or more over large areas, with the largest changes in spring. For reasons stated above, we also examined an 18K run with ocean temperatures reduced by 2°C; this reduced the number of gridboxes with rainforest from 10 to 5 in South America and from 7 to 2 in Africa, compared to Fig. 12b, in better agreement with paleoclimate evidence cited above. This additional vegetation change did not significantly change the global albedo or flux at the top of the atmosphere.

We conclude that the vegetation feedback factor between 18K and today is $f_{vegetation}$ ~ 1.05–1.1. This is much smaller than suggested by Cess (1978), but consistent with the analysis of Dickinson (1984). We find a somewhat larger feedback than Dickinson obtained, 0.003 change of planetary albedo compared to his 0.002, apparently due to the change of vegetation masking of snow-covered ground.

18K radiation balance

The simulated 18K climate (experiment 7) is close to radiation balance, the imbalance (Table 3) being 1.6 W m^{-2} at the top of the atmosphere, compared to the control run (model II) for today's climate. This imbalance is small compared to the amount of solar energy absorbed by the planet (~240 W m^{-2}). However, in reality even more precise radiation balance must have existed averaged over sufficient time, because the ice age lasted much longer than the thermal relaxation time of the ocean. (Melting the ice sheets in 10K years would require a global mean imbalance of only ~0.1 W m^{-2}.) Although the model calculations contain imprecisions comparable in magnitude to the radiation imbalance, we expect these to be largely cancelled by the procedure of differencing with the control run. This type of study should become a powerful tool in the future, as the accuracy of the reconstructed ice age boundary conditions improves and as the climate models become more realistic.

Taken at face value, the radiation imbalance in the 18K experiment 7 implies an imprecision in either some of the assumed boundary conditions for 18K or in the climate model sensitivity. The sense of the imbalance is such that the planet would cool further (to −4.8°C, based on the ΔF in Table 3), if the ocean temperature were computed rather than specified. Before studying this imbalance further, we make three modifications to the 18K simulation. First, the Southern Hemisphere sea ice cover is reduced as discussed above; this reduces the radiation imbalance. Second, the vegetation is replaced by the 18K vegetation of Fig. 12b; this slightly reduces the radiation imbalance. Third, the amount of atmospheric CO_2 is reduced in accord with evidence (Neftel et al., 1982) that the 18K CO_2 amount was only ~200 ppm; this significantly aggravates the radiation imbalance.

These three changes are all included in experiment 13, the sea ice and vegetation changes being those tested in experiments 11 and 12. The CO_2 decrease was 75 ppm from the control run value of 315 ppm; this is equivalent to the change from an estimated preindustrial abundance of 270 ppm to an ice age abundance of ~200 ppm. With these changes the radiation imbalance with space becomes 2.1 W m^{-2}. This imbalance would carry the surface temperature to −5.3°C if the constraint on ocean temperature were released.

Two principal candidates we can identify for redressing the 18K radiation imbalance are the CLIMAP sea surface temperatures and the cloud feedback in the climate model. The imbalance is removed if the CLIMAP ocean temperature is 1.5°C too warm (experiment 8, Table 3). The possibility that the CLIMAP sea surface temperatures may be too warm is suggested by the discussion above. The imbalance is also removed if it is assumed that clouds provide no feedback, rather than the positive feedback which they cause in this model; this conclusion is based on the estimate that the clouds cause 30–40 percent of the combined water vapor/cloud feedback (experiment 8), as is the case in the S_0 and CO_2 experiments.

One plausible solution is the combination of a reduction of low latitude ocean temperature by −1°C and a cloud feedback factor between 1 and 1.3. An alternative is a reduction of low latitude ocean temperature by ~1°C and a greater value for the 18K CO_2 abundance; indeed, recent analyses of Shackleton et al. (1983) suggest a mean 18K CO_2 abundance ~240 ppm. It is also possible that there were other presently unsuspected changes of boundary conditions.

There are presently too many uncertainties in the climate boundary conditions and climate model to permit identification of the cause of the radiation imbalance in the 18K simulation.

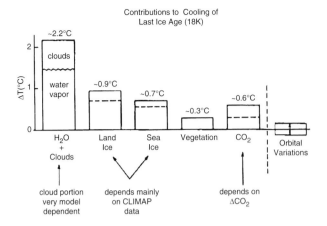

Fig. 13. Contributions to the global mean temperature difference between the Wisconsin ice age and today's climate as evaluated with a 3-D climate model and assumed boundary conditions. The cloud and water vapor portions were not separated, but based on other 3-D experiments the cloud part is estimated as 30–40 percent of their sum. The dashed line for land ice refers to the 'minimal extent' model of CLIMAP, and the dashed line for sea ice refers to the reduced sea ice cover discussed in the text. The solid line for CO_2 refers to $\Delta CO_2 \sim 100$ ppm (300 ppm $\rightarrow$ 200 ppm) and the dashed line to $\Delta CO_2 \sim 50$ ppm.

However, as the boundary conditions and climate models become more accurate, this approach should yield valuable checks on paleoclimate data and climate models. In the meantime, the data permits some general conclusions about the strength of climate feedback processes.

Conclusions from 18K experiments

The above calculations suggest the following major contributions to the global cooling at 18K, as shown schematically in Fig. 13:

$$
\begin{aligned}
\Delta T_{\text{water vapour + clouds}} &\equiv \Delta T_{WC} \sim 1.4 - 2.2°C \\
\Delta T_{\text{land ice}} &\equiv \Delta T_{l} \sim 0.7 - 0.9°C \\
\Delta T_{\text{sea ice}} &\equiv \Delta T_{S} \sim 0.6 - 0.7°C \\
\Delta T_{CO_2} &\sim 0.3 - 0.6°C \\
\Delta T_{\text{vegetation}} &\equiv \Delta T_{v} \sim 0.3°C
\end{aligned}
\tag{17}
$$

These estimates are the product of the gain for each process and the total cooling at 18K. But note that the uncertainty in the total ΔT cancels in obtaining ΔT_l, ΔT_s, ΔT_{CO_2} and ΔT_v; thus these ΔT_i are more fundamental and accurate than the corresponding g_i. The ΔT_i represent the isolated radiative forcings, which can be computed accurately, for the assumed changes in these radiative constituents between 18K and today. $\Delta T_{WC} = 2.2°C$ is

obtained from experiment 8 which yielded $g_{WC} \sim 0.6$. The cloud portion of ΔT_{WC} is uncertain because of the rudimentary state of cloud modeling; however even with no cloud feedback the water vapor contribution ($\sim 1.4°C$) is a large part of the total ice age cooling. $\Delta T_l = 0.9°C$ is based on the CLIMAP maximal ice sheet extent; it is $\sim 0.7°C$ for the minimal extent model. $\Delta T_S = 0.7°C$ is based on CLIMAP sea ice; it is $0.6°C$ with the reduced sea ice cover in the Southern Hemisphere in experiments 11 and 11*. $\Delta T_{CO_2} = 0.6°C$ refers to a CO_2 change from 200 ppm (at 18K) to 300 ppm (at say 1900); this is reduced to $0.3°C$ if the CO_2 amount was 225 ppm at 18K and 275 ppm at 1900.

The sum of the temperature contributions in Fig. 13 slightly exceeds the computed cooling $\Delta T = 3.7°C$ at 18K. This is a restatement of the radiation imbalance which exists in the model when the CLIMAP boundary conditions are used with ΔCO_2 of 50–100 ppm. If the model ocean temperature were allowed to change to achieve radiation balance, it would balance at a global mean 18K cooling of $\sim 5.3°C$ [model sensitivity $= 0.76°C/$ (W m^{-2})]. We conclude that either the CLIMAP 18K ocean temperatures are too warm by $\sim 1.5°C$ or we have overestimated one or more of the contributions to the 18K cooling in (17).

It is apparent from Fig. 13 that feedback processes account for most of the 18K cooling. The water vapor, cloud and sea ice contributions represent at least half of the total cooling. On long time scales the land ice portion of the cooling also may be regarded as a feedback, though it operates on a very regional scale and may be a complex function of a variety of factors such as the position of land areas, ocean currents and the meteorological situation. Even the CO_2 portion of the cooling, or at least part of it, may be a feedback, i.e., in response to the change of climate.

Variations in the amount of absorbed solar radiation due to Milankovitch (earth orbital) changes in the seasonal and latitudinal distributions of solar irradiance, which occur on time scales of several thousand years, can provide a global mean forcing of up to a few tenths of a degree. In view of the strength of the climate feedbacks, it is plausible for the Milankovitch and CO_2 forcings to 'drive' glacial to interglacial climate variations. However, discussion of the sequence of causes and mechanisms of glacial to interglacial climate change is beyond the scope of this paper.

We can use the contributions to the 18K cooling summarized in Fig. 13 to obtain an empirical estimate of the climate feedback factor due to the processes operating on 10–100 year time scales, taking the land ice, CO_2 and vegetation changes from 18K to today as slow (or at least specified). The global mean radiative forcing due to the difference in 18K and today's orbital parameters is negligible compared to the other forcing summarized in Fig. 13. The feedback factor for the fast (water vapor, clouds, sea ice) processes is

$$f(\text{fast processes}) = \frac{\Delta T(\text{total})}{\Delta T(\text{slow processes})} \quad (18)$$

$\Delta T(\text{total})$ is ~3.7°C for the CLIMAP boundary conditions, but may be ~5°C, if CLIMAP low latitude ocean temperatures are 1–2°C too warm. Using the nominal CLIMAP boundary conditions and intermediate estimates for $\Delta T_1 \sim 0.4$, $\Delta T_{CO2} \sim 0.45$ and $\Delta T_V \sim 0.3$, yields $f(\text{fast processes}) \sim 2.4$. Using $\Delta T(\text{total}) \sim 5°C$ and these nominal radiative forcings yields $f \sim 3.2$. Allowing for the more extreme combinations of the forcings and $\Delta T(\text{total})$, we conclude that

$$f(\text{fast processes}) \sim 2 - 4 \quad (19)$$

This feedback factor, $f \sim 2$–4, corresponds to a climate sensitivity of 2.5–5°C for doubled CO_2. Note that this result is independent of our climate model sensitivity: it depends on the total ΔT at 18K (fixed by CLIMAP) and on the assumption that land ice, CO_2 and vegetation are the only major slowly changing boundary conditions. Of course some vegetation and CO_2 feedbacks may occur in less than 100 years but for projecting future climate it is normal to take these as specified boundary conditions.

Finally, note that a given sensitivity for fast feedback processes, say 4°C for doubled CO_2, does not mean that the climate necessarily would warm by 4°C in 10 or even 100 years. Although water vapor, cloud and sea ice feedbacks respond rapidly to climate change, the speed of the climate response to a changed forcing depends on the rate at which heat is supplied to the ocean and on transport processes in the ocean.

TRANSIENT RESPONSE

Surface response time

The time required for the surface temperature to approach its new equilibrium value in response to a change in climate forcing depends on the feedback factor, f. The following example helps clarify this relationship.

Let the solar flux absorbed by a simple blackbody planet ($f \equiv 1$) change suddenly from $F_0 = \sigma T_0^4$ to $F_0 + \Delta F \equiv \sigma T_1^4$, with $\Delta F \ll F_0$. The rate of change of heat in the climate system is

$$\frac{d(cT)}{dt} = \sigma T_1^4 - \sigma T^4 \sim -4\sigma T_0^3 (T - T_1) \quad (20)$$

where c is the heat capacity per unit area and T is the time varying temperature. Since $T_1 - T_0 \ll T_0$, the solution is

$$T - T_0 = (T_1 - T_0)[1 - \exp(-t/\tau_b)] \quad (21)$$

where the blackbody no-feedback e-folding time is

$$\tau_b = c/4\sigma T_o^3. \quad (22)$$

For a planet with effective temperature 255 K and heat capacity provided by 63 m of water (as in our 3-D experiments), τ_b is approximately 2.2 years. Thus, this planet with $f = 1$ exponentially approaches its new equilibrium temperature with e-folding time 2.2 years.

Feedbacks modify the response time since they come into play only gradually as the warming occurs, the initial flux of heat into the ocean being independent of feedbacks. It is apparent that the actual e-folding time for a simple mixed layer heat capacity is

$$\tau = f\tau_b \quad (23)$$

An analytic derivation of (23) is given in Appendix A. The proportionality of the mixed layer response time to f is apparent in Fig. 3; the e-folding time for that model, which has $f \sim 3.5$ and a 63 m mixed layer, is ~8 years.

The 63 m mixed layer depth in our 3-D experiments was chosen as the minimum needed to obtain a realistic seasonal cycle of temperature, thus minimizing computer time needed to reach equilibrium. However, the global-mean annual-maximum mixed layer depth from our compilation of observations (see above) is ~110 m, and thus the isolated ocean mixed layer has a thermal response time of ~15 years if the climate sensitivity is 4°C for doubled CO_2. Even if the climate sensitivity is 2–3°C for doubled CO_2, the (isolated) mixed layer response time is about 10 years.

In order to determine the effect of deep ocean layers on the surface response time, it is useful to express the heat flux into the ocean as a function of the difference between current surface temperature and the equilibrium temperature for current atmospheric composition. In Appendix A we show that

$$F(W\,m^{-2}) = \frac{F_o}{\Delta T_{eq}(2*CO_2)}(\Delta T_{eq} - \Delta T) \qquad (24)$$

where ΔT is the ocean surface temperature departure from an arbitrary reference state and ΔT_{eq} is the equilibrium temperature departure for current atmospheric composition. $\Delta T_{eq}(2*CO_2)$ is the equilibrium sensitivity for doubled CO_2; for our 3-D climate model it is 4.2°C. F_o is the flux into the ocean in the model when CO_2 is doubled and the stratospheric temperature has equilibrated; our 3-D model yields $F_o = 4.3\,W\,m^{-2}$.

The long response time of the isolated mixed layer allows a portion of the thermal inertia of the deeper ocean to come into play in delaying surface temperature equilibrium. Exchange between the mixed layer and deeper ocean occurs by means of convective overturning in the North Atlantic and Antarctic oceans and principally by nearly horizontal motion along isopycnal (constant density) surfaces at lower latitudes. Realistic modeling of heat perturbations is thus rather complex, especially since changes of surface heating (and other climate variables) may alter the ocean mixing. However, we can obtain a crude estimate for the surface response time by assuming that *small* positive heat perturbations behave as a passive tracer; numerical experiments of Bryan *et al.* (1984) support this assumption. Measurements of transient tracers in the ocean, such as the tritium sprinkled on the ocean surface by atmospheric atomic testing, provide a quantitative indication of the rate of exchange of water between the mixed layer and the upper thermocline (see, e.g., Ostlund *et al.*, 1976).

We estimate an effective thermocline diffusion coefficient (k) at each GEOSECS measurement station from the criterion that the modeled and observed penetration depths (Broecker *et al.*, 1980) be equal at each station. The resulting diffusion coefficients are well correlated (inversely) with the stability at the base of the winter mixed layer (Fig. 14). In particular, we find a correlation coefficient of 0.85 between k and $1/N^4$, where N is the Brunt-Vaisala frequency at the base of the mixed layer. The global distribution of N^2 was obtained from the ocean data set of Levitus (1982).

The empirical relation between k and stability,

$$k = 5 \times 10^{-8}/N^4, \qquad (25)$$

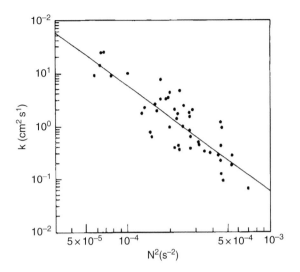

Fig. 14. Relationship between the effective diffusion coefficient (k) and the stability (N^2, where N is the Brunt-Vaisala frequency) at the base of the winter mixed layer for the GEOSECS tritium stations north of 20°S. The regression line fit (Eq. 25) has correlation coefficient 0.85 with the points for individual stations.

and the global ocean data set of Levitus (1982) yield the global distribution of k at the base of the mixed layer shown in Fig. 15a. There is a low rate of exchange ($k < 0.2\,cm^2\,s^{-1}$) in the eastern equatorical Pacific where upwelling and the resulting high stability at the base of the mixed layer inhibit vertical mixing, but rapid exchange ($k > 10\,cm^2\,s^{-1}$) in the Greenland – Norwegian Sea area of convective overturning.

The e-folding time for the mixed layer temperature (time to reach 63 percent of the equilibrium response) is shown in Fig. 15b. This is calculated from the geographically varying k and annual-maximum mixed layer depth, assuming a sudden doubling of CO_2 and an equilibrium sensitivity of 4.2°C everywhere. The (63 percent) response time is about 20–50 years at low latitudes, where the shallow mixed layer and small k allow the mixed layer temperatures to come into equilibrium relatively quickly. At high latitudes, where the deep winter mixed layer and large k result in a larger thermal inertia, the response time is about 200–400 years. The time for the global area-weighted mixed layer temperature to reach 63 percent of its equilibrium response is 124 years.

We estimate an equivalent k for use in a global 1-D model by choosing that value of k which fits the global (area-weighted) mean perturbation of the mixed layer temperature as a function of time obtained with the above calculation in which

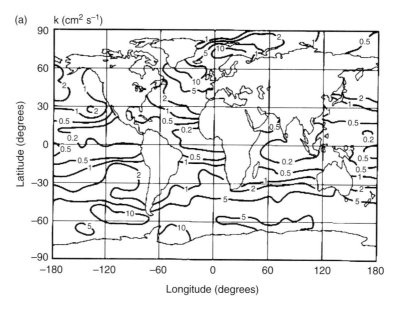

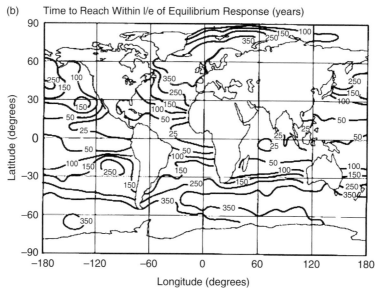

Fig. 15. (a) Geographic distribution of effective thermocline diffusion coefficient, k (cm²s⁻¹). k is derived from Eq. 25 and the global distribution of N^2, the latter obtained from the ocean data of Levitus (1982). (b) Geographic distribution of the 63 percent response time for surface temperature response to doubled CO_2 in the atmosphere. Only geographic variability of k and mixed layer depth are accounted for. ΔT_{eq} (2*CO_2) is taken as 4.2°C everywhere. The flux into the ocean is from Eq. (24).

k and mixed layer depth vary geographically. We find that k ~ 1 cm² s⁻¹ provides a reasonable global fit to the area-weighted local calculations for either a step function change of CO_2 or exponentially increasing CO_2 amount. Other analyses of the tracer data yield empirical values of 1–2 cm² s⁻¹ for the effective rate of exchange between the mixed layer and deeper ocean (Broecker *et al.*, 1980).

The delay time due to the ocean thermal inertia is graphically displayed in Fig. 16. Equation (24)

provides a good approximation of the time dependence of the heat flux into the ocean in our 3-D climate experiment with doubled CO_2, as shown by comparison of Figs. 3b and 16. Note that in our calculation with a mixed layer depth of 110 m, k = 1 cm² s⁻¹, and ΔT_{eq} = 4.2°C. The time required to reach a response of 2.65°C is 102 years. This is in rough agreement with the 124 years obtained above with the 3-D calculation.

The ocean delay time is proportional to f for an isolated mixed layer [eq. (23) and Appendix A], but

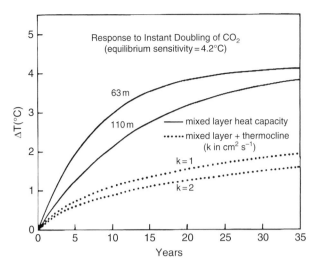

Fig. 16. Transient response to step function doubling of atmospheric CO_2 from 315 ppm to 630 ppm, computed from (24) with three representations of the ocean. The 63 m mixed layer corresponds to the mean mixed layer depth in the 3-D experiments; 110 m is the global-mean annual maximum mixed layer depth obtained from global ocean data. The curves including diffusion beneath the mixed layer are not exponentials (Appendix A).

depends more strongly on f if mixing into the deeper ocean is included. Our 1-D calculation with $k = 1 \, cm^2 \, s^{-1}$ and mixed layer depth 110 m yields an e-folding time of 55 years for $\Delta T_{eq} = 3°C$ and 27 years for $\Delta T_{eq} = 2°C$. Thus our ocean response time is consistent with that of Bryan *et al.* (1982), who obtained a response time of about 25 years for a climate model with sensitivity ~2°C for doubled CO_2.

Although our calculations were made with a simple diffusion model, the conclusion that the ocean surface temperature response time is highly nonlinear in ΔT_{eq} (or f) is clearly more general. The surface response time increases faster than linearly with f when the deeper ocean is included, because as f increases greater ocean depths come into play. Thus more realistic modeling of ocean transport processes should not modify these conclusions for small climate perturbations.

Our calculations of ocean response time neglect ocean circulation feedbacks on climate. The relationship between k and stability, equation (25), provides one way to examine the temperature feedback. By using that relation with our 1-D ocean diffusion model, we find that the time required to reach a given transient response is decreased, typically by several percent. Real ocean transports may be more sensitive to surface warming, as well as to related mechanisms such as melting of sea ice and ice sheets and changing

winds, precipitation and evaporation. It is easy to construct scenarios in which the ocean feedbacks are much greater, especially in the areas of deep water formation, but not enough information is available for reliable calculation of ocean/climate feedbacks.

Finally, we note that the ocean surface thermal response time reported in the literature generally has been 8–25 years (Hunt and Wells, 1979; Hoffert *et al.*, 1980; Cess and Goldenberg, 1981; Sohneider and Thompson, 1981; Bryan *et al.*, 1982). The 3-D ocean model result of Bryan *et al.* is consistent with our model when we employ a climate sensitivity of 2°C for doubled CO_2, as noted above. The discrepancy between our model response time and that of the other models arises from both the climate sensitivities employed and the choice of ocean model parameters. Key parameters are: mixed layer depth (we use 110 m since any depth mixed during the year should be included), rate of exchange with deeper ocean we use diffusion with $k = 1 \, cm^2 \, s^{-1}$, the minimum global value suggested by transient tracers, of., Broecker *et al.*, 1980) and the atmosphere-ocean heat flux [we use (24) which has initial value $4.3 \, W \, m^{-2}$ over the ocean area for doubled CO_2 and is consistent with other 3-D models]. Obviously the use of a 1-D box-diffusion model is a gross oversimplification of ocean transports. As an intermediate step between this and a 3-D ocean general circulation model, it may be valuable to study the problem with a model which ventilates the thermocline by means of transport along isopycnal surfaces. The agreement between the results from the 3-D ocean model of Bryan *et al.* and our model with a similar climate sensitivity suggests that our approach yields a response time of the correct order.

Impact on empirically-derived climate sensitivity

The delay in surface temperature response due to the ocean must be included if one attempts to deduce climate sensitivity from empirical data on time scales of order 10^2 years or less. Furthermore, in such an analysis it must be recognized that the lag caused by the ocean is not a constant, independent of climate sensitivity.

We computed the expected warming due to increase of CO_2 between 1850 and 1980 as a function of the equilibrium climate sensitivity. Results are shown in Fig. 17a for five choices of the 1850 CO_2 abundance (270±20 ppm), with CO_2 increasing linearly to 315 ppm in 1958 and then based on

(a)

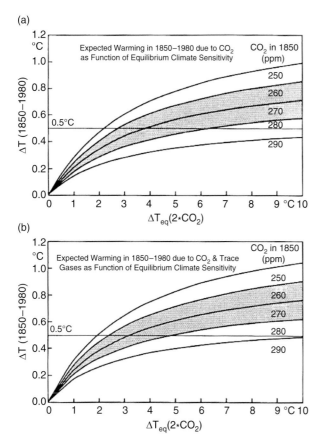

(b)

Fig. 17. Computed global warming between 1850 and 1980 as a function of the equilibrium climate sensitivity for doubled CO_2 (315 ppm → 630 ppm), ΔT_{eq} ($2*CO_2$). Results are shown for five values of the assumed abundance of CO_2 in 1850; the shaded area covers the range 270±10 ppm recommended by WMO (1983). (a) includes only CO_2 growth, while (b) also includes the trace gas growths of Table 4. In all cases CO_2 increases linearly from the 1850 abundance to 315 ppm in 1958 and then according to measurements of Keeling *et al.* (1982).

TABLE 4. Trace gas abundances employed in our calculations of the transient climate response for Figs. 17b and 18. CO_2 increases linearly for 1850–1958 and as observed by Keeling *et al.* (1982) for 1958–1980; ΔCO_2 increases about 2 percent yr^{-1} in the future. The chlorofluorocarbon abundances are based on estimated release rates to date, 150 year and 75 year lifetimes for CCl_2F_2 and CCl_3F, respectively, and constant future emissions at the mean release rate for 1971–1980. The CH_4 increase is about 1 percent yr^{-1} for 1970–1980 and 1.5 percent yr^{-1} after 1980. The N_2O increase is 0.2 percent yr^{-1} for 1970–1980 and 0.3 percent yr^{-1} after 1980.

Date	CO_2 (ppm)	CCl_2F_2 (ppt)	CCl_3F (ppt)	CH_4 (ppb)	N_2O (ppb)
1850	270	0	0	1400	295
1900	291	0	0	1400	295
1950	312	7	1	1400	295
1960	317	33	11	1416	295
1970	326	126	62	1500	295
1980	338	308	178	1650	301
1990	353	479	280	1815	307
2000	372	638	369	1996	313
2010	396	787	447	2196	320

Undoubtedly some other greenhouse gases also have increased in the past 130 years. Chloroflourocarbons, for example, are of recent anthropogenic origin. CH_4 and N_2O are presently increasing at rates of 1–2 percent yr^{-1} and 0.2–0.3 percent per yr^{-1}, respectively (Ehhalt, *et al.*, 1983; Weiss, 1981; CDAC, 1983). We estimate the influence of these gases on the empirical climate sensitivity by using the trace gas scenarios in Table 4. Although the CH_4 and N_2O histories are uncertain, the chlorofluorocarbons provide most of the non-CO_2 greenhouse effect, at least in the past 10–20 years (Lacis *et al.*, 1981), and their release rates are known. CH_4 may have increased slowly for the past several hundred years (Craig and Chou, 1982), but the reported rate of increase would not affect the results much. O_3 is also a potent greenhouse gas, but information on its past history is not adequate to permit its effect to be included.

The climate sensitivity implied by the assumed global warming since 1850, including the effect of trace gases in addition to CO_2, is shown in Fig. 17b. If the 1850 CO_2 abundance was 270±10 ppm, as concluded by WMO (1983), a warming of 0.5°C requires a climate sensitivity 2.5–5°C for doubled CO_2. The range for the implied climate sensitivity increases if uncertainty in the amount of warming is also included. For example, a warming of 0.4–0.6°C and an 1850 CO_2 abundance of 270±10 ppm yield a climate sensitivity of 2–7°C.

Keeling *et al.* (1982) measurments to 338 ppm in 1980. For simplicity a one-dimensional ocean was employed with mixed layer depth 110 m and k = 1 cm² s⁻¹. However, we obtained a practically identical graph when we used a simple three-dimensional ocean with the mixed layer depth varying geographically according to the data of Levitus (1982) and k varying as in Fig. 15a.

Use of Fig. 17a is as follows. If we take 270 ppm as the 1850 CO_2 abundance (WMO, 1983) and assume that the estimated global warming of 0.5°C between 1850 and 1980 (CDAC, 1983) was due to the CO_2 growth, the implied climate sensitivity is 4°C for doubled CO_2 (f = 3–4). Results for other choices of the 1850 CO_2 abundance or global warming can be read from the figure.

Although other climate forcings, such as volcanic aerosols and solar irradiance, may affect this analysis, we do not have information adequate to establish substantially different magnitudes of these forcings prior to and subsequent to 1850.

The climate sensitivity we have inferred is larger than obtained by CDAC (1983) from analysis of the same time period (1850–1980) with the same assumed temperature rise. The chief reason is that CDAC did not account for the dependence of the ocean response time on climate sensitivity [equation (23) and Appendix A]. Their choice of a 15 year response time, independent of ΔT_{eq} or f, biased their result to low sensitivities.

We conclude that the commonly assumed empirical temperature increase for the period 1850–1980 (0.5°C) suggests a climate sensitivity of 2.5–5°C (f = 2–4) for doubled CO_2. The significance of this conclusion is limited by uncertainties in past atmospheric composition, the true global mean temperature change and its cause, and the rate at which the ocean takes up heat. However, knowledge of these factors may improve in the future, which will make this a powerful technique for investigating climate sensitivity.

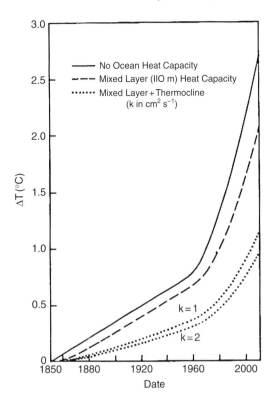

Fig. 18. Global mean warming computed for the CO_2 and trace gas scenarios in Table 4.

Growing gap between current and equilibrium climate

One implication of the long surface temperature response time is that our current climate may be substantially out of equilibrium with current atmospheric composition, as a result of the growth of atmospheric CO_2 and trace gases during recent decades. For example, in the last 25 years CO_2 increased from 315 ppm to 340 ppm and the chloroflourocarbons from near zero to their present abundance. Since the growth rates increased during the period, the gas added during the past 25 years has been present on the average about 10 years. 10 years is short compared to the surface temperature response time, even if the climate sensitivity is only 2.5°C for doubled CO_2.

We illustrate the magnitude of this disequilibrium by making some calculations with the 1-D model specified to give the climate sensitivity of our 3-D model, 4.2°C for doubled CO_2, and with the changing atmospheric composition of Table 4. Fig. 18 shows the modeled surface temperature during the past century (1) for instant equilibrium with changing atmospheric composition, (2) with thermal lag due to the mixed layer included, and

(3) with the thermocline's heat capacity included via eddy diffusion.

We infer that there is a large and growing gap between current climate and the equilibrium climate for current atmospheric composition. Based on the estimate in Fig. 18, we already have in the pipeline a future additional warming of almost 1°C, even if CO_2 and trace gases cease to increase at this time. A warming of this magnitude will elevate global mean temperature to a level at least comparable to that of the Altithermal (NAS, 1975, adapted in Fig. 1 of Hansen et al., 1984) about 6,000 years ago, the warmest period in the past 100,000 years.

The rate of warming computed after 1970 is much greater than in 1850–1980. This is because (1) ΔCO_2 is ~0.4 ppm yr^{-1} in 1850–1960, but >1 ppm yr^{-1} after 1970, and (2) trace gases, especially chlorofluorocarbons, add substantially to the warming after 1970. The surface warming computed for the period 1970–1990 is ~0.25°C; this is almost twice the standard deviation of the 5-year-smoothed global temperature (Hansen et al., 1981). But note that the equilibrium temperature increases by 0.75°C in the period

1970–1990, if the climate sensitivity is ~4°C for doubled CO_2. Thus our calculations indicate that the gap between current climate and the equilibrium climate for current atmospheric composition may grow rapidly in the immediate future, if greenhouse gases continue to increase at or near present rates.

As this gap grows, is it possible that a point will be reached at which the current climate "jumps" to the equilibrium climate? If exchange between the mixed layer and deeper ocean were reduced greatly, the equilibrium climate could be approached in as little as 10–20 years, the relaxation time of the mixed layer. Indeed, the stability of the upper ocean layers seems likely to increase as the greenhouse warming heats the ocean surface, especially if the warming leads to an increased melting of ice which adds fresh water to the mixed layer. Regions of deep-water formation, such as the North Atlantic Ocean, may be particularly sensitive to changes in surface climate. However, it is difficult to predict the net effect of greenhouse warming on ocean mixing, because changes of precipitation, evaporation and atmospheric winds, in addition to temperature, will affect ocean mixing and transport. If possible, it would be useful to examine paleoclimate records for evidence of sudden climate warmings on 10–20 year time scales, since there may have been cases in the past when the long thermal response time of the ocean allowed gaps between actual and equilibrium climates to build up.

Even if it does not lead to a dramatic jump to a new climate state, the gap between current climate and the equilibrium climate for current atmospheric composition may have important climatic effects as it grows larger. For example, it seems possible that in the summer, when zonal winds are weak, continental regions may tend partly toward their equilibrium climate, thus causing a relatively greater warming in that season. Also, in examining the climate effects of recent and future large volcanoes, such as the 1982 El Chichon eruption, the cooling effect of stratospheric aerosols must be compared to the warming by trace gases which have not yet achieved their equilibrium effect; it is not obvious that a global cooling of several tenths of a degree (Robock, 1983) should actually be expected. These problems should be studied by using a global model in which the atmospheric composition changes with time in accord with measurements, and in which the atmosphere, land and ocean each have realistic response times.

SUMMARY

Climate sensitivity inferred from 3-D models

Our analysis of climate feedbacks in 3-D models points strongly toward a net climate feedback factor of f ~ 2–4 for processes operative on 10–100 year time scales. The water vapor and sea ice feedbacks, which are believed to be reasonably well understood, together produce a feedback f ~ 2. The clouds in our model produce a feedback factor ~1.3, increasing the net feedback to f = 3–4 as a result of the nonlinear way in which feedbacks combine.

Present information on cloud processes is inadequate to permit confirmation of the cloud feedback. However, some aspects of the cloud changes in the model which contribute to the positive feedback appear to be realistic, e.g., the increase in tropical cirrus cloud cover and the increase of mean cloud altitude in conjunction with more penetrating moist convection in a warmer climate. It seems likely that clouds provide at least a small positive feedback. More realistic cloud modeling, as verified by detailed global cloud observations, is crucial for improving estimates of climate sensitivity based on climate models.

Climate sensitivity inferred from paleoclimate data

Analysis of the processes contributing to the cooling of the last ice age shows that feedbacks provide most of the cooling. The paleoclimate studies serve as proof of the importance of feedback processes and permit quantitative evaluation of the magnitude of certain feedbacks. The CLIMAP data allow us to evaluate individually the magnitudes of the land ice (f ~ 1.2–1.3) and sea ice (f ~ 1.2) feedbacks for the climate change from 18K to today, and to establish that the vegetation feedback was smaller but significant (f ~ 1.05–1.1).

We obtain an empirical estimate of f = 2.5–5 for the fast feedback processes (water vapor, clouds, sea ice) at 18K by assuming that the major radiative feedback processes have been identified (as seems likely from consideration of the radiative factors which determine the planetary energy balance with space) and grouping the slow or specified changes of the ice sheets and CO_2 as the principal climate forcings. This estimate for the fast feedback factor is consistent with the feedback in our 3-D model experiments, providing support that the model sensitivity is of the correct order.

The strength of the feedback processes at 18K implies that only relatively small climate forcings or fluctuations are needed to cause glacial to interglacial climate change. We do not try to identify the sequence of mechanisms of the glacial to interglacial changes, but it seems likely that both the direct effect of solar radiation (Milan-kovitch) changes on the planetary energy balance and induced changes of atmospheric composition, especially CO_2, are involved.

Climate sensitivity inferred from recent temperature trends

The temperature increase believed to have occurred in the past 130 years (~0.5°C) implies a climate sensitivity 2.5–5°C for doubled CO_2 (f = 2–4), if (1) added greenhouse gases are responsible for the temperature increase, (2) the 1850 CO_2 abundance was 270±10 ppm, and (3) the heat perturbation is mixed like a passive tracer in the ocean. This technique inherently yields a broad range for the inferred climate sensitivity, because the response time for the ocean increases with increasing climate sensitivity.

Thus the 3-D climate model, the 18K study and the empirical evidence from recent temperature trends yield generally consistent estimates of climate sensitivity. Our best estimate of the equilibrium climate sensitivity for processes occurring on the 10–100 year time scale is a global mean warming of 2.5–5°C for doubled CO_2.

Transient climate response

The rate at which the ocean surface can take up or release heat is limited by the fact that feedbacks come into play in conjunction with climate change, not in conjunction with climate forcing. Thus the (isolated) ocean mixed layer thermal relaxation time, commonly taken as 3–5 years, must be multiplied by the feedback factor f. This, in turn, allows the thermal inertia of deeper parts of the ocean to be effective. If the equilibrium climate sensitivity is ~4°C for doubled CO_2 and if small heat perturbations behave like observed passive tracers, the response time of surface temperature to a change of climate forcing is of order 100 years. If the equilibrium sensitivity is 2.5°C, this response time is about 40 years.

We conclude, based on the long surface temperature response time, that there is a large growing gap between current climate and the equilibrium climate for current atmospheric composition. Our projections indicate that within a few decades the equilibrium global temperature will reach a level well above that which has been experienced by modern man.

Is there a point at which the perturbation of surface climate will be large enough to substantially affect the rate of exchange of heat between the mixed layer and deeper ocean, possibly causing a rapid trend toward the equilibrium climate? This question is similar to one asked by Representative Gore (1982): "Is there a point where we trigger the dynamics of this (greenhouse) process, and if so, when do we reach that stage?". With present understanding of the climate system, particularly physical oceanography, we can not answer these questions.

APPENDIX A: INFLUENCE OF FEEDBACKS ON TRANSIENT REPONSE

Consider a planet for which the absorbed fraction of incident solar radiation (1 minus albedo) is a linear function of the temperature, say x + yT. If the planet emits as a blackbody its temperature is determined by the condition of radiative equilibrium

$$s_o a_o = \sigma T_o^{\,4} \qquad (A1)$$

with s_o the mean solar irradiance and $a_o = x + yT_o$.

Now suppose the solar irradiance changes suddenly by a small amount Δs. At the new equilibrium

$$(s_o + \Delta s)(a_o + \Delta a_{eq}) = \sigma(T_o + \Delta T_{eq})^4. \qquad (A2)$$

Neglecting second order terms (since $\Delta s << s_o$) and using (A1) yields

$$\Delta s a_o + s_o \Delta a_{eq} = 4\sigma T_o^{\,3} \Delta T_{eq}. \qquad (A3)$$

If there were no feedbacks ($\Delta a_{eq} = 0$), the temperature change at equilibrium would be

$$\Delta T_{eq}(\text{no feedbacks}) = \Delta T_o = \frac{\Delta s a_o}{4\sigma T_o^3}. \qquad (A4)$$

Thus we can rewrite (A3) as

$$\Delta T_{eq} = \Delta T_o + g\Delta T_{eq} \qquad (A5)$$

where

$$g = \frac{s_o \Delta a_{eq}}{4\sigma T_o^3 \Delta T_{eq}} = \frac{s_o y}{4\sigma T_o^3} \qquad (A6)$$

Using the relation between gain g and feedback factor f, $f = 1/(1-g)$, equation (A5) becomes

$$\Delta T_{eq} = f\Delta T_o \qquad (A7)$$

i.e., the equilibrium temperature change exceeds the no-feedback equilibrium temperature change by the factor f.

The heat flux into the planet as a function of time is

$$\begin{aligned} F &= \Delta sa_o + s_o \Delta a - 4\sigma T_o^3 \Delta T \\ &= 4\sigma T_o^3 (\Delta T_o + g\Delta T - \Delta T) \\ &= \frac{4\sigma T_o^3 \Delta T_o}{\Delta T_{eq}}(\Delta T_{eq} - \Delta T) \\ &= \frac{F_o}{\Delta T_{eq}}(\Delta T_{eq} - \Delta T) \end{aligned} \qquad (A8)$$

where

$$F_o = 4\sigma T_o^3 \, \Delta T_o = \Delta sa_o \qquad (A9)$$

is the flux into the planet at $t = 0$ (i.e., when $\Delta T = 0$). Thus the initial rate of warming is independent of the feedbacks.

The temperature of the planet as a function of time is determined by the equation

$$\frac{dcT}{dt} = \frac{dc\Delta T}{dt} = F \qquad (A10)$$

where c is the heat capacity per unit area. If c is constant (e.g. a mixed layer without diffusion into deeper layers), the solution is

$$\Delta T = \Delta T_{eq}(1 - \exp(-t/\tau)), \qquad (A11)$$

$$\tau = f\frac{c}{4\sigma T_o^3} = f\tau_b \qquad (A12)$$

where τ_b is the no-feedback e-folding time [Equation (22)].

Finally, note that these results are much more general than the specific mechanism we chose for the feedback, which was only used as a concrete example. It is apparent from the above that the only assumption required is the linearization of the feedback as a function of temperature.

Acknowledgements

This work depended on essential scientific contributions from several people, in particular: K. Prentice applied the Koppen classification scheme to the control and 18K simulations, H. Brooks and L. Smith derived the 18K monthly sea surface temperatures and sea ice distributions from the February and August CLIMAP data, S. Lebedeff developed the chlorofluorocarbon scenarios, F. Abramopoulos performed calculations with the 1-D ocean model, D. Peteet provided advice and references on paleoclimate vegetation, and L.C. Tsang compiled mixed layer depths from ocean data tapes. We thank R. Dickinson, L. Ornstein, W. Ruddiman, S. Schneider and C. Wunsch for critically reviewing the manuscript, J. Mendoza and L. DelValle for drafting the figures, and A. Calarco and C. Plamenco for typing several versions of the manuscript. Our climate model development was supported principally by the NASA Climate Program managed by Dr. Robert Schiffer; the applications to CO_2 studies were supported by a grant in 1982–1983 from EPA, for which we are indebted especially to John Hoffman, John Topping and Joseph Cannon.

REFERENCES

Alexander, R.C., and R.L. Mobley, Monthly average sea-surface temperatures and ice-pack limits on a 1° global grid, *Mon. Wea. Rev.*, *104*, 143–148, 1976.

Berger, A.L., Long-term variations of daily insolation and quaternary climatic changes, *J. Atmos. Sci.*, *35*, 2362–2367, 1978.

Bode, H.W., *Network Analysis and Feedback Amplifier Design*, 551 pp., Van Nostrand, New York, 1945.

Broecker, W.S., T.H. Peng and R. Engh, Modeling the carbon system, *Radiocarbon*, *22*, 565–598, 1980.

Bryan, K., F.G. Komro, S. Manabe and M.J. Spelman, Transient climate response to increasing atmospheric carbon dioxide, *Science*, *215*, 56–58, 1982.

Bryan, K., F.G. Komro and C. Rooth, The ocean's transient response to global surface temperature anomalies, *Maurice Ewing Series*, *5*, 1984.

Budyko, M.I., The effect of solar radiation variations on the climate of the earth, *Tellus*, *21*, 611–619, 1969.

Burckle, L.H., D. Robinson and D. Cooke, Reappraisal of sea-ice distribution in Atlantic and Pacific sectors of the Southern Ocean at 18,000 yr BP, *Nature*, *299*, 435–437, 1982.

CDAC, *Changing Climate*, Report of the Carbon Dioxide Assessment Committee, 496 pp., National Academy Press, Washington, D.C., 1983.

Cess, R.D., Radiative transfer due to atmospheric water vapor: global considerations of the earth's energy balance, *J. Quant. Spectrosc. Radiat. Transfer*, *14*, 861–871, 1974.

Cess, R.D., Biosphere-albedo feedback and climate modeling, *J. Atmos. Sci.*, *35*, 1765–1769, 1978.

Cess, R.D. and S.D. Goldenberg, The effect of ocean heat capacity upon global warming due to increasing atmospheric carbon dioxide, *J. Geophys. Res.*, *86*, 498–502, 1981.

CLIMAP Project Members, A. McIntyre, project leader: Seasonal reconstruction of the earth's surface at the last glacial maximum, *Geol. Soc. Amer.*, Map and Chart Series, No. 36, 1981.

Craig, H., and C.C. Chou, Methane: the record in polar ice cores, *Geophys. Res. Lett.*, *9*, 1221–1224, 1982.

Denton, G.H. and T.J. Hughes, *The Last Great Ice Sheets*, 484 pp., John Wiley, New York, 1981.

Dickinson, R.E., Vegetation albedo feedbacks, *Maurice Ewing Series*, *5*, 1984.

DiLabio, R.N.W. and R.A. Klassen, Book review of The Last Great Ice Sheets, *Bull. Amer. Meteorol. Soc.*, *64*, 161, 1983.

Ehhalt, D.H., R.J. Zander and R.A. Lamontagne, On the temporal increase of tropospheric CH_4, *J. Geophys. Res.*, *88*, 8442–8446, 1983.

Flenley, J., *The Equatorial Rain Forest: A Geological History*, 162 pp., Butterworths, London, 1979.

Gates, W.L., Modeling the Ice-Age Climate, *Science*, *191*, 1138–1144, 1976.

Gordon, A.L., *Southern Ocean Atlas*, 34 pp., 248 plates, Columbia Univ. Press, New York, 1982.

Gore, A., Record of Joint Hearing on Carbon Dioxide and Climate: The Greenhouse Effect, March 25, 1982, Science and Technology Committee (available from Goddard Institute for Space Studies, New York, 10025), 1982.

Hansen, J., D. Johnson, A. Lacis, S. Lebedeff, P. Lee, D. Rind and G. Russell, Climate impact of increasing atmospheric carbon dioxide, *Science*, *213*, 957–966, 1981.

Hansen, J., D. Johnson, A. Lacis, S. Lebedeff, P. Lee, D. Rind and G. Russell, Climatic effects of atmospheric carbon dioxide, *Science*, *220*, 873–875, 1983a.

Hansen, J., A. Lacis and S. Lebedeff, in *Carbon Dioxide Review 1982*, Oxford Univ. Press, New York, 284–289, 1982.

Hansen, J., A. Lacis, D. Rind and G. Russell, Climate sensitivity to increasing greenhouse gases, in *Sea Level Rise to the Year 2100*, Van Nostrand, New York, 1984.

Hansen, J., G. Russell, D. Rind, P. Stone, A Lacis, S. Lebedeff, R. Ruedy and L. Travis, Efficient three-dimensional global models for climate studies: models I and II, *Mon. Wea. Rev.*, *111*, 609–662, (paper 1), 1983b.

Hoffert, M.I., A.J. Callegari and C.T. Hsieh, The role of deep sea heat storage in the secular response to climatic forcing, *J. Geophys. Res.*, *85*, 6667–6679, 1980.

Hunt, B.G. and N.C. Wells, An assessment of the possible future climatic impact of carbon dioxide increases based on a coupled one-dimensional atmospheric-oceanic model, *J. Geophys. Res.*, *84*, 787–791, 1979.

Keeling, C.D., R.B., Bacastow and T.P. Whorf, in *Carbon Dioxide Review 1982*, ed. W.C. Clark, Oxford Univ. Press, New York, 377–385, 1982.

Kellogg, W.W. and S.H. Schneider, Climate stabilization: for better of worse?, *Science*, 1163–1172, 1974.

Koppen, W., Das geographische system der klimate, in *Handbuch der Klimatologie 1*, part C. ed. W. Koppen and G. Geiger, Boentraeger, Berlin, 1936.

Lacis, A., J. Hansen, P. Lee, T. Mitchell and S. Lebedeff, Greenhouse effect of trace gases, 1970–1980, *Geophys. Res. Lett.*, *8*, 1035–1038, 1981.

Levitus, S., *Climatological Atlas of the World Ocean*, NOAA Prof. Paper, No. 13, U.S. Government Printing Office, Washington, D.C., 1982.

Lian, M.S. and R.D. Cess, Energy-balance climate models: a reappraisal of ice-albedo feedback, *J. Atmos. Sci.*, *34*, 1058–1062, 1977.

Manabe, S., Carbon dioxide and climate change, *Adv. Geophys.*, *25*, 39–82, 1983.

Manabe, S. and D.G. Hahn, Simulation of the tropical climate of an ice age, *J. Geophys. Res.*, *82*, 3889–3911, 1977.

Manabe, S., J. Smagorinsky and R.F. Strickler, Simulated climatology of a general circulation model with a hydrologic cycle, *Mon. Wea. Rev.*, *98*, 175–212, 1965.

Manabe, S. and R.J. Stouffer, Sensitivity of a global climate model to an increase of CO_2 concentration in the atmosphere, *J. Geophys. Res.*, *85*, 5529–5554, 1980.

Manabe, S. and R.T. Wetherald, Thermal equilibrium of the atmosphere with a given distribution of relative humidity, *J. Atmos. Sci.*, *24*, 241–259, 1967.

Manabe, S. and R.T. Wetherald, The effects of doubling the CO_2 concentration on the climate of a general circulation model, *J. Atmos. Sci.*, *32*, 3–15, 1975.

Manabe, S. and R.T. Wetherald, On the distribution of climate change resulting from an increase of CO_2 content of the atmosphere, *J. Atmos. Sci.*, *37*, 99–118, 1980.

Matthews, E., Global vegetation and land use: new high-resolution data bases for climate studies, *J. Clim. Appl. Meteorol.*, *22*, 474–487, 1983.

Miller, J.R., G.L. Russell and L.C. Tsang, Annual oceanic heat transports computed from an atmospheric model, *Dynam. Atmos. Oceans*, *7*, 95–109, 1983.

NAS, *Understanding Climatic Change*, 239 pp., National Academy of Sciences, Washington, D.C., 1975.

Neftel, A., H. Oeschger, J. Swander, B. Stauffer and R. Zumbrunn, Ice core sample measurements give atmospheric CO_2 content during the past 40,000 years, *Nature*, *295*, 220–223, 1982.

NOAA, User's Guide to NODC Data Services, Environmental Data Service, Washington, D.C., 1974.

North, G.R., Theory of energy-balance climate models, *J. Atmos. Sci.*, *32*, 2033–2043, 1975.

Ostlund, H.G., H.G. Dorsey and R. Brescher, GEOSECS Atlantic radiocarbon and tritium results, Univ. Miami Tritium Laboratory Data Report No. 5, 1976.

Robock, A., The dust cloud of the century, *Nature*, *301*, 373–374, 1983.

Sarnthein, M., Sand deserts during glacial maximum and climatic optimum, *Nature*, *272*, 43–46, 1978.

Schneider, S.H. and S.L. Thompson, Atmospheric CO_2 and climate: importance of the transient response, *J. Geophys. Res.*, *86*, 3135–3147, 1981.

Sellers, W.D., A global climate model based on the energy balance of the earth-atmosphere system, *J. Appl. Meteorol.*, *8*, 392–400, 1969.

Shackleton, N.J., M.A. Hall, J. Line and C. Shuxi, Carbon isotope data in core V19–30 confirm reduced carbon dioxide concentration in the ice age atmosphere, *Nature*, *306*, 319–322, 1983.

Street, F. and A. Grove, Global maps of lake-level fluctuations since 30,000 yr BP, *Quat. Res.*, *12*, 83–118, 1979.

Stone, P.H., Feedbacks between dynamical heat fluxes and temperature structure in the atmosphere, *Maurice Ewing Series*, *5*, 1984.

Vuilleumier, B., Pleistocene changes in the fauna and flora of South America, *Science*, *173*, 771 780, 1971.

Walsh, J. and C. Johnson, An analysis of Arctic sea ice fluctuations, *J. Phys. Ocean.*, *9*, 580–591, 1979.

Wang, W.C. and P.H. Stone, Effect of ice-albedo feedback on global sensitivity in a one-dimensional radiative-convective climate model, *J. Atmos. Sci., 37*, 545–552, 1980.

Webster, P.J. and N.A. Streten, Late Quaternary ice age climates of tropical Australia: interpretations and reconstructions, *Quat. Res., 10*, 279–309, 1978.

Weiss, R.F., The temporal and spatial distribution of tropospheric nitrous oxide, *J. Geophys. Res., 86*, 7185–7195, 1981.

Wetherald, R.T. and S. Manabe, The effects of changing the solar constant on the climate of a general circulation model, *J. Atmos. Sci., 32*, 2044–2059, 1975.

Williams, J., R.G. Barry and W.M. Washington, Simulation of the atmospheric circulation using the NCAR global circulation model with ice age boundary conditions, *J. Appl. Meteorol., 13*, 305–317, 1974.

WMO, World Meteorological Organization project on research and monitoring of atmospheric CO_2, Report no. 10, R.J. Bojkov ed., Geneva, 42 pp., 1983.

7

Aerosols

Mitchell, J.F.B, Johns, T.C., Gregory J.M., *et al.* (1995). Climate response to increasing levels of greenhouse gases and sulfate aerosols. *Nature*, **376** (6540), 501–504.

Of the human-caused changes to climate forcing, greenhouse gases warm the planet, while aerosols, tiny suspended particles or droplets in the atmosphere, tend to cool. Haze in the air scatters sunlight, allowing some of it to be lost to space without ever depositing its energy as heat in the Earth system, increasing the albedo of the planet and cooling it down. Aerosols differ from greenhouse gases, especially CO_2 and nitrous oxide N_2O, in that their lifetime in the atmosphere is short, only a few weeks, before they settle or wash out (as acid rain, for the aerosols resulting from sulfur emissions). The cooling influence from aerosols is therefore regional rather than global for the gases. Another difference is that the amount of climate forcing from the aerosols is much less well constrained than the warming impact of the gases. Aerosols scatter sunlight, but they also apparently influence the microphysics of cloud formation, leading to more numerous, smaller, and longer-lived cloud droplets than one would find in the cleanest air. This so-called indirect effect may be as important as the direct scattering of the bare aerosols, but it is difficult to know how the clouds would be different in the clean air, or even how clean the air would be without human intervention.

The potential cooling impact of aerosols is the kernel of truth at the heart of the "scientists claimed global cooling in the 1970's" myth that survives in popular perceptions of climate change science. It was quickly realized, however, that greenhouse gases are stronger than aerosols, so warming wins. Another piece of the story is the observation from sediment cores that interglacial periods such as our own do not last forever, and that our current interglacial has lasted about as long as the last interglacial period did, which could imply that a new ice age might start sometime in the next few millennia. This is of course an entirely different time scale than the threat of human-induced warming, and as it turns out the interglacial most analogous to ours, when the Earth's orbit around the sun was nearly circular as it is today, took place about 400 000 years ago, and lasted for 50 000 years, suggesting that the next ice age might have been tens of thousands of years in the future even in the absence of human releases of CO_2.

Climate Response to Increasing Levels of Greenhouse Gases and Sulphate Aerosols

J. F. B. MITCHELL, T. C. JOHNS, J. M. GREGORY & S. F. B. TETT

Hadley Centre for Climate Prediction and Research, Meteorological Office, Bracknell RG12 2SY, UK

Climate models suggest that increases in greenhouse-gas concentrations in the atmosphere should have produced a larger global mean warming than has been observed in recent decades, unless the climate is less sensitive than is predicted by the present generation of coupled general circulation models.[1,2] After greenhouse gases, sulphate aerosols probably exert the next largest anthropogenic radiative forcing of the atmosphere,[3] but their influence on global mean warming has not been assessed using such models. Here we use a coupled ocean–atmosphere general circulation model to simulate past and future climate since the beginning of the near-global instrumental surface-temperature record,[4] and include the effects of the scattering of radiation by sulphate aerosols. The inclusion of sulphate aerosols significantly improves the agreement with observed global mean and large-scale patterns of temperature in recent decades, although the improvement in simulations of specific regions is equivocal. We predict a future global mean warming of 0.3 K per decade for greenhouse gases alone, or 0.2 K per decade with sulphate aerosol forcing included. By 2050, all land areas have warmed in our simulations, despite strong negative radiative forcing in some regions. These model results suggest that global warming could accelerate as greenhouse-gas forcing begins to dominate over sulphate aerosol forcing.

The general circulation model (GCM) used is the Hadley Centre climate model, a development from an earlier model.[5] Modified formulations of the atmospheric dynamics,[6] convection,[7] land surface, boundary layer[8] and cloud[9] schemes have been used. The horizontal resolution is 2.5° × 3.75° (latitude × longitude), with 20 layers in the ocean and 19 layers in the atmosphere. We apply calibrated seasonal flux adjustments to ocean surface temperatures and salinities,[5] to bring about a faithful representation of present mean climate. A simple parametrization of ice-drift[10] is included, obviating the need for flux adjustments to sea ice. The equilibrium sensitivity to a doubling of CO_2 concentration is estimated to be 2.5 K, lower than most GCMs.[2]

The model was brought to near equilibrium through a total of 470 years coupled simulation, after which the control simulation commenced. There is no detectable trend in global mean surface temperature in the 300 model years of the control run, although there is a slow warming of the deeper ocean layers amounting to a maximum of 0.07 K per century in the global mean at 1,500 m depth. The net heating of the total system is less than 0.2 W m⁻².

Three experiments were performed, each starting at model year 1860: a control with constant CO_2 concentrations, an experiment GHG in which the concentration of CO_2 was increased gradually to give the changes in forcing due to all greenhouse gases, both in the past and to 2050 under a given scenario, and an experiment SUL in which both greenhouse gases and the direct radiative effect of sulphate aerosols were represented. The concentrations of sulphate aerosols and greenhouse gases after 1990 were based on IPCC scenario IS92a[2] which assumes a slow reduction in the rate of economic growth and gradual increase in conservation measures (Fig. 1).

The greenhouse-gas forcing increases slowly from 0.4 W m⁻² relative to the control in 1860 to 1.2 W m⁻² in 1960, then more rapidly reaching 2.5 W m⁻² in 1990, and thereafter at 0.6 W m⁻² per

Mitchell, J. F. B. (1995) Climate response to increasing levels of greenhouse gases and sulphate aerosols. *Nature* 376 (6540): 501–504. Reprinted by permission from Macmillan Publishers Limited.

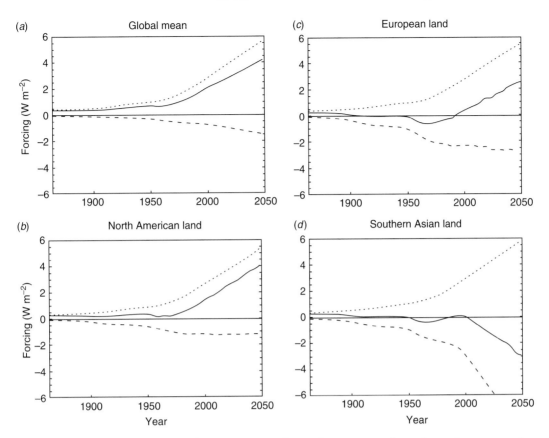

Fig. 1. Area-average annual mean radiative forcing owing to: increasing concentrations of greenhouse gases (experiment GHG, dotted curve); the direct effect of sulphate aerosols represented by increasing surface albedo (dashed curve); net forcing (experiment SUL, solid curve). a, The global mean; b, North America—land between 30° and 60° N and 40° to 140° W; c, Europe—land between 35° and 70° N and 15° W to 60° E; d, southern Asia—land between 7.5° and 42.5° N and 60° to 130° E.

METHODS. The concentration $C(t)$ of CO_2 at time t was chosen to give the estimated forcing (F) relative to the control (fixed concentration C_0) $F = 6.3 \ln (C(t)/C_0)$ W m^{-2} due to increases in all greenhouse gases from 1765 (ref. 2, Table 2.6). After 1990, $C(t)$ was increased by 1% yr^{-1}, which is within 0.2 W m^{-2} of IPCC scenario IS92a[2]. There is an initial increment of 0.4 W m^{-2} at 1860 arising from changes in greenhouse gas concentration from 1765 to 1860. The direct effect of sulphate aerosols was added in experiment SUL by increasing the surface albedo in the clear-sky fraction of each grid box.[8,22] The pattern of aerosol loading to 1990 was based on the calculated annual mean distribution for the 1980s,[23] scaled by the estimated annual mean sulphate emissions.[24,25] The global mean forcing of 0.6 W m^{-2} at 1990 lies within recent estimates of 0.3–0.9 W m^{-2} (refs 18, 19). We have ignored seasonal variations of the pattern and the indirect effect of sulphate aerosols on cloud brightness.[9] A sulphate distribution for 2050 (H. Rodhe and U. Hansson, personal communication) was obtained using a sulphur-cycle model[23] with the sulphur emissions of scenario IS92a. From 1990 to 2050, the loading pattern was interpolated between 1990 and 2050 values, with the field scaled to give the global loading of scenario IS92a.

decade (Fig. 1a). The global mean sulphate aerosol forcing increases continuously, and most rapidly, from 1950 to 1990 (Fig. 1a). Note that in recent decades, the percentage increase per year of emissions has grown faster than the global average over southern Asia, and slower over western countries[11] where in the last decade there has been a decrease.

Reproduction of the past observational record within the limits of natural variability is a necessary, though not sufficient, condition for a model to produce reliable estimates of climate change. The differences in global mean temperature between simulations and observations[4] are within the range of simulated internal variability until about 1940 (Fig. 2a). The model's variability is generally greater than observed on timescales up to several decades. (On longer timescales, the estimates of internal variability based on the observations will be exaggerated by any long-term trend due to external forcing.) In the 1940s and 1950s, SUL is significantly cooler than the

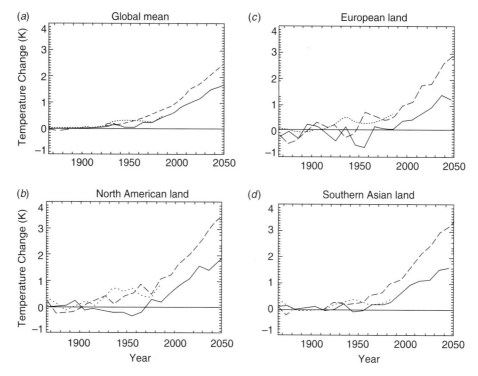

Fig. 2. Changes in area-average decadal mean temperature at 1.5 m, relative to the 1880–1920 mean. Observed[4] (dotted curve), GHG (dashed curve), SUL (solid curve). *a*, Global mean; *b*, mean over North America; *c*, mean over Europe; *d*, mean over southern Asia. Areas defined as in Fig. 1.

observations at the 5% significance level (two-tailed test assuming that decadal means are normally distributed with the same standard deviation, 0.073 K, as the control simulation). There is a 14% probability that at least two of the decades from 1860 to 1990 would be significantly different at this level by chance. Alternatively, the prescribed forcing (Fig. 1*a*) may be incorrect. The response to the 1.6 W m⁻² change in forcing by 1990 in SUL is only 0.4 K, suggesting that it would require an increase in forcing of the order of 1 W m⁻² to produce the extra warming in the 1940s. This seems unlikely, even with the large uncertainties in forcing[3]. For example, the optical depth of volcanic aerosols may have been a minimum[12], but the forcing anomaly was probably less than 0.5 W m⁻². A combination of natural fluctuations[13] and small errors in forcing seems the most likely explanation.

From 1970, the difference between GHG and observations[4] is significant in each decade at both the 5% (and 1%) levels whereas SUL is not significantly different from the observations at the 5% level after the 1930s and 1940s. Thus including the cooling attributed to sulphate aerosols

(SUL) gives a simulation closer to the observations in recent decades, as found in energy-balance models[14–16] or idealized GCMs[17]. Note the rapid warming after 1970 in the observations and in SUL. In SUL, this is the response to accelerated greenhouse warming and a slower rate of increase in cooling from sulphate aerosols (Fig. 1*a*).

Comparison of the simulated spatial distribution of changes provides a potentially more stringent test of the model's credibility. We show decadal results starting with the 1860s, but focus mainly on the changes in recent decades when the observed data coverage is better and the forcing and signal-to-noise in the GHG and SUL experiments is largest (Figs 1, 2). We first compute decadal anomalies for GHG, SUL and the observations in the same manner by subtracting the respective 1860–1990 means from each decadal mean. Then we compute a centred spatial correlation for each successive decade between the model experiments and the observations. For the decades since 1950, the magnitude of the pattern correlation between SUL and the observations increases steadily, rising above the 10% significance level in the two most recent decades

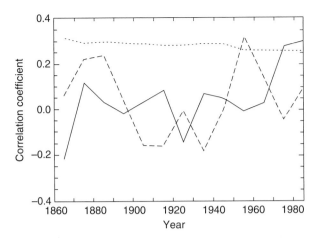

Fig. 3. Spatial correlation between simulated and observed decadal temperature changes relative to the 1860–1990 mean. Dashed line, GHG; solid line, SUL. The dotted line gives the 10% (one-tailed) level of significance, which varies with data coverage.
METHODS. Observed annual means were computed where there is at least one value in every month in a 5° grid-box. Observed and simulated data were averaged on a 15° × 15° grid, and decadal means formed in grid-boxes containing at least one annual observed value. The significance level was estimated from the distribution of correlations between decadal means in the control, assumed to be gaussian, using the observational mask for each decade.

(Fig. 3). This recent trend is consistent with what could also be an emerging greenhouse gas//sulphate aerosol signal in the observations. The GHG correlation in the decade beginning in 1950 rises above the 10% significance level, but subsequent values are much smaller despite increasing forcing in the more recent decades. Thus the spatial patterns of surface temperature anomalies in SUL, compared to GHG, more closely resemble those in the observations in recent decades when the forcing is largest.

The distinction between the simulations is less evident in specific regions because of the high level of variability of temperature on smaller scales. GHG is substantially warmer than observed over southern Asia after 1960 (Fig. 2*d*). However, SUL is generally cooler than observed over Europe and North America after the 1940s (Fig. 2*b–d*). Possible explanations include excessive sulphate aerosol forcing in these regions and natural variability.

After 1990, the rate of global warming in GHG increases to 0.3 K per decade (Fig. 2*a*). In SUL, the rate increases rapidly to 0.2 K per decade, as the net forcing increases (Fig. 1*a*). Southern Asia continues to warm despite increasingly negative

forcing (Fig. 1*d*), although at less than half the rate in GHG (Fig. 2*d*). Over North America and Europe, the aerosol forcing levels off and the greenhouse warming is more evident, though still less than in GHG. The increase in mean sulphate aerosol forcing still has a global effect, even where the local loading is substantially unchanged. The extension of these sensitivity studies to 2050 under a prescribed scenario indicates the potential strong influence of sulphate aerosols on regional climate change.

The patterns of forcing and response, averaged over 20 years from 2030, are shown in Fig. 4. The radiative forcing in GHG shows relatively little regional variation[18] (Fig. 4*a*). The response is enhanced in high latitudes by sea-ice feedbacks, and slowed in the Southern Ocean and the northern North Atlantic by deep mixing in the ocean (Fig. 4*b*), as in other transient experiments[2]. In SUL, there are areas of both positive and negative forcing (Fig. 1), but by 2030–50, areas of substantial net negative forcing are restricted to southern Asia (Fig. 4*c*). Even so, this region remains warmer than in the control simulation, due to the movement of warmer air from surrounding areas (Fig. 4*d*). The largest decreases in temperature on adding aerosols occur in northern mid-latitudes, where the forcing is largest, and in the Arctic, where the global-scale cooling is amplified by increases in sea ice[8,19,20] (Fig. 4*e, f*). There is considerable variability on decadal and longer timescales—the standard deviation of 20-year means from the control simulation is 0.2–0.4 K over the extratropical continents—but the differences between GHG and SUL are several times larger than this and thus statistically significant. A more rigorous statistical assessment of the geographical distribution of changes will form a separate study.

We have shown that inclusion of sulphate aerosol forcing improves the simulation of global mean temperature over the last few decades, although further work is needed to clarify why the simulation over North America and Europe is not improved. There remain uncertainties in model sensitivity, particularly associated with clouds[21], and in the external forcing due to sulphate and other aerosols, tropospheric ozone and solar variability[3], and natural variability on long (greater than decadal) timescales. In particular, this study suggests that if we are to improve model predictions of climate change on global and regional scales it is not sufficient to consider greenhouse gases alone; the effects of aerosols and perhaps other forcing factors must be included.

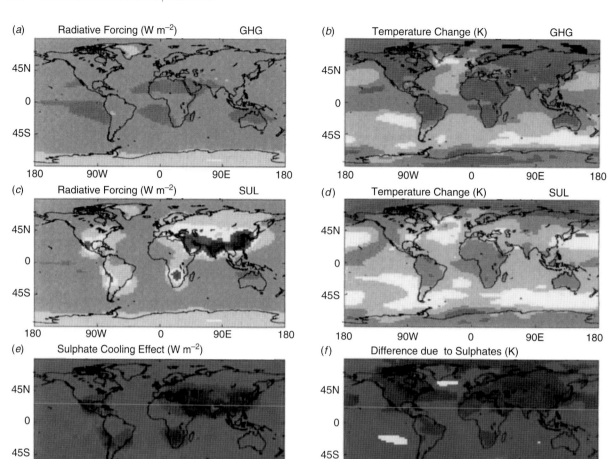

Fig. 4. Annual mean changes, averaged over 2030–50 relative to the control simulation. Temperature contours (*b, d, f* and right-hand scale bar) every 1 K, forcing contours (*a, c, e* and left-hand scale bar) every 2 W m^{-2}. *a, b*, Forcing (*a*) and temperature change (*b*) in GHG. *c, d*, Forcing (*c*) and temperature change (*d*) in SUL. *e, f*, Radiative cooling (*e*) and difference in response (*f*) due to adding sulphate aerosols. The greenhouse-gas forcing at year *t* is obtained by scaling the change at the top of the troposphere on instantaneously doubling CO_2 by ln $(C(t)/C_0)/$ln 2. The sulphate aerosol forcing is evaluated at the top of the atmosphere timestep by timestep.

ACKNOWLEDGEMENTS

We thank H. Rodhe and U. Hansson for providing the estimates of sulphate distribution, D. E. Parker for observational data, and G. Meehl and our colleagues for helpful comments. R. Davis, A. Brady and J. Lavery assisted with running the experiments and processing data. This work was supported by the UK Department of Environment.

NOTES

1 Houghton, J. T., Jenkins, G. J. & Ephraums, J. J. (eds) *Climate Change. The IPCC Scientific Assessment.* (Cambridge Univ. Press, 1990).

2 Houghton, J. T., Callander, B. A. & Varney, S. K. (eds) *Climate Change 1992. The Supplementary Report to the IPCC Scientific Assessment* (Cambridge Univ. Press, 1992).

3 Houghton, J. T. *et al.* (eds) *Climate Change 1994, Radiative Forcing of Climate Change, and an Evaluation of the IPCC IS92 Emission Scenarios* (Cambridge Univ. Press, 1995).

4 Parker, D. E., Jones, P. D., Folland, C. K. & Bevan, A. *J. geophys. Res.* **99**, 14373–14399 (1994).

5 Murphy, J. M. *J. Clim.* **8**, 36–56 (1995).

6 Cullen, M. J. P. & Davies, T. R. *Q. J. R. met. Soc.* **117**, 993–1002 (1991).

7 Gregory, D. & Allen, S. in *Preprints 9th Conf. on Numerical Weather Prediction* 122–123 (American Meteorological Soc., Boston, 1991).

8 Mitchell, J. F. B., Davis, R. A., Ingram, W. J. & Senior, C. A. *J. Clim.* **8**, 2364–2386 (1995).

9   Jones, A., Roberts, D. L. & Slingo, A. *Nature* **369**, 450–453 (1994).
10 Bryan, K. *Mon. Weath. Rev.* **97**, 806–827 (1969).
11 Engardt, M. & Rodhe, H. *Geophys. Res. Lett.* **20**, 117–120 (1993).
12 Sato, M., Hansen, J. E., McCormick, M. P. & Pollack, J. B. *J. geophys. Res.* **98**, 22987–22994 (1993).
13 Schlesinger, M. E. & Ramankutty, N. *Nature* **367**, 723–726 (1994).
14 Wigley, T. M. L. & Raper, S. C. B. *Nature* **357**, 293–300 (1992).
15 Schlesinger, M. E., Jiang, X. & Charlson, R. J. in *Climate Change and Energy Policy* (eds Rosen, L. & Glasser, R.) 75–108 (Am. Inst. Phys., New York, 1993).
16 Murphy, J. M. *J. Clim.* **8**, 496–514 (1995).
17 Hansen, J., Lacis, A., Reudy, R. & Wilson, H. *Nat. Geogr. Explor.* **9**, 142–158 (1993).
18 Kiehl, J. T. & Briegleb, P. B. *Nature* **260**, 311–314 (1993).
19 Taylor, K. & Penner, J. E. *Nature* **369**, 734–737 (1994).
20 Roeckner, E., Siebert, T. & Feichter, J. in *Proc. Dahlem Workshop* on *Aerosol Forcing of Climate* (eds Charlson, R. J. & Heintsenberg, J.) (Wiley, in the press).
21 Senior, C. A. & Mitchell, J. F. B. *J. Clim.* **6**, 393–418 (1993).
22 Charlson, R. J., Langner, J., Rodhe, H., Leovy, C. B. & Warren, S. G. *Tellus* **43AB**, 152–163 (1991).
23 Langner, J. & Rodhe, H. *J. atmos. Chem.* **13**, 225–263 (1991).
24 Dignon, J. & Hameed, S. *J. Air Pollut. Control Ass.* **39**, 180–186 (1989).
25 Hameed, S. & Dignon, J. *J. Air Pollut. Control Ass.* **42**, 159–186 (1992).

8

Ocean Heat Uptake and Committed Warming

Hansen J., Nazarenko, L., Ruedy, R., *et al.* (2005). Earth's energy imbalance: confirmation and implications. *Science*, **308** (5727), 1431–1435. Times Cited: 93. 4 pages.

The oceans play a major role in the global warming forecast, both in absorbing fossil fuel CO_2 (as discussed by the Revelle, Bolin, and Broecker papers in this volume), and also by absorbing heat, the topic of Hansen *et al.* (2005). The oceans are the largest heat store in the climate system, requiring more energy to warm them up to a new equilibrium than the land surface or even the great ice sheets. The ideas presented by Hansen *et al.* in 2005 were not really revolutionary or new; the role of the oceans in slowing the change in climate were explicitly discussed in the Charney report, for example, and indeed by Hansen in 1985 (both in this volume). By absorbing heat as they transiently warm, the oceans delay the full expression of the equilibrium warming that rising CO_2 concentrations in the atmosphere will eventually produce. "We may not be given a warning until the CO_2 loading is such that an appreciable climate change is inevitable," according to Charney.

Hansen *et al.* (2005) is the first assembly of observations of ocean heat uptake, laboriously compiled by statistical analysis of subsurface ocean temperature data, with the top-of-atmosphere energy imbalance from climate model results. The climate system is complex and subtle, and it could be that decadal or century time-scale variations ocean heat storage might provide a source of variability in the climate of the Earth's surface. However, if the ocean were responsible for the warming of the last few decades, instead of the greenhouse effect of human-released CO_2, we would expect to see the oceans cooling as they released heat to the atmosphere, rather than warming as is observed. Also, the oceans tend to average over decades, smoothing out any short-term excursions or wiggles in the climate system. It turns out the deep oceans are a good place to look for global warming.

Earth's Energy Imbalance: Confirmation and Implications

JAMES HANSEN,[1,2]* LARISSA NAZARENKO,[1,2] RETO RUEDY,[3] MAKIKO SATO,[1,2]
JOSH WILLIS,[4] ANTHONY DEL GENIO,[1,5] DOROTHY KOCH,[1,2] ANDREW LACIS,[1,5]
KEN LO,[3] SURABI MENON,[6] TICA NOVAKOV,[6] JUDITH PERLWITZ,[1,2] GARY RUSSELL,[1]
GAVIN A. SCHMIDT,[1,2] NICHOLAS TAUSNEV[3]

[1]NASA Goddard Institute for Space Studies, New York, NY 10025, USA.
[2]Columbia Earth Institute, Columbia University, New York, NY 10025, USA.
[3]SGT Incorporated, New York, NY 10025, USA.
[4]Jet Propulsion Laboratory, Pasadena, CA 91109, USA.
[5]Department of Earth and Environmental Sciences, Columbia University, New York, NY 10025, USA.
[6]Lawrence Berkeley National Laboratory, Berkeley, CA 94720, USA.

Our climate model, driven mainly by increasing human-made greenhouse gases and aerosols among other forcings, calculates that Earth is now absorbing 0.85 ± 0.15 W/m^2 more energy from the Sun than it is emitting to space. This imbalance is confirmed by precise measurements of increasing ocean heat content over the past 10 years. Implications include: (i) expectation of additional global warming of about 0.6°C without further change of atmospheric composition; (ii) confirmation of the climate system's lag in responding to forcings, implying the need for anticipatory actions to avoid any specified level of climate change; and (iii) likelihood of acceleration of ice sheet disintegration and sea level rise.

Earth's climate system has considerable thermal inertia. This point is of critical importance to policy and decision-makers who seek to mitigate the effects of undesirable anthropogenic climate change. The effect of the inertia is to delay Earth's response to climate forcings, i.e., changes of the planet's energy balance that tend to alter global temperature. This delay provides an opportunity to reduce the magnitude of anthropogenic climate change before it is fully realized, if appropriate action is taken. On the other hand, if we wait for more overwhelming empirical evidence of climate change, the inertia implies that still greater climate change will be in store, which may be difficult or impossible to avoid.

The primary symptom of Earth's thermal inertia, in the presence of an increasing climate forcing, is an imbalance between the energy absorbed and emitted by the planet. This imbalance provides an invaluable measure of the net climate forcing acting on Earth. Improved ocean temperature measurements in the past decade, along with high precision satellite altimetry measurements of the ocean surface, permit an indirect but precise quantification of Earth's energy imbalance. We compare observed ocean heat storage with simulations of global climate change driven by estimated climate forcings, thus obtaining a check on the climate model's ability to simulate the planetary energy imbalance.

The lag in the climate response to a forcing is a sensitive function of equilibrium climate sensitivity, varying approximately as the square of the sensitivity (1), and it depends on the rate of heat exchange between the ocean's surface mixed layer and the deeper ocean (2–4). The lag could be as short as a decade, if climate sensitivity is as small as ¼°C per W/m^2 of forcing, but it is a century or longer if climate sensitivity is 1°C per W/m^2 or larger (1, 3). Evidence from Earth's history (3–6) and climate models (7) suggests that climate sensitivity is ¾ ± ¼ °C per W/m^2, implying that 25–50 years are needed for Earth's surface temperature to reach 60 percent of its equilibrium response (1).

We investigate Earth's energy balance via computations with the current global climate model of the NASA Goddard Institute for Space Studies (GISS). The model and its simulated climatology have been documented (8), as has its response to a wide variety of climate forcing mechanisms (9). The climate model's equilibrium sensitivity to

* To whom correspondence should be addressed. E-mail: jhansen@giss.nasa.gov

Hansen, J. *et al.* (2005) Earth's energy imbalance: Confirmation and implications. *Science* 308: 1431–1435. Reprinted with permission from AAAS.

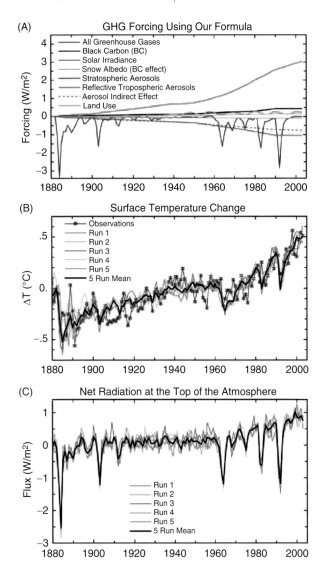

Fig. 1. (A) Forcings (9) used to drive global climate simulations. (B) Simulated and observed temperature change; prior to 1900 the observed curve is based on observations at meteorological stations and the model is sampled at the same points, while after 1900 the observations include SSTs for the ocean area and the model is the true global mean (21). (C) Net radiation at the top of the atmosphere in the climate simulations. Five climate simulations are carried out differing only in initial conditions.

doubled CO_2 is 2.7°C (~2/3°C per W/m²). The climate simulations for 1880-2003 used here will be included in the Intergovernmental Panel on Climate Change (IPCC) 2007 report and are available with other IPCC runs at http://www-pcmdi.llnl.gov/ipcc/about_ipcc.php or via www.giss.nasa.gov/data/imbalance.

Climate forcings. Figure 1A summarizes the forcings that drive the simulated 1880–2003 climate change. Among alternative definitions of climate forcing (9), we use the effective forcing, F_e. F_e differs from conventional climate forcing definitions (10) by accounting for the fact that some forcing mechanisms have a lesser or greater "efficacy" in altering global temperature than an equal forcing by CO_2 (9). F_e is an energy flux change arising in response to an imposed forcing agent. It is constant throughout the atmosphere, as it is evaluated after atmospheric temperature has been allowed to adjust to the presence of the forcing agent.

The largest forcing is due to well-mixed greenhouse gases, CO_2, CH_4, N_2O, CFCs (chlorofluorocarbons) and other trace gases, totaling 2.75 W/m² in 2003 relative to 1880 (Table 1). Ozone (O_3) and stratospheric H_2O from oxidation of increasing CH_4 make the total greenhouse gas (GHG) forcing 3.05 W/m² (9). Estimated uncertainty in the total GHG forcing is ~15% (10, 11).

Atmospheric aerosols cause climate forcings by reflecting and absorbing radiation, as well as via indirect effects on cloud cover and cloud albedo (10). The aerosol scenario in our model uses estimated anthropogenic emissions from fuel use statistics and includes temporal changes in fossil fuel use technologies (12). Our parameterization of aerosol indirect effects (13, 9) is constrained by empirical evidence that the aerosol indirect forcing is ~ −1 W/m² (9). The effective aerosol forcing in 2003 relative to 1880, including positive forcing by absorbing black carbon (BC) aerosols, is −1.39 W/m², with subjective estimated uncertainty ~ 50%.

Stratospheric aerosols from volcanoes cause a sporadically large negative forcing, with an uncertainty that increases with age from 15% for 1991 Mount Pinatubo to 50% for 1883 Krakatau (9). Land use and snow albedo forcings are small on global average and uncertain by about a factor of two (9). Solar irradiance is taken as increasing by 0.22 W/m² between 1880 and 2003, with estimated uncertainty a factor of two (9). All of these partly subjective uncertainties are intended as 2σ error bars. The net change of effective forcing between 1880 and 2003 is +1.8 W/m², with formal uncertainty ±0.85 W/m² due almost entirely to aerosols (Table 1).

Climate simulations. The global mean temperature simulated by the GISS model driven by this forcing agrees well with observations (Fig. 1B). An ensemble of five simulations was obtained by using initial conditions at intervals of 25 years of the climate model control run, thus revealing the model's inherent unforced variability. The spatial distribution of the simulated warming (see fig. S1) is slightly excessive in the tropics, as much as a few

Table 1. Effective climate forcings (W/m²) used to drive the 1880–2003 simulated climate change in the GISS climate model (9).

Forcing Agent*	Forcing (W/m²)	
Greenhouse Gases (GHGs)	–	–
Well-Mixed GHGs	2.75	–
Ozone¶§	0.24	–
CH₄–Derived Stratospheric H₂O	0.06	–
Total: GHGs		3.05 ± 0.4
Solar Irradiance		0.22 (×2)
Land Use		–0.09 (×2)
Snow Albedo		0.14 (×2)
Aerosols		
Volcanic Aerosols	0.00	–
Black Carbon§	0.43	–
Reflective Tropospheric Aerosols	–1.05	–
Aerosol Indirect Effect	–0.77	–
Total: Aerosols	–	–1.39 ± 0.7
Sum of Individual Forcings	–	1.93
All Forcings at Once	–	1.80 ± 0.85

* Effective forcings are derived from 5-member ensembles of 120-year simulations for each individual forcing and for all forcings acting at once (see ref. 9 and Supporting Online Material). The sum of individual forcings slightly from all forcings acting at once because on non-linearities in combined forcings and unforced variability in climate simulations.

¶ This is the ozone forcing in our principal IPCC simulations; it decreases from 0.24 to 0.22 W/m² when the stratospheric ozone change of Randel and Wu (SI) is employed (see ref. 9 and Supporting Online Material).

§ Ozone and black forcings are less than they would be for conventional forcing definitions (10), because their "efficacy" is only ~75% (9).

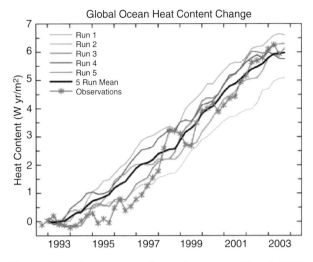

Fig. 2. Ocean heat content change between 1993 and 2003 in top 750 m of world ocean. Observations are from (19). Five model runs are for the GISS coupled dynamical ocean-atmosphere model (8, 9).

tenths of a degree, and on average the simulated warming is a few tenths of a degree less than observed in middle latitudes of the Northern Hemisphere, but there is substantial variation from one model run to another (fig. S1).

Discrepancy in the spatial distribution of warming may be partly a result of the uncertain aerosol distribution, specifically the division of aerosols between fossil fuel and biomass burning aerosols (9). However, excessive tropical warming in our model is primarily in the Pacific Ocean, where our primitive ocean model is unable to simulate climate variations associated with ENSO processes.

The planetary energy imbalance in our model (Fig. 1C) did not exceed a few tenths of 1 W/m² prior to the 1960s. Since then, except for a few years following each large volcanic eruption, the simulated planetary energy imbalance has grown steadily. According to the model, Earth is now absorbing 0.85 ± 0.15 W/m² more solar energy than it radiates to space as heat.

Ocean heat storage. Confirmation of the planetary energy imbalance can be obtained by measur-ing the heat content of the ocean, which must be the principal reservoir for excess energy (3, 14). Levitus *et at.* (14) compiled ocean temperature data that yielded increased ocean heat content of about 10 W yr m⁻², averaged over the Earth's surface, during 1955–1998 [1 W yr/m² over the full Earth ~1.61×10²² joules; see table S1 for conversion factors of land, air, water and ice temperature changes and melting to global energy units]. Total ocean heat storage in that period is consistent with climate model simulations (15–18), but the models do not reproduce reported decadal fluctuations. The fluctuations may be a result of variability of ocean dynamics (16) or, at least in part, an artifact of incomplete sampling of a dynamically variable ocean (17, 18).

Improved definition of Earth's energy balance is possible for the past decade. First, the predicted energy imbalance due to increasing greenhouse gases has grown to 0.85 ± 0.15 W/m² and the past decade has been uninterrupted by any large volcanic eruption (Fig. 1). Second, more complete ocean temperature data is available, including more profiling floats and precise satellite altimetry that permits improved estimates in data sparse regions (19).

Figure 2 shows that the modeled increase of heat content in the past decade in the upper 750 m of the ocean is 6.0 ± 0.95 (std. dev.) W yr/m², averaged over the surface of Earth, varying from 5.0 to 7.15 W yr/m² among the five simulations. The observed annual mean rate of ocean heat gain between 1993 and mid 2003 was $0.86 \pm 0.12\,\mathrm{W/m^2}$ per year for the 93.4% of the ocean that was analyzed (19). Assuming the same rate for the remaining 6.6% of the ocean yields a global mean heat storage rate of $0.7 \times 0.86 = 0.60 \pm 0.10\,\mathrm{W/m^2}$ per year or $6 \pm 1\,\mathrm{W\,yr/m^2}$ for 10 years, 0.7 being the ocean fraction of Earth's surface. This agrees well with the $5.5\,\mathrm{W\,yr/m^2}$ in the analysis of Levitus *et al.* (20) for the upper 700 meters based only on in situ data.

Figure 3 compares the latitude-depth profile of the observed ocean heat content change with the five climate model runs and the mean of the five runs. There is a large variability among the model runs, revealing the chaotic "ocean weather" fluctuations that occur on such a time scale. This variability is even more apparent in maps of change in ocean heat content (fig. S2). Yet the model runs contain essential features of observations, with deep penetration of heat anomalies at middle to high latitudes and shallower anomalies in the tropics.

The modeled heat gain of $\sim 0.6\,\mathrm{W/m^2}$ per year for the upper 750 meters of the ocean differs from the decadal mean planetary energy imbalance of $\sim 0.75\,\mathrm{W/m^2}$ primarily because of heat storage at greater depths in the ocean. 85% of the ocean heat storage occurred above 750 meters on average for the five simulations, with the range from 78% to 91%. The mean heat gain below 750 m was $\sim 0.11\,\mathrm{W/m^2}$. The remaining $0.04\,\mathrm{W/m^2}$ warmed the atmosphere and land and melted sea ice and land ice (see supplementary information).

Earth's energy imbalance. We infer from the consistency of observed and modeled planetary energy gains that the forcing still driving climate change, i.e., the forcing not yet responded to, averaged $\sim 0.75\,\mathrm{W/m^2}$ in the past decade and was $\sim 0.85 \pm 0.15\,\mathrm{W/m^2}$ in 2003 (Fig. 1C). This imbalance is consistent with the total forcing $\sim 1.8\,\mathrm{W/m^2}$ relative to 1880 and climate sensitivity $\sim 2/3°\mathrm{C}$ per $\mathrm{W/m^2}$. The observed 1880-2003 global warming is 0.6–0.7°C (*10, 21*), which is the full response to nearly $1\,\mathrm{W/m^2}$ of forcing. $0.85\,\mathrm{W/m^2}$ of the $1.8\,\mathrm{W/m^2}$ forcing remains, i.e., additional global warming of $0.85 \times 2/3 \sim 0.6°\mathrm{C}$ is "in the pipeline" and will occur in the future even if atmospheric composition and other climate forcings remain fixed at today's values (3, 4, 22).

The present planetary energy imbalance is large by standards of Earth's history. For example, an imbalance of $1\,\mathrm{W/m^2}$ maintained for the last 10,000 years of the Holocene is sufficient to melt ice equivalent to 1 km of sea level (if there were that much ice), or raise the temperature of the ocean above

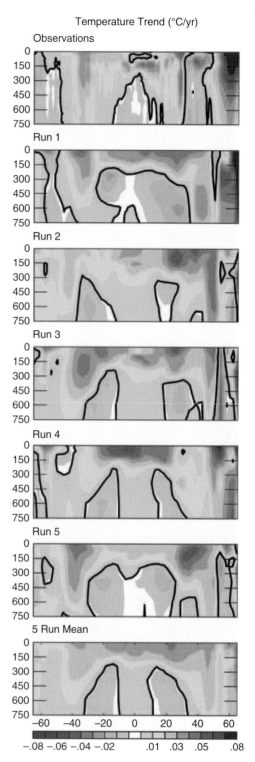

Fig. 3. Trend of zonally averaged temperature versus depth and latitude. Observations are from (19). Five model runs are as in Fig. 2.

the thermocline by more than 100°C (table S1). Clearly on long time scales the planet has been in energy balance to within a small fraction of $1\,\mathrm{W/m^2}$.

An alternative interpretation of the observed present high rate of ocean heat storage might be that it results, not from climate forcings, but from unforced atmosphere-ocean fluctuations. However, if a fluctuation had brought cool water to the ocean surface, as needed to decrease outgoing heat flux, the ocean surface would have cooled, while in fact it warmed (21). A positive climate forcing, anticipated independently, is the more viable interpretation.

The present $0.85 \, \text{W/m}^2$ planetary energy imbalance, its consistency with estimated growth of climate forcings over the past century (Fig. 1A), and its consistency with the temporal development of global warming based on a realistic climate sensitivity for doubled CO_2 (Fig. 1B) offer strong support for the inference that the planet is out of energy balance because of positive climate forcings. If climate sensitivity, climate forcings, and ocean mixing are taken as arbitrary parameters (23), one may find other combinations that yield warming comparable to that of the past century, but (1) climate sensitivity is constrained by empirical data, (2) our simulated depth of penetration of ocean warming anomalies is consistent with observations (fig. S2), thus supporting the modeled rate of ocean mixing, and (3) despite ignorance about aerosol changes, there is sufficient knowledge to constrain estimates of climate forcings (9).

The planetary energy imbalance and implied warming "in the pipeline" complicate the task of avoiding any specified level of global warming. For example, it has been argued, based on sea level during prior warm periods, that global warming of more than 1°C above the level of 2000 would constitute "dangerous anthropogenic interference" with climate (24, 25). With 0.6°C global warming in the pipeline and moderate growth of non-CO_2 forcings, a 1°C limit on further warming limits peak CO_2 to about 440 ppm (11). Given current CO_2 ~378 ppm, annual growth ~1.9 ppm (11), and a still expanding worldwide fossil fuel energy infrastructure, it may be impractical to avoid 440 ppm CO_2. A conceivable, though difficult, reduction of non-CO_2 forcings could increase the peak CO_2 limit for 1°C warming to a more feasible 520 ppm (11). This example illustrates that the 0.6°C unrealized warming associated with the planet's energy imbalance implies the need for near-term anticipatory actions, if a low limit on climate change is to be achieved.

Sea level. Sea level change includes steric (mainly thermal expansion) and eustatic (mainly changes of continental ice and other continental water storage) components. Observed temperature change in the upper 700–750 m yield a steric sea level rise of 1.4–1.6 cm (19, 20). The full ocean temperature changes in our five simulations yield a mean steric 10-year sea level increase 1.6 cm. Our climate model does not include ice sheet dynamics, so we cannot calculate eustatic sea level change directly. Sea level measured by satellite altimeters since 1993 increased 2.8 ± 0.4 cm/decade (26), but as a measure of the volume change (steric + eustatic) of ocean water this must be increased ~0.3 cm to account for the effect of global isostatic adjustment (27). We thus infer a eustatic contribution to sea level rise of ~1.5 cm in the past decade.

Both the total sea level rise in the past decade and the eustatic component, which is a critical metric for ice melt, are accelerations over the rate of the preceding century. IPCC (*10*) estimated sea level rise of the past century as 1.5 ± 0.5 cm/decade, with a central estimate of only 0.2 cm/decade for the eustatic component, albeit with large uncertainty. Decadal variability limits the significance of sea level change in a single decade (27, 28). However, we suggest that both the steric and eustatic increases are a product of the large, unusual, persistent planetary energy imbalance that overwhelms normal variability and as such may be a harbinger of accelerating sea level change (25).

The estimated ~1.5 cm eustatic sea level rise in the past decade, even if entirely ice melt, required only 2% of the Earth's present energy imbalance (table S1). Much more rapid melt is possible if iceberg discharge is accelerating, as some recent observations suggest (29, 30) and has occurred in past cases of sharp sea level rise that accompanied rapid global warming (31). Unlike ice sheet growth, which is limited by the snowfall rate in cold dry regions, ice sheet disintegration can be a wet process fed by multiple radiative and dynamical feedbacks (25). Thus the portion of the planetary energy imbalance used for melting is likely to rise as the planet continues to warm, summer melt increases, and melt-water lubricates and softens the ice sheets. Other positive feedbacks include reduced ice sheet albedo, a lowering of the ice sheet surface, and rising sea level (25).

Implications. The thermal inertia of the ocean, with resulting unrealized warming "in the pipeline", combines with ice sheet inertia and multiple positive feedbacks during ice sheet disintegration to create the possibility that the climate system could reach a point where large sea level change is practically impossible to avoid. If the ice sheet response time is millennia, the ocean thermal inertia and ice sheet dynamical inertia are relatively independent matters. However, it has been suggested, based on the saw-toothed shape of glacial-interglacial global temperature and qualitative arguments about positive feedbacks, that substantial ice sheet change could occur on the time scale of a century (25).

The destabilizing impact of comparable ocean and ice sheet response times is apparent. Say initial stages of ice sheet disintegration are detected. Before action to counter this trend could be effective it would be necessary to eliminate the positive planetary energy imbalance, now ~0.85 W/m², which exists due to the ocean's thermal inertia. Given energy infrastructure inertia and trends in energy use, that task could require of order a century. If the time for significant ice response is as short as a century, the positive ice-climate feedbacks imply the possibility of a system out of our control.

A caveat accompanying our analysis concerns the uncertainty in climate forcings. Good fit of observed and modeled temperatures (Fig. 1) also could be attained with smaller forcing and larger climate sensitivity, or with the converse. If climate sensitivity were higher (and forcings smaller), the rate of ocean heat storage and warming "in the pipeline" or "committed" would be greater, e.g., models with sensitivity 4.2–4.5°C for doubled CO_2 yield ~1°C "committed" global warming (3, 4). Conversely, smaller sensitivity and larger forcing yields lesser committed warming and ocean heat storage. The agreement between modeled and observed heat storage (Fig. 2) favors an intermediate climate sensitivity, as in our model. This test provided by ocean heat storage will become more useful as the period with large energy imbalance continues.

Even if the net forcing is confirmed by continued measurement of ocean heat storage, there will remain much room for trade-offs among different forcings. Aerosol direct and indirect forcings are the most uncertain. The net aerosol forcing that we estimate, −1.39 W/m², includes a large positive forcing by black carbon and a negative aerosol indirect forcing. Both of these aerosol forcings reduce sunlight reaching the surface, and may be the prime cause of observed "global dimming" (32) and reduced pan evaporation (33).

Given the unusual magnitude of the current planetary energy imbalance and uncertainty about its implications, careful monitoring of key metrics is needed. Continuation of the ocean temperature and altimetry measurements is needed to confirm that the energy imbalance is not a fluctuation and

determine the net climate forcing acting on the planet. The latter is a measure of the changes that will be needed to stabilize climate. Understanding of the forcings that give rise to the imbalance requires more precise information on aerosols (34). The high rate of recent eustatic sea level rise that we infer suggests positive contributions from Greenland, alpine glaciers, and West Antarctica. Quantification of these sources is possible using precise satellite altimetry and gravity measurements as initiated by the IceSat (35) and GRACE satellites (36), which warrant follow-on missions.

REFERENCES AND NOTES

1   J. Hansen et al., Science 229, 857 (1985).
2   M.I. Hoffert, A.J. Callegari, C.T. Hsieh, J. Geophys. Res. 85, 6667 (1980).
3   J. Hansen et al., J. Lerner, Amer. Geophys. Union Geophys. Mono. 29, 130 (1984).
4   T.M.L. Wigley, M.E. Schlesinger, Nature 315, 649 (1985).
5   M.I. Hoffert, C. Covey, Nature 360, 573 (1992).
6   J. Hansen, A. Lacis, R. Ruedy, M. Sato, H. Wilson, Natl. Geogr. Res. Explor. 9, 143 (1993).
7   R. Kerr, Science 305, 932 (2004).
8   G.A. Schmidt, et al., J. Climate, in review (2005).
9   J. Hansen, et al., J. Geophys. Res., in review (2005).
10 IPCC, Climate Change 2001: The Scientific Basis, J.T. Houghton et al., Eds. (Cambridge Univ. Press, New York, 2001).
11 J. Hansen, M. Sato, Proc. Natl. Acad. Sci. 101, 16,109 (2004).
12 T. Novakov et. al, Geophys. Res. Lett. 30, 1324, doi:10.1029/2002GL016345, (2003).
13 S. Menon and A. Del Genio, in Human-Induced Climate Change: An Interdisciplinary Assessment, M. Schlesinger et al., Eds., Cambridge Univ. Press, in review (2005).
14 S. Levitus, J.I. Antonov, T.P. Boyer, C. Stephens, Science 287, 2225 (2000).
15 S. Levitus et al., Science 292, 267 (2001).
16 T.P. Barnett, D.W. Pierce, R. Schnur, Science 292, 270 (2001).
17 S. Sun, J.E. Hansen, J. Climate 16, 2807 (2003).
18 J.M. Gregory, H.T. Banks, P.A. Stott, J.A. Lowe, M.D. Palmer, Geophys. Res. Lett. 31, L15312, doi:10.1029/2004GL020258 (2004).
19 J.K. Willis, D. Roemmich, B. Cornuelle, J. Geophys. Res., 109, C12036, doi:10.1029/2003JC002260 (2004).
20 S. Levitus, J.I. Antonov, T.P. Boyer, Geophys. Res. Lett., 32, L02604 (2004).
21 J. Hansen, R. Ruedy, M. Sato, M. Imhoff, W. Lawrence, D. Easterling, T. Peterson, T. Karl, J. Geophys. Res. 106, 23,947 (2001).
22 R.T. Wetherald, R.J. Stouffer, K.W. Dixon, Geophys. Res. Lett., 28, 1535 (2001).
23 R.S. Lindzen, Geophys. Res. Lett. 29(8), 10.1029/ 2001GL014360 (2002).
24 J. Hansen, Sci. Amer. 290, 68 (2004).

25 J. Hansen, *Ctim. Change* **68,** 269 (2005).
26 E.W. Leuliette, R.S. Nerem, G.T. Mitchum, *Mar. Geodesy* **27,** 79 (2004).
27 B.C. Douglas, W.R. Peltier, *Phys. Today* **55,** 35 (2002).
28 W. Munk, *Proc. Natl. Acad. Sci.* **99,** 6550 (2002).
29 I. Joughin, W. Abdalati, M. Fahnestock, *Nature* **432,** 608 (2004).
30 R. Thomas *et al.*, *Science* **306,** 255 (2004).
31 G. Bond *et al.*, *Nature* **360,** 245 (1992).
32 B. Liepert, *Geophys. Res. Lett.* **29(10),** 1421, 10.1029/2002GL014910 (2002).
33 M.L. Roderick, G.D. Farquhar, *Science* **298,** 1410 (2002)
34 M.I. Mishchenko *et al.*, *J. Quan. Spec. Rad. Trans.* **88,** 149 (2004).
35 H.J. Zwally, *et al. J. Geodyn.* **34,** 405 (2002).
36 B.D. Tapley, S. Bettadpur, J.C. Ries, P.F. Thompson, M.M. Watkins, *Science* **305,** 503 (2004).
37 We thank Waleed Abdalati, Ben Chao, Jean Dickey, Walter Munk, and Jay Zwally for helpful information, Darnell Cain for technical assistance, and NASA Earth Science Research Division managers Jack Kaye, Don Anderson, Phil DeCola, Tsengdar Lee, and Eric Lindstrom, and Hal Harvey of the Hewlett Foundation for research support.

SUPPORTING ONLINE MATERIAL

www.sciencemag.org/cgi/content/full/1110252/DC1
SOM Text
Figs. S1 and S2
Table S1
References

26 January 2005; accepted 19 April 2005
Published online 28 April 2005; 10.1126/science.1110252
Include this information when citing this paper.

9

Taking Earth's Temperature

Jones, P.D., Raper, S.C.B., Bradley, R.S., *et al.* (1986). Northern-hemisphere surface air-temperature variations – 1851–1984. *Journal of Climate and Applied Meteorology*, **25** (2): 161–179.

Fu, Q., Johanson, C.M., Warren S.G., *et al.* (2004). Contribution to stratospheric cooling to satellite-inferred tropospheric temperature trends. *Nature*, **429** (6987), 55–58.

Mann, M.E., Bradley, R.S., Hughes, M.K. (1998). Global-scale temperature patterns and climate forcing over the past six centuries. *Nature*, **392** (6678), 779–787.

Theoretical predictions are OK for academic questions, but when expensive decisions hang in the balance, the public generally requires a smoking gun, an observation from the real world to test the prediction. Jones *et al.* (1986) were not the first to compile a global average temperature of the Earth. The paper by Callendar (this volume) is another example. With a few early exceptions, the various compilations of temperature through time come to similar conclusions (Fig. 9.1), except that there are more years in "the past" by the time of the later studies. The agreement between the studies is interesting, in that it defangs the possibility that the global temperatures could be heavily impacted by the urban heat island effect, the bubbles of higher temperature often produced by the dry cement, roads, and roofing material of the urban environment. The urban heat island effect is a real phenomenon, but it is not global warming and it would be a mistake to attribute an urban heat warming to increases in atmospheric CO_2 concentration. The various studies represented in Fig. 9.1 took different strategies for eliminating it from the global average temperature, and the bottom line is the urban heat island has a negligible effect on the global average temperature. We chose Jones *et al.* for your consideration in this volume because it was by happenstance the first to encompass what in retrospect turned out to be the smoking gun for global warming: the temperature rise beginning in the 1970s.

Fu and Warren clear up a fly in the ointment of the Earth's temperature changes: the estimates of global temperature from the Microwave Sounding Unit (MSU) satellites. Satellites seem intuitively attractive as a means of taking the Earth's temperature, because they see the whole Earth nearly at once. However, satellites are not ideally suited for constructing records of the Earth's temperature over long spans of time. The orbits of the satellites decay, changing their readings, and the entire satellite era is covered with not one but many satellites, which have to be calibrated against each other to see any long-term trends. The satellites estimate the temperature of the atmosphere using microwaves emitted by molecular oxygen. The higher the temperature, the brighter the microwaves. However, the signal that the satellite sees is a composite of different altitudes. The closest you can get to the ground is a computational produce called LT, which still contains a substantial contribution from temperatures aloft. The fundamental discovery of Fu and Warren is that cooling in the stratosphere was masking some of the warming of the lower troposphere signal that we would like to compare with thermometer measurements. The corrected satellite temperature trends corroborate the warming measured by ground thermometers just fine.

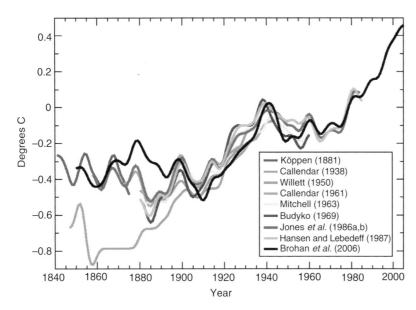

Fig. 9.1 **Reconstructions of changes in Earth's surface temperature from various studies. (From IPCC, 2007,** *The Scientific Basis,* **Figure 1.3, Le Treut, H., Somerville, R., Cubasch, U.,** *et al.* **(2007) Historical overview of climate change. In:** *Climate Change 2007: The Physical Science Basis. Contribution of Working Group I to the Fourth Assessment Report of the Intergovernmental Panel on Climate Change* **(eds S. Solomon, D. Qin, M. Manning, Z. Chen, M. Marquis, K.B. Averyt, M. Tignor H.L. Miller). Cambridge University Press, Cambridge, United Kingdom and New York.)**

Mann *et al.* (1998) pioneered the use of "proxy" measurements of temperature to extend the climate record back through the historical era of the last millennium and a bit further. Because of its generally flat shape through time, followed by an uptick of temperatures since about the year 1800, the Mann *et al.* record is often referred to as the "hockey stick." Many proxies for temperature have been developed: chemical or physical manifestations of past temperatures that have been preserved to this day, from which past changes in temperature can be inferred. The main source of information in the hockey stick is the width of tree rings, annual growth rings preserved in wood from temperate locations. Most of the data come from the northern hemisphere, so Mann *et al.* and many of its successor papers only apply to the northern hemisphere, rather than globally. Of course, other factors affect tree growth, such as shade, soil fertility, the age of the tree, and so on. Mann *et al.* "purified" the parts of each tree-ring record that really was driven by temperature by assuming that the spatial patterns of temperature variability in the deep past are the same as the patterns in the instrumental record, and tossing away apparent temperature changes from the tree rings that do not fit these observed spatial patterns. A dozen or more other authors have revisited the question since Mann *et al.* The newer studies tend to find more century-timescale changes in the Earth's temperature, but the conclusion stands that present-day temperatures are warmer than they have been in over a 1000 years.

Global Temperature Variations Between 1861 and 1984

P. D. JONES, T. M. L. WIGLEY & P. B. WRIGHT*

Climatic Research Unit, School of Environmental Sciences, University of East Anglia, Norwich NR4 7TJ, UK

Recent homogenized near-surface temperature data over the land and oceans of both hemispheres during the past 130 years are combined to produce the first comprehensive estimates of global mean temperature. The results show little trend in the nineteenth century, marked warming to 1940, relatively steady conditions to the mid-1970s and a subsequent rapid warming. The warmest 3 years have all occurred in the 1980s.

Global mean surface air temperature is the most commonly used measure of the state of the climate system. When general issues of climatic change are addressed, global mean temperature change is often used as a yardstick; the age of the dinosaurs was warmer than today, the ice ages were colder, and so on. Paradoxically, in the present era of instrumental meteorology, with data coverage far better than at any earlier time, our knowledge of global mean temperature changes is still uncertain. Variations in global mean air temperature are of considerable importance, as they are a measure of the sensitivity of the climate system to external forcing factors such as changes in carbon dioxide concentration, solar output and the frequency of explosive volcanic eruptions. Quantifying the response of the climate to external forcing changes is a major goal of climatology and a prerequisite for predicting future climatic change. As a step towards this goal, we present here the first global synthesis of near-surface temperature measurements over the land and oceans.

Most earlier estimates of global and hemispheric mean temperature (see refs 1, 2) were based solely on data from land-based meteorological stations. Since >70% of the globe is ocean, one might suspect the global representativeness of such estimates, although on long timescales (≥decades) the thermal coupling between land and ocean should ensure that the land data largely mirror changes occurring over the oceans.[1] Recently, data from ships at sea collected for routine weather forecasting purposes, have been compiled by groups in the United Kingdom[3,4] and the United States,[5,6] and these data give us the potential to calculate improved estimates of global mean temperature. Apart from our own work,[7] the only previous attempt to analyse both land and marine data is that of Paltridge and Woodruff.[8,9] These authors, however, failed to account for inhomogeneities in the marine data, which are substantial (see below and also refs 4 and 10). The quality and coverage of the land data they used was also less than adequate, but this is understandable because they were primarily interested in sea-surface temperature variations.

The land data we use are those from refs 11, 12. These have been carefully examined to detect and correct for non-climatic errors that may result from station shifts or instrument changes, changes in the methods used for calculating means, urban warming, and so on. Although problems still exist,[13,14] the quality of these data is much better than that of material used in earlier studies. Area averages based on these data show medium to long timescale trends (≥10yr) whose spatial consistency provides a strong pointer to the data's overall reliability.[1,11,12] The marine data we employ are those in the COADS (Comprehensive Ocean Atmosphere Data Set) compilation[6] which extends to 1979, and data from the Climate Analysis Center, NOAA, for 1980–84. We use both sea surface temperatures (SST) and marine air temperatures (MAT).

MARINE DATA PROBLEMS

Both SST and MAT data contain 'inhomogeneities', variations resulting from non-climatic factors.[4,10,15] For example, early SSTs were measured

* Present address: Max-Planck-Institut für Meteorologie, Bundesstrasse 55, 2000 Hamburg 13, FRG.

Jones, P. D. (1986) Global temperature variations between 1861 and 1984, *Nature* 322: 430–434. Reprinted by permission from Macmillan Publishers Limited.

using water collected in uninsulated, canvas buckets, while more recent data come either from insulated bucket or cooling water intake measurements, with the latter considered to be 0.3–0.7 °C warmer than uninsulated bucket measurements.[10] For marine air temperatures, changes in the size and speed of ships, especially those increases associated with the sail to steam transition, are both thought to have influenced data homogeneity. In addition, many early air temperature observations were not taken in screened locations. Because of these non-climatic factors, both SST and MAT data must be corrected (or 'homogenized') to remove their effects.

Folland *et al.*[4,16] and Folland and Kates,[17] using the UK Meteorological Office (UKMO) data bank,[3] attempted to overcome these problems by identifying specific sources of error, attempting to quantify these and using this information to make corrections to the raw gridded data. Such corrections have inherent uncertainties because of difficulties in their *a priori* quantification and a lack of knowledge of how most measurements were taken. Information on whether bucket or intake measurements were made has, in most cases, apparently been lost or never recorded. It has also been shown[18] that supposedly homogeneous (that is bucket-only or intake-only) SST data series appear to have non-climatic changes that are similar to those found in mixed data series, suggesting that all historical data sets contain a mix of measurement types. Since 1945, however, it is generally assumed that available SST data contain a reasonably consistent mix of intake and bucket measurements.[18]

The Folland *et al.*[4] corrected MAT and SST series have been compared with averages of land-based data by Jones *et al.*[11,12] Agreement is reasonable since the start of the twentieth century, although MAT values for the years 1942–45 appear to be too warm in both hemispheres. Before 1900, the marine and land series diverge markedly, with both marine series being about 0.3 °C warmer than the land data.

CORRECTING THE COADS DATA

The COADS compilation contains some 63.25 million non-duplicated SST observations, of which 0.96 million have been 'trimmed' to remove extreme outliers.[5] While these are more data than in the UKMO SST set (which has about 46 million non-duplicated observations),[4] the effective area and density of coverage is very similar in both data sets. However, unlike the UKMO data set used by Folland *et al.*,[4] none of the data in COADS have been corrected for non-climatic effects. Our first task, therefore, was to homogenize the COADS data. We did this by comparing marine and land data in areas where the two abut or overlap (coastal areas and around ocean islands).

The trimmed COADS data include monthly means and medians on a 2° × 2° grid, together with the number of observations in a month and the mean observation date. We compressed the data onto a more manageable grid (5° × 5° for MAT, 4° × 10° for SST) after first eliminating values where the number and distribution of observations was likely to have produced unrepresentative monthly means, and expressed the values as anomalies from a 1950–79 reference period. As a test of data quality at this stage we calculated hemispheric mean values by appropriately weighting the gridded MAT and SST data (NH, Northern Hemisphere; SH, Southern Hemisphere). Year-to-year variations for these uncorrected data were found to be in excellent agreement with the UKMO corrected data (NHSST, $r = 0.86$; NHMAT, $r = 0.87$; SHSST, $r = 0.88$; SHMAT, $r = 0.75$ over 1856–1979: correlation coefficients calculated using residuals from a 10-yr gaussian filter), but, as expected, the long-term (≥ 10 yr) fluctuations showed marked differences. Similar high frequency correlations between SST and MAT for the uncorrected COADS data (NH, $r = 0.91$; SH, $r = 0.89$) were higher than in the corrected UKMO data (NH, $r = 0.81$; SH, $r = 0.80$).

Because of the high SST-MAT correlation (see also ref. 19), SST data can be corrected by comparison with MAT data, once the latter have been corrected. For the MAT data, any attempt to assess, *a priori*, the magnitudes of errors arising from instrumental changes, changes in observation methods, and the effects of changes in ships' thermal inertia, speed and size (the latter determines the height at which observations were taken), must be fraught with uncertainty. Data reliability and long-term homogeneity can be far more convincingly demonstrated for the gridded land data than for the marine data because land station data homogeneities can be more easily identified, explained and corrected.[11,12] We therefore use these data directly to correct the marine data. Fifteen regions (see Fig. 1) were chosen in which land and marine data are in close proximity. Area averages of annual mean MAT and land air temperature were calculated for each region using the

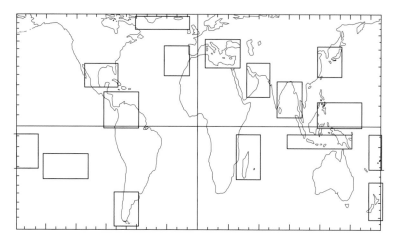

Fig. 1. Map showing the 15 regions where marine air temperatures and land-based temperatures were compared (Peters equal-area rectangular projection).

uncorrected COADS data and the homogenized land data produced by Jones *et al.*[11,12] No attempt was made to consider night-time observations only, as used by Folland *et al.*[4] In addition to the 15 pairs of area averages, annual mean coastal land time series were produced for both hemispheres and compared with the uncorrected hemispheric-mean MAT series.

The 17 land minus MAT time series were then examined for systematic differences between the land and marine data. For the period 1861–1979 (both marine and Southern Hemisphere land data are unrepresentative before 1861 because of poor data coverage), five distinct periods could be discerned in all 15 regional land minus MAT time series and in the two hemispheric land minus MAT time series. The latter are shown in Fig. 2. The three main periods are: the period up to the 1880s when the MAT data appear to be too warm by 0.4–0.5 °C; the period from the 1900s to 1941 when the MAT data are too cold by 0.1–0.2 °C; and 1946–79 when there is no obvious bias. There is a strong upward trend in the land-minus-MAT difference between the mid 1880s and the late 1900s, and the war years, 1942–45, are marked by anomalously warm MAT values. The consistency between the hemispheres is clear from Fig. 2, and the land minus MAT data for the individual smaller regions, although showing greater inter-annual variability, all show the same features.

The nineteenth century land minus MAT data also show differences between the values before and after about 1873 (see Fig. 2). By examining land, MAT and SST data it can be shown that this difference is also likely to reflect a non-climatic

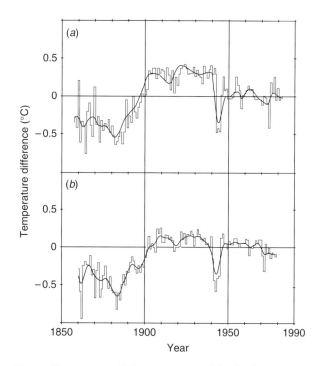

Fig. 2. Temperature differences: coastal land values minus uncorrected COADS marine air temperature values for the Northern (*a*) and Southern (*b*) Hemispheres. Smooth curves show 10-yr gaussian filtered values, padded at each end as described in ref. 11.

inhomogeneity in either the MAT data or the land data, probably the former.

The means and standard deviations of the land minus MAT values are shown in Table 1. The consistency of these values strongly suggests that these land/MAT discrepancies are not climatic in origin. They may, therefore, be used to estimate

Table 1. Comparison between coastal land and MAT data.

		1861–73	1874–89	1903–41	1942–45	1946–79
NH	$\bar{X}$	−0.35	−0.50	0.23	−0.49	−0.02
	s	0.26	0.11	0.09	0.02	0.12
SH	$\bar{X}$	−0.36	−0.53	0.10	−0.44	0.03
	s	0.23	0.14	0.09	0.09	0.10
NH (9 region average)	$\bar{\bar{X}}$	−0.36	−0.42	0.17	−0.54	−0.03
	$s(\bar{X})$	0.40	0.21	0.10	0.10	0.05
SH (6 region average)	$\bar{\bar{X}}$	−0.61	−0.52	0.17	−0.44	0.05
	$s(\bar{X})$	0.57	0.36	0.22	0.15	0.08
Correction		−0.40	−0.48	0.17	−0.54	0.00

$\bar{X}$ = mean land minus MAT value; s = corresponding standard deviation defined by $s^2 = (Y-1)^{-1} \sum (X_j = \bar{X})^2$ where X_j is the value in year j, and Y is the number of years; $s(\bar{X})$ = standard deviation of the means defined by $(s(\bar{X}))^2 = (n-1)^{-1} \sum (\bar{X}_i - \bar{\bar{X}})^2$ where n is the number of regions (6 or 9), $\bar{X}_i$ is the mean for region i and $\bar{\bar{X}}$ is the average value of $\bar{X}_i$. The last line shows the inferred correction which was added to the uncorrected annual MAT data.

annual correction factors for the MAT data in order to make these data compatible with the existing homogenized land data. Except for the 1942–45 period, when war conditions apparently prompted observers to measure temperature in unconventional locations,[4] the specific reasons for these non-climatic MAT fluctuations are not known. Although their reality cannot be questioned, there is clearly some uncertainty in the magnitude of the implied corrections.

The correction values we have used (added to the raw MAT data) are (°C): 1861–73, −0.40; 1874–89, −0.48; 1903–41, 0.17; 1942–45, −0.54; 1946–79, 0.0; with linear interpolation between 1889 and 1903. Slightly different corrections were judged necessary for Southern Hemisphere data between 1941 and 1945: 1941, −0.14; 1942–45, −0.44. Most of the transition dates for these correction factors, which are based on a number of considerations, could be altered slightly with no appreciable effect on the resulting corrected MAT values. Although the 0.08 °C difference in the MAT corrections before and after 1873 may be inappropriate if it arises from a land data inhomogeneity, we judge this to be unlikely. It has the effect of slightly reducing the magnitude of the long-term MAT warming between the period before 1873 and today. The corrections generally reflect the mean land minus MAT values shown in Table 1, but the precise values used and the transition dates also take MAT-SST comparisons into account. Our corrections differ markedly from those applied by Folland *et al.*[4] to their night-time MAT data. This is a clear indication of incompatibilities between the corrected UKMO MAT data and the homogenized land data (see also refs 11 and 12).

Table 2. Comparison between corrected MAT data and uncorrected SST data.

		1861–89	1903–41	1942–45	1946–79
NH	$\bar{X}$	0.08	0.49	−0.07	0.02
	s	0.08	0.08	0.07	0.09
SH	$\bar{X}$	0.07	0.50	−0.14	0.02
	s	0.04	0.05	0.08	0.08
Correction		0.08	0.49	−0.10	0.00

$\bar{X}$ = mean MAT minus SST value; s = corresponding standard deviation. The correction is the number added to the uncorrected annual SST data.

Having corrected the MAT data, we can now estimate the SST corrections required to ensure overall compatibility between the land, MAT and SST data by comparing the corrected MAT and raw SST values. Table 2 and Fig. 3 show the hemispheric mean differences between the corrected MAT data and the raw SST data. As with the MAT analysis, three distinct periods can be discerned: pre-1890 when the SST data are slightly but consistently cooler than the MAT data; 1903–41 when SSTs are markedly cooler than MATs; and post-1945 when there is no consistent difference. Rather complex transitions exist between these three phases. The MAT-SST difference curves are essentially the same in both hemispheres. This is a strong indication that the differences reflect non-climatic effects, and it provides a valuable consistency check on the MAT corrections.

The implied SST corrections, are (°C): 1861–89, 0.08; 1903–41, 0.49; 1942–45, −0.10; 1946–79, 0.0; with linear interpolation between 1889 and 1903. For 1941 we applied a slightly different

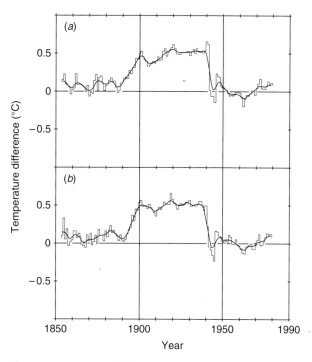

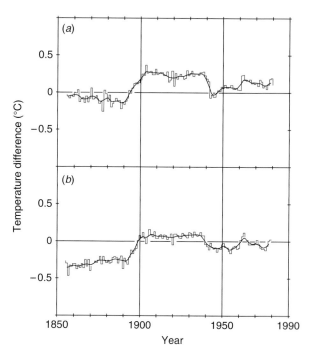

Fig. 3. Temperature differences: corrected marine air temperatures minus uncorrected sea surface temperatures for the Northern (*a*) and Southern (*b*) Hemispheres. Smooth curves show 10-yr gaussian filtered values.

Fig. 4. Differences between the hemispheric-mean sea surface temperature values produced in the present work and those of Folland *et al.*;[4] Northern Hemisphere (*a*), Southern Hemisphere (*b*). Smooth curves show 10-yr gaussian filtered values. The implied warmth of the Folland *et al.* SH data relative to the NH (by ~0.2 °C), is due to their use of 1951–60 as a reference period. Conditions during this decade differed noticeably from the mean conditions during the reference period, 1950–79, used here (see Fig. 5).

correction in the Southern Hemisphere, 0.19 °C. As for MAT, these corrections also differ somewhat from those used by Folland *et al.*[4] In their analysis, SST values were adjusted to ensure compatibility with corrected MAT values, just as we have done. However, since their corrected MAT values must differ noticeably from those produced here, differences in the SST corrections will, in part, reflect these MAT differences.

In our analysis, the difference between the twentieth century SST correction factor before 1941 and after 1946 is 0.49 °C. This difference is in the range (0.3–0.7 °C) generally accepted for the difference between uninsulated bucket and intake SST measurements.[18,20,21] The precise reasons for the differences that we obtain between the nineteenth century and early twentieth century MAT and SST corrections are uncertain. For MAT, the change is likely to be related to the transition from sail to steam. Between 1880 and 1910, the percentage of steamship tonnage as a fraction of total shipping tonnage rose from ~25 to 75% (ref. 22). Noticeable increases in ship speed occurred over the period 1880–1900, and in ship size over the period 1890–1910 (ref. 22). These dates should be compared with the duration of the rising trends in land minus uncorrected-MAT data in both

hemispheres shown in Fig. 2. Changes in MAT may be related to exposure changes attendant on the above, and to other changes in instrument exposure procedure which occurred over the same period. For SST, the main reasons for the change may be the standardization of the measuring technique and the introduction of more reliable instruments.[23] It is also possible that, in the mid to late nineteenth century, many bucket temperatures were not taken in the shade.[24] In addition, some of the earlier measurements may have been made with wooden rather than canvas buckets. The latter, being uninsulated and subject to evaporative cooling, produce lower temperature readings.

The overall differences between the hemispheric mean SST values produced here and those of Folland *et al.*[4] are shown in Fig. 4. The results for an MAT comparison are similar. The discrepancies are large and comparable in magnitude to either set of corrections. The reasons for these differences stem mainly from the different correction factors applied to what are essentially similar raw data. Because there are several sources of data inhomogeneity, we have not attempted to correct for these individually. The result should be more complete than Folland *et al.*[4] who attempted to make specific corrections for identified sources of inhomogeneity based on physical arguments. Our corrections synthesize the effects of several different factors. However, while they ensure compatibility between the marine and land data, the fact that the reasons for these corrections are uncertain must point towards some remaining uncertainty in our corrected marine data, especially in the nineteenth century.

GLOBAL MEAN TEMPERATURES

It is a relatively simple matter to produce estimates of annual global mean surface air temperature using the available (corrected) marine data and the most recent compilations of land data.[11,12] There are three different ways in which global or hemispheric (land plus marine) averages can be calculated. The first method is to average only those grid point values (with appropriate cosine weighting) for which data exist. This is the way hemispheric means have been produced for the land data.[11,12] The second and third methods assume that each of the four independent time series (NH and SH land and NH and SH marine, either SST or MAT) are, at all times, representative either of their maximum coverage or of the total areas of the four domains. The results obtained differ but little, and the use of either SST or MAT to represent the marine domains produces only minor differences. We therefore show only results using the second method based on SST data, obtained using

$$T \text{ global} = 0.25\text{NH land} + 0.25\text{NH SST} \\ + 0.2\text{SH land} + 0.3\text{SH SST}$$

(Fig. 5) where, after 1957, SH land includes Antarctic data from Raper *et al.*,[25] updated. The insensitivity to the precise method of weighting arises because all time series are quite strongly correlated.

The reliability of the time series given in Fig. 5 as true hemispheric and global averages can be questioned because the spatial coverage, even at best, is less than 75% and because the coverage changes with time. Coverage is always much better in the Northern Hemisphere. Coverage before 1900 is generally less than one third of the globe, down to <20% in the 1860s. The question of representativeness of

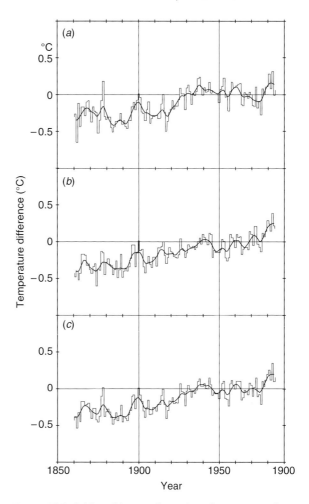

Fig. 5. Global (*c*) and hemispheric (Northern, *a*; Southern, *b*) annual mean temperature variations since 1861, based on sea-surface temperature data to represent the marine domain and using weights corresponding approximately to the maximum coverage for the four domains (method two in the text). Smooth curves show 10-yr gaussian filtered values. 1980–84 values are based on SST data obtained from the Climate Analysis Center, U.S. National Oceanic and Atmospheric Administration (see ref. 29 for information about this data source). These data were adjusted to be compatible with the values in earlier years by comparing values in both hemispheres over the overlap period, 1970–79. The CAC data correlate highly with the COADS data (*r* = 0.984 for the Northern Hemisphere mean and *r* = 0.991 for the Southern Hemisphere mean).

the land data has been considered in detail in refs 1, 11 and 12. Although marine coverage before 1900 is sparse, the spatial correlation length over the oceans is large and limited coverage should still give results representative of a much larger area. Nevertheless, there are large parts of the Southern Hemisphere that nearly always lack data, especially the southern oceans south of 45 °S and the whole of the southeastern Pacific (except near the South

American coast). Before 1957, when most Antarctic data first became available, there are essentially no data at all for the globe south of 45 °S (refs 25, 26). Although this represents only ~15% of the area of the globe, temperature fluctuations at high latitudes are known to be larger than at lower latitudes and so can have a disproportionate effect on the global average.[12,27] Any interpretation of Fig. 5 must bear in mind both these basic data deficiencies and the marine data uncertainties implied by Fig. 4. We note, however, that the latter do not affect the gross features of the global mean changes observed this century.

The global curve is extremely interesting when viewed in the light of recent ideas of the causes of climatic change.[1,2] The data show a long timescale warming trend, with the three warmest years being 1980, 1981 and 1983, and five of nine warmest years in the entire 134-yr record occurring after 1978. With regard to the hypothesized warming due to increasing concentrations of carbon dioxide and other greenhouse gases, the overall change is in the right direction and of the correct magnitude.[1,7,28] However, the relatively steady conditions maintained between the late 1930s and mid 1970s requires either the existence of some compensating forcing factor or, possibly, a lower sensitivity to greenhouse gas changes than is generally accepted.

We thank particularly C. K. Folland, D. E. Parker, P. M. Kelly, T. P. Barnett and D. J. Shea for useful comments. We thank D. E. Parker and C. K. Folland (UK Meteorological Office) for access to unpublished data used in Fig. 4 Scott Woodruff (Environmental Research Laboratories, US National Oceanic and Atmospheric Administration) for an early copy of the trimmed COADS data set, and the Climate Analysis Center, US National Oceanic and Atmospheric Administration, for the recent marine data used in Fig. 5. This work was funded by the Carbon Dioxide Research Division, US Department of Energy, grant no. DE-FGO2-86-ER60397.

NOTES

1 Wigley, T. M. L., Angell, J. K. & Jones, P. D. in *Detecting the Climatic Effects of Increasing Carbon Dioxide* (eds MacCracken, M. C. & Luther, F. M.) 55–90 (U.S. Dept of Energy, 1985).

2 Hansen, J. *et al. Science* **213**, 957–966 (1981).

3 Shearman, R. J. *Met. Mag.* **112**, 1–10 (1983).

4 Folland, C. K., Parker, D. E. & Kates, F. E. *Nature* **310**, 670–673 (1984).

5 Woodruff, S. D. in *Proc. 3rd Conf. on Climate Variations and Symp. on Contemporary Climate: 1850–2100*, 14–15 (American Meteorological Society, Boston, 1985).

6 Slutz, R. J. *et al. Comprehensive Ocean–Atmosphere Data Set Release 1* (NOAA Environmental Research Laboratories, Boulder, 1985).

7 Wigley, T. M. L., Jones, P. D. & Kelly, P. M. in *The Greenhouse Effect, Climatic Change, and Ecosystems* (eds Bolin, B., Döös, B. R., Jäger, J. & Warrick, R. A.) (SCOPE series, Wiley, in the press).

8 Paltridge, G. W. & Woodruff, S. *Mon. Weath. Rev.* **109**, 2427–2434 (1981).

9 Paltridge, G. W. *Mon. Weath. Rev.* **112**, 1093–1095 (1984).

10 Barnett, T. P. in *Detecting the Climatic Effects of Increasing Carbon Dioxide* (eds MacCracken, M. C. & Luther, F. M.) 91–107 (U.S. Dept of Energy, 1985).

11 Jones, P. D. *et al. J. Clint, appl. Met.* **25**, 161–179 (1986).

12 Jones, P. D., Raper, S. C. B. & Wigley, T. M. L. *J. Clim. appl. Met.* **25** (in the press).

13 Bradley, R. S., Kelly, P. M., Jones, P. D., Diaz, H. F. & Goodess, C. *A Climatic Data Bank for the Northern Hemisphere Land Areas, 1851–1980, DoE Tech. Rep. No. TR017* (U.S. Dept of Energy, 1985).

14 Bradley, R. S. & Jones, P. D. in *Detecting the Climatic Effects of Increasing Carbon Dioxide* (eds MacCracken, M. C. & Luther, F. M.) 29–53 (U.S. Dept of Energy, 1985).

15 Wright, P. B. *Mon. Weath. Rev.* **114** (in the press).

16 Folland, C. K., Parker, D. E. & Newman, M. in *Proc. 9th Annual Climate Diagnostics Workshop*, 70–85 (U.S. Dept. of Commerce, NOAA, Washington DC, 1984).

17 Folland, C. K. & Kates, F. E. in *Milankovitch and Climate* (eds Berger, A., Imbrie, J., Hays, J., Kukla, G. & Saltzman, B.) 721–727 (Reidel, Dordrecht, 1984).

18 Barnett, T. P. *Mon. Weath. Rev.* **112**, 303–312 (1984).

19 Cayan, D. R. *Mon. Weath. Rev.* **108**, 1293–1301 (1980).

20 Saur, J. F. T. *J. appl. Met.* **2**, 417–425 (1963).

21 James, R. W. & Fox, P. T. *Comparative sea-surface temperature measurements. Marine Science Affairs Rep. No. 5*, WMO Publ. No. 336 (Geneva, 1972).

22 Kirkaldy, A. W. *British Shipping, Its History, Organization and Importance* (Kegan Paul, Trench, Trubner, London, 1919).

23 Krümmel, O. *Handbuch der Ozeanographie*, Vol. 1 (Engelhorn, Stuttgart, 1907).

24 Brooks, C. F. *Mon. Weath. Rev.* **54**, 241–253 (1926).

25 Raper, S. C. B., Wigley, T. M. L., Mayes, P. R., Jones, P. D. & Salinger, M. J. *Mon. Weath. Rev.* **112**, 1341–1353 (1984).

26 Raper, S. C. B., Wigley, T. M. L., Jones, P. D., Kelly, P. M., Mayes, P. R. & Limbert, D. W. S. *Nature* **306**, 458–459 (1984).

27 Kelly, P. M., Jones, P. D., Sear, C. B., Cherry, B. S. G. & Tavakol, R. K. *Mon. Weath. Rev.* **110**, 71–83 (1982).

28 Wigley, T. M. L. & Schlesinger, M. E. *Nature* **315**, 649–652 (1985).

29 Reynolds, R. W. & Gemmill, W. H. *Trop. Ocean-Amos. Newslett.* No. 23, 4–5 (1984).

Contribution of Stratospheric Cooling to Satellite-Inferred Tropospheric Temperature Trends

QIANG FU,[1] CELESTE M. JOHANSON,[1] STEPHEN G. WARREN[1] & DIAN J. SEIDEL[2]

[1]Department of Atmospheric Sciences, University of Washington, Seattle, Washington 98195, USA
[2]NOAA Air Resources Laboratory, Silver Spring, Maryland 20910, USA

From 1979 to 2001, temperatures observed globally by the mid-tropospheric channel of the satellite-borne Microwave Sounding Unit (MSU channel 2), as well as the inferred temperatures in the lower troposphere, show only small warming trends of less than 0.1 K per decade (refs 1–3). Surface temperatures based on *in situ* observations however, exhibit a larger warming of ~0.17 K per decade (refs 4, 5), and global climate models forced by combined anthropogenic and natural factors project an increase in tropo-spheric temperatures that is somewhat larger than the surface temperature increase.[6–8] Here we show that trends in MSU channel 2 temperatures are weak because the instrument partly records stratospheric temperatures whose large cooling trend[9] offsets the contributions of tropospheric warming. We quantify the stratospheric contribution to MSU channel 2 temperatures using MSU channel 4, which records only stratospheric temperatures. The resulting trend of reconstructed tropospheric temperatures from satellite data is physically consistent with the observed surface temperature trend. For the tropics, the tropo-spheric warming is ~1.6 times the surface warming, as expected for a moist adiabatic lapse rate.

The inconsistency between the trends at the surface and in the troposphere, traceable to the pioneering work of ref. 10, has raised questions about the ability of current global climate models (GCMs) to predict climate changes, the reliability of the observational data used to derive temperature trends, and the reality of human-induced climate change.[4,11–15] It is generally agreed that the warming trend in global-mean surface temperature observations during the past 20 years is real and at least partly of anthropogenic origin.[4,12] This increase of temperature is supported by observations of a reduction of snow cover and sea ice, thawing of permafrost, changes in freeze/thaw dates of lake and river ice, ocean warming and sea-level rise, and other related environmental changes.[4] However, the situation is less clear for tropospheric temperatures. Balloon-borne radiosondes have been the principal tool for atmospheric profiling. From 1979 to 2001, the trends of tropospheric-layer temperature between 850 and 300 hPa, as derived from different radiosonde data sets[3], range from −0.03 to +0.04 K per decade. The radiosondes have limited spatial coverage, particularly over large parts of the oceans, and are subject to a host of complications, including changing instrument types and observation practices,[16,17] which confound analyses of climate trends.

The MSU, since 1979, and its successor, the Advanced MSU (AMSU), from 1998, provide a global measure of temperature for several atmospheric layers from NOAA polar-orbiting satellites. Although the original purpose of MSU measurements was to improve weather forecasts, a continuing data-analysis effort has been made to satisfy climate research requirements of homogeneity and calibration.[1,2,10,18–24] Several important non-climatic influences have been identified and removed, including diurnal temperature biases related to local sampling times of the satellite and their changes over its lifetime, errors in the MSU calibration, and biases due to decay of the satellite orbits. Recent analyses of MSU channel 2 (most sensitive to the mid-troposphere) by the University of Alabama at Huntsville (UAH) and the Remote Sensing Systems (RSS) teams find temperature trends of 0.01 K per decade[1] and 0.1 K per decade,[2] respectively, during 1979–2001. This trend difference is mainly due to differences in data adjustments related to instrument calibration and diurnal drift correction.[2] The purpose of this Letter is not to reconcile the trend

Fu, Q. (2004) Contribution of stratospheric cooling to satellite-inferred tropospherice temperature trends. *Nature* 429 (6987): 55–58. Reprinted with permission of Macmillan Publishers Limited.

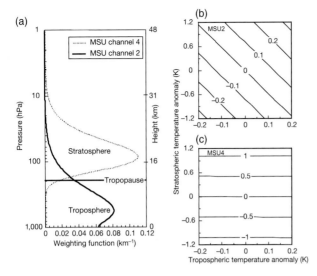

Fig. 1. Atmospheric weighting functions and brightness temperature responses. (a), Weighting function profiles for MSU channels 2 and 4 over ocean[1,2]. The boundary between the troposphere and the stratosphere (the tropopause) is shown at 200 hPa. The satellite-observed brightness temperature, T_b, can be expressed in the form:

$$T_b = T_s W_s + \int_0^\infty T(z)W(z)\mathrm{d}z$$

where T_s is the surface temperature, W_s the surface contribution factor, $T(z)$ the atmospheric temperature profile, and $W(z)$ the weighting function. Thus the weighting function describes the relative contributions of atmospheric temperatures at different heights to the brightness temperatures observed by the satellite. (b), Responses of MSU channel 2 brightness temperature to changes in stratospheric and tropospheric temperatures assuming a US Standard Atmosphere and a surface emissivity of 0.5. Contours and axes are both labelled in K. (c), Same as (b) but for MSU channel 4.

differences between these two research teams, but to address the question of whether the MSU data really imply small or negligible tropospheric warming over the past two decades. We argue that the trends reported by both teams for the 'mid-tropospheric' channel are substantially smaller than the actual trend of the mid-tropospheric temperature.

To infer the temperature of the mid-troposphere, we use two microwave channels: MSU channels 2 and 4 (or AMSU channels 5 and 9). Their weighting functions are shown in Fig. 1a. The weighting function for MSU channel 2 (or AMSU channel 5) peaks at ~550 hPa (~4.5 km). Thus the MSU Channel 2 brightness temperatures (T_2) have often been used to represent mid-tropospheric temperatures.[1,2,3,8,15,22,24] The MSU channel 4 (or AMSU channel 9), whose weighting function peaks

at ~85 hPa (~18 km), has been used to represent stratospheric temperatures.[1,3,8,15]

As ~85% of the signal for T_2 comes from the troposphere and surface, it is not a bad approximation to say that the seasonal and interannual variations of mean deep-layer temperature in the troposphere can be well represented by T_2. However, this might not be the case for trends. Figure 1b shows simulated changes of T_2 owing to changes of tropospheric and stratospheric temperatures, and indicates that T_2 remains constant when the changes in tropospheric and stratospheric temperatures have a ratio of about −1/5. This is because the vertical integral of the weighting function from the surface to tropopause, near 200 hPa, is about 5 times the integral above the tropopause. For example, if the tropospheric temperature trend were 0.15 K per decade and the stratospheric trend were −0.8 K per decade, the trend of T_2 would be close to zero. Intriguingly, this scenario resembles what may actually be the case in the atmosphere. The temperature trend in the lower stratosphere (15 to 23 km), as obtained from radiosondes and satellite observations, is about −0.5 to −0.9 K per decade for the last 20 years.[1,3,4,9] Therefore, because of the combined influence of stratospheric and tropospheric changes, T_2 trends are not an ideal indicator of global climate change. To derive the tropospheric temperature trends, the effects of stratospheric cooling on T_2 must be taken into account.

Although a stratospheric influence on the T_2 trend has long been recognized,[20,25,26] it has never been well quantified. The UAH team created a synthetic channel called T_{2LT}, where LT means 'low-middle troposphere',[1,19,25] by subtracting signals at different view-angles of MSU channel 2. However, this approach amplifies noises, increases satellite inter-calibration biases, and may introduce other complications involving effects of changes in surface emissivity and of mountainous terrain.[2,19,20,24–28] For these reasons, the T_{2LT} record is now receiving less attention than the better-calibrated T_2 record.[2,15,22,24] Here we develop an alternative method to remove the stratospheric contribution, which should be free of the complications afflicting T_{2LT}, by making use of data from MSU channel 4.

The MSU channel 4 brightness temperature (T_4) is sensitive mainly to stratospheric temperature changes (Fig. 1c), so it can be used to remove the contribution of the stratosphere to T_2. We define the free-tropospheric temperature as the mean temperature between 850 and 300 hPa ($T_{850-300}$; ref. 3). We derive this temperature from the measured brightness temperatures of MSU channels 2 and 4, as:

$$T_{850-300} = a_0 + a_2 T_2 + a_4 T_4 \qquad (1)$$

To obtain these three coefficients, we use global-, hemispheric- and tropical-average monthly temperature anomaly profiles from radiosonde observations at 87 stations, for the period 1958–97 (ref. 17). The radiosonde data at the surface and at 15 pressure levels between 1,000 and 10 hPa are used to derive temperature anomalies for the 850-300-hPa layer as well as for MSU channels 2 and 4 (ref. 3). The coefficients in equation (1) are then obtained by least-squares regression (see Supplementary Table 1 for the values). For global-average anomalies, a_2 is 1.156 and a_4 is −0.153. The effective vertical weighting function for $T_{850-300}$ (that is, $a_2 W_2 + a_4 W_4$, where $W_{2,4}$ are the physical weighting functions for $T_{2,4}$) peaks at the same level as T_2 but is 15% larger. In the stratosphere it is negative above ~100 hPa and positive below, so that the integrated contribution of the stratosphere becomes near-zero. The effective weighting function may have a negative part[25]; this is different from the physical weighting function, which must be positive everywhere.

The success of equation (1) in predicting $T_{850-300}$ from T_2 and T_4 is shown in Supplementary Fig. 1. The global-average anomalies of 850-300-hPa layer temperature, as derived from the radiosonde-simulated T_2 and T_4, closely follow those directly observed by radiosondes for the period 1958–79. The correlation coefficient is 0.984, with a root-mean-square error of 0.065 K. The trend differences are only about 0.001 K per decade. The T_4 time series can therefore be used to remove nearly all of the contribution of the stratosphere to T_2 in the trend analyses. This is because temperature variations in the stratosphere are vertically coherent and well correlated with T_4.

We now apply equation (1) to satellite-observed time series of T_2 and T_4 from 1979 to 2001, as reported by UAH[1] and RSS[2], to derive $T_{850-300}$. The global anomaly time series of $T_{850-300}$ using RSS data is shown in Fig. 2, along with those of T_2 and T_4. It is evident that the $T_{850-300}$ trend is more positive than the T_2 trend. (Similar results are obtained using UAH data, not shown here.)

Figure 3 shows the trends for T_2 (Fig. 3a) and MSU-derived $T_{850-300}$ (Fig. 3b) for the globe, Northern Hemisphere, Southern Hemisphere, and tropics (30° N-30° S) using both the UAH and RSS datasets, as well as surface temperature trends

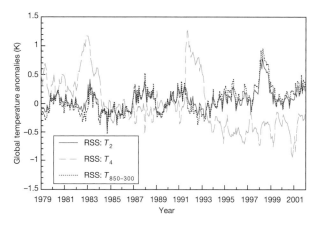

Fig. 2. Time series of monthly mean, global-average temperature anomalies. Shown are the MSU channel 2 brightness temperature (T_2), MSU channel 4 brightness temperature (T_4) and MSU-derived 850-300-hPa layer temperature ($T_{850-300}$), based on the MSU data as analysed by the RSS team. All data are expressed as anomalies relative to climatological monthly means over 1985–94.

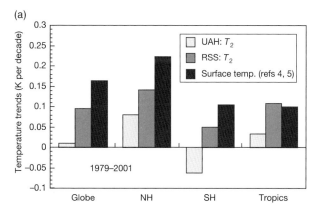

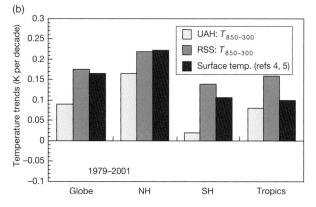

Fig. 3. Trends in monthly mean temperature anomalies. (a), MSU channel 2 brightness temperature trend and (b), MSU-derived 850-300-hPa layer temperature trend for the globe, Northern Hemisphere (NH), Southern Hemisphere (SH) and tropics (30° N–30° S). The results using both UAH (version 5) and RSS data sets are shown. See Supplementary Table 2 for these trend values and their error estimates. The surface temperature trends are also shown for comparison. The surface temperature trends for the globe, Northern Hemisphere and Southern Hemisphere are from ref. 5; the surface trend for the tropical region is from ref. 4.

based on *in situ* observations.[4,5] The trends of 0.01 K per decade (UAH) and 0.1 K per decade (RSS) for global mean T_2 are substantially smaller than the surface temperature trend of 0.17 K per decade. In the Southern Hemisphere, the T_2 trend from UAH is actually negative. However, as shown in Fig. 3b, the global trends of $T_{850\text{-}300}$ are 0.09 K per decade (UAH) and 0.18 K per decade (RSS), which are about 0.08 K per decade larger than the corresponding T_2 trends. The trend difference between $T_{850\text{-}300}$ and T_2 for the tropics is smaller (~0.05 K per decade) because there the tropopause is higher and the stratospheric cooling is smaller, so the stratospheric influence is smaller.

Our analysis of the RSS data set suggests that over the past 22 years the global free troposphere has warmed at close to the same rate as the surface. The ratio of free-tropospheric temperature trend to surface temperature trend is ~1.1 for the globe and 1.6 for the tropics. In the tropics where the temperature follows the moist adiabatic lapse rate,[29] this ratio should be larger than unity.[14,28,30] GCM studies have predicted a global ratio of ~1.2 (ref. 8) and a tropical ratio of ~ 1.54 (ref. 14). Note that the RSS T_2 trend is also statistically consistent with a GCM prediction of T_2 (ref. 15).

Applying the stratospheric corrections to the UAH data set also enhances the mid-tropospheric temperature trends, but they are still smaller than the surface warming rate, particularly over the Southern Hemisphere (Fig. 3b). In addition, the UAH-reported T_{2LT} (bulk temperatures for the low-middle troposphere) seems to be inconsistent with the $T_{850\text{-}300}$ obtained from the UAH data. For example, over the period from 1979 to 2001, the UAH-reported T_{2LT} trend in the tropics is −0.01 K per decade, which is substantially smaller than the $T_{850\text{-}300}$ trend of 0.08 K per decade. (The trend of the difference time series between $T_{850\text{-}300}$ and T_{2LT} (that is, $T_{850\text{-}300} - T_{2LT}$), which is 0.09 ± 0.05 K per decade (the 95% confidence interval), is significantly different from zero at less than 0.1% significance level.) Also note that in the tropics the UAH T_{2LT} is cooling at −0.04 k per decade relative to the UAH T_2. This apparent inconsistency may be attributed to complications involving the T_{2LT} retrieval, as well as to the techniques used by UAH to analyse the MSU channel 2 data, which could also explain the lack of agreement between GCM simulations[8] and UAH results for the trend differences of either $T_s - T_{2LT}$, where T_s is the surface temperature, or $T_s - T_2$, despite their agreement for T_2 trends.[8]

In an independent analysis of MSU data, Vinnikov and Grody[24] found a large positive global trend of 0.22–0.26 K per decade for T_2, which they used to represent the tropospheric temperature. But irrespective of the techniques used to analyse the data, T_2 is subject to the effects of stratospheric cooling. Assuming a stratospheric temperature trend of −0.5 K per decade,[1,3,4] the Vinnikov-Grody T_2 trend translates to a $T_{850\text{-}300}$ trend of ~0.33–0.37 K per decade. This value is about twice as large as the surface warming globally. It also suggests a ratio of ~3 between tropospheric and surface temperature trends for the tropical region. These ratios seem large, and suggest that the technique Vinnikov and Grody used to analyse the satellite data may require further scrutiny.

ACKNOWLEDGEMENTS

We thank J. M. Wallace for discussions. We also thank J. M. Wallace, D. L. Hartmann, J. R. Holton, J. K. Angell and M. Free for their comments on the manuscript. This study was supported by the US DOE, NSF and NASA.

NOTES

1 Christy, J. R. *et al.* Error estimates of version 5.0 of MSU-AMSU bulk atmospheric temperatures. *J. Atmos. Ocean. Technol.* **20**, 613–629 (2003).

2 Mears, C. A., Schabel, M. C. & Wentz, F. J. A reanalysis of the MSU channel 2 tropospheric temperature record. *J. Clim.* **16**, 3650–3664 (2003).

3 Seidel, D. J. *et al.* Uncertainty in signals of large-scale climate variations in radiosonde and satellite upper-air temperature datasets. *J. Clim.* **17**, 2225–2240 (2004).

4 Houghton, J. T. *et al.* in *Climate Change 2001: The Scientific Basis* (Cambridge Univ. Press, London, 2001).

5 Jones, P. D. & Moberg, A. Hemispheric and large-scale surface air temperature variations: An extensive revision and an update to 2001. *J. Clim.* **16**, 206–223 (2003).

6 Bengtsson, L., Roeckner, E. & Stendel, M. Why is the global warming proceeding much slower than expected? *J. Geophys. Res.* **104**, 3865–3876 (1999).

7 Santer, B. D. *et al.* Interpreting differential temperature trends at the surface and in the lower troposphere. *Science* **287**, 1227–1232 (2000).

8 Hansen, J. *et al.* Climate forcings in Goddard Institute for Space Studies SI2000 simulations. *J. Geophys. Res.* **107**, doi:10.1029/2001JD001143 (2002).

9 Ramaswamy, V. *et al.* Stratospheric temperature trends: Observations and model simulations. *Rev. Geophys.* **39**, 71–122 (2001).

10 Spencer, R. W. & Christy, J. R. Precise monitoring of global temperature trends from satellites. *Science* **247**, 1558–1662 (1990).

11 Hansen, J. *et al.* Satellite and surface temperature data at odds? *Clim. Change 30*, 103–117 (1995).

12 Wallace, J. M. *et al. Reconciling Observations of Global Temperature Change* (National Academy Press, Washington DC, 2000).

13 Singer, S. F. Difficulty in reconciling global warming data. *Nature* **409**, 281 (2001).

14 Hegerl, G. C. & Wallace, J. M. Influence of patterns of climate variability on the difference between satellite and surface temperature trends. *J. Clim.* **15**, 2412–2428 (2002).

15 Santer, B. D. *et al.* Influence of satellite data uncertainties on the detection of externally forced climate change. *Science* **300**, 1280–1284 (2003).

16 Gaffen, D. J., Sargent, M. A., Habermann, R. E. & Lanzante, J. R. Sensitivity of tropospheric and stratospheric temperature trends to radiosonde data quality. *J. Clim.* **13**, 1776–1796 (2000).

17 Lanzante, J. R., Klein, S. A. & Seidel, D. J. Temporal homogenization of monthly radiosonde temperature data. *J. Clim.* **16**, 224–262 (2003).

18 Christy, J. R., Spencer, R. W. & McNider, R. T. Reducing noise in the MSU daily lower-tropospheric global temperature dataset. *J. Clim.* **8**, 888–896 (1995).

19 Christy, J. R., Spencer, R. W. & Lobl, E. S. Analysis of the merging procedure for the MSU daily temperature time series. *J. Clim.* **11**, 2016–2041 (1998).

20 Wentz, F. J. & Schabel, M. Effects of orbital decay on satellite-derived lower-tropospheric temperature trends. *Nature* **394**, 661–664 (1998).

21 Christy, J. R., Spencer, R. W. & Braswell, W. D. MSU tropospheric temperature: Dataset construction and radiosonde comparisons. *J. Atmos. Ocean. Technol.* **17**, 1153–1170 (2000).

22 Prabhakara, C., Iacovazzi, J. R., Yoo, J. M. & Dalu, G. Global warming: Evidence from satellite observations. *Geophys. Res. Lett.* **27**, 3517–3520 (2000).

23 Mo, T., Goldberg, M. D. & Crosby, D. S. Recalibration of the NOAA microwave sounding unit. *J. Geophys. Res.* **106**, 10145–10150 (2001).

24 Vinnikov, K. Y. & Grody, N. C. Global warming trend of mean tropospheric temperature observed by satellites. *Science* **302**, 269–272 (2003).

25 Spencer, R. W. & Christy, J. R. Precision and radiosonde validation of satellite gridpoint temperature anomalies. Part II: A tropospheric retrieval and trends during 1979–1990. *J. Clim.* **5**, 858–866 (1992).

26 Hurrell, J. W. & Trenberth, K. E. Spurious trends in satellite MSU temperature from merging different satellite records. *Nature* **386**, 164–167 (1997).

27 Trenberth, K. E. & Hurrell, J. W. Reply to "How accurate are satellite 'thermometers'?". *Nature* **389**, 342–343 (1997).

28 Hurrell, J. W. & Trenberth, K. E. Difficulties in obtaining reliable temperature trends: Reconciling the surface and satellite Microwave Sounding Unit records. *J. Clim.* **11**, 945–967 (1998).

29 Stone, P. H. & Carlson, J. H. Thermal equilibrium of the atmosphere with a given distribution of relative humidity. *J. Atmos. Sci.* **36**, 415–423 (1979).

30 Wentz, F. J. & Schabel, M. Precise climate monitoring using complementary satellite data sets. *Nature* **403**, 414–416 (2000).

Supplementary Information accompanies the paper on **www.nature.com/nature**.

Competing interests statement The authors declare that they have no competing financial interests.

Correspondence and requests for materials should be addressed to Q.F. (qfu@atmos.washington.edu).

Northern Hemisphere Temperatures During the Past Millennium: Inferences, Uncertainties, and Limitations

MICHAEL E. MANN AND RAYMOND S. BRADLEY

Department of Geosciences, University of Massachusetts, Amherst Massachusetts

MALCOLM K. HUGHES

Laboratory of Tree-Ring Research, University of Arizona, Tucson, Arizona

ABSTRACT

Building on recent studies, we attempt hemispheric temperature reconstructions with proxy data networks for the past millennium. We focus not just on the reconstructions, but the uncertainties therein, and important caveats. Though expanded uncertainties prevent decisive conclusions for the period prior to AD 1400, our results suggest that the latter 20th century is anomalous in the context of at least the past millennium. The 1990s was the warmest decade, and 1998 the warmest year, at moderately high levels of confidence. The 20th century warming counters a millennial-scale cooling trend which is consistent with long-term astronomical forcing.

INTRODUCTION

Estimates of climate variability during past centuries must rely upon indirect "proxy" indicators–natural archives that record past climate variations. Trends Over several centuries are evident in the recession of glaciers [Grove and Switsur, 1994], and the sub-surface information from boreholes [Pollack *et al*, 1998]. Annual climate estimates, however, require proxies such as tree rings, varved sediments, ice cores, and corals (combined with any available instrumental or historical records), which record seasonal/annual variations. Studies based on such "multiproxy" data networks [e.g., Bradley and Jones, 1993; Hughes and Diaz, 1994; Mann *et al*, 1995] have allowed the 20th century climate to be placed in a longer-term perspective, thus allowing for improved estimates of the influence of climate forcings [Lean *et al*, 1995; Crowley and Kim, 1996; Overpeck *et al*, 1997], and validation of the low-frequency behavior exhibited by climate models [e.g., Jones *et al*, 1998].

Recently, Mann *et al* [1998–henceforth "MBH98"] reconstructed yearly global surface temperature patterns back in time through the calibration of multiproxy networks against the modern temperature record. Skillful reconstruction of Northern Hemisphere mean annual surface temperature ("NH") was possible back to AD 1400, as the pattern of surface temperature most readily calibrated by the available multiproxy network corresponds largely to synchronous large-scale temperature variation. It has been speculated that temperatures were warmer even further back, ~1000 years ago–a period described by Lamb [1965] as the Medieval Warm Epoch (though Lamb, examining evidence mostly from western Europe, never suggested this was a global phenomenon). We here apply the methodology detailed by MBH98 to the sparser proxy data network available prior to AD 1400, to critically revisit this issue, extending NH reconstructions as far back as is currently feasible. We also reevaluate earlier estimates of uncertainties in the NH series.

DATA AND METHOD

The multiproxy data network and instrumental temperature data used to calibrate it are discussed in detail by MBH98 (see supplementary information therein). Before AD 1400, only 12 indicators of the more than 100 described by MBH98 are available. This includes the first 3

Table 1. 12 Proxy Indicators Available Back to AD 1000. Description ("SERIES" – see MBH98 for details regarding data and reference), location ("LOC" – region or lat/lon coordinates, start year ("y_0") AD, and type ("TYPE") of series is indicated. These data (and the NH series discussed in the text) are available over the internet through the World Data Center-A for Paleoclimatology (http://www.ngdc.noaa.gov/paleo/paleo.html).

Series	LOC	y_0	TYPE
ITRDB (PC #1)	N. Amer	1000	T. Ring width
ITRDB (PC #2)	N. Amer	1000	T. Ring width
ITRDB (PC #3)	N. Amer	1000	T. Ring width
Fennoscandia	68N 23E	500	T. Ring density
Polar Urals	67N 65E	914	T. Ring density
Tasmania	43S 148E	900	T. Ring width
N. Patagonia	38S 68W	869	T. Ring width
Morocco	33N 5W	984	T. Ring width
France	44N 7E	988	T. Ring width
Greenland stacked core	77N 60W	553	ice core $\delta^{18}O$
Quelccaya (2)	14S 71W	488	ice core $\delta^{18}O$
Quelccaya (2)	14S 71W	488	ice accum.

principal components (PCs) of the (28) dendroclimatic series available back to AD 1000 in the International Tree Ring Data Bank ("ITRDB")–all from North America. The 12 indicators (14 counting two nearby ice core sites) are summarized in Table 1.

The calibration procedure (see MBH98) invokes the assumptions (1) that a linear relationship exists between proxy climate indicators and and some combination of large-scale temperature patterns, and (2) that patterns of surface temperature in the past can be suitably described in terms of some linear combination of the dominant present-day surface temperature patterns. MBH98 performed extensive cross-validation experiments to verify the reliability of the reconstruction using global temperature data from 1854–1901 withheld from (1902–1980) calibration, and, further back, by the small number of instrumental temperature series available back through the mid-18th century.

In using the sparser dataset available over the entire millennium (Table 1), only a relatively small number of indicators are available in regions (e.g., western North America) where the primary pattern of hemispheric mean temperature variation has significant amplitude (see Fig. 2 in MBH98), and where regional variations appear to be closely tied to global-scale temperature variations in model-based experiments [Bradley, 1996]. These few indicators thus take on a particularly important role (in fact, as discussed below, one

such indicator – PC #1 of the ITRDB data–is found to be essential), in contrast with the post AD 1400 reconstructions of MBH98 for which indicators are available in several key regions [e.g., the North American northern treeline ("NT") dendroclimatic chronologies of Jacoby and D'Arrigo, 1989].

Due to the leverage of ITRDB PC #1 in the millennial reconstruction, any non-climatic influence must first be removed before it can meaningfully be used in the reconstructions. Spurious increases in variance back in time associated with decreasing sample sizes [see e.g. Jones *et al*, 1998] are not an issue with this series, owing to the high degree of replication in the underlying chronologies back to AD 1000. A number of the highest elevation chronologies in the western U.S. do appear, however, to have exhibited long-term growth increases that are more dramatic than can be explained by instrumental temperature trends in these regions. Gray bill and Idso [1993] suggest that such high-elevation, CO_2-limited trees, in moisture-stressed environments, should exhibit a growth response to increasing CO_2 levels. Though ITRDB PC #1 shows significant loadings among many of the 28 constituent series, the largest are indeed found on high-elevation western U.S. sites. The ITRDB PC#1 is shown along with that of the composite NT series, during their 1400–1980 period of overlap (Figure 1). The low-frequency coherence of the ITRDB PC#1 series and composite NT series during the initial four centuries of overlap (1400–1800)is fairly remarkable, considering that the two series record variations in entirely different environments and regions. In the 19th century, however, the series diverge. As there is no *a priori* reason to expect the CO_2 effect discussed above to apply to the NT series, and, furthermore, that series has been verified through cross-comparison with a variety of proxy series in nearby regions [Overpeck *et al*, 1997], it is plausible that the divergence of the two series, is related to a CO_2 influence on the ITRDB PC #1 series. The residual is indeed coherent with rising atmospheric CO_2 (Figure 1b), until it levels off in the 20th century, which we speculate may represent a saturation effect whereby a new limiting factor is established at high CO_2 levels. For our purposes, however, it suffices that we consider the residual to be non-climatic in nature, and consider the ITRDB PC #1 series "corrected"' by removing from it this residual, forcing it to align with the NT series at low frequencies throughout their mutual interval of overlap. This

(a)

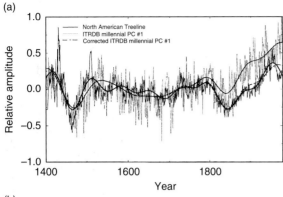

(b)

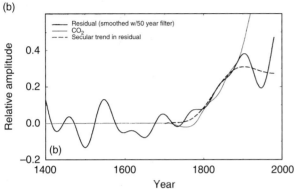

Fig. 1. Comparison of ITRDB PC#1 and NT series, (a) composite NT series vs. ITRDB PC #1 series during AD 1400–1980 overlap. Thick curves indicate smoothed (75 year low-passed) versions of the series. The smoothed "corrected" ITRDB PC #1 series (see below) is shown for comparison, (b) Residual between the smoothed NT and ITRDB series, and its secular trend (retaining timescales longer than 150 years). Relative variations in atmospheric CO_2 since AD 1700 are shown for comparison.

correction is independently justified by the fact that temperatures averaged over the NT region and western U.S. region dominating ITRDB PC #1 exhibit very similar low-frequency trends this century (not shown).

VERIFICATION AND CONSISTENCY CHECKS

The calibration/verification statistics for reconstructions based on the 12 indicators available back to AD 1000, are, as expected, somewhat degraded relative to those for the post AD 1400 period. The calibration and verification resolved variance (39% and 34% respectively) are consistent with each other, but lower than for reconstructions back to AD 1400 (42% and 51% respectively–see MBH98). Results further back than a millennium, based on even sparser data

Spectrum of calibration residuals

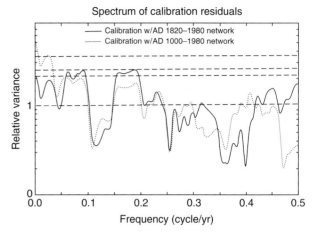

Fig. 2. Spectrum of NH series calibration residuals from 1902–1980 for post-AD 1820 (solid) and AD 1000 (dotted) reconstructions (scaled by their mean white noise levels). Median and 90%,95%,and 99% significance levels (dashed lines) are shown.

(see Table 1) are yet further degraded. With only a single eigenvector of the instrumental temperature data (#1– see Figure 2 in MBH98) skillfully resolved by the network available back to AD 1000, the total *spatial* variance calibrated is far more modest than that for the NH mean ($\approx$ 5% in calibration and verification). Thus, the NH series, but not the spatial details, are most meaningful in the millennial reconstructions.

Further consistency checks are required. The most basic involves checking the potential resolvability of long-term variations by the underlying data used. An indicator of climate variability should exhibit, at a minimum, the red noise spectrum the climate itself is known to exhibit [see Mann and Lees, 1996 and references therein]. A significant deficit of power relative to the median red noise level thus indicates a possible loss of true climatic variance, with a deficit of zero frequency power indicative of less trend than expected from noise alone, and the likelihood that the longest ("secular") timescales under investigation are not adequately resolved. Only 5 of the indicators (including the ITRDB PC #1, Polar Urals, Fennoscandia, and both Quelccaya series) are observed to have at least median red noise power at zero frequency for the pre-calibration (AD 1000–1901) period. It is furthermore found that only one of these series–PC #1 of the ITRDB data–exhibits a significant correlation with the time history of the dominant temperature pattern of the 1902–1980 calibration period. Positive calibration/variance scores for the NH series cannot be obtained if this indicator is removed from the network of 12 (in contrast with post-AD 1400 reconstructions for which a variety of indicators are available which correlate against the instrumental record). Though, as discussed earlier, ITRDB PC#1 represents a vital region for resolving hemispheric temperature trends, the assumption that this relationship holds up overtime nonetheless demands circumspection. Clearly, a more widespread network of quality millennial proxy climate indicators will be required for more confident inferences.

A further consistency check involves examining the calibration residuals. In Figure 2 we show the power spectrum of the residuals of the NH calibration from 1902–1980 for both the calibrations based on all indicators in the network available back to 1820 (see MBH98), and the calibrations based on the 12 indicators available back to AD 1000. Not only (as indicated earlier) is the calibrated variance lower for the millennial reconstruction, but there is evidence of possible bias. While the residuals for the post-AD 1820 reconstructions are consistent with white noise (at no frequency does the spectrum of the residuals breach the 95% significance level for white noise– this holds in fact back to AD 1600), a roughly five-fold increase in unresolved variance is observed at secular frequencies (>99% significant) for the millennial reconstruction. In contrast to MBH98 where uncertainties were self-consistently estimated based on the observation of Gaussian residuals, we here take account of the spectrum of unresolved variance, separately treating unresolved components of variance in the secular (longer than the 79 year calibration interval in this case) and higher-frequency bands. To be conservative, we take into account the slight, though statistically insignificant inflation of unresolved secular variance for the post-AD 1600 reconstructions. This procedure yields composite uncertainties that are moderately larger than those estimated by MBH98, though none of the primary conclusions therein are altered.

TEMPERATURE RECONSTRUCTION

The reconstructed NH series and estimated uncertainties are shown in Figure 3, along with its associated power spectrum. The substantial secular spectral peak is highly significant relative to red noise, associated with a long-term cooling trend in the NH series prior to industrialization ($\delta T = -0.02°C$/century). This cooling is

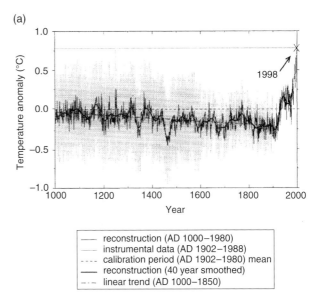

(a)

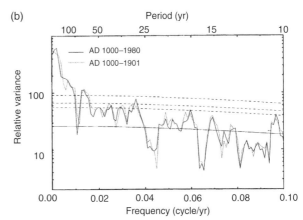

(b)

Fig. 3. Millennial temperature reconstruction. (a) NH reconstruction (solid) and raw data (dotted) from AD 1000–1998. Smoothed version of NH series (thick solid), linear trend from AD 1000–1850 (dot-dashed) and two standard error limits (shaded) are also shown. (b) Power spectrum of the NH series based on full (AD 1000–1980) and pre-calibration (AD 1000–1901) intervals. Robustly estimated median and 90%, 95%, and 99% significance levels relative to red noise are shown [see Mann and Lees, 1996].

possibly related to astronomical forcing, which is thought to have driven long-term temperatures downward since the mid-Holocene at a rate within the range of −0.01 to −0.04°C/century [see Berger, 1988]. In addition, significant century-scale variability may be associated with solar irradiance variations [see Lean *et al*, 1995; MBH98], and a robust spectral peak centered at 50–70 year period seems to correspond to a multidecadal climate signal discussed by Mann *et al* [1995].

The 20th century (1900–1998) (anomaly of $\bar{T}$ = 0.07°C relative to the 1902–1980 calibration period mean) is nominally the warmest of the millennium (11–12th: –0.04; 13th: –0.09; 14th: –0.07; 15th: –0.19; 16th: –0.14; 17th: –0.18; 18th : –0.14; 19th: –0.21). Expanded uncertainties in centennial means prior to AD 1600, and warmer conditions during the earlier centuries of the millennium, however, preclude a definitive statement prior to AD 1400–the 11th and 12th centuries are within a (centennial) standard error of the 20th century. The late 11th, late 12th, and late 14th centuries rival *mean* 20th century temperature levels (see Figure 3a). Our reconstruction thus supports the notion of relatively warm hemispheric conditions earlier in the millennium, while cooling following the 14th century could be viewed as the initial onset of the Little Ice Age *sensu lato*. Considerable spatial variability is evident however [see Hughes and Diaz, 1994] and, as in in Lamb's [1965] original concept of a Medieval Warm Epoch, there are episodes of cooler as well as warmer conditions punctuating this period. Even the warmer intervals in our reconstruction pale, however, in comparison with *modern* (mid-to-late 20th century) temperatures. For the NH series, both the past year (1998) and past decade (1989–1998) are well documented as the warmest in the 20th century instrumental record. Furthermore, the past decade ($\bar{T}$ = 0.45°C) is nearly two (decadal) standard errors warmer than the next warmest decade prior to the 20th century (1166–1175: $\bar{T}$ = 0.11), and 1998 (T = 0.78°C) more than two standard errors warmer than the next warmest year (1249 with an anomaly T = 0.27°C; 1253 and 1366 with $T \approx 0.25$°C are the only other two years approaching typical modern warmth), supporting the conclusion that both the past decade and past year are likely the warmest for the Northern Hemisphere *this millennium*. The recent warming is especially striking if viewed as defying a long-term cooling trend associated with astronomical forcing.

CONCLUSIONS

Although NH reconstructions prior to about AD 1400 exhibit expanded uncertainties, several important conclusions are still possible. While warmth early in the millennium approaches *mean* 20th century levels, the late 20th century still appears anomalous: the 1990s are likely the warmest decade, and 1998 the warmest year, in at least a millennium. More widespread high-resolution data which can resolve millennial-scale variability are needed before more confident conclusions can be reached with regard to the spatial and temporal details of climate change in the past millennium and beyond.

ACKNOWLEDGMENTS

We thank P.D. Jones and two anonymous reviewers for their comments. We gratefully acknowledge the numerous researchers who have contributed to the ITRDB. This research was supported by grants from the NSF (ATM-9626833) and DOE. M.E.M. acknowledges support through the Alexander Hollaender Distinguished Postdoctoral Fellowship Program (DOE).

REFERENCES

Berger, A., Milankovitch theory and climate, *Rev. of Geophys., 26,* 624–657, 1988.

Bradley, R.S., and Jones, P.D., 'Little Ice Age' summer temperature variations: their nature and relevance to recent global warming trends, *Holocene, 3,* 367–376, 1993.

Bradley, R.S. in *Climate Variations and Forcing Mechanisms of the Last 2000 Years* (eds Jones, P.D., Bradley, R.S., and Jouzel, J.), 603–624 (Springer, Berlin), 1996.

Crowley, T.J., and K.Y. Kim, Comparison of proxy records of climate change and solar forcing, *Geophys. Res. Lett., 23,* 359–362, 1996.

Graybill, D.A., and S.B. Idso, Detecting the aerial fertilization effect of atmospheric CO_2 enrichment in tree-ring chronologies, *Glob. Biogeochem. Cycles, 7,* 81–95, 1993.

Grove, J.M. and R. Switsur, Glacial geological evidence for the Medieval Warm Period, *Climatic Change, 26,* 143–169, 1994.

Hughes, M.K., and H.F. Diaz, Was there a 'Medieval Warm Period' and if so, where and when?, *Climatic Change, 26,* 109–142, 1994.

Jacoby, G.C. and R. D'Arrigo, Reconstructed Northern Hemisphere annual temperature since 1671 based on high-latitude tree-ring data from North America, *Clim. Change, 14,* 39–59 1989.

Jones, P.D. *et al*, High-resolution palaeoclimatic records for the last millennium: interpretation, integration and comparison with General Circulation Model control run temperatures, *Holocene, 8,* 477–483, 1998.

Lamb, H.H., The early Medieval warm epoch and its sequel, *Palaeogeography, Palaeoclimatology, Palaeoecology,* **1,** 13–37, 1965.

Lean, J., Beer, J., and R. Bradley, Reconstruction of solar irradiance since 1610: Implications for climate change, *Geophys. Res. Lett., 22,* 3195–3198, 1995.

Mann, M.E., Bradley, R.S., and Hughes, M.K., Global-Scale Temperature Patterns and Climate Forcing Over the Past Six Centuries, *Nature, 392,* 779–787, 1998.

Mann, M.E. and Lees, J., Robust estimation of background noise and signal detection in climate time series, *Climatic Change, 33*, 409–445, 1996.

Mann, M.E., Park, J., and R.S. Bradley, Global Interdecadal and Century-Scale Climate Oscillations During the Past Five Centuries, *Nature, 378*, 266–270, 1995.

Overpeck, J. *et al*, Arctic environmental changes of the last four centuries, *Science, 278*, 1251–1256, 1997.

Pollack, H., Huang, S., and P.Y. Shen, Climate change revealed by subsurface temperatures: A global perspective, *Science, 282*, 279–281, 1998.

M.E. Mann, Department of Geosciences, Morrill Science Building, University of Massachusetts, Amherst MA 01003–5820.

(Received October 14, 1998; revised January 21, 1999; accepted January 27,1999.)

10

Ice Sheets and Sea Level

Zwally, H.J., Abdalati, W., Herring, T., *et al.* (2002). Surface melt-induced acceleration of Greenland ice-sheet flow. *Science*, **297** (5579), 218–222. 4 pages.

The IPCC Fourth Assessment Report, released in 2007, concluded that sea level might rise by half a meter or so by the year 2100, due to thermal expansion of warming seawater, and melting mountain glaciers. This forecast explicitly excludes any contribution from the great ice sheets in Greenland and Antarctica. Although the ultimate change in sea level from fossil fuel combustion will come from these ice sheets, the available models of ice sheet flow and melting did not predict much response in a mere 100 years; they predicted that it would take thousands of years for the ice sheets to respond to changing climate.

Recent observations from the flowing perimeter of the ice sheets seem to be telling us that ice sheets know a few tricks about melting that glaciologists have yet to coax from the ice sheet models. There are also indications from the geologic past, such as the Heinrich events in which the Laurentide ice sheet collapsed on a century time scale into the Atlantic Ocean (Hemming, S.R. (2004). Heinrich events: Massive late pleistocene detritus layers of the North Atlantic and their global climate imprint. Review of Geophysics, **42** (1), RG1005), and the rate of change of sea level as the ice sheets partially or completely melted at the end of the last glacial time, in particular an event called Meltwater Pulse 1-A (Clark, *et al.* (2002). *Science*, **295**, 5564).

The Warming Papers, 1ˢᵗ edition. Edited by David Archer & Raymond Pierrehumbert. Editorial matter © 2011 Blackwell Publishing Ltd

Surface Melt–Induced Acceleration of Greenland Ice-Sheet Flow

H. JAY ZWALLY,[1]* WALEED ABDALATI,[2] TOM HERRING,[3] KRISTINE LARSON,[4] JACK SABA,[5] KONRAD STEFFEN[6]

[1]Oceans and Ice Branch, Code 971, NASA Goddard Space Flight Center, Greenbelt, MD 20771, USA.
[2]Code YS, NASA Headquarters, 300 E Street, SW, Washington, DC 20546, USA.
[3]Department of Earth, Atmospheric, and Planetary Sciences, MIT Room 54-618, 77 Massachusetts Avenue, Cambridge, MA 02139, USA.
[4]Department of Aerospace Engineering Sciences, University of Colorado, Boulder, CO 80309, USA.
[5]Raytheon Inc., Code 971, NASA Goddard Space Flight Center, Greenbelt, MD 20771, USA.
[6]CIRES, University of Colorado, CB 216, Boulder, CO 80309, USA.
*To whom correspondence should be addressed. E-mail: jay.zwally@gsfc.nasa.gov

Ice flow at a location in the equilibrium zone of the west-central Greenland Ice Sheet accelerates above the midwinter average rate during periods of summer melting. The near coincidence of the ice acceleration with the duration of surface melting, followed by deceleration after the melting ceases, indicates that glacial sliding is enhanced by rapid migration of surface meltwater to the ice-bedrock interface. Interannual variations in the ice acceleration are correlated with variations in the intensity of the surface melting, with larger increases accompanying higher amounts of summer melting. The indicated coupling between surface melting and ice-sheet flow provides a mechanism for rapid, large-scale, dynamic responses of ice sheets to climate warming.

The time scale for dynamic responses of ice sheets to changes in climate (e.g., snow accumulation and surface temperature) is typically considered to be hundreds to thousands of years (1). Because most ice-sheet motion occurs by ice deformation in the lower layers, basal sliding, or deformation in basal till, the effects of changes in surface climate must be transmitted deep into the ice to affect the ice flow markedly. In particular, changes in the surface-mass balance alter the ice thickness slowly, and therefore the driving stresses in the deforming layers, as thickness changes accumulate. Changes in surface temperature can also affect the rate of ice deformation or basal sliding, but only after the very slow conduction of heat to the lower layers (2). In contrast to the flow of grounded ice, both floating glacier tongues (3) and Antarctic ice shelves (4) respond quickly to changes in basal heat fluxes and melting, as well as to the effects of surface meltwater trapping in crevasses. Although the transfer of surface meltwater to the base of a grounded ice sheet can provide a rapid mechanism for transferring heat and lubricating fluid to the bottom, this mechanism has not been given much consideration in studies of ice-sheet dynamics. In particular, the Greenland Ice Sheet is grounded above sea level and is generally believed to respond gradually to climate warming (5, 6), mainly by melting at the surface.

Increases in ice velocity occur in alpine glaciers during periods of surface melt in summer (7–9) and have also been observed in Greenland outlet glaciers (10–12). For example, Bindschadler et al. (8) measured a velocity increase from 10 cm/day in winter to 12 cm/day in summer in the Variegated Glacier, a surge-type glacier in Alaska. However, short-term velocity variations have not been observed in the flow of ice sheets away from ice streams and outlet glaciers. Even inland of the fast-flowing Jakobshavn Isbrae in west-central Greenland, seasonal variations were not found (13).

VELOCITY AND MELT OBSERVATIONS

To examine the possibility of seasonal or interannual changes in ice-sheet flow, we initiated year-round global positioning system (GPS) measurements at the Swiss Camp (14, 15) at 1175 m (69.57°N, 49.31°W) near the equilibrium line (16) in west-central Greenland in June 1996 (Fig. 1). The camp is about 35 km from the ice edge and

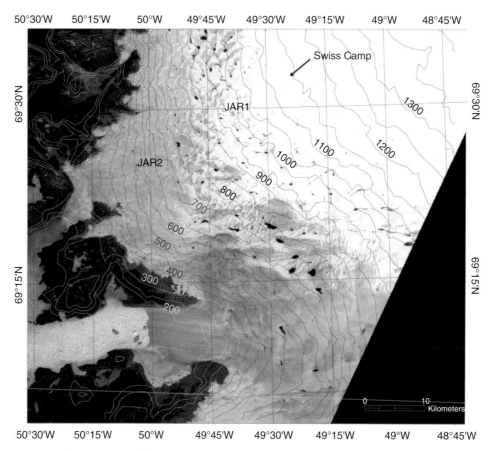

Fig. 1. Elevation contours (50 m) on a Landsat Thematic Mapper image (channel 3) taken on 22 June 1990, which is typically about one-third of the way through the melt season. Locations of the Swiss Camp and the Automated Weather Station at JAR-1 and JAR-2 are marked. The indicated flowline direction of 235°E at the camp curves westward toward the grounded ice edge. The grayer areas at lower elevations in the image are bare ice, with some whiter patches of remaining winter snow near the ice-snow line. By the end of the melt season in late August to early September, the firn-ice boundary usually retreats to around the average location of the equilibrium line (15) near the Swiss Camp. The dark patches are melt lakes, some of which show dark lines of inflow channels. Later in the season, melt lakes also form above the equilibrium line. Jacobshavn Isbrae is in the lower part.

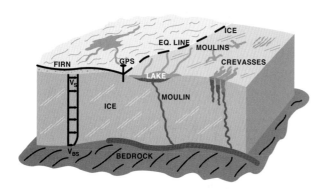

Fig. 2. Schematic of glaciological features in the equilibrium and ablation zones, including surface lakes, inflow channels, crevasses, and moulins. Ice flow for basal ice at the pressure melting point is partly from basal sliding and partly from shear deformation, which is mostly in a near-basal boundary layer.

about 510 km downstream from the ice divide. Characteristic glaciological features of the region are illustrated in Fig. 2. The ice thickness at the camp is 1220 m (17), and a numerical ther-modynamic-dynamic ice model suggests that the basal temperature is at the pressure-melting point (PMP) of −1.0°C (18). The horizontal ice velocity is nearly constant from the surface to the lower boundary layer, where ice shearing occurs. For basal ice at the PMP, part of the surface velocity is typically from ice sliding at the base, but the magnitude of sliding is difficult to estimate without borehole measurements (19).

Our ice velocities are derived from GPS-measured positions of a 4-m pole embedded 2 m in the ice beginning in June 1996 (20). The receiver recorded automatically for 12-hour periods at intervals of 10 or 15 days (21). GPS data were analyzed using GAMIT and GLOBK software (22). To examine changes in ice velocity with respect to the direction of ice flow, a smoothed line of motion was derived as a function of time. The time series of north and east positions [N(*t*), E(*t*)] were each fitted to parabolas as a function of time (*t*). The direction of motion was computed from the derivatives (dN/dE = dN/dt/dE/dt) and is approximately a linear function of time. The derived direction of motion is toward the southwest at an azimuth of 234.963°E on 1 January 1997 and curving toward the west at the rate of 0.106°/year. Although most of the motion is along-track, systematic cross-track displacements up to about ±20 cm occur annually, as discussed below.

Figure 3 shows the horizontal along-track velocity from June 1996 through mid-November 1999. Data from periods during the winter months with little or no change in velocity were used to compute a constant base velocity of 31.33 ± 0.02 cm/day (*23*). In 1996, the velocity increased slightly from the winter value to a maximum of 32.8 cm/day on August 9, and then returned to the winter value. A small temporary increase to 32.8 cm/day also occurred in March of 1997. In subsequent summers, the velocity increased markedly to a maximum of 35.1 cm/day around 9 August 1997, of 40.1 cm/day around 10 July 1998, and of 38.6 cm/day around 30 July 1999. After these periods of accelerating flow, the velocity decreased markedly to a minimum of 28.9 cm/day around 18 September 1997, of 29.8 cm/day around 19 August 1998, and of the lowest value of 27.6 cm/day around 29 August 1999. After the velocity minima, the ice slowly accelerated again over several fall months to return to the midwinter values.

The cumulative additional motion (Fig. 3) caused by the summer accelerations was computed as the difference between the measured positions and the calculated along-track positions that would have occurred under a constant velocity of 31.33 cm/day. At the transition from accelerating flow to decelerating flow in early September 1997, the ice had moved an additional 3.0 m relative to the baseline rate. At the 1998 and 1999 transitions, the respective net additional displacements were 4.7 m and 6.0 m. During periods of slower flow in the fall, after the transitions, the additional displacement was reduced by about 45 to 65% (*24*).

From a 21-year record of Greenland surface melting from passive microwave data, the summer of 1996 was the second-lowest melt year since 1979, the summer of 1997 was slightly below the 21-year average, and the summers of 1998 and 1999 were well above average (*25*). We investigated relationships between the availability of summer meltwater for basal lubrication and changes in

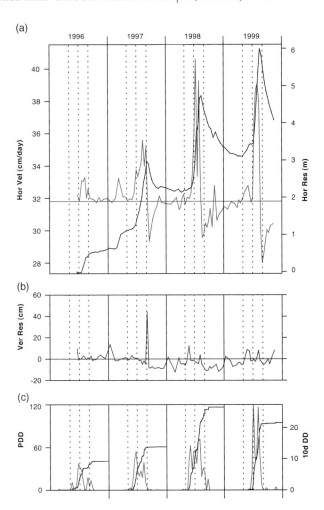

Fig. 3. (a) Horizontal ice velocity (red curve) along a smoothed line of motion showing ice accelerations during the summer melt seasons and the abrupt transitions to deceleration around the times of melt cessation. The cumulative additional motion (horizontal residual, black) relative to a wintertime-average velocity of 31.33 cm/day is 6.0 m by the time of the maximum velocity in 1999. (b) The vertical residual (blue) indicates a 50-cm uplift at the time of the 1997 transition from accelerating flow to decelerating flow. (c) Cumulative PDDs and PDDs for 10-day intervals (10d DD, red) from temperatures measured at the Swiss Camp, showing correlations of the melting with the intensity and timing of the ice accelerations and decelerations (units are degree-days). Vertical dotted lines mark May 1, July 1, and September 1 for each year.

the ice velocity using the cumulative positive degree-days (PDDs) (*26*) at the camp, as shown in Fig. 3C. For the four summers (1996 to 1999), the respective total PDDs were 47.6, 60.8, 116.5, and 94.7. The corresponding mean areas of melt from June through September in the Jacobshavn region from passive microwave data were 28.2, 52.0, 79.6, and 74.5 × 10³ km², which show similar year-to-year variations as the PDDs (*27*). The quality of PDDs as an indicator of ablation and consequent meltwater production was shown by Braithwaite and Olesen (*28*), who obtained a 0.96 correlation coefficient between annual ice ablation and PDDs, even though variations in radiation and other energy-balance factors such as wind speed also affect the rate of ice ablation (*29*).

CORRELATIONS BETWEEN ACCELERATION AND SURFACE MELTING

In all years, correlations are evident between the changes in ice velocity and both the intensities and timings of surface melting. For the summers of 1998 and 1999, the respective ratios of the increases in velocity of 8.8 and 7.3 cm/day to the PDDs of 116.5 and 94.7 are nearly the same (0.076 and 0.077), whereas the ratios in 1996 and 1997 are somewhat smaller (0.032 and 0.063). For the summer of 1999, the melting period was shorter in duration, but of greater intensity in the month of July than in the other summers.

Although melting downstream of the equilibrium line begins a few days to several weeks earlier than at the camp, comparison of the timing of the acceleration and melting is based on the melt record at the camp (*30*). Although some melting usually occurs in May, more-continuous melting does not usually begin until the beginning of June. In 1997, the onsets of ice acceleration and melting were nearly simultaneous in mid-June. In 1998, the acceleration onset (also in mid-June) lagged the melting onset by about 2 weeks. In 1999, although a small temporary increase in velocity appeared in May, the major ice acceleration and the melting onset were both delayed until the beginning of July. Correlations between the timing of the transitions from acceleration to deceleration and the cessations of melting were also good, particularly in 1997 and 1999 when the respective transitions at the beginning of September and mid-August were nearly coincident with the end of melting. In 1998, some melting continued after the transition in early August. In 1996, the year of minimal acceleration, the ice accelerated in July and early August until the end of a period of minimal melting in late August when the transition to deceleration occurred.

ACCELERATION MECHANISM

In the ablation zone of Greenland, surface meltwater runs along the surface and collects in surface lakes or flows directly into moulins (Fig. 2).

Although the internal or subglacial pathways for transit of the meltwater to the margins are generally not known, Thomsen *et al.* (*29*) assumed that water flowing into moulins quickly flows to the bottom and drains subglacially. Whether the drainage pathways tend to be vertical and channel meltwater to the base of the ice sheet, or tend to be horizontal and remain englacial, markedly affects the local availability of water for basal lubrication. One indication that the water flow is largely subglacial, at least near the margins, is that the meltwater primarily leaves the ice sheet in subglacial streams, and not in surface flow over the ice edges. Theoretically, water-filled crevasses are expected to propagate to the bottom due to the over-burden pressure of water, as compared to ice (*31*), and could provide numerous pathways for meltwater drainage throughout the ablation zone and for the initiation of moulins. Mapping of moulins and surface lakes (*29*) in the ablation zone below the camp shows that the areal density of moulins is about 0.2 moulins/km^2 and most moulins are not associated with the less-numerous surface lakes.

Therefore, flow of meltwater through moulins, and perhaps through the numerous crevasses throughout the ablation zone, is likely to provide a widespread and continual drainage from the surface to the ice base during the melt season. The surface lakes usually drain during summer, but their drainage tends to be episodic, depending on the irregular timing of the opening of drainage channels from the lakes (*15, 32*). We believe that the observed correlations between the changes in ice velocity and the timing and intensity of the surface melting show that there is widespread and continual drainage of meltwater from the surface to the ice-sheet base during summer ablation. Because the ice base is at the PMP (*18*), a wet base maintained by basal melting is expected to be a normal condition throughout the year. Water may be maintained during winter in subglacial conduits that expand during periods of increased water pressure and accelerated flow (*33*). The probable cause of the summer acceleration is an increase in the water pressure at the bedrock interface, which is a well-known mechanism for velocity variations in alpine glaciers. For example, Iken *et al.* (*33*) found that the summer increase in subglacial water pressure actually raised the glacier by as much as 0.6 m and partially decoupled the ice from the bed. The horizontal velocity increased by three to six times and was largest at times of maximum upward velocity.

The magnitude of the acceleration at the Swiss Camp was small in the low-melt year of 1996 and increased in later years as the melting increased. Apparently, more meltwater produces larger accelerations, and the timings of melting and acceleration are closely related. The ice accelerations commenced a few weeks after the onset of melting. Later, the transition from accelerating to decelerating ice flow occurred near the cessation or slowing of melting at the end of the melt season. The short time between melt onset and acceleration, as well as the close timing between the transition and the ending of ablation, suggests that storage of meltwater after production is not an important factor affecting the acceleration.

The rapid ice deceleration after the peak velocity is reached could be associated with an increase in basal friction and reduced basal sliding caused by changes at the ice-bedrock interface that occur during the period of increased water flow. As noted above, the ice decelerates to velocities about 10% below the average winter velocity before returning to the winter velocity in late fall. This suggests that the basal sliding gradually recovers to normal winter values. Presumably, the ice acceleration begins in the lower elevations of the ablation zone, with changes in basal sliding, and propagates upstream past the Swiss Camp as sliding increases. The process of deceleration is likely to be a combination of changes in basal sliding and the dynamic response of the ice to the changing patterns of driving stresses during acceleration.

In early September 1997, during the one GPS measurement period just at the transition from accelerating to decelerating flow, the vertical residuals show a 50-cm uplift of the ice sheet (Fig. 3A). Around the time of the 1998 and 1999 transitions, the excursions of the vertical position were <10 cm, which is on the order of other occasional excursions in the data throughout the year. The 50-cm uplift in the 1997 data could be indicative of changes in the basal water pressure similar to that in small glaciers (*34*), or of dynamic effects in the ice occurring as the ice flow at the camp changes from accelerating to decelerating.

Around the time of the maximum accelerations in 1997, 1998, and 1999, the direction of the ice motion changed by about 1° to the right (more westward) over about 30 to 45 days, as shown by 15- to 20-cm cross-track displacements. The peak cross-track velocities are 0.5 to 1 cm/day to the right. After the rightward displacement, an equal displacement to the left slowly occurs until the next summer. These changes in direction are in addition to the 0.106°/year average rotation to the right of the smoothed line of motion. The change in direction during the acceleration could be indicative of changes in the basal friction and the ratio of the sliding velocity to the velocity from ice deformation.

CONCLUSIONS

The interaction among warmer summer temperatures, increased surface meltwater production, water flow to the base, and increased basal sliding provides a mechanism for rapid response of the ice sheets to climate change. In general, a direct coupling between increased surface melting and ice-sheet flow has been given little or no consideration in estimates of ice-sheet response to climate change (*35*). In addition to the direct effect of increased water pressure on the basal sliding, the flow of surface water at approximately 0°C to basal ice, at the PMP of −1.0°C, transfers heat for additional basal melting. The occurrence of this melt-driven acceleration in the equilibrium zone implies that the mechanism may be occurring throughout much of the ablation zone of the ice sheet, or at least where the basal temperature is at the PMP. Therefore, the rate of retreat of the ice-sheet margin under climate warming is probably faster than predicted by estimates based only on the direct increase in surface ablation (*36*). Enhanced basal sliding from surface meltwater may have contributed to the rapid demise of the Laurentide Ice Sheet during increased summer insolation and surface ablation circa 10,000 years ago (*37*) and to extensive melting of the Greenland Ice Sheet during the last interglacial (*38*), by causing a faster flow of ice to the margins, an increase in the thinning rate, and more rapid inward migration of the ablation zone.

REFERENCES AND NOTES

1 R. B. Alley, I. M. Whillans, *J. Geophys. Res.* **89**, 6487 (1984).
2 Another long-term dynamic-response mechanism is caused by variation in the dust impurities in snowfall that affects the ice hardness, so the upper layers of Holocene ice with lower dust content are harder than the lower layers of ice-age ice with higher dust content (*39*).
3 E. Rignot, W. B. Krabill, S. P. Gogineni, I. Joughin, *J. Geophys. Res.* **106**, 34007 (2001).
4 T. Scambos, C. Hulbe, M. Fahnestock, J. Bohlander, *J. Glaciol.* **46**, 516 (2000).

5   C. Ritz, A. Fabre, A. Letreguilly, *Clim. Dyn.* **13**, 11 (1997).

6   P. Huybrechts, J. de Wolde, *J. Clim.* **12**, 2169 (1999).

7   A. Iken, *Z. Gletscherkd. Glazialgeol.* **13**, 23 (1978).

8   R. Bindschadler, W. D. Harrison, C. F. Raymond, R. Crosson, *J. Glaciol.* **18**, 181 (1977).

9   G. H. Gudmundsson *et al.*, *Ann. Glaciol.* **31**, 63 (2000).

10 N. Reeh and O. B. Olesen (*40*) found summer velocity fluctuations of 15% on time scales of 2 to 10 days in the Daugaard-Jensen Glacier, an outlet glacier on the east coast of Greenland. Also, an observed 50% velocity increase for about 7.5 hours was inferred to be the result of a subglacial increase in water pressure caused by the drainage of an ice-dammed lake.

11 I. Joughin, S. Tulaczyk, M. Fahnestock, and R. Kwok (*41*) found a threefold velocity increase during the 1995 melt season in the Ryder outlet glacier in northwest Greenland, which they described as a short-lived mini-surge.

12 On Storstrommen Glacier in northeast Greenland, GPS-determined velocities for summer are generally greater than winter velocities and annual averages (*42*).

13 K. Echelmeyer, W. D. Harrison, *J. Glaciol.* **36**, 82 (1990).

14 K. Steffen, J. E. Box, *J. Geophys. Res.* **106**, 33951 (2001).

15 A. Ohmura *et al.*, *ETH Greenland Expedition Progress Report No. 1* (Department of Geography, ETH, Zurich, Switzerland, 1991).

16 The equilibrium line is defined as the boundary between annual net ice accumulation and net ablation at the surface and is located at the upper elevation of bare ice (without refrozen firn or superimposed ice) at the end of the summer melt season. The empirical equation for the equilibrium-line altitude (ELA) is a decreasing quadratic function of latitude (*43*) that gives an ELA of 1418 m at the latitude of the Swiss Camp. The camp was established in 1990 at the nominal location of the equilibrium line, based on surface observations in previous years (*29*). It lies on a small ridge, and local variations in the altitude of the equilibrium zone range from around 1200 m at the camp to 1400 m south of the camp.

17 Measured by University of Kansas ice-penetrating radar. Data are available online at http://tornado.rsl.ku.edu/1998thick.htm.

18 W. L. Wang, H. J. Zwally, W. Abdalati, S. Luo, *Ann. Glaciol.*, in press, using the measured ice thickness of 1220 m. The modeled basal temperature is slightly below the PMP at the ice divide and is essentially at the PMP along a flowline from 400 km above the camp to the ice edge.

19 M. Luthi, M. Funk, A. Iken, S. Gogineni, M. Truffer, *J. Glaciol.*, in press, obtained a ratio of basal sliding of 60% in the lateral shear zone of Jakobshavn Isbrae, but noted that such a high ratio is probably not representative for the ice sheet.

20 GPS data were recorded by a geodetic-quality dual-frequency Trimble 4000 SSi receiver (Trimble, Sunnyvale, CA), which is inside an insulated box in a camp tent and powered by four to six 100-a-h batteries and two 18-W solar panels. The 2-m extension of the antenna above the ice keeps the antenna above the seasonal snow accumulation, which ranges from about 0.5 to 1.8 m interannually. Yearly data were downloaded when the camp was occupied each May. Data collection ended in the fall of 1999 because of anomalous behavior of the receiver.

21 During the spring occupation of the Swiss Camp, data were usually recorded continuously instead of once every 10 to 15 days. To avoid large velocity errors for short time intervals, the following position data for multiple 12-hour time periods were averaged: 23 to 29 May 1998 (13 periods), 23 to 28 May 1999 (10 periods), and 9 and 10 June 1999 (4 periods).

22 R. W. King, *Documentation of the GAMIT GPS Analysis Software Version 10.05* (2001); and T. A. Herring, *Documentation of the GLOBK Software Version 5.1* (2001), MIT, Cambridge, MA. See also http://www-gpsg.mit.edu/~simon/gtgk The Cartesian coordinates of the antenna were calculated, relative to bedrock sites at Thule Air Force Base, Kellyville, and Kulusuk in Greenland, from 12-hour increments of data. The average standard deviations of the coordinates calculated with GLOBK is 1.2 cm horizontally and 3.3 cm vertically. The standard deviations of the north and east components of the positions are somewhat higher during the winter periods (e.g., the standard deviation of the north component went from a base value of about 0.5 cm to a high of nearly 6 cm during the winter of 1999). The large uncertainties occurred on days when only the bedrock site at Thule was operating and when less than 12 hours of data were recorded at Swiss Camp. Times assigned to the points are the midpoints of the times of the data increments. The overall average standard deviation of the horizontal velocity is 0.14 cm/day. Analysis techniques are also described in (*44*).

23 The horizontal velocity at markers 2.15 km downstream from Swiss Camp is 0.67 cm/day slower than at the camp. Therefore, as the camp moved from 1996 to 1999, the estimated decrease in horizontal velocity was 0.036 cm/day/year, or 0.14 cm/day during the 4-year period.

24 The respective advancements and reductions for 1997, 1998, and 1999 were: 1.8 and 0.8 m, 2.5 and 1.6 m, and 2.9 and 1.9 m.

25 W. Abdalati, K. Steffen, *J. Geophys. Res.* **106**, 33983 (2001).

26 $\text{PDD} = \sum_i \alpha_i T_i / 24$, where T_i are hourly averaged near-suface air temperatures, $\alpha_i = 1$ if $T_i > 0$, and $\alpha_i = 0$ if $T_i \le 0$.

27 The year-to-year ratios of cumulative PDDs to melt areas (*19*) differ from unity by less than 19%. The Jacobshavn melt region is defined from about 65.5 to 76°N and from the western ice edge to the divide. The respective cumulative melt-days obtained from passive microwave for the area near the camp are 30, 60, 97, and 47. The year-to-year correspondence for the camp location is not as good (ratios differ from unity by up to 33%), probably because the passive-microwave index of melt days records only the occurrence of surface melting, whereas the deg-days index registers more for warm days (e.g., 5 times more for +5°C) than for days that are just above the melting point. In contrast, the melt-area index used for the Jacobshavn region is also sensitive to the intensity of melting, which expands the area of melting to higher elevations.

28 R. J. Braithwaite, O. B. Olesen, in *Glacier Fluctuations and Climate Change*, J. Oerlemans, Ed. (Kluwer, Dordrecht, Netherlands, 1989), pp. 219–233.

29 H. H. Thomsen, L. Thorning, R. J. Braithwaite, *Glacier-Hydrological Conditions on the Inland Ice North-East of Jakobshavn/Ilusissat, West Greenland: Report 138* (Gronlands Geologiske Undersogelse, Copenhagen, Denmark, 1998).

30 For example, data for 1997 through 1999 from the Automated Weather Station (AWS) (*15*) at JAR-1 at a 960-m elevation and 16.3 km downstream from the camp show that the lag in melting onset between the sites varies from only 1 to 14 days until about 30 deg-days are cumulated. The end of melting may occur even closer in time at different elevations in the ablation zone. In 1999 at least, the cessation was almost simultaneous at Swiss Camp, JAR-1, and JAR-2, which is at a 528-m elevation and 17.1 km downstream from JAR-1.

31 G. de Q. Robin, *J. Glaciol.* **13**, 543 (1974).

32 At the JAR-1 AWS site, which was located near the side of a lake hundreds of meters in diameter, the sonic surface-height data show that the surface level dropped 1.5 m in a few hours on 17 July 1996 as the water drained from the ice. A similar drop of 0.8 m

occurred on 25 June 1997, when the station had moved 70 m closer to the lake edge.

33 A. Iken, H. Rothlisberger, A. Flotron, W. Haberli, *J. Glaciol.* **29**, 28 (1983).

34 C. F. Raymond, *J. Geophys. Res.* **92**, 9121 (1987).

35 J. T. Houghton *et al.*, Eds., *Climate Change 2001: The Scientific Basis* (IPCC Report, Cambridge, UK, 2001).

36 R. J. Braithwaite, O. B. Olesen, *Ann. Glaciol.* **14**, 20 (1990).

37 P. U. Clark, R. B. Alley, D. Pollard, *Science* **286**, 1103 (1999).

38 R. M. Koerner, *Science* **244**, 964 (1989).

39 N. Reeh, *Nature* **317**, 797 (1985).

40 ———, O. B. Olesen, *Ann. Glaciol.* **8**, 146 (1986).

41 I. Joughin, S. Tulaczyk, M. Fahnestock, R. Kwok, *Science* **274**, 228 (1996).

42 J. J. Mohr, N. Reeh, S. Madsen, *Nature* **391**, 273 (1998).

43 H. J. Zwally, M. B. Giovinetto, *J. Geophys. Res.* **106**, 33717 (2001).

44 K. Larson, J. Plumb, J. Zwally, W. Abdalati, *Polar. Geogr.* **25**, 22 (2002).

45 Supported by NASA's ICESat (Ice Cloud and Land Elevation Satellite) science activities and Cryospheric Sciences Program.

11

The Public Statement

Sawyer, J.S. (1972). Man-made carbon dioxide and the "greenhouse" effect. *Nature*, **239**, 23–26.

Charney, J.G., Arakawa, A., James Baker, D., *et al.* (1979). Carbon dioxide and climate: a scientific assessment. National Academy of Sciences, Washington, DC, 22pp.

The papers in this section are not new, seminal scientific findings like the other papers in this volume, but were instead intended to convey the implications of the scientific literature to the general public. The editorial piece by Sawyer contains a clear prediction of warming that would be observed by the year 2000, 0.6°C, which according to the Jones *et al.* temperature reconstruction, was spot on. Note that human-caused warming was masked in the year 1972 by cooling due to sulfate aerosols and natural climate forcings. This was a forecast in advance.

The Charney report, commissioned by the National Research Council, a branch of the National Academy of Sciences, addresses the question of whether the Earth's climate can and would change from fossil fuel CO_2 release. After examining possible stabilizing feedback mechanisms of clouds and ocean circulation, and reviewing results from climate modeling from Hansen and Manabe (this volume), the report finds no reason to doubt that doubling atmospheric CO_2 would lead to a significant change in the global average temperature of the Earth.

It is interesting to contrast the Charney report with its successors, the various IPCC reports, most recently the Fourth Assessment Report published in 2007. The conclusions have not changed except to become more specific, but the number of person-years of effort contained in the underlying science has exploded since 1979. Charney *et al.* had results from two global climate models to think about, and they were rather fledgling models at that, whereas there were about 20 global models from around the world that participated in the Fourth Assessment Report. The climate sensitivity of the Earth to doubling CO_2 is also constrained now by paleoclimate studies such as the proxy temperature records of the last millennium and the last glacial maximum, and by analysis of the instrumental temperature record, which in retrospect only began to show signs of human-induced warming, the smoking gun, beginning in the 1970s. The bottom line of the Charney report was an estimated climate sensitivity of 1.5–4.5°C, which is very similar to the IPCC current range of 2–4.5°C.

The evidence as it stood in the 1970s was adequate to pronounce a relatively strong conclusion, according to the usual standards of scientific discourse. If the question of radiative forcing and climate response were of purely academic interest, it would have been considered provisionally settled at that time. The extraordinary effort represented by the IPCC since then may be scientifically over the top, but it reflects the extraordinary importance of the question.

The Warming Papers, 1ˢᵗ edition. Edited by David Archer & Raymond Pierrehumbert. Editorial matter
© 2011 Blackwell Publishing Ltd

Man-Made Carbon Dioxide and the "Greenhouse" Effect

J. S. SAWYER

Meteorological Office, London Road, Bracknell, Berkshire RG12 2SZ

In spite of the enormous mass of the atmosphere and the very large energies involved in the weather systems which produce our climate, it is being realized that human activities are approaching a scale at which they cannot be completely ignored as possible contributors to climate and climatic change.

THE first thing that has to be recognized is that significant effects on the climate are only likely where human activity impinges on a particularly sensitive factor among those that control climate. The output of human industry is still very much less than the total mass of the atmosphere and man-made energy is still small compared with the energy of meteorological systems. The total industrial output of heat each day is, for example, considerably less than 0.1% of the total kinetic energy of the atmosphere, which itself is destroyed by friction and replaced naturally within a few days. Another useful comparison is that of the total man-made heat output in Britain with natural processes over the same area. Even over this area of relatively intense human activity man's efforts are relatively quite small—man-made heat is less than 1% of the energy received from the Sun.

It must also be remembered that the mass of the atmosphere is enormous compared with the products of human activity. The total mass of the atmosphere is more than 500 times the mass of the known coal reserves, for example, and human activities will not significantly change its chief constituents. Nevertheless there are certain minor constituents of the atmosphere which have a particularly significant effect in determining the world climate. They do this by their influence on the transmission of heat through the atmosphere by radiation. Carbon dioxide, water vapour and ozone all play such a role, and the quantities of these substances are not so much greater than the products of human endeavour that the possibilities of man-made influences can be dismissed out of hand.

INFLUENCE OF CO_2

This article is concerned with the part played by carbon dioxide in determining climate and the way in which it may be affected by human activity. There are several other possible ways in which man might affect climate on a global scale, but the carbon dioxide effect is probably the one about which the most is known, and at the same time it illustrates clearly the inherent difficulties in assessing whether such activity can have a significant effect on climate and how much that effect might be.

First, it is necessary to consider the natural behaviour of carbon dioxide in the atmosphere and the evidence for changes produced by human activity. Carbon dioxide is, of course, a product of the combustion of nearly all fuels and is discharged to the atmosphere through the chimneys or exhaust of the power or heating plant in which the fuel is consumed. The carbon dioxide content of the atmosphere was first measured in the earlier part of the nineteenth century and found to be fairly uniform, both geographically and with season. By the earlier part of the century it had been pointed out that the combustion of fossil fuel was discharging carbon dioxide, which might be increasing the carbon dioxide content of the atmosphere, and which might have an effect on the heat balance of the Earth. The nineteenth century measurements of carbon dioxide were naturally of a somewhat uncertain accuracy, but by comparison with more recent observations Callendar[1] was able to demonstrate that there is reasonable evidence for an increase attributable to the carbon dioxide added to the atmosphere from the burning of fuel.

Sawyer, J. S. (1972) Man-made carbon dioxide and the 'greenhouse' effect. *Nature* 239: 23–26. Reprinted by permission from Macmillan Publishers Limited.

The reality of this increase has been confirmed in a remarkable way in the past decade by two series of measurements made at two locations specially chosen to be far away from local sources of pollution. These were at the South Pole and on the summit of the volcanic mountain Mauna Loa in Hawaii. Fig. 1 (taken from ref. 2) shows the trend of carbon dioxide concentration over the past ten years or so at Mauna Loa. The upward trend is apparent, but the diagram also illustrates some other aspects of the problem. The upward trend at the South Pole is closely similar to that at Mauna Loa.

The upward trend amounts to about 0.7 parts per million (p.p.m.) by volume per year over the eleven year period. The value of 312 p.p.m. in 1958 rising to 319 p.p.m. by 1969 compares with the value of around 292 p.p.m. measured by observers in the nineteenth century. Fig. 1 also shows a marked annual fluctuation, emphasizing that carbon dioxide is essential to the growth of vegetation, and is taken up by plants as they grow. (Some is then returned to the atmosphere when they rot and some incorporated in humus, only to be subsequently returned to the atmosphere.) The annual fluctuation in the concentration of carbon dioxide arises because plant growth is greater in the northern hemisphere than in the southern hemisphere (where the land mass is smaller) and plant growth and the uptake of CO_2 is thus a maximum in the northern summer. The carbon dioxide content of the atmosphere is a maximum in the northern spring and a minimum in the northern autumn.

Fig. 1 also shows the rate at which the concentration of CO_2 would have increased if all the man-made carbon dioxide had remained in the atmosphere. The observed increase is only about half of this, the remainder has clearly been removed from the atmosphere by natural processes, and an assessment of the future rise in carbon dioxide content requires a knowledge of what the processes are and where the carbon dioxide is going to. There is also a suggestion in Fig. 1 that the rate of increase of carbon dioxide was rather slower in the mid-1960s than before and after—an indication that these natural processes may vary in their effectiveness from time to time.

STORAGE AND RESERVOIRS

Fig. 2, which is based on a diagram due to Craig[3], shows the natural reservoirs of the carbon which takes part in the carbon dioxide cycle and the relative size of these reservoirs. On land the carbon dioxide is taken up by vegetation and stored in plants and humus. This reservoir is of similar magnitude to that of the atmosphere, and the exchange time is probably of the order of 30 to 40 yr. The ocean provides a much larger reservoir and has the potential for storing some sixty times as much carbon dioxide as the atmosphere. The upper layers of the sea (above the thermocline) must, however, be distinguished from the deeper layers of the ocean. The upper layers are well mixed, and are in contact with the atmosphere, but they can hold only about as much carbon dioxide as exists in the atmosphere. Studies of the concentration of ^{14}C, which is produced by cosmic rays in the atmosphere and subsequently decays

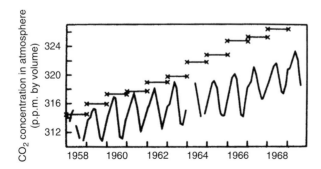

Fig. 1. Increase in carbon dioxide concentration from burning of fossil fuels. ____, Mean monthly atmospheric CO_2 concentration at Mauna Loa; ×—×, potential annual increase from burning fossil fuels. (Reproduced from ref. 2.)

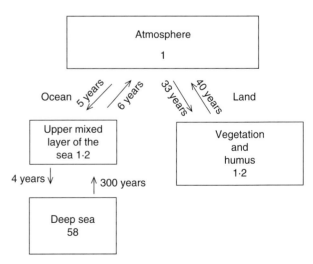

Fig. 2. Natural reservoirs of carbon dioxide (based on ref. 3). Figures indicate content as multiples of atmospheric content.

to ^{12}C, suggest that the rate of transfer of carbon dioxide from the atmosphere to the upper layers of the ocean is such as to require some 5 to 10 yr for the transfer of a quantity equivalent to that in the atmosphere. Transfer to the deep ocean from the upper layers is a slower process, and as a result it would be a matter of centuries before the deep ocean reached equilibrium with any new level of concentration in the atmosphere.

Industrial development has recently been proceeding at an increasing rate so that the output of man-made carbon dioxide has been increasing more or less exponentially. So long as the carbon dioxide output continues to increase exponentially, it is reasonable to assume that about the same proportion as at present (about half) will remain in the atmosphere and about the same amount will go into the other reservoirs. On this basis Bolin[4] has estimated that the concentration of carbon dioxide will be about 400 p.p.m. by the year 2000. A recent conference[5] put the figure somewhat lower (375 p.p.m.)

On the other hand, there must ultimately be a levelling off in industrial output of carbon dioxide, if only on account of the limitations of fuel supply. At this stage a larger proportion of the carbon dioxide will be absorbed by the oceans because, on the longer time scale, the deep sea will have opportunity to come nearer to equilibrium with the atmosphere. If the carbon dioxide were to be shared between the various reservoirs in proportion to their capacities, only one-sixtieth of the man-made carbon dioxide would remain in the atmosphere—but unfortunately the situation is more complex than that.

CHEMICAL COMPLICATIONS

Kanwisher[6] has pointed out that only a small proportion of the carbon dioxide entering the sea remains as dissolved CO_2 directly available for exchange with the atmosphere. The remainder forms magnesium and sodium carbonates which provide a chemical buffering solution for carbon dioxide. In consequence an increase of 0.6% in the carbon dioxide content of the sea corresponds to a 10% increase in the partial pressure of CO_2 in the atmosphere above, and on this basis one might expect that the ultimate sharing of carbon dioxide between the atmosphere and the whole ocean might still leave more than 20% of the extra carbon dioxide in the atmosphere. On a still longer time scale one might expect some of the oceanic

carbon to be deposited as carbonates on the sea bed, but the typical times associated with this process is probably too long to be of much relevance to the fate of industrially produced CO_2.

There is little doubt that in assessing the future level of carbon dioxide in the atmosphere, it is important to understand fully the balance between the carbon dioxide in the atmosphere and in the ocean. There are several other complications which are not fully understood. The solubility of carbon dioxide is greater at lower temperatures and the tropical oceans are thus continuously giving up carbon dioxide to the atmosphere; it is then reabsorbed in the oceans of higher latitudes. In most parts of the ocean there is a relatively warm layer of water lying above colder and denser water beneath— the transition layer, known as the thermocline, lies at a depth of 100 to 200 m. This stable layer is a barrier to the mixing of the lower and upper water of the ocean, but the barrier disappears in certain parts of the polar ocean when the surface water is cooled in winter, and such areas may provide a pathway by which carbon dioxide absorbed from the atmosphere can be transferred to the deep ocean more readily. Such areas may play a significant role in determining the carbon dioxide balance between ocean and atmosphere, and an understanding of this balance will require a better understanding of the long term circulation of the ocean than is available at present.

INDIRECT EFFECTS OF INCREASED CO_2

The direct effect of a small increase in carbon dioxide on mankind would be negligible (except that it might make some vegetation grow a little faster) and I shall now consider its possible indirect effect on the world climate.

The temperature of the Earth is, of course, maintained by the energy received from the Sun as radiation in a band of wavelengths centred in the visible region. Some of this radiation is reflected by the Earth's surface, and more especially by clouds, but most of the remainder penetrates the atmosphere and heats the Earth's surface including the ocean. Some of the heat is radiated back by the surface at longer wavelengths corresponding to its lower temperature than the Sun, part is communicated to the overlying air by conduction, and part is used to evaporate water and becomes available to heat the air when the water condenses as rain. The atmosphere is not transparent to the long wavelength radiation which is emitted by the Earth

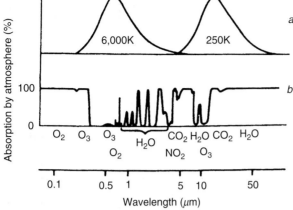

Fig. 3. Radiation spectrum illustrating absorption by principal atmospheric absorbing gases. *a*, Black body curves; *b*, absorption by atmosphere.

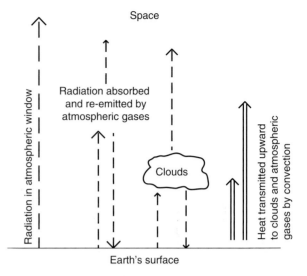

Fig. 4. Pathways of heat loss from the Earth and atmosphere.

and its atmosphere, as it is to the incoming short wave radiation. Thus certain atmospheric gases, principally water vapour and carbon dioxide, absorb a significant part of the outgoing radiation and reradiate it both upwards and downwards. (The significant aspects of the radiation spectrum are illustrated in Fig. 3.) The outgoing radiation from the Earth-atmosphere system is thus made up of, first, some radiation emitted by the Earth's surface at wavelengths to which the Earth's atmosphere is transparent—the so-called "window", primarily between 7 and 14 μm; second, some radiation which has been emitted from the surface (or from clouds), absorbed by atmospheric gases and reradiated outward by the same (or other) gases; and, third, radiation from clouds which themselves may be receiving heat from below. Some of the heat radiated outward by the gases and clouds is transported from below to the level where it is radiated. These aspects of the radiation balance of the atmosphere are illustrated in Fig. 4.

As carbon dioxide is one of the principal gases taking part in radiation exchange in the atmosphere and in the radiation of the Earth's heat content, a change in the content of carbon dioxide within the atmosphere is likely to influence the process. The chief effect of increasing carbon dioxide is that the gas which is radiating heat to space is found at a higher level in the atmosphere than before—the radiation from lower down in the atmosphere is absorbed by the extra carbon dioxide above and then reradiated to space. In the troposphere, at least, temperature decreases with height, so the effective radiating temperature of the carbon dioxide becomes lower if the amount

of the gas is increased, and therefore less heat is radiated to space. Thus the additional carbon dioxide tends to act as a blanket which keeps the Earth warmer—the Earth has to get rid of the incoming radiation from the Sun, and the same amount can only be removed if the temperature of the atmosphere rises a little.

The calculation of the effect on the radiation budget is not easy because the absorption and emission of heat through the whole of the long and short wave radiation spectum has to be taken into account, and the heat transfer has to be calculated by integration over the very complicated spectrum resulting from the complex pattern of absorption bands introduced by the various atmospheric constituents. Neverthless modern computing resources allow reasonably accurate computations to be made. Probably the most reliable calculations performed so far are those reported by Manabe and Wetherald[7], who point out and overcome two important deficiencies of earlier calculations. The deficiencies arise, first, because the vertical distribution of temperature in the atmosphere is not determined by radiation transfer alone; and, second, because, if world temperature rises due to an increase in carbon dioxide, it is almost certain that there will be more evaporation of water—the water vapour content of the atmosphere will also increase and will have its own effect on the radiation balance.

The first difficulty is overcome by incorporating a restriction in the calculations, namely, that the temperature must not fall off more rapidly with height than −6.5°C km⁻¹, the rate normally found in the troposphere. If this rate is exceeded during the calculations a so-called "convective adjustment" is made.

The second effect is probably more important in the present context. An atmosphere at a higher temperature can hold more water vapour, and the additional water vapour produces a similar blanketing effect to that produced by carbon dioxide. Manabe and Wetherald[7] calculate that an increase of 100% in the content of carbon dioxide would increase the world temperature by 1.3° C if the water content of the atmosphere remained constant, but by 2.4° C if the water vapour increased to retain the same relative humidity. The increase of 25% in CO_2 expected by the end of the century therefore corresponds to an increase of 0.6° C in world temperature—an amount somewhat greater than the climatic variations of recent centuries. Rasool and Schneider[8] give a more recent and substantially smaller estimate of the effect of carbon dioxide, but their scheme makes no allowance for adjustments of stratospheric temperature or absorption of incoming radiation by carbon dioxide.

OTHER VARIABLES

The water vapour content is by no means the only variable factor in the atmosphere which would change as a result of a general warming. The increased water vapour would probably lead to the formation of more clouds because evaporation increases much faster than temperature, and substantially more condensed water would be available. The additional cloud would reflect incoming solar radiation and tend to produce a lowering of temperature—a negative feedback effect to counter balance the positive feedback arising from water vapour. Other calculations[7] show that world temperature is likely to be remarkably sensitive to the average global cloudiness. A change of only 1% in the average cloudiness would produce a change in temperature of almost 1° C. As cloudiness varies greatly from day to day and from place to place, it is really rather remarkable that the fluctuations in world temperature are not greater than those observed. There is thus a suggestion that the negative feedback from cloud cover is a real stabilizing effect on world temperature, increasing cloud keeping the temperature down and cutting off evaporation until the average cloudiness is restored. No quantitative estimate of this effect has, however, yet been made.

On the other hand, a destabilizing feedback which has been given some prominence recently arises from the changes in the Earth's reflectivity which would accompany any change in its permanent cover of ice and snow. A lowering of temperature would lead to an extension of glaciers and permanent snow fields, and these would reflect more of the incoming solar radiation. Budyko[9] has stressed the destabilizing influence of such positive feedback and has suggested that a decrease of incoming solar radiation of only about 2% might lead to another ice age and indeed to complete glaciation of the Earth. The effect of the decrease of ice cover which would accompany an increase of world temperature is, however, less significant. The present limits of permanent snow and ice are such that no extensive recession is possible with increasing temperature without a thawing of the ice cover of the Arctic Ocean. If this took place it would undoubtedly have a significant effect on the climate of the adjacent coasts and hinterland, but the indications are that the effect on the mean temperature in other latitudes would be limited to 1° or 2° C. Regional climatic effects might, however, be significant.

The oceans also have an influence on climatic changes through their large mass and thermal inertia. The atmosphere cannot settle down to a new temperature regime until the temperature of the oceans has come into equilibrium with that regime. Calculations show that this would take of the order of 100 yr, and in consequence the oceans impose a substantial lag on the response of world temperature to such changes as I have discussed here. On the other hand, a rise in temperature of the oceans will itself release additional carbon dioxide to the atmosphere and provide a positive feedback tending to enhance slightly the carbon dioxide effect.

The response of the atmosphere to a change in its heat balance is far from simple. Almost any change in the heat supply is likely to result not only in a change in the overall temperature, but also in the system of winds and weather which derive their energy from the heating and cooling of the atmosphere. Such a change in the winds produces of itself a change in climate, and also modifies the cloud and temperature distribution with a possible feedback to the heat supply and heat loss by radiation.

In spite of the enormous complications of attempting to calculate the whole circulation of atmospheric winds and the resulting cloud and rain distributions, it is probable that only in this way can a soundly based estimate of possible man-made climatic changes be made. Numerical models of the atmospheric circulation have been developed and the computations are just practicable on the largest modern computers. The calculations reproduce the chief features of the world climate, but do not have sufficient precision to differentiate two regimes which would produce a difference of temperature of 1° or 2° C. Possibly such calculations will be feasible in the future, but the sophisticated models required will have to take into account the complicated feedback processes which I have already discussed. Most difficult is to devise a method of calculating the amount of cloud to be expected in a particular circulation regime because individual clouds are too small to be treated; the statistical behaviour of assemblies of clouds covering a region will have to be calculated, and it is not yet clear how this may be achieved.

POSSIBLE CHANGES AND NATURAL FLUCTUATIONS

The climate has undergone many changes in the past. Some of these have been associated with the formation of the atmosphere itself, the generation of oxygen by vegetation and so on, but large scale geographical changes, the moving of the continents, have also played a part. Even since geography and atmospheric composition have been more or less as they are known at present, there have been ice ages and periods considerably milder than the present—the causes of these changes are unknown. The last ice age came to an end over a period of centuries about 10,000 yr ago, but even since then there have been changes. Since the temperature recovered to more or less current values the climatic fluctuation in England has at most been a matter of 2° C or so. The biggest swing in recent decades has probably been about 1° C from the medieval optimum to the little ice age around the seventeenth century. Individual years, even in the past two decades or so, have, however, covered a wider range than 2° C. Even global mean temperatures have varied by 0.6° C from a minimum around 1880 to the last maximum around 1940. Against this background a change of 0.6° C by the end of the century will not be easy to distinguish from natural fluctuations and certainly is not a cause for alarm. Even a doubling of the amount of carbon dioxide in the atmosphere, which would probably require the burning of a large part of the known fuel reserves, would appear to result in a rise of temperature little above that experienced in the climatic optimum which followed the last ice age. Nevertheless it must not be overlooked that variations in climate of only a fraction of a degree centigrade have considerable economic importance, as experience of natural fluctuations has already shown. The more frequent incidence of severe winters or of frost can readily affect the economies of sensitive crops.

Although there may be no immediate cause for alarm about the consequences of carbon dioxide increase in the atmosphere, there is certainly need for further study. We need to have a better assessment of where the carbon dioxide goes after it has been dispersed from our chimneys, and in particular the long term balance with the ocean. We also need a better assessment of the various feedback effects which enter into the control of climate, and the development of increasingly sophisticated numerical simulation of the global climate seems the only possible approach in spite of the effort involved.

NOTES

1 Callendar, G. S., *Quart. J. Roy. Met. Soc.*, **66**, 395 (1940).
2 *Man's Impact on the Global Environment* (MIT Press, Cambridge, Massachusetts, 1970).
3 Craig, H., *Tellus*, **9**, 1 (1957).
4 Bolin, B., *Stockholm, Stat. Naturvetensk. Furskningsgr., Sartryck. Svensk Naturevetensk.*, **134** (1966).
5 *Inadvertent Climate Modification* (MIT Press, Cambridge, Massachusetts, 1971).
6 Kanwisher, J., *Tellus*, **12**, 209 (1960).
7 Manabe, S., and Wetherald, R. T., *J. Atmos. Sci.*, **24**, 241 (1967).
8 Rasool, I., and Schneider, S. H., *Science*, **173**, 138 (1971).
9 Budyko, M. I., *Tellus*, **21**, 611 (1969).

Carbon Dioxide and Climate: A Scientific Assessment

Ad Hoc Study Group on Carbon Dioxide and Climate

JULE G. CHARNEY
Massachusetts Institute of Technology, Chairman

AKIO ARAKAWA
University of California, Los Angeles

D. JAMES BAKER
University of Washington

BERT BOLIN
University of Stockholm

ROBERT E. DICKINSON
National Center for Atmospheric Research

RICHARD M. GOODY
Harvard University

CECIL E. LEITH
National Center for Atmospheric Research

HENRY M. STOMMEL
Woods Hole Oceanographic Institution

CARL I. WUNSCH
Massachusetts Institute of Technology

STAFF
John S. Perry
Robert S. Chen
Doris Bouadjemi
Theresa Fisher

Foreword

Each of our sun's planets has its own climate, determined in large measure by the planet's separation from its mother star and the nature of its atmospheric blanket. Life on our own earth is possible only because of its equable climate, and the distribution of climatic regimes over the globe has profoundly shaped the evolution of man and his society.

For more than a century, we have been aware that changes in the composition of the atmosphere could affect its ability to trap the sun's energy for our benefit. We now have incontrovertible evidence that the atmosphere is indeed changing and that we ourselves contribute to that change. Atmospheric concentrations of carbon dioxide are steadily increasing, and these changes are linked with man's use of fossil fuels and exploitation of the land. Since carbon dioxide plays a significant role in the heat budget of the atmosphere, it is reasonable to suppose that continued increases would affect climate.

These concerns have prompted a number of investigations of the implications of increasing carbon dioxide. Their consensus has been that increasing carbon dioxide will lead to a warmer earth with a different distribution of climatic regimes. In view of the implications of this issue for national and international policy planning, the Office of Science and Technology Policy requested the National Academy of Sciences to undertake an independent critical assessment of the scientific basis of these studies and the degree of certainty that could be attached to their results.

In order to address this question in its entirety, one would have to peer into the world of our grandchildren, the world of the twenty-first century. Between now and then, how much fuel will we burn, how many trees will we cut? How will the carbon thus released be distributed between the earth, ocean, and atmosphere? How would a changed climate affect the world society of a generation yet unborn? A complete assessment of all the issues will be a long and difficult task.

It seemed feasible, however, to start with a single basic question: If we were indeed certain that atmospheric carbon dioxide would increase on a known schedule, how well could we project the climatic consequences? We were fortunate in securing the cooperation of an outstanding group of distinguished scientists to study this question. By reaching outside the membership of the Climate Research Board, we hoped to find unbiased viewpoints on this important and much studied issue.

The conclusions of this brief but intense investigation may be comforting to scientists but disturbing to policymakers. If carbon dioxide continues to increase, the study group finds no reason to doubt that climate changes will result and no reason to believe that these changes will be negligible. The conclusions of prior studies have been generally reaffirmed. However, the study group points out that the ocean, the great and ponderous flywheel of the global climate system, may be expected to slow the course of observable climatic change. A wait-and-see policy may mean waiting until it is too late.

In cooperation with other units of the National Research Council, the Climate Research Board expects to continue review and assessment of this important issue in order to clarify further the scientific questions involved and the range of uncertainty in the principal conclusions. We hope that this preliminary report covering but one aspect of this many-faceted issue will prove to be a constructive contribution to the formulation of national and international policies.

We are grateful to Jule Charney and to the members of the study group for agreeing to undertake this task. Their diligence, expertise, and critical judgment has yielded a report that has significantly sharpened our perception of the implications of the carbon dioxide issue and of the use of climate models in their consideration.

Verner E. Suomi, *Chairman*
Climate Research Board

Preface

In response to a request from the Director of the Office of Science and Technology Policy, the President of the National Academy of Sciences convened a study group under the auspices of the Climate Research Board of the National Research Council to assess the scientific basis for projection of possible future climatic changes resulting from man-made releases of carbon dioxide into the atmosphere. Specifically, our charge was

1. To identify the principal premises on which our current understanding of the question is based,
2. To assess quantitatively the adequacy and uncertainty of our knowledge of these factors and processes, and
3. To summarize in concise and objective terms our best present understanding of the carbon dioxide/climate issue for the benefit of policymakers.

The Study Group met at the NAS Summer Studies Center at Woods Hole, Massachusetts, on July 23–27, 1979, and additional consultations between various members of the group took place in subsequent weeks. We recognized from the outset that estimates of future concentrations of atmospheric carbon dioxide are necessarily uncertain because of our imperfect ability to project the future workings of both human society and the biosphere. We did not consider ourselves competent to address the former and recognized that the latter group of problems had recently been reviewed in considerable detail by the Scientific Committee on Problems of the Environment (SCOPE) of the International Council of Scientific Unions. We therefore focused our attention on the climate system itself and our ability to foretell its response to changing levels of carbon dioxide. We hope that the results of our study will contribute to a better understanding of the implications of this issue for future climate and human welfare.

In our review, we had access not only to the principal published studies relating to carbon dioxide and climate but also to additional unpublished results. For these contributions, we gratefully acknowledge the assistance of the following scientists:

A. Gilchrist, British Meteorological Office
J. Hansen, Goddard Institute for Space Studies, NASA
S. Manabe, R. T. Wetherald, and K. Bryan, Geophysical Fluid Dynamics Laboratory, NOAA

We also had the benefit of discussions with a number of other scientists in the course of the review. We wish to thank the following individuals for their helpful comments:

R. S. Lindzen, Harvard University
C. G. Rooth, University of Miami
R. J. Reed, University of Washington
G. W. Paltridge, Commonwealth Scientific and Industrial Research Organization (CSIRO), Australia
W. L. Gates, Oregon State University

Finally, I wish to express my appreciation to the members of the Study Group for their contributions. In particular, the report benefited greatly from Akio Arakawa's careful examination of the results of general circulation model studies. Our group is also grateful to the staff of the Climate Research Board for their support.

Jule G. Charney, *Chairman*
Ad Hoc Study Group on
Carbon Dioxide and Climate

1 Summary and Conclusions

We have examined the principal attempts to simulate the effects of increased atmospheric CO_2 on climate. In doing so, we have limited our considerations to the direct climatic effects of steadily rising atmospheric concentrations of CO_2 and have assumed a rate of CO_2 increase that would lead to a doubling of airborne concentrations by some time in the first half of the twenty-first century. As indicated in Chapter 2 of this report, such a rate is consistent with observations of CO_2 increases in the recent past and with projections of its future sources and sinks. However, we have *not* examined anew the many uncertainties in these projections, such as their implicit assumptions with regard to the workings of the world economy and the role of the biosphere in the carbon cycle. These impose an uncertainty beyond that arising from our necessarily imperfect knowledge of the manifold and complex climatic system of the earth.

When it is assumed that the CO_2 content of the atmosphere is doubled and statistical thermal equilibrium is achieved, the more realistic of the modeling efforts predict a global surface warming of between 2°C and 3.5°C, with greater increases at high latitudes. This range reflects both uncertainties in physical understanding and inaccuracies arising from the need to reduce the mathematical problem to one that can be handled by even the fastest available electronic computers. It is significant, however, that none of the model calculations predicts negligible warming.

The primary effect of an increase of CO_2 is to cause more absorption of thermal radiation from the earth's surface and thus to increase the air temperature in the troposphere. A strong positive feedback mechanism is the accompanying increase of moisture, which is an even more powerful absorber of terrestrial radiation. We have examined with care all known negative feedback mechanisms, such as increase in low or middle cloud amount, and have concluded that the oversimplifications and inaccuracies in the models are not likely to have vitiated the principal conclusion that there will be appreciable warming. The known negative feedback mechanisms can reduce the warming, but they do not appear to be so strong as the positive moisture feedback. We estimate the most probable global warming for a doubling of CO_2 to be near

3°C with a probable error of ±1.5°C. Our estimate is based primarily on our review of a series of calculations with three-dimensional models of the global atmospheric circulation, which is summarized in Chapter 4. We have also reviewed simpler models that appear to contain the main physical factors. These give qualitatively similar results.

One of the major uncertainties has to do with the transfer of the increased heat into the oceans. It is well known that the oceans are a thermal regulator, warming the air in winter and cooling it in summer. The standard assumption has been that, while heat is transferred rapidly into a relatively thin, well-mixed surface layer of the ocean (averaging about 70 m in depth), the transfer into the deeper waters is so slow that the atmospheric temperature reaches effective equilibrium with the mixed layer in a decade or so. It seems to us quite possible that the capacity of the deeper oceans to absorb heat has been seriously underestimated, especially that of the intermediate waters of the subtropical gyres lying below the mixed layer and above the main thermocline. If this is so, warming will proceed at a slower rate until these intermediate waters are brought to a temperature at which they can no longer absorb heat.

Our estimates of the rates of vertical exchange of mass between the mixed and intermediate layers and the volumes of water involved give a delay of the order of decades in the time at which thermal equilibrium will be reached. This delay implies that the actual warming at any given time will be appreciably less than that calculated on the assumption that thermal equilibrium is reached quickly. One consequence may be that perceptible temperature changes may not become apparent nearly so soon as has been anticipated. We may not be given a warning until the CO_2 loading is such that an appreciable climate change is inevitable. The equilibrium warming will eventually occur; it will merely have been postponed.

The warming will be accompanied by shifts in the geographical distributions of the various climatic elements such as temperature, rainfall, evaporation, and soil moisture. The evidence is that the variations in these anomalies with latitude, longitude, and season will be at least as great as the globally averaged changes themselves, and it would be misleading to predict regional climatic changes on the basis of global or zonal averages alone. Unfortunately, only gross globally and zonally averaged features of the present climate can now be reasonably well simulated. At present, we cannot simulate accurately the details of regional climate and thus cannot predict the locations and intensities of regional climate changes with confidence. This situation may be expected to improve gradually as greater scientific understanding is acquired and faster computers are built.

To summarize, we have tried but have been unable to find any overlooked or underestimated physical effects that could reduce the currently estimated global warmings due to a doubling of atmospheric CO_2 to negligible proportions or reverse them altogether. However, we believe it quite possible that the capacity of the intermediate waters of the oceans to absorb heat could delay the estimated warming by several decades. It appears that the warming will eventually occur, and the associated regional climatic changes so important to the assessment of socioeconomic consequences may well be significant, but unfortunately the latter cannot yet be adequately projected.

2 Carbon in the Atmosphere

A brief account of the key features of the exchange of carbon between the atmosphere, the living and dead organic matter on land (the terrestrial biosphere), and the oceans is essential as a basis for the discussion that follows. The intermediate layers (100–1000 m) of the oceans also play a central role both as a sink for excess atmospheric CO_2 and for heat. For these reasons some basic features of the carbon cycle will be outlined, based primarily on the recently published review by the Scientific Committee on Problems of the Environment (SCOPE) of the International Council of Scientific Unions (Bolin *et al.*, 1979).

The CO_2 concentration in the atmosphere has risen from about 314 ppm (parts per million, volume) in 1958 to about 334 ppm in 1979, i.e., an increase of 20 ppm, which is equivalent to 42×10^9 tons of carbon. During this same period, about 78×10^9 tons of carbon have been emitted to the atmosphere by fossil-fuel combustion. It has further been estimated that more than 150×10^9 tons of carbon have been released to the atmosphere since the middle of the nineteenth century, at which time the CO_2 concentration in the atmosphere most likely was less than 300 ppm, probably about 290 ppm.

By reducing the extent of the world forests (at present about 30 percent of the land surface) and increasing the area of farmland (at present about 10 percent of the land surface) man has also transformed carbon in trees and in organic matter in the soil into CO_2. The magnitude of this additional emission into the atmosphere is poorly known. Estimates range between 40×10^9 tons and more than 200×10^9 tons for the period since early last century.

Since these emissions are not known with any degree of accuracy during the period for which accurate observations of atmospheric CO_2 are available (1958–1979), we know only approximately the ratio between the net increase of CO_2 in the atmosphere and the total man-induced emissions. However, at least 50 percent of the emissions and perhaps more than 70 percent have been transferred into other natural reservoirs for carbon. We need to consider three possible sinks for this transfer:

1. The remaining forests of the world (because of more effective carbon assimilation as a result of higher CO_2 levels in the atmosphere);

2. The surface and intermediate waters of the oceans (above about 1000 m);
3. The deep sea (below about 1000 m).

The distribution of past emissions of CO_2 between these sinks is not entirely clear. On the basis of the radiocarbon concentration in the deep sea, it has been concluded that only a rather small part of the emissions so far have been transferred into the deep sea. However, the proper role of the deep sea as a potential sink for fossil-fuel CO_2 has not been accurately assessed. As indicated in Section 3.3 on the oceans, theoretical estimates of mass transfer from the mixed layer into the intermediate waters indicate that this part of the ocean may have been a more important sink for carbon dioxide emitted into the atmosphere than has so far been considered. This conclusion is also in accord with observations of the penetration of radioactive trace substances produced by nuclear-weapons testing into the intermediate waters. Whether some increase of carbon in the remaining world forests has occurred is not known.

Our limited knowledge of the basic features of the carbon cycle means that projections of future increases of CO_2 in the atmosphere as a result of fossil-fuel emissions are uncertain. It has been customary to assume to begin with that about 50 percent of the emissions will stay in the atmosphere. The possibility that the intermediate waters of the oceans, and maybe also the deep sea, are in more rapid contact with the atmosphere may reduce this figure to 40 percent, perhaps even to a somewhat smaller figure. On the other hand, a continuing reduction of the world forests will further add to any increase due to fossil-fuel combustion. The ability of the oceans to serve as a sink for CO_2 emissions to the atmosphere is reduced as the concentrations increase because of the chemical characteristics of the carbonate system of the sea.

If all the fossil-fuel reserves were used for combustion, the airborne fraction would increase considerably above the values of 30 to 50 percent mentioned above. Global fossil-fuel resources contain at least 5000×10^9 tons of carbon, of which oil and gas together represent about 10 percent. The maximum conceivable amount of future releases from the land biosphere due to deforestation and other changes in land use is of the order of 500×10^9 tons. An emission of 5000×10^9 tons of carbon as CO_2 (i.e., about eight times the pre-industrial amount of CO_2 in the atmosphere) during the next few centuries probably would lead to four to six times higher CO_2 concentration than at present, i.e., 1300–2000 ppm. In view of the huge amounts involved, it seems unlikely that increases in carbon stored in the terrestrial biosphere could reduce these values substantially.

Decline of CO_2 levels in the atmosphere will take centuries because of the slow turnover of the deep sea. However, as the more CO_2-rich waters reach the calcium carbonate deposits on the continental slopes, dissolution may increase the capacity of the oceans to absorb CO_2. Since this process fundamentally depends on the rate with which ocean water can get in contact with the bottom sediments, it is not likely to proceed quickly, although our knowledge is inadequate to assess the role of this process more than qualitatively at present.

Considering the uncertainties, it would appear that a doubling of atmospheric carbon dioxide will occur by about 2030 if the use of fossil fuels continues to grow at a rate of about 4 percent per year, as was the case until a few years ago. If the growth rate were 2 percent, the time for doubling would be delayed by 15 to 20 years, while a constant use of fossil fuels at today's levels shifts the time for doubling well into the twenty-second century.

There are considerable uncertainties about the future changes of atmospheric CO_2 concentrations due to burning of fossil fuels. It appears, in particular, that the role of intermediate waters as a sink for CO_2 needs careful consideration. Predictions of CO_2 changes on time scales of 50 to 100 years may be significantly influenced by the results of such studies. However, considerable changes of atmospheric CO_2 levels will certainly occur as a result of continuing use of fossil fuels. This conclusion is a sufficient basis for the following discussion of possible climatic changes.

3 Physical Processes Important for Climate and Climate Modeling

In order to assess the climatic effects of increased atmospheric concentrations of CO_2, we consider first the primary physical processes that influence the climatic system as a whole. These processes are best studied in simple models whose physical characteristics may readily be comprehended. The understanding derived from these studies enables one better to assess the performance of the three-dimensional circulation models on which accurate estimates must be based.

3.1 RADIATIVE HEATING

3.1.1. Direct Radiative Effects

An increase of the CO_2 concentration in the atmosphere increases its absorption and emission of infrared radiation and also increases slightly its absorption of solar radiation. For a doubling of atmospheric CO_2, the resulting change in net heating of the troposphere, oceans, and land (which is equivalent to a change in the net radiative flux at the tropopause) would amount to a global average of about $\Delta Q = 4\,\mathrm{W\,m^{-2}}$ if all other properties of the atmosphere remained unchanged. This quantity, ΔQ, has been obtained by several investigators, for example, by Ramanathan *et al.* (1979), who also compute its value as a function of latitude and season and give references to other CO_2/climate calculations. The value $4\,\mathrm{W\,m^{-2}}$ is obtained by several methods of calculating infrared radiative transfer. These methods have been directly tested against laboratory measurements and, indirectly, are found to be in agreement with observation when applied to the deduction of atmospheric temperature profiles from satellite infrared measurements. There is thus relatively high confidence that the direct net heating value ΔQ has been estimated correctly to within ± 25 percent. However, it should be emphasized that the accurate calculation of this term has required a careful treatment of the thermal radiative fluxes with techniques that have been developed over the past two decades or more. Crude estimates may easily be in error by a large factor. Thus, in an interim report, MacDonald *et al.* (1979) obtain a ΔQ of 6 to $8\,\mathrm{W}$ $\mathrm{m^{-2}}$, a value about 1.5 to 2 times too large.

Greater uncertainties arise in estimates of the resulting change in global mean surface temperature, ΔT, for this quantity is influenced by various feedback processes that will increase or decrease the heating rate from its direct value. These processes will influence the feedback parameter λ in the expression $\Delta T = \Delta Q / \lambda$. For the simplest case in which only the temperature change is considered, and the earth is assumed to be effectively a blackbody, the value of $\lambda = 4\sigma T^3$ is readily computed to be about $4\,\mathrm{W\,m^{-2}\,K^{-1}}$. For such a case, doubled CO_2 produces a temperature increase of 1 °C.

3.1.2 Feedback Effects

The most important and obvious of the feedback effects arises from the fact that a higher surface temperature produces a much higher value of the surface equilibrium water-vapor pressure through the highly nonlinear Clapeyron-Clausius relation. This, in turn, leads to increased water vapor in the atmosphere. A plausible assumption, borne out qualitatively by model studies, is that the relative humidity remains unchanged. The associated increase of absolute humidity increases the infrared absorptivity of the atmosphere over that of CO_2 alone and provides a positive feedback. There is also increased absorption of solar radiation by the increased water vapor, which further increases the infrared feedback by about 10 percent. As with CO_2, the radiative transfer calculation of water-vapor effects is relatively reliable, and the consequence is that λ is decreased and ΔT increased by about a factor of 2. For doubled CO_2, the temperature increase would be 2 °C.

One-dimensional radiative-convective models that assume fixed relative humidity, a fixed tropospheric lapse rate of $6.5\,\mathrm{K\,km^{-1}}$, and fixed cloud cover and height give $\lambda = 2.0\,\mathrm{W\,m^{-2}\,K^{-1}}$ (Ramanathan and Coakley, 1978). This value is uncertain by at least $\pm 0.5\,\mathrm{W\,m^{-2}\,K^{-1}}$ because of uncertainties in the possible changes of relative humidity, temperature lapse rate, and cloud cover and cloud height.

Snow and ice albedo provide another widely discussed positive feedback mechanism (see, for example, Lian and Cess, 1977, and additional references therein). As the surface temperature increases, the area covered by snow or ice decreases; this lowers the mean global albedo and increases the fraction of solar radiation absorbed. Estimates of this effect lead to a further decrease of λ by between 0.1 and 0.9 W m^{-2} K^{-1} with 0.3 a likely value. Some uncertainty in albedo feedback also arises from cloud effects discussed in the next section. Taking into consideration all the above direct effects and feedbacks, we estimate λ to be 1.7 ± 0.8 W m^{-2} K^{-1} and hence ΔT for doubled CO_2 to lie in the range of 1.6 to 4.5 K, with 2.4 K a likely value.

3.2   CLOUD EFFECTS

Most clouds are efficient reflectors of solar radiation and at the same time efficient absorbers (and emitters) of terrestrial infrared radiation. Clouds thus produce two opposite effects: as cloud amount and hence reflection increase, the solar radiation available to heat the system decreases, but the decreased upward infrared radiation at the tropopause and downward radiation from the base of the clouds raises the temperature of the earth's surface and troposphere.

Because the change of solar absorption dominates, the *net* result of increased low cloudiness, and very likely also middle cloudiness, is to lower the temperature of the system. The net effect of an increased amount of high cirrus clouds is less certain because their radiative characteristics are sensitive to height, thickness, and microphysical structure. Present estimates are that they raise the temperature of the earth's surface and the troposphere.

It follows that if a rise in global temperature results in an increased amount of low or middle clouds, there is a negative feedback, and if a rise in global temperature results in an increased amount of high clouds, there is a positive feedback. The effect of cloud albedo by itself gives a negative feedback. Thus if clouds at all levels were increased by 1 percent, the atmosphere-earth system would absorb about 0.9 m^{-2} less solar radiation and lose about 0.5 W m^{-2} less thermal radiation. The net effect would be a cooling of about 0.4 W m^{-2}, or, if this occurred together with a doubling of CO_2, a decrease of ΔQ from 4.0 to 3.6 W m^{-2}.

How important the overall cloud effects are is, however, an extremely difficult question to answer. The cloud distribution is a product of the entire climate system, in which many other feedbacks are involved. Trustworthy answers can be obtained only through comprehensive numerical modeling of the general circulations of the atmosphere and oceans together with validation by comparison of the observed with the model-produced cloud types and amounts. Unfortunately, cloud observations in sufficient detail for accurate validation of models are not available at present.

Since individual clouds are below the grid scale of the general circulation models, ways must be found to relate the total cloud amount in a grid box to the grid-point variables. Existing parameterizations of cloud amounts in general circulation models are physically very crude. When empirical adjustments of parameters are made to achieve verisimilitude, the model may appear to be validated against the present climate. But such tuning by itself does not guarantee that the response of clouds to a change in the CO_2 concentration is also tuned. It must thus be emphasized that the modeling of clouds is one of the weakest links in the general circulation modeling efforts.

The above uncertainties, and others such as those connected with the modeling of ground hydrology and snow and ice formation, create uncertainties in the model results that will be described in Chapter 4.

3.3   OCEANS

Existing numerical models of the atmosphere, which treat the ocean as having no meridional heat transports of its own, may give somewhat improper accounts of the CO_2 impact. It is currently estimated that at some latitudes the ocean transports as much as 50 percent of the poleward heat flux in the existing climatic system. A proper accounting for oceanic dynamics has several possible consequences as levels of CO_2 continue to rise.

The role of the ocean as an active transporter of heat meridionally leads one to consider several possible feedback mechanisms. Atmospheric models suggest that the warming at high latitudes will be larger than at low latitudes. If this reduced atmospheric baroclinicity reduces the wind stress at the ocean surface (and there are not good estimates of the anticipated size of such a reduction), it is possible that oceanic meridional heat flux might be reduced. Because of the required overall radiative heat balance of the total system, the atmosphere would then be required to compensate for reduced oceanic heat transport by steepening the equator-to-pole temperature gradient, thus ameliorating somewhat the predicted polar warming. However, the total atmospheric warming would not likely be greatly affected, merely its distribution in latitude.

The only part of the ocean that has been included in the general circulation modeling of the CO_2 effects is the mixed layer. The rationale for this simplification is that only the mixed layer needs to be modeled in order to deal with the annual cycle, while the heat capacity of the deeper ocean does not matter once thermal equilibrium has been reached.

On time scales of decades, however, the coupling between the mixed layer and the upper thermocline must be considered. The connections between upper and lower ocean are generally presumed to have response times of the order of 1000 years, the essential coupling being local vertical diffusion and formation of bottom water at high latitudes. This ignores the mechanism of Ekman convergence of the surface mixed layers in the large subtropical gyres, which pumps water down into the upper thermocline over more than half the ocean surface area, a reservoir much larger than that of the mixed layer alone. The connections between the upper-thermocline reservoir and the deep ocean may indeed require very long time constants, but the carbon and heat budgeting on the decadal time scale must account properly for the potentially large reservoir directly beneath the mixed layer.

Simple model calculations involving Ekman pumping from the mixed layer into the intermediate waters of the order of 10–20 cm/day^{-1} and estimates of mixing coefficients for the intermediate waters from tracer studies (Östlund *et al.*, 1974; National Science Foundation, 1979) suggest that the upper-thermocline reservoir communicates effectively with the mixed layer on time scales of several decades. Therefore, the effective thermal capacity of the ocean for absorbing heat on these time scales is nearly an order of magnitude greater than that of the mixed layer alone.[1] If this reservoir is indeed involved, it could delay the attainment of ultimate global thermal equilibrium by the order of a few decades. It would also increase the rate at which the ocean can take up carbon from the air and might at least partially account for the current discrepancies between the observed rise in atmospheric CO_2 and the estimated rise due to the anthropogenic input of CO_2 into the air.

NOTES

1   The existence of the Ekman pumping underlies all the generally accepted ideas about the physics of the general circulation of the oceans. The order of magnitude estimated above (10–20 cm/day) is consistent with a variety of oceanographic data, including wind stress, chemical tracers, and local heat-budget calculations.

4   Models and Their Validity

The independent studies of the CO_2/climate problem that we have examined range from calculations with simple radiative-convective models to zonally and vertically averaged heat-balance models with horizontally diffusive heat exchange and snow-ice albedo feedbacks to full-fledged three-dimensional general circulation models (GCM's) involving most of the relevant physical processes. Our confidence in our conclusion that a doubling of CO_2 will eventually result in significant temperature increases and other climate changes is based on the fact that the results of the radiative-convective and heat-balance model studies can be understood in purely physical terms and are verified by the more complex GCM's. The last give more information on geographical variations in heating, precipitation, and snow and ice cover, but they agree reasonably well with the simpler models on the magnitudes of the overall heating effects.

The radiative-convective models have been reviewed by Ramanathan and Coakley (1978). The latitudinally varying energy-balance models were originally developed by Budyko (1969) and Sellers (1969) for studies of climatic change. More recently they have been employed by many authors, including Ramanathan *et al.* (1979) and MacDonald *et al.* (1979), for CO_2/climate-change determinations. These models prescribe the infrared feedback but calculate the snow-ice albedo feedback by coupling to a simple parameterized horizontal heat transport; the snow and ice occur poleward of the latitude at which the temperature has an empirically prescribed value. The principal value of these models lies in their inclusion of the snow-ice albedo feedback. However, they do not deal with real geography or explicit dynamics and therefore can yield only crude approximations to the latitudinal variations of the CO_2-induced temperature changes.

4.1   THREE-DIMENSIONAL GENERAL CIRCULATION MODELS

We proceed now to a discussion of the three-dimensional model simulations on which our conclusions are primarily based. Some of the existing general circulation models have been used to predict the climate for doubled or quadrupled CO_2 con-

centration. The results of several such predictions were available to us: three by S. Manabe and his colleagues at the NOAA Geophysical Fluid Dynamics Laboratory (hereafter identified as M1, M2, and M3) and two by J. Hansen and his colleagues at the NASA Goddard Institute for Space Studies (hereafter identified as H1 and H2). Some results obtained with the British Meteorological Office model (Mitchell, 1979) were also made available to us but will not be described here because both the sea-surface temperature and the sea-ice distribution were prescribed in this model, thus placing strong constraints on the surface ΔT, whereas it is just the surface ΔT that we wish to estimate.

The only one of the five predictions available in published form is M1. M2 is described in a pre-publication manuscript, and H1 in a research proposal. We learned of M3 and H2 through personal communication.

The Geophysical Fluid Dynamics Laboratory and the Goddard Institute for Space Studies general circulation models, which are the basic models used in the M and H series, respectively, were independently constructed and differ from one another in a number of physical and mathematical aspects. They also differ in respect to their geographies, seasonal changes, cloud feedbacks, snow and ice properties, and horizontal and vertical grid resolutions. These differences are summarized in Table 1. In this table "swamp" means that the model ocean has no heat capacity though it provides a water surface for evaporation, and "mixed layer" means that the model ocean has a heat capacity corresponding to that of an oceanic mixed layer of constant depth. Heat transport by ocean currents is neglected in both model oceans. This is one of the weaknesses of all the predictions, as discussed in Section 3.3.

The horizontal resolution of the H series is rather coarse and perhaps only marginal for meaningful climate prediction. On the other hand, these models take into account more physical factors, such as ground heat storage, sea-ice leads, and dependence of snow-ice albedo on snow age, than do the models of the M series.

The models M1, H1, and H2 were run for doubled CO_2 concentrations, M2 for both doubled and quadrupled concentrations, and M3 for quadrupled concentrations. The temperature changes for

Table 1. Characteristics of General-Circulation Models Examined (λ, Longitude; ϕ, Latitude; T, Temperature).

Model Characteristics	M1[a]	M2[a]	Model Predictions M3[a]	H1[b]	H2[b]
Domain	$0° < \lambda < 120°C$ $0° < \phi < 81.7°$	$0° < \lambda < 120°C$ $0° < \phi < 90°$	Global	Global	Global
Land-ocean distribution	Ocean for $60° < \lambda < 120°$ $0° < \phi < 66.5°$	Ocean for $60° < \lambda < 120°$ $0° < \phi < 90°$	Realistic	Realistic	Realistic
Ocean	Swamp	Swamp	Mixed layer	Mixed layer	Swamp
Seasonal change	No	No	Yes	Yes	No
Cloud feedbacks	No	Yes	No	Yes	Yes
Snow and ice albedo	When $T < -25°C$ 0.7 When $T > -25°C$ 0.45 for snow 0.35 for ice	When $T < -10°C$ 0.7 When $T > -10°C$ 0.45 for snow 0.35 for ice	Depends on depth and underlying surface albedo For deep snow, 0.8 For thick ice, 0.7	For snow, depends on snow age, snow depth, underlying surface albedo, etc. For ice, 0.45	Same as H1
Horizontal resolution	About 500 km on a mercator projection	5° in longitude 4.5° in latitude	Spectral model with the maximum zonal wave number 15	10° in longitude 8° in latitude	Same as H1
Vertical resolution	9 layers	9 layers	9 layers	7 layers	7 layers

[a] Models developed by S. Manabe and colleagues at the NOAA Geophysical Fluid Dynamics Laboratory, Princeton, N.J.
[b] Models developed by J. Hansen and colleagues at the NASA Goddard Institute for Space Studies, New York, N.Y.
[c] Cyclic continuity assumed at boundaries.

doubled CO_2 in M2 were approximately half of those for quadrupled CO_2. Since it can be expected that a similar result would have been obtained for M3, we have halved the M3 temperature changes.[1]

At low latitudes, the predicted values of the mean surface ΔT for doubled CO_2 concentration were slightly more than 1.5°C in the M series, 2.5°C in H1, and 3.0°C in H2. Both series predict larger ΔT at upper levels, primarily because of added heating by cumulus convection. The discrepancy in the surface ΔT may well be due to differences in the respective parameterizations of cumulus convection.

The hemispheric mean surface ΔT is about 3°C in M1 and M2, and the global mean about 2°C in M3, 3.5°C in H1, and 3.9°C in H2. The 1°C difference between M3 and M1/M2 has been ascribed partly to the exclusion of seasonal changes and southern hemisphere effects in M1/M2 and their inclusion in M3; in the southern hemisphere the area covered by land, and therefore the snow-ice albedo feedback, is smaller than in the northern hemisphere, and there is no albedo feedback over Antarctica. The differences between the M series and the H series may be at least partially attributed to differences in the areas covered by snow and ice.

All the GCM's predict larger surface ΔT at high latitudes. This is partly due to the snow-ice albedo feedbacks and also to the fact that the strong gravitational stability produced by cooling from below suppresses convective and radiative transfer of heat and thereby concentrates the CO_2-enhanced heating in a thin layer near the ground. Although the magnitudes and locations of the temperature increases vary significantly, all the predictions give a maximum of between 4°C and 8°C in polar or subpolar regions for the annual mean surface ΔT. More detailed descriptions of the model predictions for high latitudes are given in the Appendix.

With regard to clouds, M2 gives a decrease of high clouds in low latitudes, whereas H1 and H2 give an increase. This discrepancy may well be due to differences in the parameterization of cumulus convection. The M series relies on an adjustment process for distributing heat and moisture by cumulus convection. This process takes place when and only when a layer of air is both saturated and moist-convectively unstable. In contrast, the H series permits cumulus convection to extend through nonsaturated and stable layers; but because it does not allow for entrainment of noncloud air, the penetrating cumuli extend higher than they otherwise would. At high latitudes, M2, which has no seasons, predicts an increase of both high and low clouds; in comparison, H1, which does have seasons, predicts an increase of high clouds throughout the year but an increase of zonally averaged low cloud amount only in spring. It may be shown from data presented by Manabe and Wether-ald that the M2 cloud radiative feedback effects are relatively small intrinsically and are rendered even smaller by the tendency of their short and long wave components to compensate. This tendency is not apparent in H1, but there the negative and middle cloud feedback is on the average weak or nonexistent.

For comparison purposes, the convective adjustment parameterization was introduced into an H model with fixed sea-surface temperatures and was found to reduce appreciably the penetration of water vapor and cloud to high levels (J. Hansen, NASA Goddard Institute for Space Studies, personal communication). Since the original penetration was probably too high because of lack of noncloud air entrainment, we conclude that the surface ΔT's due to the upper water-vapor-cloud feedback may very well have been overestimated in the H series, whereas, because of insufficient penetration, they were probably underestimated in the M series. Since, moreover, the snow-ice boundary is too far equatorward in H1 and too far poleward in M1 and M2 (see Appendix), we believe that the snow-ice albedo feedback has been over-estimated in the H series and underestimated in M1 and M2. For the above reasons, we take the global or hemispheric surface warmings to approximate an upper bound in the H series and a lower bound in the M series (with respect to positive water-vapor-cloud and snow-ice albedo feedback effects). These are at best informed guesses, but they do enable us to give rough estimates of the probable bounds for the global warming. Thus we obtain 2°C as the lower bound from the M series and 3.5°C as the upper bound from H1, the more realistic of the H series. As we have not been able to find evidence for an appreciable negative feedback due to changes in low- and middle-cloud albedos or other causes, we allow only 0.5°C as an additional margin for error on the low side, whereas, because of uncertainties in high-cloud effects, 1°C appears to be more reasonable on the high side. We believe, therefore, that the equilibrium surface global warming due to doubled CO_2 will be in the range 1.5°C to 4.5°C, with the most probable value near 3°C. These estimates may be compared with those given in our discussion of feedback effects in one-dimensional, radiative-convective models. There the range was 1.6°C to 4.5°C, with 2.4°C estimated as a likely value.

We recall that the snow-ice albedo feedback is greater in the northern than in the southern hemisphere because of the greater land area and the lack of albedo change over Antarctica. Hence we estimate that the warming will be somewhat greater in the northern hemisphere and somewhat less in the southern hemisphere.

The existing general circulation models produce time-averaged mean values of the various meteorological parameters, such as wind, temperature, and rainfall, whose climate is reasonably accurate in global or zonal mean. Their inaccuracies are revealed much more in their regional climates. Here physical shortcomings in the treatments of cloud, precipitation, evaporation, ground hydrology, boundary-layer turbulent transport phenomena, orographic effects, wave-energy absorption and reflection in the high atmosphere, as well as truncation errors arising from lack of sufficient resolution combine to produce large inaccuracies. Two models may give rather similar zonal averages but, for example, very different monsoon circulations, positions, and intensities of the semipermanent centers of action and quite different rainfall patterns. It is for this reason that we do not consider the existing models to be at all reliable in their predictions of regional climatic changes due to changes in CO_2 concentration.

We conclude that the predictions of CO_2-induced climate changes made with the various models examined are basically consistent and mutually supporting. The differences in model results are relatively small and may be accounted for by differences in model characteristics and simplifying assumptions. Of course, we can never be sure that some badly estimated or totally overlooked effect may not vitiate our conclusions. We can only say that we have not been able to find such effects. *If the CO_2 concentration of the atmosphere is indeed doubled and remains so long enough for the atmosphere and the intermediate layers of the ocean to attain approximate thermal equilibrium, our best estimate is that changes in global temperature of the order of 3°C will occur and that these will be accompanied by significant changes in regional climatic patterns.*

NOTES

1   It should, however, be pointed out that the snow-ice albedo feedback may not be linear. For example, quadrupled CO_2 in M3 melts the arctic ice altogether in summer.

REFERENCES

General Circulation Models Examined in the Study

M1: Manabe, S., and R. T. Wetherald (1975). The effects of doubling the CO_2 concentration on the climate of a general circulation model, *J. Atmos. Sci. 32*, 3.

M2: Manabe, S., and R. T. Wetherald (1980). On the distribution of climate change resulting from an increase in CO_2 content of the atmosphere (accepted for publication in *J. Atmos. Sci.*, January 1980).

M3: Manabe, S., and R. Stouffer (1979). Study of climatic impact of CO_2 increase with a mathematical model of global climate (submitted to *Nature*).

H1: Goddard Institute for Space Studies (1978). *Proposal for Research in Global Carbon Dioxide Source/Sink Budget and Climate Effects*, Institute for Space Studies, 2880 Broadway, New York, N.Y. 10025.

H2: Hansen, J. E. Private communication. Paper in preparation for submission to *J. Atmos. Sci.* Information available from Institute for Space Studies, 2880 Broadway, New York, N.Y. 10025.

OTHER REFERENCES

Bolin, B., E. T. Degens, S. Kempe, and P. Ketner, eds. (1979). *The Global Carbon Cycle*, SCOPE 13, Scientific Committee on Problems of the Environment, International Council of Scientific Unions, Wiley, New York, 491 pp.

Budyko, M. I. (1969). The effect of solar radiation variations on the climate of the earth, *Tellus 21*, 611.

Lian, M. S., and R. D. Cess (1977). Energy balance climate models—a reappraisal of ice-albedo feedback, *J. Atmos. Sci. 34*, 1058.

MacDonald, G. F., H. Abarbanel, P. Carruthers, J. Chamberlain, H. Foley, W. Munk, W. Nierenberg, O. Rothaus, M. Ruderman, J. Vesecky, and F. Zachariasen (1979). The long term impact of atmospheric carbon dioxide on climate, *JASON Technical Report JSR-78-07*, SRI International, Arlington, Virginia.

Mitchell, J. F. B. (1979). Preliminary report on the numerical study of the effect on climate of increasing atmospheric carbon dioxide, *Meteorological Office Technical Note No. II/137*, Meteorological Office, Bracknell, Berkshire, United Kingdom.

National Science Foundation (1979). *GEOSECS Atlas* (in press).

Östlund, H. G., H. G. Dorsey, and C. G. Rooth (1974). GEOSECS North Atlantic radiocarbon and tritium results. *Earth Planet. Sci. Lett. 23*, 69–86.

Ramanathan, V., and J. A. Coakley, Jr. (1978). Climate modeling through radiative-convective models, *Rev. Geophys. Space Phys. 16*, 465.

Ramanathan, V., M. S. Lian, and R. D. Cess (1979). Increased atmospheric CO_2: zonal and seasonal estimates of the effect on the radiation energy balance and surface temperature, *J. Geophys. Res. 84*, 4949–4958.

Sellers, W. D. (1969). A global climatic model based on the energy balance of the earth-atmosphere system. *J. Appl. Meteorol. 8*, 392.

Part II

The Carbon Cycle

12

The Sky is Rising!

Callendar, G.S. (1938). The artificial production of carbon dioxide and its influence on climate. *Quarterly Journal of the Royal Meteorological Society*, **64**, 223–240. 17 pages.

This is essentially three papers in one, which combine to assert for the first time that human activity could change the climate of the Earth.

The first leg of the paper is a claim that atmospheric CO_2 is already rising due to human activity. Callendar cited measurements of the CO_2 concentration from around the year 1900, and again from about 1935. The values available to him were very close to modern measurements from fossil air samples preserved in ice cores, and the CO_2 concentration had indeed risen in that time interval. Callendar compared the atmospheric CO_2 rise with industrial emission of carbon, which he underestimated somewhat and so ended up with an airborne fraction of 75%, a bit higher than the current best guess, for his time and ours, of about 50% airborne fraction of emitted CO_2 during a CO_2-emitting era.

Callendar apparently constructs a conceptual picture of the carbon cycle dominated by ocean uptake, but unfortunately does not elaborate very much on how he imagined the carbon cycle to work. He projected the future trajectory of atmospheric CO_2 concentration. Callendar showed the sensitivity of the CO_2 rise to the overturning time scale of the ocean, which he bracketed by values of 2000 and 5000 years (a modern estimate is about 1000 years). Presumably he imagined that the ocean would absorb essentially all of the CO_2 on this time scale, so that a slow mixing time would mean slower CO_2 uptake, and higher atmospheric concentration. It cannot have been obvious that ocean uptake would be slow; gazing at a blue globe of the mostly water-covered Earth, it would be easy to imagine that the ocean would equilibrate quickly, even continuously, with the atmosphere. In this realization, Callendar appears to be ahead of Revelle and Suiss two decades later, who concluded that the ocean would take up CO_2 in only a decade (Fig. 12.1).

Callendar's projections of future evolution of atmospheric CO_2 suffered from his assumption of constant emissions with time, rather than the exponential growth in CO_2 emissions, which has persisted from his day to ours. Arrhenius stated that it would take 1000 years for human activity to double the CO_2 concentration of the atmosphere, and Callendar is not far off from this trajectory either. The two authors spun the conclusion entirely differently, however. Arrhenius figured that there is nothing to worry about because the CO_2 rise is so slow, and that anyway a little warming would be OK. Callendar was astonished that human activity could change CO_2 at all, but thought that anyway a little warming would be OK.

The second leg of Callendar's paper was a recalculation of the radiative impact of rising CO_2 concentration, and its impact on the Earth's temperature. The spectral data were probably better than Arrehnius', but in a few conceptual ways his calculation was a step backward from what Arrhenius had done. Like Plass, Callendar held the temperature of the atmosphere fixed, and calculated the change in the ground temperature by adding to the initial steady state energy balance an additional source of heat from the downward radiation from higher CO_2 concentration. Arrhenius also reasoned out the water vapor feedback, but neither Plass nor Callendar took advantage of his insight.

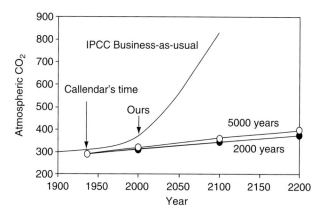

Fig. 12.1   **Projections of atmospheric CO$_2$ concentration, from Callendar (lines with symbols) and the present (to 2000, followed by IPCC 1992 Business-as-usual Scenario A, line with no symbols). The difference is largely due to exponentially increasing emissions, which Callendar did not anticipate.**

The third leg of the paper was a reconstruction of northern hemisphere temperature variations through time. Callendar considered the urban heat island effect, which turns out to be small but that has been a source of concern as a potential effect to this day. Callendar's reconstruction agrees well with subsequent reconstructions including those presented in the IPCC Fourth Assessment Report. The slight rise in temperature from 1900 to his time looked to Callendar like the signature of global warming. In this regard, Callendar is now believed to be wrong; the human influence on climate is not thought to have risen above natural variability until the 1970s.

John Callendar is probably best known as the first to piece together the carbon cycle of the Earth as it pertains to the question of global warming. This is the thinnest part of the paper. One of the reviewers, whose comments are printed at the end of the paper, asked for more details on Callendar's thinking on the carbon cycle, to which Callendar replied that his musings on this subject were fully eight times longer than the rest of the paper, and he was unable to publish this for lack of space. He might have done better to skimp on some of the more traditional work in the rest of the paper.

The Artificial Production of Carbon Dioxide and its Influence on Temperature

G. S. CALLENDAR

(Steam technologist to the British Electrical and Allied Industries Research Association.)
(Communicated by Dr. C. M. B. Dobson, F.R.S.)
[Manuscript received May 19, 1937—read February 16, 1938.]

SUMMARY

By fuel combustion man has added about 150,000 million tons of carbon dioxide to the air during the past half century. The author estimates from the best available data that approximately three quarters of this has remained in the atmosphere.

The radiation absorption coefficients of carbon dioxide and water vapour are used to show the effect of carbon dioxide on "sky radiation." From this the increase in mean temperature, due to the artificial production of carbon dioxide, is estimated to be at the rate of 0.003°C. per year at the present time.

The temperature observations at 200 meteorological stations are used to show that world temperatures have actually increased at an average rate of 0.005°C. per year during the past half century.

Few of those familiar with the natural heat exchanges of the atmosphere, which go into the making of our climates and weather, would be prepared to admit that the activities of man could have any influence upon phenomena of so vast a scale.

In the following paper I hope to show that such influence is not only possible, but is actually occurring at the present time.

It is well known that the gas carbon dioxide has certain strong absorption bands in the infra-red region of the spectrum, and when this fact was discovered some 70 years ago it soon led to speculation on the effect which changes in the amount of the gas in the air could have on the temperature of the earth's surface. In view of the much larger quantities and absorbing power of atmospheric water vapour it was concluded that the effect of carbon dioxide was probably negligible, although certain experts, notably Svante Arrhenius and T. C. Chamberlin, dissented from this view.

Of recent years much new knowledge has been accumulated which has a direct bearing upon this problem, and it is now possible to make a reasonable estimate of the effect of carbon dioxide on temperatures, and also of the rate at which the gas accumulates in the atmosphere. Amongst important factors in such calculations may be mentioned the temperature-pressure-alkalinity-CO_2 relation for sea water, determined by C. J. Fox (1909), the vapour pressure-atmospheric radiation relation, observed by A. Angstrom (1918) and others, the absorption spectrum of atmospheric water vapour, observed by Fowle (1918), and a full knowledge of the thermal structure of the atmosphere.

This new knowledge has been used in arriving at the conclusions stated in this paper, but for obvious reasons only those parts having a meteorological character will be referred to here.

1. THE RATE OK ACCUMULATION OF ATMOSPHKRIC CARBON DIOXIDE

I have examined a very accurate set of observations (Brown and Escombe, 1905), taken about the year 1900, on the amount of carbon dioxide in the free air, in relation to the weather maps of the period. From them I concluded that the amount of carbon dioxide in the free air of the North Atlantic region, at the beginning of this century, was 2.74 ± 0.05 parts in 10,000 by volume of dry air.

Callendar, G. S. (1938) The artificial production of carbon dioxide and its influence on climate. *Quarterly Journal of the Royal Meteorological Society* 64: 223–240. Reprinted with permission of John Wiley & Sons.

Table I. The effect of the artificial production of carbon dioxide upon its pressure in the atmosphere.

Annual net addition of CO_2 to the air = 4,300 million tons.
Total pressure from CO_2 = 0.000274 atmosphere in the year 1900.
See surface at 15°C. and total alkalinity = 40 mg. of negative hydroxyl ions per litre; it is this quantity which is maintained neutral by the dissolved CO_2.
P (CO_2) stands for the pressure of CO_2 in the air at normal barometric pressure.

Sea equilibrium time, years	Date: 1936	2000	2100	2200
	P (CO_2) in atmos./10,000			
2000	2.89	3.14	3.46	3.73
5000	2.90	3.17	3.58	3.96
All CO_2 to the air	2.96	3.35	3.96	4.58

From 1900 to 1936 the increase should be close to 6 per cent.

A great many factors which influence the carbon cycle in nature have been examined in order to determine the quantitative relation between the natural movements of this gas and the amounts produced by the combustion of fossil fuel. Such factors included the organic deposit of carbon in swamps, etc., the average rate of fixation of the gas by the carbonisation of alkalies from igneous rocks, and so on. The general conclusion from a somewhat lengthy investigation on the natural movements of carbon dioxide was that there is no geological evidence to show that the *net* offtake of the gas is more than a small fraction of the quantity produced from fuel. (The artificial production at present is about 4,500 million tons per year.)

The effect of solution of the gas by the sea water was next considered, because the sea acts as a giant regulator of carbon dioxide and holds some sixty times as much as the atmosphere. The rate at which the sea water could correct an excess of atmospheric carbon dioxide depends mainly upon the fresh volume of water exposed to the air each year, because equilibrium with the atmospheric gases is only established to a depth of about 200 m. during such a period.

The vertical circulation of the oceans is not well understood, but several factors point to an equilibrium time, in which the whole sea volume is exposed to the atmosphere, of between two and five thousand years. Using Fox's solution coefficients for sea water of known total alkalinity and average surface temperature, it is possible to calculate the change in atmospheric CO_2 pressure over a given period, when the rate of addition of the gas is known and the equilibrium time for the sea water is assumed. A few such figures are given in Table I, and it will be seen that when periods of a few centuries are considered the sea equilibrium time is less important.

Since calculating the figures in Table I, I have seen a report of a great number of observations on atmospheric CO_2, taken recently in the eastern U.S.A. The mean of 1,156 "free air" readings taken in the years 1930 to 1936 was 3.10 parts in 10,000 by volume. For the measurements at Kew in 1898 to 1901 the mean of 92 free air values was 2.92, including a number of rather high values effected by local combustion, etc. ; and assuming that a similar proportion of the American readings are affected in the same way, the difference is equal to an increase of 6 per cent. Such close agreement with the calculated increase is, of course, partly accidental.

2.   INURA-RED ABSORPTION BY CARBON DIOXIDE AND WATER VAPOUR

The loss of heat from the earth's surface and atmosphere is nearly all carried upon wave lengths greater than 4μ, the maximum intensity being at about $IO\mu$.

There have been a great many careful and accurate measurements of the absorption and radiation by various gases in this part of the spectrum, but owing to the very great difficulties attending these observations most of the earlier values were highly conflicting. However, considerable accuracy has now been attained, and from a number of considerations, which cannot be detailed, the values observed by Rubens and Aschkinass (1898), of Germany, are used here for the absorption by carbon dioxide on the longer wave lengths.

For water vapour I have made many comparisons between the measurements of F. E. Fowle (1918),

Table II. Absorption exponents for carbon dioxide.

Band 4 to 4.6 μ by H. Schmidt (1913).
Band 13 to 16 μ by Rubens and Aschkinass (1898).

Wave length μ	4.0	4.1	4.2	4.25	4.3	4.4	4.45	4.5	4.6
$10^3 k$	4.1	12.0	33.5	50.0	61.0	38.0	27.2	19.7	11.0
Wave length μ	13	13.5	14	14.25	14.5	15	15.25	15.5	16
$10^3 k$	2.5	6.4	17.8	44.6	58.5	59.3	39.7	15.7	1.0

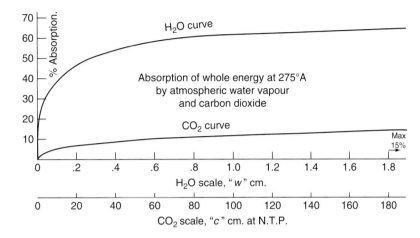

Fig. 1.

who observed the absorption by atmospheric water vapour, and those of Rubens and of Hettner (1918), who used steam at 1 atmos. pressure for their measurements. These comparisons fully support the conclusion arrived at by Fowle, that the absorption by water vapour as it occurs in the air is less than half as great as that found for steam under laboratory conditions. Perhaps the most powerful support for this conclusion comes from a comparison between the observed and calculated atmospheric radiation, which shows that, for dry and cold air conditions, the absorption exponents found for steam lead to much too high a value for atmospheric radiation.

To return to the absorption by carbon dioxide, the three primary bands given by this gas are at 2.4 to 3.0 μ, 4 to 4.6 μ, and 13 to 16 μ, the latter being much the most important for atmospheric conditions because very little low temperature radiation is carried on the small bands.
The relation between absorption and quantity of gas is usually expressed as

$$A_\lambda = I - e^{-kl} \qquad (1)$$

where A_λ is the absorption by l units of gas in the path of radiation of wave length λ, k being the absorption exponent for the gas at this wave length. Some values of this exponent for carbon dioxide are given in Table II.

The lower curve on Fig. 1 shows the absorption by carbon dioxide when the radiation comes from a source at 275° A, as given by the exponents in Table II.

The absorption by water vapour in the infra-red region of the spectrum is very complicated as it consists of innumerable fine bands. As already remarked, it is apparently affected by the presence of air, which may account for the observed difference between the values found for steam and for atmospheric water vapour.[1]

Owing to the difficulties of measurement and to the presence of carbon dioxide in the air, Fowle was not able to extend his observations on water vapour beyond about 13.5 μ. He observed strong CO_2, absorption at about 14.5 to 15 μ.

For the longer wave lengths I have used the values found for steam, and corrected these for the conditions in the atmosphere from a comparison of the values at wave lengths where each are available.

The absorption/quantity relation for atmospheric water vapour, which is shown by Fowle's experiments, can be written:—

$$A_\lambda = I - e^{-K_\lambda w^4} \qquad (2)$$

where A_λ is the absorption on band λ by a quantity of water vapour which would give w cm. of liquid water when condensed on a surface of equal area.

Table III. Absorptions exponents for water vapour, and the percentage of energy on the wave bands.

Wave band μ	3–4	4–5	5–6	6–7	7–8	8–13	13–16	>16
K	1.1	2.1	1.9	6.2	2.2	0.20	1.0	2.0
Temp. abs.			Percentage of energy on band					
250	0	0.3	1.0	2.1	3.4	25.8	15.2	52.4
275	0.1	0.6	1.8	3.2	4.5	29.2	14.9	45.7
300	0.2	1.1	2.8	4.4	5.7	31.7	14.5	39.8

The values obtained from this expression, (2), fit Fowle's observations remarkably well over a wide range of water vapour.

For the important band 13 to 16μ Rubens and Aschkinass found that steam equivalent to 0.045ω cm. gave 35 per cent absorption (Hettncr 38 per cent) for the whole band; this amount of steam is equivalent to 0.1ω cm. of atmospheric water vapour, from a careful comparison over the band 11 to 13.5μ where each are available, hence the mean exponent K in expression (2) becomes 1.0 for band 13 to 16μ.

Table III shows the approximate percentage of radiation energy on various wave bands for a range of temperature to cover surface conditions. The water vapour exponents are also shown for the same bands. With the aid of expression (2) and these figures the absorption of any quantity of water vapour may be calculated in terms of the whole energy at these temperatures. The value of K for wave lengths greater than 16μ has been adjusted to suit the observed values of atmospheric radiation for different quantities of water vapour Many of these comparisons have been made and show that the absorption/quantity relation, deduced from Fowle's measurements, leads to excellent agreement with the observed variation of atmospheric radiation with water vapour pressure.

3. SKY RADIATION

The downward radiation from the sky, excluding the direct and scattered short wave radiation from the sun, is usually called the "sky radiation." Valuable papers on this subject have been published by A. Angstrom (1918), W. H. Dines (1927), Simpson (1928), Brunt (1932), and others, and it is not proposed to refer to it at any length here.

For normal conditions near the earth's surface, with a clear sky the downward radiation varies between three and four fifths of that from the surface, the proportion being greatest when the air is warm and carries much water vapour.

The method used to calculate the sky radiation from the absorption coefficients of water vapour and carbon dioxide is simple but laborious. It consists of dividing the air into horizontal layers of known mean temperature, water vapour, and carbon dioxide content, and summing the absorbing power of these layers on the different wave bands, in conjunction with the spectrum distribution of energy at the surface temperature. In this way the perpendicular component of sky radiation is obtained.

A long series of observations by W. H. Dines show that the effective radiation from the hemisphere of the sky, when free from clouds, to a horizontal surface bears a very constant relation to the perpendicular component ("zenith" sky radiation); it is 7 per cent greater for low level conditions in England.

Table IV shows the zenith sky radiation, as calculated from the absorption coefficients of water vapour and carbon dioxide, for a variety of air conditions. The effective sky radiation to a horizontal surface would be 7 per cent greater, for temperate and tropical conditions, than the zenith values in Table IV. For arctic conditions it would be about 5 per cent greater.

For normal conditions about three fourths of the zenith sky radiation comes from the lowest 400 m. of air, but when the air is very cold and dry a much greater depth takes part in this radiation.

4. THE EFFECT OF CARBON DIOXIDE UPON SKY RADIATION

For atmospheric conditions the sky radiation on wave band 13 to 16μ comes from a mixture of water vapour and carbon dioxide. This is true also for band 4 to 4.7μ, but the energy here is so small that it may be disregarded in relation to the probable error of the large band absorption.

In the case of mixed gases the absorption for the mixture is equal to the difference between the sum and product of their respective absorptions:—

Table IV. Calculated zenith sky radiation for different air conditions.

$Sz\%$ = sly radiation as percentage of that from the surface.
F = vertical lapse rates used °C. /km.
P (H_2O) = surface vapour pressure.
St = the temperature of a "black body" which would give the same radiation as that from the sky. As the minimum surface temperature cannot fall below this, values below about −70°C. (−93°F.) are not to be expected at low levels.

Location	Antarctic plateau	Mountain top	Polar surface	Temperate	Tropical
Season	Summer	Any	Equinox	Equinox	Any
Altitude, km.	3	5	0	0	0
Surface, °abs	250°	257°	257°	283°	296°
P (H_2O)	0.5	1.0	1.0	7.5	15.0
F	8°	7°	6°	5-6–6-7°	4-6–6-7°
$Sr.$ %	40	48	52	68	73.5
$Sr.$ cal/Cm²/day	186	246	267	514	664
Sky temp., °abs	200°	214°	219°	257°	274°

Under the stratosphere only Sz = 30, and sky temp. 163° Abs.

$$Acw = Ac + Aw - AcAw \qquad (3)$$

This relation is true if the respective absorptions are symmetrical in relation to the energy distribution over the wave band to which they refer.

The observed sky radiation results from the infra-red absorption by variable amounts of water vapour and from 3 parts in 10,000 of carbon dioxide, which are normally present in the air. For temperate conditions at vapour pressure 7.5 mm.Hg. I calculate that 95 per cent of the radiation comes from the water vapour; for arctic conditions the carbon dioxide may supply as much as 15 per cent of the total.

For the purpose required here it is necessary to consider the effect of a change in the amount of carbon dioxide, firstly upon sky radiation, and secondly the effect of changes in the latter upon temperatures.

When radiation takes place from a thick layer of gas, the average depth within that layer from which the radiation comes will depend upon the density of the gas. Thus if the density of the atmospheric carbon dioxide is altered it will alter the altitude from which the sky radiation of this gas originates. An increase of carbon dioxide will lower the mean radiation focus, and because the temperature is higher near the surface the radiation is increased, without allowing for any increased absorption by a greater total thickness of the gas.

The change of sky radiation with carbon dioxide depends largely upon this change in the alti-

Table V. The effect of changes in atmospheric carbon dioxide upon the amount and vertical distribution of sky radiation on band 13 to 16 μ.

Temperate air section. P (H_2O) = 7.5 mm. Hg. T surface = 283° Abs.

Altitude of air layer, km.	0–1	1–2	>2	Total on band 13 to 16 μ
	Radiation from air layer, % of surface radiation			
P (CO_2) Atmos.				
0.0001	9.35	1.98	0.74	12.06
0.0003 (normal)	10.93	1.73	0.62	13.25
0.0006	11.93	1.43	0.47	13.83

tude of the radiation focus, because the present quantity in the atmosphere (equal to a layer of 2 m. at N.T.P.) can absorb nearly the maximum of which this gas is capable. The latter assumption depends upon the exponents used, but it is probable that great thicknesses of carbon dioxide would absorb on other wave lengths besides those of the primary bands.

Table V shows the effect upon that part of the sky radiation which comes on band 13 to 16 μ, of changing the amount of atmospheric carbon dioxide from one third to double the present quantity. For these values twelve air layers were used, and the surface vapour pressure was 7.5 mm.

From the figures in Table V it will be seen that an increase of carbon dioxide causes the radiation to be concentrated from the lowest air layers, whilst the amount from the cold upper layers is still further screened off, the net effect being a small increase in the total sky radiation. For these values the equivalent thickness of carbon dioxide in the lowest kilometre is changed from 10 to 55 cm. "c," and these are quantities for which the exponents of Rubens and Aschkinass should be reliable.

5.   THE RELATION BETWEEN SKY RADIATION AND TEMPERATURE

If the whole surface of the earth is considered as a unit upon which a certain amount of heat falls each day, it is obvious that the mean temperature will depend upon the rate at which this heat can escape by radiation, because no other type of heat exchange is possible. For simplicity the reflection loss from clouds and ice surfaces is assumed to be a constant factor.

The radiation loss from the surface and clouds depends upon the fourth power of the absolute temperature and is proportional to the difference between the surface and sky radiation:—

$$H = \sigma . T^4 . (1 - S) \tag{4}$$

where:

H = radiation heat loss from surface.

σ = radiation constant. $10^{-7} \times 1.18$ cal/cm^2/day.

T = temperature of surface, Abs.

S = sky radiation, as proportion of that from the surface.

Suppose that the sky radiation is changed from S_1 to S_2 whilst H remains constant. Then:—

$$T_2 = T_1 \sqrt[4]{[(1 - S_1) / (1 - S_2)]} \tag{5}$$

From this relation it will be seen that the change of temperature for a given change of sky radiation increases rapidly as the latter approaches that from the surface, it being always assumed that the heat supply and loss are constant.

On the earth the supply of water vapour is unlimited over the greater part of the surface, and the actual mean temperature results from a balance reached between the solar "constant" and the properties of water and air. Thus a change of water vapour, sky radiation and temperature is corrected by a change of cloudiness and atmospheric circulation, the former increasing the reflection loss and thus reducing the effective sun heat.

There is also a further loss owing to the scattering of solar energy by the water molecule.

Small changes of atmospheric carbon dioxide do not affect the amount of sun heat which reaches the surface, because the CO_2 absorption bands lie well outside the wave lengths, 0.25 to 1.5 μ, on which nearly all the sun energy is carried. Consequently a change of sky radiation due to this gas can have its full effect upon low level temperatures, provided it does not increase the temperature *differences* on which the atmospheric circulation depends.

An increase of temperature due to sky radiation will be different from that caused by an increase of sun heat; the latter would tend to increase the temperature differences and atmospheric circulation, and the ultimate rise of temperature should not be in proportion to the change of sun heat.

From the change of sky radiation with carbon dioxide, and from expression (5), the resulting change in surface temperature can be obtained. The relation between atmospheric carbon dioxide

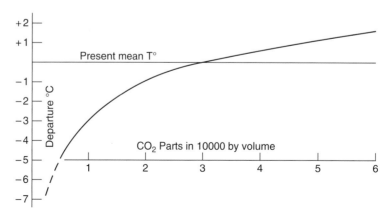

Fig. 2. Change of surface temperature with atmospheric carbon dioxide (H$_2$O vapour pressure, 7.5 mm. Hg.)

Table VI. Increase of main temperature from the artificial production of carbon dioxide.

Annual excess of CO_2 to the air = 4,300 million tons.
$P(CO_2)$ is expressed in units of a ten-thousandth of an atmosphere.
ΔT = increase from mean temperature of 19th century.
Sea water equilibrium time 2,000 years.

Period	1910–1930	20th century	21st century	22nd century
Mean $P(CO_2)$	2.82	2.92	3.30	3.60
Mean ΔT. °C	+0.07°	+0.16°	+0.39°	+0.57°
Polar displacement of climate zones	15	36	87	127 km.

and surface temperature is shown in Fig. 2 for the temperate air section.

At first sight one would expect that carbon dioxide would have far more effect upon the temperature of arctic regions where the amount of water vapour is very small; this is true as regards its effect on sky radiation, but not on temperatures, because the ratio dT/dS increases rapidly as the sky radiation approaches that from the surface. The result of these opposing changes is to make the quantitative influence of carbon dioxide on temperature remarkably uniform for the different climate zones of the earth.

There are a great many other points in connection with the influence of this gas upon temperatures which cannot be referred to here, but sufficient has been said to indicate the method used and the actual changes given by the best absorption coefficients, and also by the absorption/quantity relation for water vapour given by Fowle's measurements in the atmosphere.

In Table I the increase in atmospheric carbon dioxide for the present rate of production is shown, and from the change in temperature so caused it is possible to make a reasonable estimate of the change of mean temperature to be expected during the next few centuries.

It may be supposed that the artificial production of this gas will increase considerably during such a period, but against this is the ever-increasing efficiency of fuel utilisation, which has tended to stabilise carbon dioxide production at around 4,000 million tons during the last 20 years, in spite of the greatly increased number of heat units turned to useful purpose.

The last line in Table VI, showing the polar displacement of climate zones, is for a horizontal gradient of 1°C. for each 2° of latitude between equator and poles. This is an average figure for temperate latitudes.

The estimate of the effect of carbon dioxide on temperatures made by S. Arrhenius (1903) about 40 years ago gave changes about twice as large as those shown in Table VI, but he had taken the maximum energy absorption by this gas as 30 per cent, whereas the Rubens and Aschkinass exponents give a maximum of only 15 per cent.

6.   THE OBSERVED TEMPERATURE VARIATIONS ON THE EARTH

Coming now to the actual temperatures which have been observed near the earth's surface during the recent past, these measurements have provided an almost overwhelming mass of statistical detail, including many millions of accurate and standardized readings of temperature. The period to which these standardized observations refer is generally not more than 65 years and often less. It is a matter of opinion whether such a period is sufficiently extended to show a definite trend in world temperatures.

I have relied principally on that valuable Smithsonian publication, "World Weather Records," to obtain the temperature readings summarized here. In all I have examined about two hundred records, but a small proportion of these are found to be defective when their period temperature departures are compared with those of neighbouring stations; also in certain cases the records are unreliable, owing to changes of conditions which are detailed in the above publication.

Out of 18 records going back more than a century I found only two which could be classed as continuous throughout; these are at the Radcliffc Observatory (1930), Oxford, and at Copenhagen. There are two or three others which comparisons show must be very reliable, and amongst these

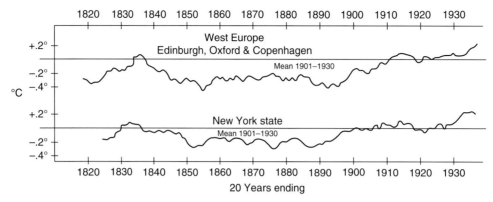

Fig. 3. The most reliable long period temperature records. Twenty-year moving departures from the mean. 1901–1930.

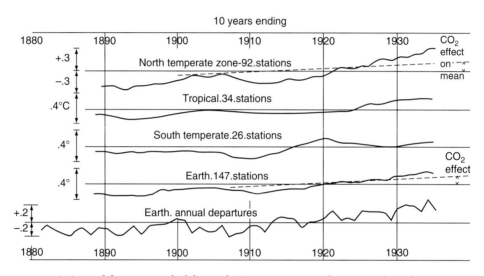

Fig. 4. Temperature variations of the zones and of the earth. Ten-year moving departures from the mean, 1901–1930, °C.

perhaps the best are those from Edinburgh (Mossman, 1902), and from New York City.

Fig. 3 shows the combined departures from the mean 1901–30, at Edinburgh, Oxford, and Copenhagen, as twenty-year "moving average departures"; the lower curve for New York State is based principally upon the New York City record, but other long records from Albany and New Haven have also been used to check the New York readings. There is a very marked agreement between the west Europe curve and that of New York State, almost throughout the period, although the principal temperature anomalies seem to occur a few years earlier in the eastern States than in west Europe.

In order to represent the temperature anomalies of great regions of the earth's surface I have grouped a number of stations together, and then weighted each group according to the area represented by its stations. In this way the curves for the different zones of the earth, shown in Fig. 4, were obtained.

I have made more than one estimate of the temperature anomalies of the north temperate zone, and the figures in Table VII compare the result of an earlier estimate with the final values obtained from nearly twice as many stations after a more careful elimination of all doubtful records. A different grouping of the stations was also used in the two cases shown.

It would appear from the values in Table VII that quite a moderate number of reliable temperature records may be used to give the period anomalies of very large areas.

I have not included the arctic stations of Upernivik and Spitsbergen in the curves, Fig. 4,

Table VII. Temperature anomalies of the north temperate zone represented by different numbers of stations.

Departures from mean. 1901–1930, °C.

Decade or period	1880–1889	1890–1899	1900–1909	1910–1919	1920–1929	1930–1934
No. of stations used						
47 (original estimate)	−.23	−.12	−.12	−.07	+.20	+.50
92 selected stations	−.22	−.12	−.10	−.06	+.16	+.40

Table VIII. Annual temperature departures for the earth, from 147 records.

Departures from mean. 1901–1930, in 1/100°C.

Date	0	1	2	3	4	5	6	7	8	9	Decade
1880	−5	−10	−4	−10	−15	−19	−11	−9	−18	−9	−11
1890	−9	−14	−18	−12	−5	−15	−6	−3	−2	0	−8
1900	+4	+2	−12	−7	−11	−11	−5	−18	−14	−12	−9
1910	−5	−3	−11	+1	+7	+6	+3	−7	0	−1	−1
1920	+6	+15	+2	+6	+2	+16	+14	+14	+13	+2	+9
1930	+20	+23	+22	+16	+30	+17					(+22)

because the variations are so large and there are no compensating stations from the antarctic regions.

As a matter of interest the annual departures for the whole earth are also plotted on Fig. 4, and the actual figures given by the 150 stations are shown in Table VIII.

From the curves in Fig. 4 it will be seen that the greater part of the warmth of recent years has occurred in the northern regions; these, with the exception of the west Pacific stations, all show a decided rising tendency since about 1920. In tropical regions departures are more local and variable, but each of the three groups used (Indian Ocean, Tropical Atlantic and Tropical West Pacific) show a distinctly higher mean temperature for the 25 years, 1910–34, than for the previous 25 years. In Australasia and South America (Kincer, 1933) temperatures are very stable, but both regions show a decided increase over the above period.

At high altitudes in Europe the warmth of the 1920's has been greater than at the surface, as the following figures show:—

Sonnblick (3 km. altitude) and Santis ($2\frac{1}{2}$ km. altitude), departure 1920–29 = +.44°C.

All Europe, 30 low-level stations (Lisbon to Kazan), 1920–29 = +.17.

It is well known that temperatures, especially the night minimum, are a little higher near the centre of a large town than they are in the surrounding country districts; if, therefore, a large number of buildings have accumulated in the vicinity of a station during the period under consideration, the departures at that station would be influenced thereby and a rising trend would be expected.

To examine this point I have divided the observations into three classes, as follows:—

(i)   First class exposures, small ocean islands or exposed land regions without a material accumulation of buildings.

(ii) Small towns which have not materially increased in size.

(iii) Large towns, most of which have increased considerably during the last half century.

As this is a matter which is open to controversy it is necessary to give examples of the stations used in each class.

In each class one quarter of the stations were in the southern hemisphere.

In order to make a more specific estimate I have examined a number of country and city records in England. The former only extend over about half a century, but during the last 55 years the 20-year moving average at the country stations of Stonyhurst College, Worksop, and Rothamsted Farm has risen slightly (0.1°F.) more than that given by the precise measurements at Greenwich and Oxford Observatories. This shows that no secular increase of temperature, due to "city influence," has occurred at these city stations, in spite of the great increase of population in the immediate neighbourhood during the period under consideration.

Table IX. The effect of town sites on period temperature departures.

(i) Best exposures.	(ii) Small towns.	(iii) Large towns.
Upernivik, Greenland.	Victoria, B.C., Canada.	St. Louis, Mo.*
Father Point, Q., Canada.	St. John's, Newfoundland.	Alexandria.
Valencia, Irish Free State.*	Sibiu, Roumania.	Moscow.
Ponta Delgada, Azores.	Yakutsk, U.S.S.R.	Nagasaki.
St. Helena.	Miyako, Japan	Bombay.
Cape Pembroke, Falklands.	Leh, 3½ km. alt., Kashmir.	Cape Town.
Petropavlovsk (lighthouse),	Antananarivo, Madagascar.	Rio de Janeiro.
Kamchatka.	Bahia Blanca, Argentine.	Santiago.
Honolulu, Hawaii.*	Magellenes (Punta Arenas),	Auckland.
Apia, Samoa.	Chile.	Adelaide.
Port Victoria, Seychelles.	Darwin, N.T., Australia.	

* See Fig. 5.
Mean increase from period 1890–1909, to period 1910-1929.
(i) 16 stations, 0.23°C.   (ii) 32 stations, 0.19°.   (iii) 34 stations, 0.21°.
In each class are quarter of the stations were in the southern hemisphere

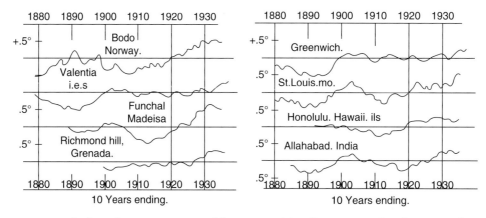

Fig. 5. Temperature records from the great oceans and from great cities. Ten-year moving departures from mean, 1901–1930, °C.

In Fig. 5 records from some of the best "ocean" stations are shown, together with those from the great cities of London and St. Louis.

Very few reliable records show a fall of temperature over the last half century, except those of the China Sea, and perhaps also the Bay of Bengal.

To return to the world temperature curve, Fig. 4, the dotted line shows the displacement of the mean due to increasing atmospheric carbon dioxide, and it is evident that present temperatures, particularly in the northern hemisphere, are running above the calculated values. The course of world temperatures during the next twenty years should afford valuable evidence as to the accuracy of the calculated effect of atmospheric carbon dioxide.

As regards the long period temperature variations represented by the Ice Ages of the geologically recent past, I have made many calculations to see if the natural movements of carbon dioxide could be rapid enough to account for the great changes of the amount in the atmosphere which would be necessary to give glacial periods with a duration of about 30,000 years. I find it almost impossible to account for movements of the gas of the required order because of the almost inexhaustible supply from the oceans, when its pressure in the air becomes low enough to give a fall of 5 to 8°C. in mean temperatures (see Fig. 2). Of course, if the effect of carbon dioxide on temperatures was considerably greater than supposed, glacial periods might well be accounted for in this way.

In conclusion it may be said that the combustion of fossil fuel, whether it be peat from the surface or oil from 10,000 feet below, is likely to prove beneficial to mankind in several ways, besides the provision of heat and power. For instance the above mentioned small increases of mean temperature would be important at the northern margin of cultivation, and the growth of favourably situated plants is directly proportional to the carbon dioxide pressure (Brown and Escombe, 1905). In any case the return of the deadly glaciers should be delayed indefinitely.

As regards the reserves of fuel these would be sufficient to give at least ten times as much carbon dioxide as there is in the air at present.

DISCUSSION

Sir George Simpson expressed his admiration of the amount of work which Mr. Callendar had put into this paper. It was excellent work. It was difficult to criticise it, but he would like to mention a few points which Mr. Callendar might wish to reconsider. In the first place he thought it was not sufficiently realised by non-meteorologists who came for the first time to help the Society in its study, that it was impossible to solve the problem of the temperature distribution in the atmosphere by working out the radiation. The atmosphere was not in a state of radiative equilibrium, and it also received heat by transfer from one part to another. In the second place, one had to remember that the temperature distribution in the atmosphere was determined almost entirely by the movement of the air up and down. This forced the atmosphere into a temperature distribution which was quite out of balance with the radiation. One could not, therefore, calculate the effect of changing any one factor in the atmosphere, and he felt that the actual numerical results which Mr. Callendar had obtained could not be used to give a definite indication of the order of magnitude of the effect. Thirdly, he thought Mr. Callendar should give a little more information as to how he had calculated the results shown in Fig. 2. These contained the crucial point of the paper, but the paper did not explain how they were obtained. In Table V Mr. Callendar had given the effect of doubling the CO_2 in one band, 13 to 16μ, which included nearly the whole of the energy connected with the CO_2. The increase of temperature obtained by calculation from these results, however, was not the same for a similar increase in CO_2 as that shown in Fig. 2. This sort of discrepancy should be cleared up. Lastly he thought that the rise in CO_2 content and temperature during the last 50 years, must be taken as rather a coincidence. The magnitude of it was even larger than Mr. Callendar had calculated, and he thought the rise in temperature was probably only one phase of one of the peculiar variations which all meteorological elements experienced.

Dr. F. J. W. Whipple expressed the hope that the author would give the Society an account of his investigation of the natural movements of carbon dioxide. It was not clear how the calculations regarding the gradual diffusion of CO_2 into the sea were carried out. The calculations embodied in Table IV depended on the assumption of high lapse rates of temperature everywhere. The inversions of temperature at night and throughout the winter in the polar regions were ignored. As an inversion implied a reversal of the new flow of radiation in the critical waveband, it seemed necessary to make additional calculations to allow for the varying circumstances. Other processes besides radiation are involved in the exchange of normal energy between ground and atmosphere, but it may be justifiable to ignore these other processes in an investigation of the effect of variations in the amount of CO_2.

Prof. D. Brunt referred to the diagrams showing the gradual rise of temperature during the last 30 years, and said that this change in the mean temperature was no more striking than the changes which appear to have occurred in the latter half of the eighteenth century, and whose reality does not appear to be a matter of defective instruments. The long series of pressure observations made in Paris showed clearly that there had been great changes in the mean path of depressions coming from the Atlantic. Prof. Brunt agreed with the view of Sir George Simpson that the effect of an increase in the absorbing power of the atmosphere would not be a simple change of temperature, but would modify the general circulation, and so yield a very complicated series of changes in conditions. He was not quite clear how the temperature changes had been evaluated. He appreciated, however, that Mr. Callendar had put a tremendous amount of work into his most interesting paper.

Dr. C. E. P. Brooks said that he had no doubt that there had been a real climatic change during the past thirty or forty years. This was shown not only by the rise of temperature at land stations, but also by the decrease in the amount of ice in arctic and probably also in antarctic regions and by the rise of sea temperatures. This rise of temperature could however be explained, qualitatively if not quantitatively, by changes in the atmospheric circulation, and in those regions where a change in the circulation would be expected to cause a fall of temperature, there had actually been a fall; moreover the rise of temperature was about ten times as great in the arctic regions as in middle or low latitudes, and he did not think that a change in the amount of carbon dioxide could cause such a differential effect. The possibility certainly merited discussion, however, and he welcomed the paper as a valuable contribution to the problem of climatic changes.

Mr. L. II. G. DINKS asked Mr. Callendar whether he was satisfied that the change in the temperature of the air which he had found was significant, and that it was not merely a casual variation.

Mr. J. H. COSTE congratulated Mr. Callendar on his courage and perseverance. He would like to raise some practical issues. Firstly, was the CO_2 in the air really increasing? It used to be given as .04%, then as methods of chemical analysis improved it went down to .03%, and he thought it was very doubtful whether the differences which Mr. Callendar made use of were real. The methods of determining CO_2 thirty or forty years ago were not sufficiently accurate for making such a comparison. A. Krogh calculated that for a constant difference in tension of $1/10^5$ atmosphere between the air and the ocean, the latter being less rich, the annual invasion of CO_2 would be equal to 3.7×10^9 metric tons, which was about the annual contribution of CO_2 to the atmosphere by the burning of fuel; to this absorption by the ocean must be added the effects of vegetation, by photosynthesis. Thermometers thirty years ago were not instruments of very high precision and one would hesitate to consider variations of fractions of a degree based on observations made with such thermometers.

In replying, Mr. G. S. CALLENDAR said he realized the extreme complexity of the temperature control at any particular region of the earth's surface, and also that radiative equilibrium was not actually established, but if any substance is added to the atmosphere which delays the transfer of low temperature radiation, without interfering with the arrival or distribution of the heat supply, some rise of temperature appears to be inevitable in those parts which are furthest from outer space.

As stated in the paper the variation of temperature with CO_2 (Fig. 2), was obtained from the values of sky radiation, calculated for different amounts of this gas, substituted in expression (5) at S_1, S_2. If the *changes* of S shown in Table V are used for expression (5), it will be found that the temperature changes lie on the curve of Fig. 2 when the total sky radiation is 7/10 of the surface radiation. The sky radiation is calculated as a proportion of that from the surface, hence, at constant heat supply, a change of sky "temperature" involves an equilibrium change of surface temperature as in expression (5).

It was found that even the minimum numerical explanation of the method used for calculating sky radiation would occupy several pages, and as a number of similar methods have been published from time to time, it was decided to use the available space for matter of more direct interest.

In reply to Dr. Whipple, the author regretted that space did not permit an account of the natural movements of CO_2; he had actually written an account of these, but it was just eight times as long as the present paper.

For the calculation of the diffusion of CO_2 into the sea the effective depth was considered to be 200 m. at any one time.

The effect of CO_2 on temperatures has been calculated for a variety of lapse rates, including large inversions. In the latter case the effect on the surface temperature is small, but the protection for the warm middle layers remains.

In reply to Prof. Brunt, the author stated that the warm periods about 1780, 1797 and 1827, appeared to be of the nature of short warm intervals of up to 10 years, with some very cold years intervening, whereas recent conditions indicated a more gradual and sustained rise of temperature; this was perhaps best shown by a 40-year moving average.

In reply to Dr. Brooks, the author agreed that the recent rise in arctic temperatures was far too large to be attributed to change of CO_2; he thought that the latter might act as a promotor to start a series of imminent changes in the northern ice conditions. On account of their large rise he had not included the arctic stations in the world temperature curve (Fig. 4).

In reply to Mr. Dines, the author said he thought the change of air temperature appeared too widespread to be a casual change due to local variations of pressure.

In reply to Mr. Coste, the author said that the early series of CO_2 measurements he had used were probably very accurate; he had only used values observed on days when strong and steady west winds were blowing at Kew. The actual CO_2 added in the last 40 years was equal to an increase of 8%; the observed and calculated values agreed in giving an effective increase of about 6%.

The author is not aware of the solution coefficients for sea water used by A. Krogh to give the stated figure which appears to be far too high. It must be remembered that less than 1/1000 of the sea volume would be replaced at the surface in one year, and the annual increase of CO_2 pressure in the air is less than $1/10^6$ atmosphere. About 98% of the CO_2 used by vegetation appears to be returned by decay oxidation and respiration.

The author thought that very accurate temperatures were taken last century; if there was any doubt on this point the introduction to the long period tables from the Radcliffe Observatory (*Met. Obs.*, Vol. **55**, 1930), should set this at rest.

551.511

The effect of vertical motion on the "geostrophic departure" of the wind

In section 7 of our paper "The importance of vertical motion in the development of tropical revolving storms," published it the *Journal* of January, 1938, a general proposition is enunciated, which reads as follows:

"If vertical motion occurs in the atmosphere in a region of horizontal temperature gradient, then ascending motion gives a component of wind from warm to cold, and descending motion a component from cold to warm."

We are indebted to Prof. Brunt for pointing out that the brief argument given in the paper does not stress at all adequately the type of air motion to which the proposition is intended to apply, and that it certainly does not apply to the case of a convection current rising by instability through an environment. By not realising this natural application of the proposition, we failed fully to appreciate Prof. Brunt's contribution to the discussion of the paper, and in order to avoid further misunderstanding we should like to amend the statement to read:

"If there is general vertical motion in a region where the winds are quasi-geostrophic, then the departure of the wind velocity from the geostrophic value has a component directed along the horizontal gradient of temperature, from warm to cold with ascending motion, and from cold to warm with descending motion."

The proposition is believed to have an important bearing on general meteorological developments, where gradual ascending motion, as, for example, in a frontal zone, is associated with convergence or divergence, and with departures of the wind velocity from the geostrophic value. Although the departures are generally small compared with the geostrophic velocities (so that the whole motion may be described as quasi-geostrophic) they are of fundamental dynamical significance.

A further paper will shortly be presented, in which the mathematical and physical basis of the proposition will receive a more adequate treatment.

C. S. DURST

R. C. SUTCLIFFE

NOTES

1 Since writing the above, my attention has been called to the recent water vapour absorption measurements by Weber and Randall (*Phys. Rev. Amer.*, **40**, 1932, p. 835) in this part of the spectrum. These new values are far lower than those found for steam, and give strong support to the general accuracy of Fowle's values for atmospheric water vapour. On band 13 to 16 μ Weber and Randall find approximately 30 per cent absorption by 0.1 cm. "w"; the exponent used here gives 33 per cent at 13 to 16 μ by this quantity of water vapour.

REFERENCES

Angstrom, A. 1918 *Smithson. Misc. Coll.*, **65**, No. 3.
Arrhenius, Svante 1903 *Kosmische Physik*, **2**.
Brown, H., and Escombe, F. 1905 *Proc. Roy. Soc.*, B, **76**.
Brunt, D. 1932 *Quart. J.R. Met. Soc.*, **58**.
Carpenter, T. M. 1937 *J. Amer. Chem. Soc.*, **59**.
Dines, W. H. 1927 *Mem. R. Met. Soc.*, **2**, No. II.
Fox, C. J. B. 1909 *International Council for the Investigation of the Sea. Publications de Circonstance*, No. **44**.
Fowle, F. E. 1918 *Smithson. Misc. Coll.*, **68**, No. 8.
Hettner, A. 1918 *Ann. Phys., Leipzig*, **55**.
Kincer, J. B. 1933 *Mon. Weath. Rev., Wash.*, **61**.
Mossman. R. C. 1902 *Trans. Roy. Soc, Edin.*, **40**.
Radcliffc Observatory 1930 *Met. Obs.*, **55**.
Rubens, H., and Aschkinass, R. 1898 *Ann. Phys. Chem.*, **64**.
Schmidt, H. 1913 *Ann. Phys., Leipzig*, **42**.
1927 and 1934 "World weather records." *Smithson. Misc. Coll.*, **79** and **90**.
Simpson, G. C. 1928 *Mem. R. Met. Soc.*, **3**, No. 21.

13

Denial and Acceptance

Bolin, B. & Eriksson, E. (1958). Changes in the carbon dioxide content of the atmosphere and sea due to fossil fuel combustion. In *The Atmosphere and the Sea in Motion: Scientific Contributions to the Rossby Memorial Volume* (ed. B. Bolin), pp. 130–142. Rockefeller Institute Press, New York.

Revelle, R. & Suess, H.E. (1957). Carbon dioxide exchange between atmosphere and ocean and the question of an increase of atmospheric CO_2 during the past decades. *Tellus*, **9**, 18–27. 9 pages.

Callendar asserted in 1938 that the CO_2 concentration in the air had risen between the turn of the century and the early 1930s. He also argued that the average temperature of the Earth had risen in that time, as a result of this rising concentration of the greenhouse gas CO_2. One scientific response to this was to refine the calculations of the potential impact of CO_2 on the Earth's temperature, and it was Manabe and Hansen that ultimately got this right. Another was to measure the CO_2 concentration in the air more accurately, which David Keeling did. The third approach is exemplified by the following two papers, by Revelle and Suess in 1957 and Bolin in 1958, who tried to figure out whether the oceans would quickly mop up any CO_2 emission, preventing an atmospheric increase. A comparison of the two papers is interesting, in that Revelle and Suess seemed almost to be in a psychological state of denial, while Bolin looked at the same problem with greater sophistication in his analysis, but also with a mind that was open to the possibility that ocean uptake of CO_2 would be slow, leaving humans a strong potential impact on global climate.

Changes in the carbon-14 concentration of the atmosphere were taken as a starting point observation by both studies. Carbon-14 is produced in the upper atmosphere by a nuclear reaction driven by cosmic rays, the transmutation of nitrogen-14 to carbon. Carbon-14 is radioactive, decaying with a half-life of 5730 years. This lifetime is long enough that the carbon-14 finds its way into trees, soil carbon, and the oceans. As the carbon dissolves in the oceans, some of the carbon-14 decays, so on average the carbon-14 concentration of the surface ocean is lower than that of the atmosphere, equivalent to an apparent "age" of the surface water of about 400 years. From this observation, and a knowledge of the relative amounts of CO_2 in the atmosphere and the ocean, Revelle and Suess calculated that the average lifetime of a CO_2 molecule in the atmosphere before it can expect to dissolve in the ocean is about 7 years. On the face of it, it seems that a slug of new CO_2 to the atmosphere should dissolve in the oceans in 7 years, preventing any buildup in the atmosphere that would lead to global warming.

Revelle and Suess assumed that the ocean is well mixed. Our CO_2 slug may be important to the atmospheric concentration, but there is so much dissolved carbon in the ocean that if the slug is mixed in, the increase in the ocean concentration would be negligible. Therefore, the slug released to the atmosphere will increase the downward CO_2 flux, dissolving into the oceans, much more than it would change the upward flux of CO_2 evaporating from the oceans. However, in a point picked up by Bolin and Calendar but missed by Revelle, the ocean takes a long time to mix. The surface ocean concentration of CO_2 increases more than the average for the whole ocean, and the rate of CO_2 degassing to the atmosphere increases as the surface ocean

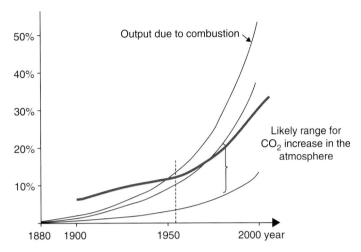

Fig 13.1 **Estimate of likely range for CO$_2$ increase in the atmosphere as a result of fossil fuel combustion according to UN estimates. From Bolin, with the addition of atmospheric CO$_2$ concentration data as it turned and in the heavy line.**

concentration of CO$_2$ builds up, almost enough to counteract the invasion of the CO$_2$ slug into the ocean. There is a difference here between net invasion and CO$_2$ exchange; invasion draws down the atmospheric CO$_2$ concentration but exchange does not.

Another complicating factor seems to have come to Revelle and Suess at a late stage in the preparation of the manuscript; the paper reads as if it was pointed out by one of the reviewers. The way the carbon chemistry of seawater works, added CO$_2$ reacts with carbonate ion, CO$_3^-$, to form the pH-neutral bicarbonate ion, HCO$_3^-$. One might naively assume that doubling the CO$_2$ concentration in the air would double the concentrations of all of the forms of dissolved carbon in the seawater, including the bicarbonate and carbonate ions. This would be the case if the acidity, or pH, of the water were held constant by some external buffer. The reality is that the pH of seawater changes as the CO$_2$ invades, and the water only holds about a 10th of the naive expectation. This factor of 10 is called, perhaps a bit unfairly, the Revelle buffer factor.

Revelle and Suess' Fig. 1 is essentially a slide rule for calculating how much fossil fuel CO$_2$ there could be in the atmosphere, based on the lifetime of CO$_2$ and the observed changes in carbon-14 in the atmosphere, but it ignores the factor of 10 from the buffer chemistry, apparently because the model was already done, the figure already drafted, when the buffer chemistry effect was tacked on to the end of the manuscript as an afterthought. Figure 2 was added to illustrate what a factor of 10 slowdown in CO$_2$ uptake should look like, but it looks hastily drawn, with straight lines for the atmospheric CO$_2$ concentration with time instead of slower versions of the curved lines from their real solution. A further problem with their Fig. 2 is that they assume that CO$_2$ emissions will be constant with time, while their Table 1 shows clearly accelerating CO$_2$ emissions.

In every instance, Revelle and Suess seem to rebel against what their analysis wants to tell them, that the oceans will not control atmospheric CO$_2$ concentration quickly enough to prevent human-induced climate change. Based on the authors' prior and subsequent careers in science, we judge that their reluctance to see the results of their analysis stems from a psychological state of denial, rather than a deliberate effort to deceive. In contrast, the analysis of Bolin, 1 year later, is a breath of fresh air, brilliant in its clarity. The slow mixing time of the ocean, the impact of the buffer chemistry, and the acceleration of CO$_2$ emission with time are all accounted for, and the result is a prediction of future CO$_2$ concentration trends in the atmosphere, in Fig. 13.1, that looks stunningly like what actually turned out to happen.

Carbon Dioxide Exchange Between Atmosphere and Ocean and the Question of an Increase of Atmospheric CO_2 during the Past Decades

ROGER REVELLE AND HANS E. SUESS
Scripps Institution of Oceanography, University of California, La Jolla, California

ABSTRACT

From a comparison of C^{14}/C^{12} and C^{13}/C^{12} ratios in wood and in marine material and from a slight decrease of the C^{14} concentration in terrestrial plants over the past 50 years it can be concluded that the average lifetime of a CO_2 molecule in the atmosphere before it is dissolved into the sea is of the order of 10 years. This means that most of the CO_2 released by artificial fuel combustion since the beginning of the industrial revolution must have been absorbed by the oceans. The increase of atmospheric CO_2 from this cause is at present small but may become significant during future decades if industrial fuel combustion continues to rise exponentially.

Present data on the total amount of CO_2 in the atmosphere, on the rates and mechanisms of exchange, and on possible fluctuations in terrestrial and marine organic carbon, are inadequate for accurate measurement of future changes in atmospheric CO_2. An opportunity exists during the International Geophysical Year to obtain much of the necessary information.

INTRODUCTION

In the middle of the 19th century appreciable amounts of carbon dioxide began to be added to the atmosphere through the combustion of fossil fuels. The rate of combustion has continually increased so that at the present time the annual increment from this source is nearly 0.4% of the total atmospheric carbon dioxide. By 1960 the amount added during the past century will be more than 15%.

CALLENDAR (1938, 1940, 1949) believed that nearly all the carbon dioxide produced by fossil fuel combustion has remained in the atmosphere, and he suggested that the increase in atmospheric carbon dioxide may account for the observed slight rise of average temperature in northern latitudes during recent decades. He thus revived the hypothesis of T. C. CHAMBERLIN (1899) and S. ARRHENIUS (1903) that climatic changes may be related to fluctuations in the carbon dioxide content of the air. These authors supposed that an increase of carbon dioxide in the upper atmosphere would lower the mean level of back radiation in the infrared and thereby increase the average temperature near the earth's surface.

Subsequently, other authors have questioned Callendar's conclusions on two grounds. First, comparison of measurements made in the 19th century and in recent years do not demonstrate that there has been a significant increase in atmospheric CO_2 (SLOCUM, 1955; FONSELIUS et al. 1956). Most of the excess CO_2 from fuel combustion may have been transferred to the ocean, a possibility suggested by S. ARRHENIUS (1903). Second, a few percent increase in the CO_2 content of the air, even if it has occurred, might not produce an observable increase in average air temperature near the ground in the face of fluctuations due to other causes. So little is known about the thermodynamics of the atmosphere that it is not certain whether or how a change in infrared back radiation from the upper air would affect the temperature near the surface. Calculations by PLASS (1956)

Contribution from the Scripps Institution of Oceanography, New Series, No. 900. This paper represents in part results of research carried out by the University of California under contract with the Office of Naval Research.

Revelle, R. (1957) Carbon dioxide exchange between atmosphere and ocean and the question of an increase of atmospheric CO_2 during the past decades. *Tellus* 9: 18–27. Reprinted with permission of John Wiley & Sons.

Table 1. CO_2 added to atmosphere by consumption of fossil fuels.

Decade	Average amount added per decade Measured		Cumulative total added			
	10^{18} gms	% Atm CO_2	10^{18} gms	% Atm CO_2	10^{18} gms	% Atm CO_2
1860–69	0.0054	0.23	0.0054	0.23		
1870–79	0.0085	0.36	0.0139	0.59		
1880–89	0.0128	0.54	0.0267	1.13		
1890–99	0.0185	0.79	0.0452	1.92		
1900–09	0.0299	1.27	0.0751	3.19		
1910–19	0.0405	1.72	0.1156	4.91		
1920–29	0.0470	2.00	0.1626	6.91		
1930–39	0.0497	2.11	0.2123	9.02		
1940–49	0.0636	2.71	0.2759	11.73		

	Forecast					
	Assuming fossil fuels are used to meet future requirements of fuel and power as estimated by UN (1955)				Assuming fossil fuel consumption remains constant at estimated 1955 rate	
1950–59	0.091	3.9	0.367	15.6	0.367	15.6
1960–69	0.128	5.4	0.495	21.0	0.458	19.5
1970–79	0.176	7.5	0.671	28.5	0.549	23.4
1980–89	0.247	10.5	0.918	39.0	0.640	27.2
1990–99	0.340	14.5	1.258	53.5	0.731	31.1
2000–09	0.472	20.0	1.730	73.5	0.822	35.0

indicate that a 10% increase in atmospheric carbon dioxide would increase the average temperature by 0.36° C. But, amplifying or feed-back processes may exist such that a slight change in the character of the back radiation might have a more pronounced effect. Possible examples are a decrease in albedo of the earth due to melting of ice caps or a rise in water vapor content of the atmosphere (with accompanying increased infrared absorption near the surface) due to increased evaporation with rising temperature.

During the next few decades the rate of combustion of fossil fuels will continue to increase, if the fuel and power requirements of our world-wide industrial civilization continue to rise exponentially, and if these needs are met only to a limited degree by development of atomic power. Estimates by the UN (1955) indicate that during the first decade of the 21st century fossil fuel combustion could produce an amount of carbon dioxide equal to 20% of that now in the atmosphere (Table 1).[1] This is probably two orders of magnitude greater than the usual rate of carbon dioxide production from volcanoes, which on the average must be equal to the rate at which silicates are weathered to carbonates (Table 2). Thus human beings are now carrying out a large scale geophysical experiment of a kind that could not have happened in

Table 2. Estimated present annual rates of some processes involving atmospheric and oceanic carbon dioxide, in part after HUTCHINSON (1954).

	10^{18} gms	In units of atm CO_2 (A_0)
Consumption of fossil fuels—CO_2 produced.	0.0091	0.0039
Bicarbonate and carbonate added to ocean by rivers (as CO_2)[1]	0.00103	0.0004
Photosynthesis on land —CO_2 consumed	0.073 ± 0.018	0.031
Photosynthesis in ocean —CO_2 consumed	0.46 ± 0.3	0.20

[1] According to CONWAY (1943) and HUTCHINSON (1954) 87% of the river borne bicarbonate and carbonate comes from weathering of carbonate rocks, the remainder from weathering of silicates.

the past nor be reproduced in the future. Within a few centuries we are returning to the atmosphere and oceans the concentrated organic carbon stored in sedimentary rocks over hundreds of millions of years. This experiment, if adequately documented, may yield a far-reaching insight into the processes determining weather and climate. It therefore becomes of prime importance to attempt to determine the way in which carbon dioxide is partitioned between the atmosphere, the oceans, the biosphere and the lithosphere.

The carbon dioxide content of the atmosphere and ocean is presumably regulated over geologic times by the tendency toward thermodynamic equilibrium between silicates and carbonates and their respective free acids, silica and carbon dioxide (UREY, 1952). The atmosphere contains and probably has contained during geologic times considerably more CO_2 than the equilibrium concentration (HUTCHINSON, 1954), although uncertainties in the thermodynamic data are too great for an accurate quantitative comparison. Equilibrium is approached through rock weathering and marine sedimentation. Estimated rates of these processes give a very long time constant of the order of magnitude of 100,000 years. Rapid changes in the amount of carbon dioxide produced by volcanoes, in the state of the biosphere, or as in our case, in the rate of combustion of fossil fuels, may therefore cause considerable departures from average conditions.

The answer to the question whether or not the combustion of coal, petroleum and natural gas has increased the carbon dioxide concentration in the atmosphere depends in part upon the rate at which an excess amount of CO_2 in the atmosphere is absorbed by the oceans. The exchange rate of isotopically labeled CO_2 between atmosphere and ocean, which, in principle, could be deduced from C^{14} measurements, is not identical with the rate of absorption, but is related to it.

Notations and geochemical constants

In our discussion of exchange and absorption rates we shall use the following notations:

S_0: Total carbon of the marine carbon reservoir at equilibrium condition, at time zero.

A_0: Atmospheric CO_2 carbon at time zero.

i: Annual amount of industrial CO_2 added to the atmosphere.

t: Time in years.

$s = S_t - S_0$: Amount of CO_2 derived from industrial fuel combustion in the sea at time t.

$r = it - s$: Amount of CO_2 derived from industrial fuel combustion in the atmosphere at time t.

r^*: Observed decrease in C^{14} activity.

$\tau(atm)$: Average lifetime of a CO_2 molecule in the atmosphere, before it becomes dissolved in the sea.

$\tau(sea)$: Average lifetime of carbon in the sea, before it becomes atmospheric CO_2.

$k_1 = 1/\tau(atm)$: Rate of CO_2 transfer per year from the atmosphere to the sea.

$k_2 = 1/\tau(sea)$: Rate of CO_2 transfer per year from the sea to the atmosphere.

P_{CO_2}: Partial CO_2 pressure in the atmosphere.

Table 3 gives the amount of carbon, expressed in CO_2 equivalents, in the various geochemical reservoirs on the surface of the Earth.

The symbols S^* and A^* shall be used for denoting respective "effective" reservoirs.

Table 3. Amount of carbon, computed as CO_2, in sedimentary rocks, hydrosphere, atmosphere and biosphere from data given by RUBEY (1951), HUTCHINSON (1954), and SVERDRUP et al. (1942).

	Total on Earth: 10^{18} gms	In units of A_0
Carbonate in sediments	67,000	28,500
Organic carbon in sediments	25,000	10,600
CO_2 in the atmosphere (A_0)	2.35	1
Living matter on land[1]	0.3	0.12
Dead organic matter on land	2.6	1.1
$CO_2 + H_2CO_3$ in ocean[2]	0.8	0.3
CO_2 as HCO_3 in ocean[2]	114.7	48.7
CO_2 as $CO_3^=$ in ocean[2]	14.2	6.0
Total inorganic carbon in ocean	129.7	55.0
Dead organic matter in ocean	10	4.4
Living organic matter in ocean	0.03	0.01
Total carbon in ocean (S_0)	140	59.4

[1] Living organic matter on land estimated from assays of standing timber in the world's forests. 30% of land surface is relatively thick forest, averaging about 5,000 board feet/acre of commercial size timber, or 0.26 gm CO_2/cm^2 of forest. Assuming that total living matter in forests is twice the amount of timber, and that other components of the biosphere are $\frac{1}{3}$ of total gives 0.34 × 10^{18} gm CO_2 in land biosphere.

[2] Carbon dioxide components in sea water are assumed to be in equilibrium with a CO_2 partial pressure of 3 × 10^{-4} atmospheres; chlorinity: 20‰; temperature: 10° C; alkalinity: 2.46 × 10^{-3} meq/L. Under these conditions pH = 8.18. Volume of the Ocean = 1.37 × 10^{24} cm^3. This is probably an underestimate, because the water below the thermocline contains CO_2 produced by oxidation of organic matter, and the average temperature of the ocean is somewhat less than 10° C.

Rate of CO_2 exchange between sea and atmosphere

Two types of C^{14} measurements independently allow calculation of the exchange rate through the sea-air interface, if assumptions are made with respect to mixing rates of the water masses in the oceans: (*a*) the apparent C^{14} age of marine materials and (*b*) the effects of industrial coal combustion on the C^{14} concentration in the atmosphere (SUESS 1953). Experimental data on these two subjects are inadequate for rigorous quantitative interpretation but are sufficiently accurate to allow estimates of the order of magnitude of the rate constant for the exchange.

Considering the combined marine and atmospheric carbon reservoir as a closed system in equilibrium, the following relation holds by definition:

$$\frac{\tau(\text{sea})}{\tau(\text{atm})} = \frac{k_1}{k_2} = \frac{S_0}{A_0}. \tag{1}$$

τ (sea), the average lifetime of a carbon atom as a member of the marine reservoir, is equal to the average apparent C^{14} age of marine material.

An apparent C^{14} age of marine material is obtained by comparison of the C^{14} activity with that of a wood standard corrected for isotopic fractionation effects in nature or in the laboratory by mass spectrometric measurement of the C^{13}/C^{12} ratios. In marine carbonate the C^{13}/C^{12} ratio is about 2.5% higher than in land plants (NIER and GULBRANSEN, 1939 and CRAIG 1953, 1954). Because of the double mass difference the effect for C^{14} should be twice as large as that for C^{13}, if the isotopic distribution is established sufficiently rapidly so that radioactive decay of C^{14} can be neglected. If, after normalizing to equal C^{13}/C^{12} ratios, a lower C^{14} concentration is found than that of the wood standard, then this difference is attributed here to the effect of radioactivity, and is expressed as apparent age. A detailed study of the expected relationship between the isotopic fractionation factors for C^{13} and C^{14} in the bio-geochemical cycle of carbon has been made by CRAIG (1954).

Assuming from the then available C^{14} measurements that shell and wood have the same specific C^{14} activity, CRAIG (1954) attributed this unexpected result to slow transfer of CO_2 across the ocean-atmosphere interface resulting in a radiocarbon age of 400 years for surface ocean bicarbonate. Since then, more precise C^{14} measurements, supplemented by mass spectroscopic C^{13} determinations, have been published by SUESS (1954, 1955) and by RAFTER (1955) (see also HAYES *et al.*,

1955). The standard error of about 0.5% corresponds to an uncertainty in the age values of about 40 years. The apparent ages calculated from the published measurements are as follows:

Atlantic: (SUESS, 1954).

Mercenaria mercenaria	Shells	440 yrs.
Nantucket sound (from under 45 ft of water)	Flesh	540 yrs.
Sargassum weed from sea surface, 36°24′ N, 69°37′ W		320 yrs.

New Zealand area: (RAFTER, 1955).

Pauna Shells	350 yrs.
Limpet shells	360 yrs.
Cockle shells	280 yrs.
Pauna flesh	290 yrs.
Limpet flesh	140 yrs.
Seaweed	250 yrs.
Seawater (East Cape area)	370 yrs.

The average of the ages given here is 430 yrs. for the Atlantic samples and 290 yrs. for the New Zealand samples. The difference of 140 yrs. is due to the fact that the ages of the first group were calculated using wood grown in the 19th century as standard, whereas contemporaneous wood was used as standard for the New Zealand samples.[2]

Assuming that these apparent ages for marine surface materials are representative for the average age of marine carbon, or in other words, that mixing times of the oceans are short compared to the ages measured, an apparent age of about 400 years for marine carbon corresponds, according to Eq. (1), to an exchange time τ (atm) of about 7 years.

This *lower* limit for the exchange time of CO_2 between the atmosphere and the sea can now be compared with computations of this quantity from the observed effect of industrial fuel combustion on the specific C^{14} activity of wood. This second way, however, will lead to an *upper* limit for the exchange time if rapid mixing in the oceans is assumed.

At present 9.1×10^{15} grams of C^{14} free CO_2, or a fraction of 3.9×10^{-3} of the atmospheric CO_2, is added per year to the atmosphere by artificial burning of fossil fuels. The total amount added during the 100 years prior to 1950 corresponds to about 12% of the atmospheric carbon reservoir.

Neglecting any effect of the industrial CO_2 on the rate constants k_1 and k_2 one obtains:

$$\frac{ds}{dt} = k_1(it - s) - k_2 s \tag{2}$$

and integrated:

$$s = it\frac{k_1}{k_1 + k_2} - i\frac{k_1}{(k_1 + k_2)^2}(1 - e^{-(k_1 + k_2)t}) \quad (3)$$

for the amount of industrial CO_2 in the sea at the time t. As $k_1 \approx 60\, k_2$, we may neglect k_2 as small compared to k_1 We then obtain:

$$s = it - \frac{i}{k_1}(1 - e^{-k_1 t}) \quad (4)$$

or

$$r = it - s = \frac{i}{k_1}(1 - e^{-k_1 t}) \quad (5)$$

Expressing i, s, and r in units of the atmospheric CO_2, we obtain with $i = 0.25\%$ (corresponding to the value during the 1940's) and $t = 40$ years, the following values of r for exchange times $\tau(atm) = \frac{1}{k_1}$ as listed:

Table 4.

$1/k_1 = \tau(atm)$ years	r%
5	1.2
10	2.5
20	4.4
30	5.5
40	6.3
100	8.3
∞	10.0

Empirical values for the decrease in the specific C^{14} activity r^* were obtained by comparing C^{14} activities of wood samples grown in the 19th century with those grown more recently, taking into account isotope fractionation effects by C^{13} measurements and correcting for the C^{14} decay by normalizing to equal age (SUESS 1955).

With the assumption that the total atmospheric carbon reservoir is only negligibly greater than the amount of CO_2 in the atmosphere, r^* will be equal to r.

Table 5.

Tree	Years of growth from annual rings	r* %
Spruce, Alaska	1945–1950*	1.77
White Pine, Massachusetts	1936–1946	3.40
	1946–1953	2.90
Incense Cedar, California	1940–1944	1.85
	1950–1953	1.05
Cedrela, Peru	1943–1946	0.05
	1948–1953	1.05

It might be tempting to assume that the effects found in the samples investigated and their individual variations are due to local contamination of air masses by industrial CO_2 and that the world-wide decrease in the C^{14} activity of wood is practically zero. This, however, implies a too fast exchange rate and is inconsistent with the lower limit of τ (atm) given above. By coincidence, it so happens that taking the average of the r^* values listed, 1.73%, an exchange time τ (atm) of 7 years is obtained as an upper limit, identical with the lower limit obtained previously from the C^{14} age of marine surface material. An exchange time of 7 years, however, makes it necessary to assume unexpectedly short mixing times for the ocean.[3]

By reconsidering the relative size of the marine and atmospheric carbon reservoirs, and the assumptions necessary for treating the combined reservoirs as a closed system, we may approximate more closely to the conditions prevailing in nature. First, it seems possible that some of the organic matter and carbonate present in soils may have to be added to the carbon that constitutes the atmospheric carbon reservoir. Practically nothing is known about the C^{14} age of soils and it may be that some of the soil carbon, partly through bacterial action, is in more rapid isotopic exchange with atmospheric CO_2 than the data on plant assimilation and biological oxidation indicate. The total amount of carbon in soils may be equal to, or larger by as much as a factor of two than that in the CO_2 of the atmosphere. Rapid exchange with such carbon would decrease the change in C^{14} activity resulting from industrial fuel combustion by a factor equal to the ratio of the total "effective" atmospheric carbon reservoir to the atmospheric CO_2. However, the overturn time of the atmospheric CO_2 through the terrestrial biological cycle and the soil is probably at least several decades and it therefore seems improbable that a mechanism exists that might account for an atmospheric reservoir of more than 1.5 times the CO_2 in the atmosphere.

Nevertheless, it is interesting, in order to demonstrate how the uncertainty in the size of the atmospheric carbon reservoir affects our results, to introduce "effective" reservoirs A^* and S^*, so that $A^* > A$, because of exchange with soil carbon, and $S^* < S$, because of incomplete mixing in the oceans. A^* and S^* are defined in such a way that Eq. (1) becomes

$$\frac{k_1}{k_2} = \frac{S^*}{A^*} \quad (1a)$$

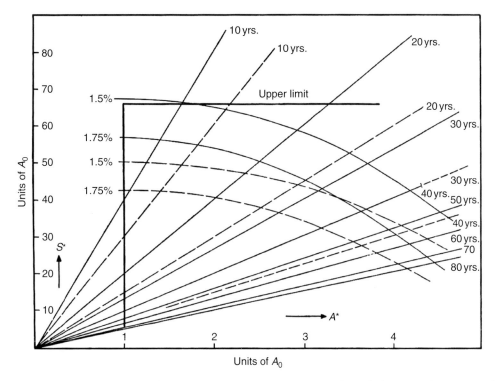

Fig. 1. Graphic solutions of Eq. (1 a) for values $1/k_1$ ranging from 10 to 80 years (straight lines), and of Eq. (5 a) for $r^* = 1.5\%$ and 1.75% (curved lines), for average apparent C^{14} age of sea water $1/k_2 = 400$ years (solid lines) and 300 years (broken lines). Points of intersection of straight lines and curves are possible solutions of the two equations for effective reservoirs A^* and S^*.

For Eq. (5):

$$r^* = \frac{A}{A^*}\frac{i}{k_1}(1-e^{-k_1t}) \qquad (5a)$$

is to be substituted.

Then the relationships of the three unknown quantities k_1, S^* and A^* according to the two equations (1 a) and (5 a) are shown graphically in figure 1.

With an average age of surface sea carbon of 400 years, i.e., $k_2 = 1/400$, one finds, even with an improbably large A^* equal to $3A$, that S^* is only 10 to 30% smaller than S, so that the resulting mixing times of the oceans cannot be larger than a few 100 years.

The overturn time for the marine carbon through rock weathering and precipitation can be estimated to be of the order of 100,000 years (table 3). During an average lifetime of sea carbon of 400 years an amount corresponding to 0.3% of the marine reservoir will be added to the oceans through rock weathering in the form of C^{14} free carbon and be precipitated after isotopic equilibration. On the average, this will cause the C^{14} age of sea carbon to appear about 35 years older. Locally, and especially along shores, the effect may well be considerably greater, so that the ages

measured so far may, on the average, be too great by as much as 100 years. In figure 1, solutions are shown for $1/k_2 = 300$ years, indicating a value of S^*/S of about $\frac{2}{3}$ for acceptable values of A^*.[4]

A possibility that the addition of C^{14} free CO_2 to the atmosphere from volcanic eruptions in the 19th century may obscure the effect from industrial fuel combustion needs further experimental investigation.

We conclude that the exchange time τ (atm) = = $1/k_1$, defined as the time it takes on the average for a CO_2 molecule as a member of the atmospheric carbon reservoir to be absorbed by the sea, is of the order of magnitude of 10 years. This corresponds to a net exchange rate of the order of 10^{-7} mol CO_2 per second and square meter of the ocean surface, larger by a factor of 100 than that postulated by PLASS (1956) and smaller by a factor of 10,000 than that deduced by DINGLE (1954) as a lower limit from numerical values of the various rate controlling constants. These are, as HUTCHINSON (1954) has forcefully pointed out, too uncertain to allow any definite conclusions. On the other hand, our exchange data give a value for the "invasion coefficient" of carbon dioxide close to that determined experimentally by BOHR (1899) for a stirred liquid surface.

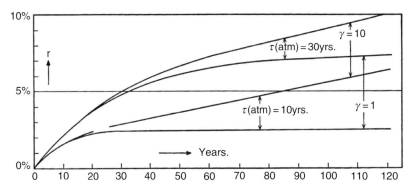

Fig. 2. Expected secular increase in the CO_2 concentration of air (r) according to Eq. (7), for average lifetimes of CO_2 in the atmosphere $\tau(atm) = 10$ years and 30 years, with and without correction for the increase in the partial CO_2 pressure with total CO_2 concentration of sea water (curves for $\gamma = 10$ and $\gamma = 1$ respectively), for a constant rate of addition of industrial CO_2 of $i = 2.5 \cdot 10^{-3} \times A_0$.

SECULAR VARIATION OF CO_2 IN THE ATMOSPHERE

In the preceding section of this paper, two simplifying assumptions were made when estimating the exchange rate of CO_2 between the atmosphere and the oceans: (1) that the rate constants k_1 and k_2 were not affected by a small increase of the exchangeable carbon reservoir such as that from industrial fuel combustion, and (2) that, except for that increase, no other changes in the sizes of the oceanic and atmospheric carbon reservoirs have taken place. If these assumptions were rigorously correct, the increase in atmospheric CO_2 due to an addition of C^{14} free CO_2 would be nearly equal to r, as given by Eq. (5) and in table 4, and equal to the decrease in the specific C^{14} activity r^*, multiplied by a factor A^*/A.

Because of the peculiar buffer mechanism of sea water, however, the increase in the partial CO_2 pressure is about 10 times higher than the increase in the total CO_2 concentration of sea water when CO_2 is added and the alkalinity remains constant (BUCH, 1933, see also HARVEY, 1955), so that under equilibrium conditions at a given alkalinity

$$\frac{r}{A_0} = \frac{\gamma s}{S_0} \quad \text{or} \quad \frac{A_0}{S_0} = \frac{k_1}{k_2} = \frac{r}{\gamma s},$$

γ being a numerical factor of the order of 10 for r and s small compared to A_0 and S_0 respectively.

As a reasonable approximation we may write instead of Eq. (3):

$$s = it\frac{k_1}{k_1 + \gamma k_2} - i\frac{k_1}{(k_1 + \gamma k_2)^2}(1 - e^{-(k_1 + \gamma k_2)t}) \quad (6)$$

or

$$r = it\frac{\gamma k_2}{k_1 + \gamma k_2} + i\frac{k_1}{(k_1 + \gamma k_2)^2}(1 - e^{-(k_1 + \gamma k_2)t}) \quad (7)$$

Figure 2 shows r as a function of time calculated from Eq. (7) for two values of k_1 corresponding to an exchange time of 10 and 30 years respectively, assuming constant addition of industrial CO_2 equal to 0.25% of the CO_2 in the atmosphere per year. At present the integrated amount of industrial CO_2 corresponds to about 40 to 50 years of addition at this constant rate. The increase in CO_2 in the atmosphere plus biosphere and soil due to industrial fuel combustion should therefore at present amount to 3 to 6%, depending on the assumptions made with respect to the size of the "effective" atmospheric carbon reservoir that exchanges with the ocean.

Eq. (7) and figure 2 show that addition of industrial CO_2 at a constant rate should eventually lead to a situation in which the secular increase in CO_2 in the atmosphere plus biosphere and soil equals $\dfrac{i\gamma k_2 / k_1}{1 + \gamma k_2 / k_1}$ per year. With $S_0/A_0 = k_1/k_2 \approx 60$ and $i = 0.25$ an increase of 3.6% per century is obtained. We have neglected the present rate of increase in alkalinity, which is small compared to the rate of CO_2 production from fossil fuels. However, over a sufficiently long period of time the alkalinity of the ocean must be expected to rise more rapidly as a consequence of the higher total CO_2 concentration in the atmosphere and ocean which will tend to increase the rate of rock weathering and to decrease the rate of deposition of $CaCO_3$. The secular increase in r should therefore be less than that calculated from Eq. (7)

with $\gamma = 10$ and somewhere between the curves shown in figure 2 for $\gamma = 10$ and $\gamma = 1$.

It seems therefore quite improbable that an increase in the atmospheric CO_2 concentration of as much as 10% could have been caused by industrial fuel combustion during the past century, as Callendar's statistical analyses indicate. It is possible, however, that such an increase could have taken place as the result of a combination with various other causes. The most obvious ones are the following:

1) Increase of the average ocean temperature of 1° C increases P_{CO_2} by about 7%. However, such increase would also raise the sea level by about 60 cm, due to thermal expansion of the ocean water. Actually, according to MUNK and REVELLE (1952), although the sea level has risen about 10 cm during the last century, this rise can be accounted for almost quantitatively by addition of melt water from retreating glaciers and ice caps. The increase in the average ocean temperature is probably not more than 0.05° C, which corresponds to an increase in P_{CO_2} of 0.35%. In the case of slow oceanic mixing, the increase could be somewhat larger, if only the top layers of the sea were affected by a rise in temperature.

2) Decrease in the carbon content of soils: HUTCHINSON (1954) considers this the most probable additional cause of the Callendar effect. The increase in arable lands of about 4 × 10^{16} cm² since the middle of the 19th century has resulted in a corresponding decrease of forest land of about 10%. Assuming that 70% of all the soil carbon is in forests (probably a considerable over-estimate), and that cultivation reduced this by 50%, the total decrease in soil carbon would correspond to 9 × 10^{16} gms of CO_2 which is 4% of that in the atmosphere. At least four-fifths of this amount should have been transferred to the ocean.

3) Change in the amount of organic matter in the oceans. About 7% of the marine carbon reservoir consists of organic material. The amount of organic matter must depend on the balance between the rates of reduction of CO_2 by photosynthesis and of production of CO_2 by oxidation. As pointed out previously, a change in the CO_2 content of sea water by a certain factor without a corresponding change in alkalinity changes P_{CO_2} by about 10 times this factor. Therefore a 1% change in the concentration of organic material in the sea will change the partial CO_2 pressure and hence the atmospheric

CO_2 by roughly 1%. During the past 50 years, the increase in marine carbon from absorption of industrial CO_2 of about 0.2% might have increased the rate of photosynthesis without a corresponding change in the rate of oxidation per unit mass of organic matter, and thus decreased P_{CO_2}. An increase in the temperature of surface water might have increased the rate of oxidation per unit mass of organic matter, and hence increased P_{CO_2}. We suspect that fluctuations in the amount of organic marine carbon might be an important cause for changes in the atmospheric CO_2 concentration.

ERIKSON and WELANDER (1956) have discussed a mathematical model of the carbon cycle between the atmosphere, the land biosphere and dead organic matter, and the ocean, in which it is assumed that the rate of input of carbon to the biosphere is directly proportional, not only to the total size of the biosphere but also to the amount of CO_2 in the atmosphere. Their estimate of the land biosphere is 7 times larger than that given in Table 3. They conclude that most of that part of the CO_2 added by fossil fuel consumption, which has not been absorbed by the ocean, has probably gone into the biosphere. Erikson and Welander's basic assumption that the amount of atmospheric carbon dioxide limits the growth of the terrestrial biosphere seems highly unlikely, in view of the fact that the principal photo-synthetic production on land is in forests, where a deficiency of plant nutrients might be expected. In any case as HUTCHINSON (1954) has shown, the amount of carbon in the biosphere and soil humus has probably decreased, rather than increased, during the past century, because of the clearing of forests.

In contemplating the probably large increase in CO_2 production by fossil fuel combustion in coming decades we conclude that a total increase of 20 to 40% in atmospheric CO_2 can be anticipated. This should certainly be adequate to allow a determination of the effects, if any, of changes in atmospheric carbon dioxide on weather and climate throughout the earth.

Present data on the total amount of CO_2 in the atmosphere, on the rates and mechanisms of CO_2 exchange between the sea and the air and between the air and the soils, and on possible fluctuations in marine organic carbon, are insufficient to give an accurate base line for measurement of future changes in atmospheric CO_2. An opportunity exists during the International Geophysical Year to obtain much of the necessary information.

ACKNOWLEDGEMENTS

A paper on the same subject by James R. Arnold and Ernest C. Anderson appears in this issue of this journal; we are grateful to Drs. Arnold and Anderson for the opportunity of discussing the problem before publication. We are happy to note that these authors have simultaneously and independently derived essentially the same conclusions as presented in this paper. We hope that the somewhat different approach will make both contributions equally valuable to the reader. We also wish to thank Dr. Carl Eckart for valuable discussions and Dr. Harmon Craig for much constructive criticism. Dr. Craig's own careful analysis of the subject appears in a separate paper in this issue.

REFERENCES

Arrhenius, Svante, 1903: *Lehrbuch der kosmischen Physik* **2**. Leipzig: Hirzel.

Bohr, C, 1899: Die Loslichkeit von Gasen in Flüssigkeiten. *Ann. d. Phys.* **68**, p. 500.

Buch, K., 1933: Der Borsäuregehalt des Meerwassers und seine Bedeutung bei der Berechnung des Kohlensäuresystems. *Rapp. Cons. Explor. Mer.* **85**, p. 71.

Callendar, G. S., 1938: The artificial production of carbon dioxide and its influence on temperature. *Quarterly Journ. Royal Meteorol. Soc.* **64**, p. 223.

Callendar, G. S., 1940: Variations in the amount of carbon dioxide in different air currents. *Quarterly Journ. Royal Meteorol. Soc.* **66**, p. 395.

Callendar, G. S., 1949: Can carbon dioxide influence climate? *Weather* **4**, p. 310.

Chamberlin, T. C, 1899: An attempt to frame a working hypothesis of the cause of glacial periods on an atmospheric basis. *J. of Geology* **7**, pp. 575, 667, 751.

Conway, E. J., 1942: Mean geochemical data in relation to oceanic evolution. *Proc. Roy. Irish Acad., B.* **48**, p. 119.

Craig, H., 1953: The geochemistry of the stable carbon isotopes. *Geochim. et Cosmochim. Acta* **3**, p. 53.

Craig, H., 1954: Carbon 13 in plants and the relationship between carbon 13 and carbon 14 variations in nature. *Journ. Geol.* **62**, p. 115.

Dingle, H. N., 1954: The carbon dioxide exchange between the North Atlantic Ocean and the atmosphere. *Tellus* **6**, p. 342.

Erikson, E., and WELANDER, P., 1956: On a mathematical model of the carbon cycle in nature. *Tellus* **8**, p. 155.

Fergusson, G. J., and RAFTER, T. A.: *New Zealand C-14 Age Measurements III*. In press.

Fonselius, S., KOROLEFF, F., and WÄRME, K., 1956: Carbon dioxide variations in the atmosphere. *Tellus* **8**, p. 176.

Harvey, H. W., 1955: *The Chemistry and Fertility of Sea Water*. Cambridge: University Press.

Hayes, F. N., ANDERSON, E. C, and ARNOLD, J. R., 1955: Liquid scintillation counting of natural radiocarbon. *Proc. of the International Conference on Peaceful Uses of Atomic Energy, Geneva*, **14**, p. 188.

Hutchinson, G. E., 1954: In *The Earth as a Planet*, G. Kuiper, ed. Chicago: University Press. Chapter 8.

Munk, W., and REVELLE, R., 1952: Sea level and the rotation of the earth. *Am. Journ. Sci.* **250**, p. 829.

Nier, A. O., and GULBRANSEN, E. A., 1939: Variations in the relative abundance of the carbon isotopes. *Journ. Am. Chem. Soc.* **61**, p. 697.

Plass, G. N., 1956: Carbon dioxide theory of climatic change. *Tellus* **8**, p. 140.

Rafter, T. A., 1955: C^{14} variations in nature and the effect on radiocarbon dating. *New Zealand Journ. Sci. Tech. B.* **37**, p. 20.

Rubey, W. W., 1951: Geologic history of sea water. *Bull. Geol. Soc. Amer.* **62**, p. 1111.

Slocum, GILES, 1955: Has the amount of carbon dioxide in the atmosphere changed significantly since the beginning of the twentieth century? *Monthly Weather Rev.* Oct., p. 225.

Suess, H. E., 1953: Natural Radiocarbon and the rate of exchange of carbon dioxide between the atmosphere and the sea. *Nuclear Processes in Geologic Settings*, National Academy of Sciences – National Research Council Publication, pp. 52–56.

Suess, H. E., 1954: Natural radiocarbon measurements by acetylene counting. *Science* **120**, p. 5.

Suess, H. E., 1955: Radiocarbon concentration in modern wood. *Science* **122**, p. 415.

Sverdrup, H. U., JOHNSON, M. W., and FLEMING, R. H., 1942: *The Oceans*. New York: Prentice-Hall, Inc.

United Nations, 1955: World requirements of energy, 1975–2000. *International Conference on Peaceful Uses of Atomic Energy, Geneva*, **1**, p. 3.

Urey, H. C., 1952: *The Planets*. New Haven: Yale Univ. Press.

NOTES

1 World production of CO_2 from the use of limestone for cement, fluxing stone and in other ways was about 1% of the total from fossil fuel combustion in 1950.

2 See FERGUSSON and RAFTER. These authors have now found that 110 years should be added to all of their previous age determinations to correct for the difference of the standards.

3 In a separate paper in this issue, H. Craig evaluates the exchange time by considering present data on the C-14 production rate by cosmic rays, and obtains a value of 7 ± 3 years. He further finds that a more detailed model of ocean mixing gives the same value. An earlier estimate of 20 to 50 years (SUESS 1953) was based on C-14 measurements of woods grown in an area of industrial contamination.

4 For a discussion of the physical significance of S^*/S in terms of mixing times in the ocean see H.Craig's paper in this issue.

Distribution of Matter in the Sea and Atmosphere: Changes in the Carbon Dioxide Content of the Atmosphere and Sea due to Fossil Fuel Combustion[1]

BERT BOLIN AND ERIK ERIKSSON

International Meteorological Institute in Stockholm

ABSTRACT

The dissociation equilibrium of carbon dioxide in the sea is discussed with particular emphasis on the buffering effect of sea water, when changes of the partial pressure of CO_2 in the gas phase take place. The results are used in a study of the changes of the carbon dioxide content of the atmosphere and the sea that occur as a result of release of CO_2 to the atmosphere by fossil fuel combustion. It is shown that the steady state considerations given by previous authors hereby are considerably modified. Thus an increase of the CO_2 content of the atmosphere of about 10% as reported by Callendar may be compatible with a Süess effect of only a few percent. Because of the small buffering effect of the sea it seems likely that the biosphere on land may play a more important role for the changes actually occurring in the atmosphere due to release of CO_2 by combustion than previously believed. This problem warrants further investigation, but already the present treatment indicates that an appreciable increase of the amount of CO_2 in the atmosphere may have occurred since last century. This increase will continue and should be detectable with present techniques for measuring CO_2 in the atmosphere within a few years in areas with little or no local pollution due to fossil fuel combustion as in the Antarctica or on Hawaii.

1. INTRODUCTION

Fossil fuel combustion has added considerable amounts of carbon dioxide to the atmosphere during the last 100 years. In view of the great importance of CO_2 in the atmosphere for maintaining a radiational balance between the earth and space it is of great interest to know whether this output of CO_2 has caused a significant increase of the total content of carbon dioxide in the atmosphere or whether most of it has been transferred into the oceans. Some twenty years ago CALLENDAR (1938) could show that most likely a noticeable increase of the CO_2 in the atmosphere had occurred and he has recently (CALLENDAR, 1958) indicated that this increase in 1955–56 amounted to about 10 % as compared with an output due to combustion of about 14 % of the total amount of CO_2 present in the atmosphere (cf. REVELLE and SÜESS, 1957). His conclusions have recently been supported by BRAY (1959) in a detailed statistical investigation.

However, by studying the C^{14} distribution in the atmosphere and the sea and its variation in the atmosphere during the last 100 years as revealed by the ratio C^{14}/C^{12} in wood one has been able to show that the exchange time between the atmosphere and the ocean is about 5 years (CRAIG, 1957, 1958; REVELLE and SÜESS, 1957; ARNOLD and ANDERSON, 1957; RAFTER and FERGUSSON, 1958). It has then been concluded (REVELLE and SÜESS, I.c.) that most of the CO_2 due to combustion has been transferred into the ocean and that a net increase of CO_2 in the atmosphere of only a few percent has actually occurred. Callendar's deduction has therefore been rejected particularly since the CO_2 measurements from the 19th century are indeed very uncertain.

Towards the end of their paper Revelle and Süess point out, however, that the sea has a buffer mechanism acting in such a way that a 10 % increase of the CO_2-content of the atmosphere need merely be balanced by an increase of about 1 % of the total CO_2 content in sea water to reach a new

1 The research reported in this paper was partly sponsored by the Office of Naval Research through the Woods Hole Oceanographic Institution. Contribution No. 1025 from the Woods Hole Oceanographic Institution.

equilibrium. The crude model of the sea they used assuming it to be one well-mixed reservoir of CO_2, did not permit them to study the effect of this process more in detail.

The low buffering capacity of the sea mentioned by Revelle and Süess is due to a change in the dissociation equilibrium between CO_2 and H_2CO_3 on one hand and HCO_3 and CO_3 ions on the other. An addition of CO_2 to the water will change the pH and thereby decrease the dissociation resulting in a larger portion of CO_2 and H_2CO_3 molecules. Since the pressure of CO_2 in the gas phase being in equilibrium with CO_2 dissolved in water is proportional to the number of CO_2 and H_2CO_3 molecules in the water, an increase of the partial pressure occurs which is much larger (about 12.5 times) than the increase of the *total content* of CO_2 in the water. The change of this equilibrium in the sea is almost instantaneous. However, in course of its circulation the ocean water gets in contact with solid $CaCO_3$ on the bottom of the sea whereby a change towards another equilibrium takes place. This latter process is extremely slow and may be disregarded when discussing changes due to fossil fuel combustion. It will, however, be indicated in section 2 how this equilibrium is of major interest when being concerned with processes with a time scale of several thousand years.

In discussing the consequences of such a shift in the dissociation equilibrium with respect to the exchange of CO_2 between the atmosphere and the sea and within the sea it is not sufficient to treat the ocean as one well-mixed reservoir. We shall instead first follow a suggestion by Craig (1957) and divide the ocean in two layers. The upper reservoir, in direct contact with the atmosphere, is the part of the ocean located above the thermocline and constitutes about 1/50 of the total sea. This part of the ocean is well mixed due to wind action and convection. The remaining part of the ocean, the deep sea, is also taken as a well-mixed water body in direct exchange with the mixed layer above. Certainly this latter assumption is a poor approximation to actual conditions but will, as we shall see, give internally consistant results. In the first instance we shall neglect the effect of living matter on the earth but some general remarks about the exchange of CO_2 between vegetation and the atmosphere will be given towards the end of the paper.

It is obvious that an addition of CO_2 to the atmosphere will only slightly change the CO_2 content of the sea but appreciably effect the CO_2 content of the atmosphere. It is possible to deduce a relation between the exchange coefficient for transfer from the atmosphere to the sea and the corresponding coefficient for the exchange between the deep sea and the mixed layer. It turns out that a 10 % increase of the CO_2 content of the atmosphere as a result of a total output due to combustion amounting to 13 % of the total CO_2 content of the atmosphere would result for a residence time of water in the deep sea of around 500 years. Only a considerably more rapid turn over of the ocean model yield appreciably lower values, while the rate of exchange between the atmosphere and the sea is much less important.

The change in the dissociation equilibrium in water resulting from a transfer of CO_2 to the sea also effects the C^{14} distribution in the three reservoirs and Revelle's and Süess' (1957) considerations in this matter are thereby appreciably changed. It is clear that a large percentage change of the CO_2 in the atmosphere and a comparatively small percentage change in the sea will yield a C^{14}/C^{12} ratio in the sea which is considerably greater than that of the atmosphere. The distribution of C^{14} between these two reservoirs is therefore not in equilibrium any longer. A transfer of C^{14} from the sea to the atmosphere will result. A more detailed study of this secondary effect reveals that

a) the steady state considerations of CO_2 exchange between the atmosphere and the sea as given by Craig (1957, 1958) are somewhat modified and a likely residence time for CO_2 in the atmosphere is 5 years.

b) the Süess effect would be 3–5 % depending on the rate of overturning of the sea.

In view of the observed values of the Süess-effect being around 3 % in 1954 (before the hydrogen bomb tests) (cf. Broecker and Walton, 1959; De Vries, 1958) the values obtained in this analysis are somewhat too large if using a value for the residence of water in the deep sea of 500 years. One possible explanation of this discrepancy would be the neglect of the exchange of CO_2 between the atmosphere and the biosphere. The main difficulty met when trying to incorporate this effect is, that we actually do not know whether a net increase of the amount of CO_2 present in vegetation may have occurred due to a transfer from the atmosphere or whether the direct influences from man's activities have had an effect in the opposite direction. It is clear, however, that even if no net transfer of CO_2 from the atmosphere into living or dead matter on land has occurred these exchange processes will modify the estimate of the "Süess effect" obtained by merely considering the atmosphere and the sea.

2. THE CO$_2$-SYSTEM IN THE SEA

The different components of CO$_2$ present in the sea are CO$_2$, H$_2$CO$_3$, HCO$_3^-$ and CO$_3^{-2}$. As CO$_2$ is difficult to distinguish from H$_2$CO$_3$ it is customary to express the sum of these species as CO$_2$. The sea is also in contact with ample amounts of solid CaCO$_3$ which should be considered in equilibrium being attained over several "turnover times" of the sea itself. Another important item in the system is the carbonate alkalinity, denoted here by A which is the sum of those cations which balance the charges of HCO$_3^-$ and CO$_3^{-2}$. The following average values for the concentrations of the different components will be used here:

$$C_{CO_2} = 0.0133 \text{ mmol} \times 1^{-1} (\text{sum of CO}_2 \text{ and H}_2\text{CO}_3)$$
$$C_{HCO_3^-} = 1.90 \text{ mmol} \times 1^{-1}$$
$$C_{CO_3^{-2}} = 0.235 \text{ mmol} \times 1^{-1}$$

The sum of all these species is denoted by ΣC_{CO_2} and becomes 2.148 mmol $\times$ 1^{-1}. A is given in mval $\times$ 1^{-1} and becomes, $A = C_{HCO_3^-} + 2C_{CO_3^{-2}} = 2.37$ mval $\times$ 1^{-1}. Now the following relationships can be derived, namely

$$\Sigma C_{CO_2} = \left(1 + \frac{K_1}{C_{H^+}} + \frac{K_1 K_2}{C_{H^+}^2}\right) C_{CO_2} \tag{1}$$

$$A = \left(\frac{K_1}{C_{H^+}} + \frac{2K_1 K_2}{C_{H^+}^2}\right) C_{CO_2} \tag{2}$$

where K_1 and K_2 are the first and second dissociation constants of H$_2$CO$_3$ in sea water and C_{H^+} is the hydrogen ion concentration. It is convenient to have the average values of the fractions within the brackets. They are

$$\frac{K_1}{C_{H^+}} = 143 \qquad \frac{K_1 K_2}{C_{H^+}^2} = 18$$

As to calcium carbonate, its solubility product L_p can be written

$$C_{Ca^{+2}} \cdot C_{CO_3^{-2}} = L_p$$

or more conveniently

$$C_{Ca^{+2}} = L_p \frac{C_{H^+}^2}{K_1 K_2} \frac{1}{C_{CO_2}} \tag{3}$$

and in sea water $C_{Ca^{2+}} = 10$ mmol $\times$ 1^{-1}.

Finally we have for the equilibrium between atmospheric CO$_2$ and that in sea water

$$P_{CO_2} = \frac{1}{\alpha} C_{CO_2} \tag{4}$$

where α is a proportionality constant and P_{CO_2} is the partial pressure of CO$_2$ in the atmosphere.

The constants K_1, K_2, L_p and α are only functions of temperature and salinity and will be regarded as constants in the following. We may therefore consider relations between small variations in P_{CO_2}, C_{CO_2}, ΣC_{CO_2}, A, C_{H^+} and $C_{Ca^{2+}}$. Using the variational method applied on equations (1) to (4) one obtains

$$\frac{\delta \Sigma C_{CO_2}}{\Sigma C_{CO_2}} = \frac{\delta C_{CO_2}}{C_{CO_2}} + \frac{\frac{\partial}{\partial C_{H^+}}\left(1 + \frac{K_1}{C_{H^+}} + \frac{K_1 K_2}{C_{H^+}^2}\right)}{1 + \frac{K_1}{C_{H^+}} + \frac{K_1 K_2}{C_{H^+}^2}} \delta C_{H^+} \tag{5}$$

$$\frac{\delta A}{A} = \frac{\delta C_{CO_2}}{C_{CO_2}} + \frac{\frac{\partial}{\partial C_{H^+}}\left(\frac{K_1}{C_{H^+}} + \frac{2K_1 K_2}{C_{H^+}^2}\right)}{\frac{K_1}{C_{H^+}} + \frac{2K_1 K_2}{C_{H^+}^2}} \delta C_{H^+} \tag{6}$$

$$\frac{\delta C_{Ca^{+2}}}{C_{Ca^{+2}}} = \frac{2\delta C_{H^+}}{C_{H^+}} - \frac{\delta C_{CO_2}}{C_{CO_2}} \tag{7}$$

$$\frac{\delta P_{CO_2}}{P_{CO_2}} = \frac{\delta C_{CO_2}}{C_{CO_2}} \tag{8}$$

First we see that if P_{CO_2} varies and the hydrogen ion concentration were kept constant, the relative changes would be the same in the sea as in the atmosphere. As the total amount of CO$_2$ in the sea is about 50 times that in the air, practically all excess CO$_2$ delivered to the atmosphere would be taken up by the sea when equilibrium has been established. One cannot, however, assume that pH is uninfluenced by changes in the ΣC_{CO_2} of the sea. We may see if any condition can be imposed upon the alkalinity. Obviously this should be kept constant if we consider changes that takes place over a relatively short time interval, less than the "turnover time" of the sea, because A is really the concentration of cations that balance the charges of HCO$_3^-$ and CO$_3^{-2}$. If CaCO$_3$ is excluded the sum of these charges must remain constant. And then we see that any change in the P_{CO_2} will also change C_{H^+}. From eq. (5), (6) and (8) we get by putting $\delta A = 0$

$$\frac{\delta P_{CO_2}}{P_{CO_2}} = \frac{\delta C_{CO_2}}{C_{CO_2}} = 12.5 \frac{\delta \Sigma C_{CO_2}}{\Sigma C_{CO_2}} \qquad (9)$$

using the numerical values listed earlier. This tells us that 1 percent change in the total CO_2 concentration in the sea would require 12.5 percent change in the atmospheric CO_2 to maintain equilibrium. If we consider only the "mixed layer" of the oceans, i.e. the surface layer which contains about as much CO_2 as the atmosphere less than 10 percent of the excess fossil CO_2 in the atmosphere should have been taken up by the mixed layer. It is therefore obvious that the mixed layer acts as a bottleneck in the transport of fossil CO_2 into the deep sea (cf. the following section).

It may be of interest also to consider the effect of the $CaCO_3$ on the bottom of the oceans; the effect this may have for the final equilibrium which is attained after a long time. Then alkalinity will change by $\delta A = 2 \; \delta C_{Ca^{+2}}$ and with this condition using the whole system of equations and $C_{Ca^{+2}} = 10$ mmol $\times 1^{-1}$,

$$\frac{\delta P_{CO_2}}{P_{CO_2}} = \frac{\delta C_{CO_2}}{C_{CO_2}} = 2.36 \frac{\delta \Sigma C_{CO_2}}{\Sigma C_{CO_2}} \qquad (10)$$

which shows that the sea, given enough time, has an appreciable buffer capacity for atmospheric CO_2. However, in the case of eq. (10) part of the change of total CO_2 comes from dissolution or precipitation of $CaCO_3$ and this has obviously to be subtracted if we want to know how much of the excess of atmospheric CO_2 the sea ultimately can consume. The amount that dissolves or precipitates is obviously equal to $\delta C_{Ca^{+2}}$ and becomes

$$\delta C_{Ca^{+2}} = 0.444 \frac{\delta C_{CO_2}}{C_{CO_2}}$$

Expressed as a part of the total CO_2 it becomes

$$\frac{\delta C_{Ca^{+2}}}{\Sigma C_{CO_2}} = 0.206 \frac{\delta C_{CO_2}}{C_{CO_2}} \qquad (11)$$

Now, rewriting (9) we find

$$\frac{\delta \Sigma C_{CO_2}}{\Sigma C_{CO_2}} = 0.424 \frac{\delta C_{CO_2}}{C_{CO_2}}$$

The part of the increase in total CO_2 that has come from the atmosphere is therefore

$$\Delta \frac{\delta \Sigma C_{CO_2}}{\Sigma C_{CO_2}} = \frac{\delta \Sigma C_{CO_2}}{\Sigma C_{CO_2}} - \frac{\delta C_{Ca^{+2}}}{\Sigma C_{CO_2}}$$

$$= 0.238 \frac{\delta C_{CO_2}}{C_{CO_2}} \qquad (12)$$

or

$$\frac{\delta P_{CO_2}}{P_{CO_2}} = \frac{\delta C_{CO_2}}{C_{CO_2}} = 4.20 \Delta \left(\frac{\delta \Sigma C_{CO_2}}{\Sigma C_{CO_2}} \right) \qquad (13)$$

Thus, in equilibrium one percent increase in ΣC_{CO_2} obtained from the atmosphere would occur for a 4.2 percent increase in the atmospheric partial pressure of CO_2. In other words, any excess CO_2 put into the atmosphere will ultimately be distributed so that about 11/12 of it goes into the sea (again assuming the sea contains, 50 times more CO_2 than the atmosphere) while about 1/12 remains in the atmosphere. Of the part that goes into the sea, 87 percent has taken part in the reaction

$$CaCO_3(s) + H_2CO_3(aq) = Ca^{+2} + 2HCO_3^{-1}$$

The rest has been used to lower the pH of sea water by the reaction

$$H_2CO_3(aq) \rightarrow H^+ + HCO_3^-$$

If the turnover time of the sea is of the order of 1,000 years, several thousands of years would be required to reach equilibrium with the $CaCO_3$ at the bottom of the sea.

It should finally be noted that, in case atmospheric CO_2 was withdrawn by some process, this would result in precipitation of $CaCO_3$ in the sea.

3. THE EXCHANGE OF INACTIVE CARBON BETWEEN THE ATMOSPHERE AND THE SEA

In order to see more clearly the effect of the shift in the dissociation equilibrium in the sea we shall first disregard the role played by the biosphere and merely consider the atmosphere and the sea, the latter composed of two reservoirs. The top one, the mixed layer, is confined to the layer above the thermocline and the lower one, the deep sea, consists of the remainder of the sea. Both these reservoirs as well as the atmosphere are considered to be well-mixed and the exchange between them takes place through first order exchange processes. Introduce the following notations:

N_i = the total amount of C^{12} and C^{13} in reservoir i in a state of equilibrium as before 1850.

N_i^* = the total amount of C^{14} in reservoir i in a state of equilibrium as before 1850.

N_i' = the amount of C^{12} and C^{13} in reservoir i present in the form of CO_2 or non-dissociated H_2CO_3 (only for the ocean) in a state of equilibrium as before 1850.

$N_i^{*'}$ = the amount of C^{14} in reservoir i present in the form of CO_2 or non-dissociated H_2CO_3 (only for the ocean) in a state of equilibrium as before 1850.

n_i, n_i^*, n_i', $n_i^{*'}$ indicate the deviations from the equilibrium values given above.

k_{i-j} = exchange coefficient for transfer of C^{12} and C^{13} from reservoir i to reservoir j.

k_{i-j}^* = exchange coefficient for transfer of C^{14} from reservoir i to reservoir j.

τ_{i-j} = $1/k_{i-j}$

τ_{i-j}^* = $1/k_{i-j}^*$

λ = decay constant for C^{14}

$\gamma(t)$ = release of C^{12} and C^{13} due to fossil fuel combustion as a function of time t.

Q = mean production of C^{14} in the atmosphere due to cosmic rays.

The indices a, m and d refer to the atmosphere, the mixed layer and the deep sea respectively. The nomenclature is in part very similar to that used by Craig (1957).

Considering now first conditions for inactive carbon we obtain the following equilibria

$$\left.\begin{array}{r} -k_{a-m}N_a + k_{m-a}N_m' = 0 \\ k_{a-m}N_a - k_{m-a}N_m' - k_{m-d}N_m + k_{d-m}N_d = 0 \\ k_{m-d}N_m - k_{d-m}N_d = 0 \end{array}\right\} \quad (14)$$

Notice here particularly that the transfer from the sea to the atmosphere is put proportional to N_m' and not to the total amount of carbon in the mixed layer, N_m. On the other hand the transfer from the mixed layer to the deep sea and *vice versa* is due to the motion of water and should therefore be proportional to the total amounts of carbon present in the two reservoirs, *i.e.* N_m and N_d respectively. It should be pointed out here that an exchange of carbon between various strata of the ocean also occurs through the motion of organisms and the settling of dead organic material and precipitated $CaCO_3$ which is gradually being dissolved. From a recent paper by Eriksson (1958) it can be estimated that this is about 1/3,000 of the total amount of CO_2 in the sea per year. Naturally it is compensated by an

upward flux of dissolved CO_2. This flux is small compared to the advective flux from deep water which is about the same as the horizontal transfer of atmospheric CO_2 in Eriksson's model giving a residence time of $\approx$ 600 years. His paper suggests anyway that the ratio between advective flux and gravitational is about 5. From (14) we now get

$$\left.\begin{array}{l} k_{m-a} = \dfrac{N_a}{N_m'} k_{a-m} = \alpha k_{a-m} \\[2mm] k_{m-d} = \dfrac{N_d}{N_m} k_{d-m} = \beta k_{d-m} \end{array}\right\} \quad (15)$$

Due to combustion a deviation from this equilibrium now has occurred, which is governed by the following set of equations

$$\left.\begin{array}{l} \dfrac{dn_a}{dt} = -k_{a-m}n_a + k_{m-a}n_m' + \gamma(t) \\[2mm] \dfrac{dn_m}{dt} = k_{a-m}n_a - k_{m-a}n_m' - k_{m-d}n_m + k_{d-m}n_d \\[2mm] \dfrac{dn_d}{dt} = k_{m-d}n_m - k_{d-m}n_d \end{array}\right\} \quad (16)$$

In the previous section we found that the following relation exists between n_m' and n_m (cf. eq. (9))

$$n_m' = 12.5 \frac{N_m'}{N_m} n_m = B_1 n_m \quad (17)$$

Introducing this expression for n_m' into (16) we obtain a system of three ordinary linear differential equations for the three dependant variables n_a, n_m and n_d. Eliminating two of the three variables we obtain

$$\frac{d^3n_i}{dt^3} + [(1 + B_1\alpha)k_{a-m} + (1 + \beta)k_{d-m}]\frac{d^2n_i}{dt^2}$$
$$+ [1 + B_1\alpha + \beta]k_{a-m}k_{d-m}\frac{dn_i}{dt} = S_i \quad (18)$$
$$i = a,\ m,\ d$$

where

$$\left.\begin{array}{l} S_a = \gamma''(t) + [B_1\alpha k_{a-m} + (1 + \beta)k_{d-m}]\gamma'(t) \\ \qquad + B_1\alpha k_{a-m}k_{d-m}\gamma(t) \\ S_m = k_{a-m}\gamma'(t) + k_{a-m}k_{d-m}\gamma(t) \\ S_d = \beta k_{a-m}k_{d-m}\gamma(t) \end{array}\right\} \quad (19)$$

The general solution of (18) is

$$n_i = C_{1i}e^{\lambda_1 t} + C_{2i}e^{\lambda_2 t} + C_{3i}e^{\lambda_3 t} + P_i \quad (20)$$

where C_{1i}, C_{2i} and C_{3i} are integration constants, λ_1, λ_2 and λ_3 solutions to the algebraic equation

$$\lambda_3 + [(1+B_1\alpha)k_{a-m} + (1+\beta)k_{d-m}]\lambda^2 \\ + [1+B_1\alpha+\beta]k_{a-m}k_{d-m}\lambda = 0 \quad (21)$$

and P_i are particular solutions depending on the functions S_i. Assuming now that [CRAIG (1957)]

$$\left.\begin{array}{l} N_m = 1.2\,N_a \\ N_d = 60\,N_a \end{array}\right\} \quad (22)$$

we obtain

$$\left.\begin{array}{l} B_1\alpha = 12.5\dfrac{N'_m}{N_m}\dfrac{N_a}{N'_m} = 10.4 \\[3mm] \beta = \dfrac{N_d}{N_m} = 50 \end{array}\right\} \quad (23)$$

The values chosen for N_m and N_d are somewhat uncertain but as we shall see later, do not influence the results significantly. The solutions to eq. (21) then are, due regard taken to the fact that $k_{a-m} \gg \gg k_{d-m}$ and thus $(1+B_1\alpha)\cdot k_{a-m} + (1+\beta)k_{d-m} \gg \gg (1+B_1\alpha+\beta)\,k_{a-m}k_{d-m}$ where $\gg$ denotes about two orders of magnitude

$$\left.\begin{array}{l} \lambda_1 = 0 \\[2mm] \lambda_2 = -\dfrac{(1+B_1\alpha+B)k_{a-m}k_{d-m}}{(1+B_1\alpha)k_{a-m} + (1+\beta)k_{d-m}} \\[3mm] \lambda_3 = -[(1+B_1\alpha)k_{a-m} + (1+\beta)k_{d-m}] \end{array}\right\} \quad (24)$$

To obtain the particular solutions we have to specify $\gamma(t)$. We shall assume that $\gamma(t)$ may be approximated by

$$\gamma(t) = \gamma_0 e^{rt} \quad (25)$$

where

$$\left.\begin{array}{l} \gamma_0 = 4.96\ N_a\ 10^{-4} \\ r = 0.029\ \text{year}^{-1} \end{array}\right\} \quad (26)$$

which fits the values given by REVELLE and SÜESS (1957) for carbon production until today and also the estimated values to year 2010 with sufficient accuracy if $t = 0$ at 1880 (see table 1).
Thus we obtain

$$\left.\begin{array}{l} S_a = \gamma_0[r^2 + \{B_1\alpha k_{a-m} + (1+\beta)k_{d-m}\}r \\ \qquad + B_1\alpha k_{a-m}k_{d-m}]e^{rt} = \gamma_0 S_{a0}e^{rt} \\[2mm] S_m = \gamma_0[k_{a-m}r + k_{a-m}k_{d-m}]e^{rt} = \gamma_0 S_{m0}e^{rt} \\[2mm] S_d = \gamma_0\beta k_{a-m}k_{d-m}e^{rt} = \gamma_0 S_{d0}e^{rt} \end{array}\right\} \quad (27)$$

One then easily finds the particular solutions

Table 1. CO_2 added to the atmosphere by fossil fuel combustion and a comparison with an analytical expression.

Decade	Average amount added per decate (% of N_a)		Cumulative total added (% of N_a)	
	Measured or estimated	$\gamma(t)$	Measured or estimated (since 1860)	$\int_0^t \gamma(t)dt$ (since 1880)
1880–89	0.54	0.57	1.13	0.57
1890–99	0.79	0.77	1.92	1.34
1900–09	1.27	1.03	3.19	2.37
1910–19	1.72	1.37	4.91	3.74
1920–29	2.00	1.83	6.91	5.57
1930–39	2.11	2.47	9.02	8.04
1940–49	2.71	3.17	11.73	11.21
1950–59	3.9	4.4	15.6	15.6
1960–69	5.4	5.8	21.0	21.4
1970–79	7.5	8.0	28.5	29.4
1980–89	10.5	10.4	39.0	39.8
1990–99	14.5	13.7	53.5	53.5
2000–09	20.0	19.0	73.5	72.5

$$P_i = \frac{\gamma_0}{r}\frac{S_{i0}}{S_{a0}+S_{m0}+S_{d0}}e^{rt} \quad (28)$$
$$i = a,\ m,\ d$$

In order to determine the three constants C_{ij} for each solution $n_i(t)$ we shall apply the initial conditions

$$n_a = n_m = n_d = 0$$

which yields

$$\left.\begin{array}{l} \dfrac{dn_a}{dt} = \gamma(t);\quad \dfrac{dn_m}{dt} = \dfrac{dn_d}{dt} = 0 \\[3mm] \hfill t = 0 \\[2mm] \dfrac{d^2n_a}{dt^2} = \gamma'(t) - k_{a-m}\gamma(t) \\[3mm] \dfrac{d^2n_m}{dt^2} = k_{a-m}\gamma(t); \\[3mm] \dfrac{d^2n_d}{dt^2} = 0 \end{array}\right\} \quad (29)$$

These equations are obtained from the system (16) and also yield

$$n_a + n_m + n_d = \frac{\gamma_0}{r}(e^{rt} - 1) \quad (30)$$

As can easily be verified the final solutions of n_a, n_m and n_d will be independant of the exact initial conditions for $t \gg 200$ years if the values for τ_{a-m} and τ_{d-m} are of the order of 5 and 500 years respectively. It also follows that $\lambda_3 = -2$ years^{-1} and that thus the term containing exp $(\lambda_3 t)$ in (20) may be neglected for $t \gg 2$–3 years. With due regard taken to this latter fact we finally obtain

$$
\left.
\begin{aligned}
n_a &= \frac{\gamma_0}{r}\left[\frac{S_{a0}}{S}(e^{rt}-1)+\frac{r}{\lambda_2}\left(1-\frac{S_{a0}}{S}\right)(e^{\lambda_2 t}-1)\right] \\
n_m &= \frac{\gamma_0}{r}\frac{S_{m0}}{S}\left[(e^{rt}-1)-\frac{r}{\lambda_2}(e^{\lambda_2 t}-1)\right] \\
n_d &= \frac{\gamma_0}{r}\frac{S_{d0}}{S}\left[(e^{rt}-1)-\frac{r}{\lambda_2}(e^{\lambda_2 t}-1)\right] \\
S &= S_{a0} + S_{m0} + S_{d0}
\end{aligned}
\right\} \quad (31)
$$

From CRAIG'S (1957, 1958) investigation of the exchange time for carbon dioxide between the atmosphere and the sea and the exchange within the sea the best estimates of τ_{a-m} and τ_{d-m} at present are $\tau_{a-m} = 5$ years and $\tau_{d-m} = 500$–1,000 years. FERGUSSON and RAFTER (1958) give a value $\tau_{a-m} = 3$ years. Fig. 1 shows the amount to be expected in the atmosphere 1954, when the total fossil fuel combustion since the middle of the last century is estimated to have been 13.2% of the previous content of the atmosphere, for values of τ_{a-m} and τ_{d-m}

in the vicinity of those quoted above. It first of all shows that *the net increase in the atmosphere is almost independant of the precise rate of exchange between the atmosphere and the sea*. This depends on the fact that the top layer of the ocean only need to absorb a small amount of CO_2 from the atmosphere as compared to the quantities released to be in approximate balance. Its capacity is therefore too small to be of any major importance. The decisive factor is instead the rate of overturning of the deep sea. Thus even using a residence time of only 200 years for the deep sea water an 8–9 percent increase of CO_2 in the atmosphere must have taken place. For a value of $\tau_{d-m} = 500$ years an increase of the atmosphere's content of CO_2 of about 10 percent would have occurred in 1954. This value compares very favourably with the value of 10 percent given by CALLENDAR (1958) as the total increase until 1955 deduced from a careful survey of all available measurements.

The results obtained above are, however, dependant on our assumptions of the size of the three reservoirs, i.e. N_a, N_m and N_d. Of course, the total amount of CO_2 in the atmosphere, N_a is quite well known, but the division of the sea into two layers is somewhat arbitrary. It is of some interest to see how sensitive the solution is to a variation of N_m and N_d. Instead of chosin the values given in (22) we shall assume

$$
\text{a)} \begin{cases} N_m = 0.8\, N_a \\ N_d = 60\, N_a \end{cases} \quad \text{b)} \begin{cases} N_m = 2\, N_a \\ N_d = 59\, N_a \end{cases} \quad (32)
$$

We then obtain the values $n_a = 10.6\%$ and $n_a = 10.0\%$ respectively as compared with $n_a = 10.3\%$ previously, using $\tau_{a-m} = 5$ years and $\tau_{d-m} = 500$ years. Our solution is quite insensitive to the exact division of the ocean. However, the value of the total amount of carbon dioxide in the sea ($\approx 61\, N_a$) effects the solution more, but here we quite accurately know the actual amounts as just mentioned.

4. THE EXCHANGE OF RADIO-CARBON BETWEEN THE ATMOSPHERE AND THE SEA

REVELLE and SÜESS (1957) assume that the exchange of radio carbon between the atmosphere and the sea takes place independantly of changes in the distribution of inactive CO_2. If this were the case the decrease of the C^{14}/C^{12} ratio in the atmosphere since 1850, the Süess effect, would be a direct

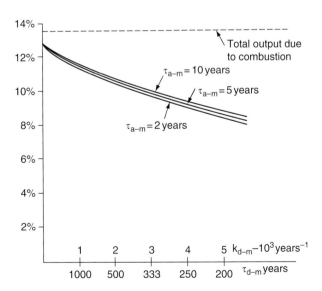

Fig. 1. Net increase of CO_2 in the atmosphere in 1954 due to release of fossil CO_2 according to UN-estimates as dependant on the rate of exchange between the atmosphere and the sea and the mixed layer and the deep sea.

measure of the increase of the inactive CO_2 during the same time, since fossil fuel contains no C^{14}. As a matter of fact, they deduce an exchange coefficient for transfer of CO_2 from the atmosphere to the sea on this bases. As was indicated in the introduction this is not correct. The change of pH in the sea will shift the dissociation equilibrium also for the carbon dioxide containing C^{14}. We may assume an equilibrium rapidly being established and have

$$\frac{N_m^{*'}}{N_m^*} = \chi \frac{N_m'}{N_m} \qquad (33)$$

where χ is dependant on the fractionation. If changes of N_m and N_m' occur as discussed in the previous section, we obtain by variation the following expression for the changes in N_m^* and $N_m^{*'}$

$$\frac{n_m^{*'}}{N_m^{*'}} - \frac{n_m^*}{N_m^*} = \frac{n_m'}{N_m'} - \frac{n_m}{N_m} \qquad (34)$$

Making use of (17) we obtain

$$\left. \begin{aligned} n_m^{*'} &= \frac{n_m^*}{N_m^*} N_m^{*'} + 11.5 \frac{n_m}{N_m} N_m^{*'} \\ &= B_2 n_m^* + B_3 n_m \end{aligned} \right\} \qquad (35)$$

We see here clearly how a change of the total amount of CO_2 in the top layer of the ocean will influence the amount of radioactive carbon in the form of CO_2 and H_2CO_3 and therefore the partial pressure, which will mean that the equilibrium with the atmosphere is disturbed. Actually the amount of C^{14} in the form of undissociated CO_2 is much more influenced by changes in the *total*

amount of CO_2 present in the water than by changes of the amount of *radioactive* carbon dioxide.

Corresponding to the system of equations in (16) we have the following set describing the transfer of radioactive carbon

$$\left. \begin{aligned} \frac{dn_a^*}{dt} &= -k_{a-m}^* n_a^* + k_{m-a}^* n_m^{*'} - \lambda n_a^* \\ \frac{dn_m^*}{dt} &= k_{a-m}^* n_a^* - k_{m-a}^* n_m^{*'} \\ &\quad - k_{m-d} n_m^* + k_{d-m} n_d^* - \lambda n_m^* \\ \frac{dn_d^*}{dt} &= k_{m-d} n_m^* - k_{d-m} n_d^* - \lambda n_d^* \end{aligned} \right\} \qquad (36)$$

Notice here that the exchange coefficients between the atmosphere and the sea are different from those used in equation (16) thereby taking into consideration the fractionation effect. Following CRAIG (1957) we have

$$\frac{k_{m-a}^*}{k_{a-m}^*} = \mu \frac{k_{m-a}}{k_{a-m}} = \mu \frac{N_a}{N_m'} \approx \frac{N_a}{N_m'} \qquad (37)$$

where $\mu = 1/1.012 = 0.988$. The deviation of μ from unity will be completely irrelevant in the following discussion and we will therefore put $\mu = 1$ as indicated by the last expression in (37). We shall also put $k_{a-m}^* = k_{a-m}$ and thus neglect fractionation. For the exchange between the top layer of the ocean and the deep sea the exchange is the same as for inactive carbon with the assumption made that it is due to the motion of sea water only.

Introducing (35) into (36), taking into consideration (15) and (37) and rearranging terms we obtain

$$\left. \begin{aligned} \left(\frac{d}{dt} + k_{a-m} + \lambda\right) n_a^* - B_2 \alpha k_{a-m} n_m^* &= B_3 \alpha k_{a-m} n_m \\ -k_{a-m} n_a^* + \left(\frac{d}{dt} + B_2 \alpha k_{a-m} + \beta k_{d-m} + \lambda\right) n_m^* - k_{d-m} n_d^* &= -B_3 \alpha k_{a-m} n_m \\ -\beta k_{d-m} n_m^* + \left(\frac{d}{dt} + k_{d-m} + \lambda\right) n_d^* &= 0 \end{aligned} \right\} \qquad (38)$$

We find thus that the changes of n_m appear as "driving forces" for the C^{14}-system. Knowing n_m as a function of time, we may calculate the changes that will occur in the distribution of radio carbon

and thus also compute the Süess effect. The problem is implicit, however, since the distribution of n_m depends on k_{a-m} and k_{d-m} as shown in the previous section and our problem is therefore to find the particular pair of values that are in agreement with observed changes in the total amount of CO_2 in the atmosphere *and* the changes of the C^{14}/C^{12} ratio. Eliminating n_m^* and n_d^* from the equations (38) and in doing so taking account of the fact that $k_{a-m} \gg k_{d-m} \gg \lambda$ we obtain

$$\frac{d^3 n_a^*}{dt^3} + [(B_2\alpha + 1)k_{a-m} + (\beta + 1)k_{d-m}]\frac{d^2 n_a^*}{dt^2} + (1 + \beta + B_2\alpha)k_{a-m}k_{d-m}\left(\frac{dn_a^*}{dt} - n_a^*\lambda\right)$$
$$= B_3\alpha k_{a-m}\left[\frac{d^2 n_m}{dt^2} + (\beta + 1)k_{d-m}\left(\frac{dn_m}{dt} + \lambda n_m\right)\right] \tag{39}$$

Now $|n_m^{-1}dn_m/dt| \gg \lambda$ and we shall also assume that our solution n_a^* satisfies the same relation, which will be verified *a posteriori*. It means that the decay of C^{14} is unimportant for the discussion of the exchange between the atmosphere and the sea. We thus finally obtain

$$\frac{d^3 n_a^*}{dt^3} + [(B_2\alpha + 1)k_{a-m} + (\beta + 1)k_{d-m}]\frac{d^2 n_a^*}{dt^2} + (1 + \beta + B_2\alpha)k_{a-m}k_{d-m}\frac{dn_a^*}{dt}$$
$$= B_3\alpha k_{a-m}\left[\frac{d^2 n_m}{dt^2} + (1 + \beta)k_{d-m}\frac{dn_m}{dt}\right] \tag{40}$$

This equation is principally the same as eq. (18) and thus possesses a solution of the character given by (20) where now λ_1^*, λ_2^* and λ_3^* are solutions of an equation similar to (24). From the three equations (38) and with due regard to the expression for dn_m/dt given by eq. (16) we obtain the initial conditions

and similar expressions for n_m^* and n_d^* if we wish to study their variations. Introducing the expression for n_m given by equation (31) and solving for n_a^* we obtain with some simplifications similar to those done previously

$$n_a^* = \frac{dn_a^*}{dt} = \frac{d^2 n_a^*}{dt^2} = 0 \tag{41}$$

$$\frac{n_a^*}{N_a^*} = Q_1\left[e^{rt} - 1 - \frac{r}{\lambda_2^*}(e^{\lambda_2^* t} - 1)\right]$$
$$- Q_2\left[e^{\lambda_2 t} - 1 - \frac{\lambda_2}{\lambda_2^*}(e^{\lambda_2^* t} - 1)\right] \tag{42}$$

where

$$Q_1 = \frac{[r + (1 + \beta)k_{d-m}]H}{r[r^2 + \{(B_2\alpha + 1)k_{a-m} + (1 + \beta)k_{d-m}\}r + (1 + \beta + B_2\alpha)k_{a-m}k_{d-m}]}$$
$$Q_2 = \frac{[(1 + \beta)k_{d-m} + \lambda_2]H}{\lambda_2[\lambda_2^2 + \{(B_2\alpha + 1)k_{a-m} + (1 + \beta)k_{d-m}\}\lambda_2 + (1 + \beta + B_2\alpha)k_{a-m}k_{d-m}]} \right\} \tag{43}$$

having introduced the symbol H and λ_2^* according to

$$H = B_3\alpha\frac{\gamma_0}{r}\frac{k_{a-m}^2(r + k_{d-m})r}{r^2 + [(B_1\alpha + 1)k_{a-m} + (1 + \beta)k_{d-m}]r + (1 + \beta + B_1\alpha)k_{a-m}k_{d-m}} \tag{44}$$

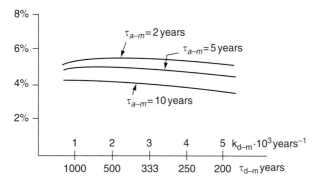

Fig. 2. The increase of C^{14} in the atmosphere in 1954 by a net transfer from the sea resulting from release of fossil CO_2 and a change of the dissociation equilibrium in the sea.

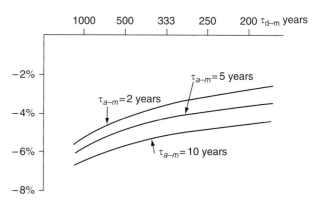

Fig. 3. The expected value of the Süess effect in 1954 for various rates of exchange between the atmosphere and the sea and within the sea.

$$\lambda_2^* = -\frac{(1+\beta+B_2\alpha)k_{a-m}k_{d-m}}{(1+B_2\alpha)k_{a-m}+(1+\beta)k_{d-m}} \qquad (45)$$

Figure 2 shows the values of n_a^* in 1954 when the total output of CO_2 into the atmosphere is estimated to have been 13.2 %. A more pronounced variation with the value of $\tau_{a-m} = k_{a-m}^{-1}$ is here obtained than was the case for n_a. This is easily understandable. The slower the exchange between the atmosphere and the sea takes place the less rapid is the response of the sea to changes of the CO_2 content of the atmosphere. If $\tau_{a-m} = 0$ and $\tau_{d-m} = \infty$ obviously $12.5 \times \left(\dfrac{n_m}{n_a}\right) : \left(\dfrac{N_m}{N_a}\right) = 1$ according to equation (17). Computing n_m from (31) and inserting we obtain here

$$12.5 \cdot \frac{n_m}{n_a} \cdot \frac{N_a}{N_m} = \begin{Bmatrix} 0.77 \\ 0.73 \\ 0.69 \end{Bmatrix} \text{ for } \tau_{a-m} \begin{Bmatrix} 2 \text{ years} \\ 5 \text{ years} \\ 10 \text{ years} \end{Bmatrix} \qquad (46)$$

which values are almost independant of the value of τ_{d-m} in the range 200–1,000 years considered here. The changes of n_m are the "driving force" for changes in the C^{14} system and again the slower the exchange between the sea and the atmosphere is, the greater the lag of the C^{14} adjustment in the atmosphere relative to the sea becomes.

The results obtained from eq. (42) are again quite independant of the exact division of the ocean into two reservoirs. Making the computations with the values of N_a and N_m given in (32) yields values of n_a^* only a few tenths of a percent different from those shown in fig. 2.

By a comparison of the results given by eq. (31) and (42) we can compute the changes of the

C^{14}/C^{12} ratio in the atmosphere, the "Süess effect". The result is shown in fig. 3. It is of special interest to compare these values with actually observed values. Süess' measurements give an average value of -1.7 % which may be considered representative for about 1946 (Revelle and Süess 1957). Broecker and Walton (1959) have found a value of -2.9% for 1938, and a lower value of -1.8% for 1954. The latter sample may, however, already have been influenced by the Castle tests in the Pacific early in 1954. Finally de Vries (1958) gives a value of -2.9% for 1954. Summarizing these measurements a value of -2.5 to -3% seems plausible for the Süess effect in 1954 before any appreciable amounts of C^{14} had been introduced into the atmosphere due to atomic bomb tests. More measurements from all over the world would, however, be desirable to determine this value more accurately. It is seen from fig. 3 that the computed value is somewhat larger particularly if the exchange between the three reservoirs is relatively slow.

Craig (1958) has given the value $\tau_{a-m} = 5$ years as the most likely value for the exchange time between the atmosphere and the sea. It is obtained by studying the difference in the C^{14}/C^{12} ratio in the atmosphere and the sea with due regard taken to fractionation. It is furthermore assumed that the Süess effect is -1.25%, which seems to be an underestimate even if it is true that some of the values later reported perhaps are not truely representative. Chosing a value of -3% would yield a value of about 3 years. However, due to the fact that a net transfer of CO_2 from the atmosphere to the sea occurs at present (N_a+n_a) $(N_m+n_m)^{-1}$ is somewhat larger than would be the case in equilibrium. On the other hand a net transfer of C^{14}

takes place from the sea to the atmosphere and therefore $(N_a^*+n_a^*)(N_m^*+n_m^*)^{-1}$ is smaller than in equilibrium. The deviations are larger the slower the exchange between the atmosphere and the sea takes place. Thus the difference of the C^{14}/C^{12} ratio in the atmosphere and the sea (corrected for fractionation) should be larger in the equilibrium case than in the case where a net transfer occurs as indicated here.

It should be pointed out at this moment that similar deviations of the C^{13}/C^{12} ratios from the equilibrium values may occur if the net transfer of C^{13} and C^{12} are different. This is important since all C^{14} measurements are corrected for fractionation with the aid of the C^{13}/C^{12} ratio. Obviously such a procedure assumes equilibrium conditions. In the case of exchange between the atmosphere and the sea this need not be the case as we have shown above. It so happens, however, that the ratio C^{13}/C^{12} is almost the same in fossil fuel as in the atmosphere, the difference being only about 2%. Thus no significant errors are therefore introduced in assuming equilibrium conditions.

With the model presented above one can now easily compute how large the deviations from equilibrium are which are due to the fact that a net transfer of both inactive and radioactive carbon occurs. The most likely value of the exchange time again becomes $\tau_{a-m} = 5$ years.

Measurements of C^{14} in the deep sea BROECKER (1957) and RAFTER and FERGUSSON (1958) in many cases indicate an age relative surface waters of 500 years or more and the interpretation of these values are not essentially influenced by considerations of the kind presented here. It is thus seen from fig. 3 that the computed value of the Süess effect should be around −5% in comparison with the observed value of about −3%. There may be many explanations for this discrepancy but first of all the accuracy of our model is not greater than it could be explained merely as due to this crudeness. Our assumption of a well-mixed deep sea is of course an unrealistic one and a more complete formulation of the problem in this respect seems very desirable. Secondly, we have completely neglected the effect of the biosphere on land. In view of the relatively small buffering effect of the sea the changes of the CO_2 (as well as $C^{14}O_2$) content of the atmosphere are here computed to be quite large. One may therefore very well expect that the biosphere also is influenced in some way (cf. ERIKSSON and WELANDER, 1956).

5. ESTIMATES OF THE EFFECT OF CO_2 EXCHANGE BETWEEN THE ATMOSPHERE AND THE BIOSPHERE

The amount of carbon stored in the biosphere on land N_b is not very well known. Different estimates give values varying between 12% (CRAIG, 1957) and about 85% (ERIKSSON and WELANDER, 1956) of the amount present in the atmosphere. Estimates of dead organic matter, humus (N_h), also vary considerably, between values of 1.1 N_a to 1.7 N_a. For the following estimates we shall assume $N_b = 0.5 N_a$ and $N_h = 1.5 N_a$.

We may now extend our previous model of the CO_2-exchange in nature to the one depicted in fig. 4, which in case of equilibrium has been studied by CRAIG (1957). In complete anology with the previous analysis we obtain in the equilibrium case

$$\left. \begin{aligned} -k_{a-m}N_a + k_{m-a}N_m' - k_{a-b}N_a & \\ + k_{b-a}N_b + k_{h-a}N_a &= 0 \\ k_{a-b}N_a - k_{b-a}N_b - k_{b-h}N_b &= 0 \\ -k_{h-a}N_h + k_{b-h}N_b &= 0 \\ k_{a-m}N_a - k_{m-a}N_m' - k_{m-d}N_m & \\ + k_{d-m}N_d &= 0 \\ k_{m-d}N_d - k_{d-m}N_d &= 0 \end{aligned} \right\} \quad (47)$$

Again denoting the deviations from equilibrium due to fossil fuel combustion by n_i we obtain

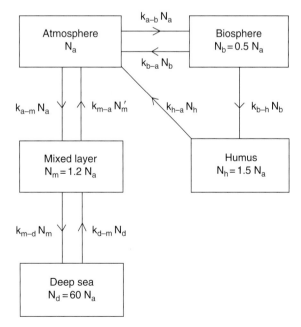

Fig. 4. Model of the CO_2 exchange between the deep sea, the mixed layer of the sea, the atmosphere, the biosphere and humus on land.

$$
\left.
\begin{array}{ll}
\dfrac{dn_a}{dt} + k_{a-m}n_a - k_{m-a}n'_m + k_{a-b}n_a & \\[1ex]
\qquad - k_{b-a}n_b - k_{h-a}n_h & = \gamma(t) \\[2ex]
\dfrac{dn_b}{dt} - k_{a-b}n_a + k_{b-a}n_b + k_{b-h}n_b & = 0 \\[2ex]
\dfrac{dn_h}{dt} + k_{h-a}n_h - k_{b-h}n_b & = 0 \\[2ex]
\dfrac{dn_m}{dt} - k_{a-m}n_a + k_{m-a}n'_m + k_{m-d}n_m & \\[1ex]
\qquad - k_{d-m}n_d & = 0 \\[2ex]
\dfrac{dn_d}{dt} - k_{m-d}n_m + k_{d-m}n_d & = 0
\end{array}
\right\} \quad (48)
$$

It is hardly justified to carry through such a complete analysis here as presented previously in view of the fact that the assumptions made about the exchange processes between the atmosphere, biosphere and humus are more doubtful. An estimate shows, however, that $n_a \approx +7\%$ if we assume $\tau_{a-m} = 5$ years, $\tau_{d-m} = 500$ years, $\tau_{a-b} = = 30$ years and $\tau_{b-h} = 30$ years compared with 10% in the case of no net increased assimilation as a result of increased CO_2 content of the atmosphere.

Obviously the distribution of $C^{14}O_2$ between the various reservoirs also is influenced by an exchange with the biosphere and this would be the case even if no net increase of inactive CO_2 in the biosphere and in the humus has occurred. We can obtain a lower limit for the Süess effect if we assume an infinitely rapid adjustment of the C^{14} content between the four reservoirs, the atmosphere, the mixed layer of the sea, the biosphere and humus, i.e. the Suess effect is the same in all reservoirs and also considering that no C^{14} is supplied from the deep sea to these reservoirs. Since $N_m + N_a + N_b + N_h \approx 4 N_a$ and since the exchange between the mixed layer and the deep sea is comparatively slow so that only a small part $(0.02–0.04\ N_a)$ of the fossil CO_2 released until 1954 $(\approx 0.13\ N_a)$ has found its way into the deep sea we estimate that the Süess effect should be 2–2.5 %. As pointed out previously the observed Süess effect was about –3 % in 1954 which is in very good agreement with this estimate.

6. FORECAST OF THE CO_2 CHANGES IN THE ATMOSPHERE DURING THE REMAINDER OF THE 20TH CENTURY

Certainly the estimates presented above are partly quite uncertain but it is of some interest to see what they imply with regard to future changes of the CO_2-content of the atmosphere. An upper limit is obtained - if neglecting the exchange with the

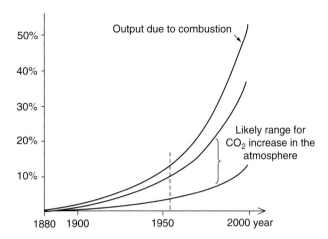

Fig. 5. Estimate of likely range for CO_2 increase in the atmosphere as a result of fossil fuel combustion according to UN estimates.

biosphere as done in section 4. A lower limit on the other hand is determined if an infinitely rapid exchange between the atmosphere and the biosphere takes place and obviously this would be equivalent to putting N_a equal to the sum of the CO_2 found in the atmosphere, the biosphere and the humus. Introducing the values for the various residence times as given in the previous section yields a forecast in between these two extremes. Fig. 5 shows the variations to be expected until year 2000 based on an output of CO_2 into the atmosphere as given by table 1. The most likely value for n_a at that time seems to be about + 25 %, it may possibly be larger but probably not exceed 40 %. These values are considerably larger than those estimated for example by REVELLE and SÜESS (1957). The implications with regard to the radiational equilibrium of the earth in such a case may be considerable but falls outside the scope of this paper.

Fig. 5 also shows that the present increase of CO_2 in the atmosphere probably is 0.1–0.3 % per year. Recent measurements in regions far away from industrial areas such as Hawaii and the Antarctica (personal communication from Rakestraw) show remarkably constant values of the CO_2 content in the atmosphere throughout the year. It should therefore be possible within a few years to observe whether an increase occurs with this computed rate or not.

REFERENCES

Arnold, J. R. and Anderson, E. C., 1957: The distribution of carbon-14 in nature. *Tellus,* **9**, p. 28.

Bray, J.R., 1959: An analysis of the possible recnt change in atmospheric carbon dioxide concentration. *Tellus,* **11**, p. 230.

Broecker, W., 1957: Application of radiocarbon to oceanography and climate chronology. *Thesis Faculty of Pure Science, Columbia Univ.*

Broecker, W. and Walton, A., 1959: The geochemistry of C14 in the fresh water system. *Geochimica et Cosmochimica Acta* **16**, p. 15–38.

Callendar, G. S., 1938: The artificial production of carbon dioxide and its influence on temperature. *Q. J. Roy. Met. Soc.*, **64**, p. 223.

Callendar, G. S., 1958: On the amount of carbon dioxide in the atmosphere. *Tellus*, **10**, p. 243.

Craig, H., 1957: The natural distribution of radio carbon and the exchange time of carbon dioxide between atmosphere and sea. *Tellus*, **9**, p. 1.

Craig, H., 1958: A critical evaluation of radio carbon techniques for determining mixing rates in the oceans and the atmosphere. *Second United Nations International Conference for the Peaceful Uses of Atomic Energy*, Geneva.

Eriksson, E., 1959: The circulation of some atmospheric constituents in the sea. *Rossby Memorial Volume.* Esselte, Stockholm, pp. 147–157.

Eriksson, E. and Welander, P., 1956: On a mathematical model of the carbon cycle in nature. *Tellus*, **8**, 155–175.

Harvey, H. W., 1955: *The chemistry and fertility of sea waters.* Cambridge University Press.

Rafter, T. A. and Fergusson, G. J., 1957: Recent increase in the C14 content of the atmosphere, biosphere and surface waters of the ocean. *N.Z. J. Science and Tech.*, B 38, p. 871.

Rafter, T. A. and Fergusson, G. J., 1958: Atmospheric radio carbon as a tracer in geophysical circulation problems. *Second United Nations International Conference on the Peaceful Uses of Atomic Energy*, Geneva (in press).

Revelle, R. and Süess, H., 1957: Carbon dioxide exchange between atmosphere and ocean and the question of an increase of atmospheric CO_2 during past decades. *Tellus*, **9**, p. 18.

14

Bookends

Keeling, C.D. (1960). The concentration and isotopic abundances of carbon dioxide in the atmosphere. *Tellus*, **XII** (2), 200–203. 5 pages.

Keeling, C.D. (1970). Is carbon dioxide from fossil fuel changing man's environment? *Proceedings of the American Philosophical Society*, **114** (1), 10–17. 8 pages.

Two papers from Keeling, published a decade apart, make a nice set of bookends to the proof that atmospheric CO_2 is rising because of human activity. Keeling was hired by Roger Revelle, following the publication of Revelle's conclusion that humans probably could not influence the CO_2 concentration of the atmosphere because ocean uptake is too quick. Regardless of that belief, Revelle wrote that "human beings are now carrying out a large scale geophysical experiment of a kind that could not have happened in the past nor be reproduced in the future." These sound like grant proposal words, apparently successful ones that stimulated the International Geophysical Year in 1957–8.

The 1960 Tellus paper would be analogous of a snapshot of a baseball batter, just as the ball has been hit by the bat, with a resounding crack that presages a game-changing home run. Atmospheric CO_2 is a noisy quantity to measure, affected by local sources and uptake of CO_2 by living things and machinery. Much of the attention in the paper is focused on making sense of this chaos, by comparing samples from various remote locations such as a Hawaiian mountaintop, remote Antarctica, and samples taken from airplanes. There is a large seasonal cycle from some of the stations, less so in others. In all locations the measurements from the second year are higher than they are in the first, consistent with expectations if the oceans were not taking up CO_2 as quickly as humans were releasing it. Keeling wrote, very tentatively, that "one might be led to conclude that the oceans have been without effect in reducing the annual increase in concentration resulting from the combustion of fossil fuel." One might, unless one is an extremely careful scientist who would prefer to wait for more definitive data.

It was not long in coming. If the 1960 paper is the bat cracking at the baseball, the 1970 paper, a transcript of a spoken talk, is the batter recounting the story of his astonishing feat in the pub later that evening. The story has been polished in its frequent retelling. Fig. 5 shows the recognizable Moana Loa CO_2 curve, one of the most iconic in climate science. Keeling also shows "typical" (i.e., presumably some of the prettiest) short-term concentration records from exemplary locations such Moanna Loa and Antarctica, illustrating the potential for local pollution of the global CO_2 signal, and also the reassuring stability of the method of analysis when times are quiet. The talk muses about the growth of the human footprint on the Earth, with alarm at the potential for its impact on human and environmental well-being. Humans tend to be myopic in time, he notes, in that, for example, all planning seems to stop at the year 2000. This observation is still true, only now we stop at the year 2100.

The Concentration and Isotopic Abundances of Carbon Dioxide in the Atmosphere

Scripps Institution of Oceanography, University of California, La Jolla, California
(Manuscript received March 25, 1960)

ABSTRACT

A systematic variation with season and latitude in the concentration and isotopic abundance of atmospheric carbon dioxide has been found in the northern hemisphere. In Antarctica, however, a small but persistent increase in concentration has been found. Possible causes for these variations are discussed.

The content of carbon dioxide in the atmosphere, in contrast to oxygen, nitrogen, and the rare gases, has been found to be significantly variable (GLUECKAUF, 1951). New data indicate, however, that the degree of variability is smaller and the variations are more systematic than previously believed (STEPANOVA, 1952, SLOCUM, 1955, FONSELIUS ET AL., 1956, CALLENDAR, 1958, BRAY, 1959).

Three gas analysers, as described by SMITH (1953), equipped with strip chart recorders, have been employed to measure the concentration of carbon dioxide continuously at stations in Antarctica, Hawaii, and California. A fourth analyser has been used in the laboratory to analyse samples of air collected in glass flasks at various places. These analysers provide direct comparisons of the partial pressure of carbon dioxide in air relative to that in prepared mixtures of carbon dioxide in nitrogen gas. The concentration of carbon dioxide in these reference mixtures is determined manometrically (KEELING, 1958).

The relative accuracy of the data presented here is approximately ±0.3 p.p.m. (parts per million by volume). The uncertainty in the absolute values is considerably larger, however, since only a preliminary calibration of the reference mixtures has been made. When an accurate manometric calibration is completed, the absolute accuracy is expected to approach ±0.1 p.p.m.

Monthly averages of the data from the continuously recording stations and from collections in flasks are presented numerically in Table 1 and plotted in figures 1 and 2. The locations of the sampling stations and tracks are shown in figure 3.

Local contamination has been found to occur at all three continuous recording stations. At Little America, Antarctica, it was evidently due solely to combustion of fuel in the immediate vicinity of the station. It could be readily spotted from the significant fluctuations in the otherwise steady trace of the recorder pen and was eliminated from consideration in the initial reading of the charts. At Mauna Loa Observatory, Hawaii, a less prominent variability has been found in approximately half of the records. This is attributed to release of carbon dioxide by nearby volcanic vents; combustion on the island associated with agricultural, industrial, and domestic activities; and lower concentration of carbon dioxide in the air transported to the station by upslope winds. The values reported here are averages of data for periods of downslope winds or strong lateral winds when the concentration remained nearly constant for several hours or more. At La Jolla, California, the concentration has been found to be highly variable. Highest concentrations occur during light winds from the north, from the direction of Los Angeles; lowest concentrations when the wind is from the west or southwest and

Contribution from the Scripps Institution of Oceanography, New Series. This paper represents preliminary results of research carried out as part of the International Geophysical Year and its extension. Support was provided by the National Science Foundation, United States Weather Bureau, and the Office of Naval Research.

Table 1. Monthly Average Concentrations of Atmospheric Carbon Dioxide at Various Stations in Parts per Million of Dry Air by Volume.

Month	Year	Continuous Recording Stations			Surface Flask Samples			Flask Samples from Aircraft[2]		
		Little America	Mauna Loa	La Jolla[1]	South Pole	Arctic Ice Floes	Down-wind Cruise	Data at Lower Air temp.	Data at Higher Air temp.	Limiting Air temp. (°C.)
Sept.	1957				311.1	306.5				
Oct.							310.9[1]			
Nov.						310.8	311.9			
Dec.					312.0	313.6	311.6			
Jan.	1958									
Feb.		310.8								
Mar.		311.0	313.4		311.9	315.5				
Apr.		311.4	314.4	314.3					314.9[1]	−27
May		311.7	315.1	315.3						
Jun.		311.9			312.5				314.9[1]	−21
Jul.		311.8	312.9	311.1						
Aug.		313.0	312.3	308.4						
Sept.		312.9	311.6	308.7	313.3			308.3[1]	309.6	−18
Oct.		313.0		310.8				312.8[1]	311.2	−27
Nov.			310.6	313.2						
Dec.			311.6	314.3						
Jan.	1959		312.5	314.9	312.8				313.2	−36
Feb.			313.5	314.4	313.4				314.1	−36
Mar.			314.0	315.4	312.7					
Apr.			314.7	315.0	314.3			316.6[1]	316.6	−27
May			315.3	315.0	314.3				317.3	−24
Jun.			315.2	315.2	314.4			314.3	314.1	−21
Jul.			313.5	311.5	314.2					
Aug.			311.9	308.1	314.2			308.1	310.7	−18
Sept.			311.1	308.7	314.0			311.0	311.1	−27
Oct.			310.5	312.1				313.0	312.1	−27
Nov.			311.8	313.7	313.9			314.2	313.2	−36
Dec.			312.5	313.5				315.6	314.3	−36
Jan.	1960		313.4	314.7				317.6	315.6	−36
Feb.			313.7	315.4					315.6	−36
Mar.			314.4	314.7						

[1] Data not shown in Figures 1 or 2.

[2] Data have been separated into two groups based upon the temperature of the air at the point of sampling. Limiting air temperature refers to the temperature separating lower air temperature from higher. It has been chosen so that, for each period of sampling, the boundary between warmer and colder air lies at approximately 50° North (see figure 3).

of moderate force or greater. Lowest weekly values usually do not differ by more than ±1 p.p.m. during any month, and, within a range of 2 p.p.m., agree with other data for the northern Pacific ocean. Monthly averages of these data, which presumably indicate nearly uncontaminated air, are cited here.

Data for air collected in flasks from aircraft flying at 5 to 6 kilometers above the Pacific and Arctic oceans and from surface stations at the South Pole and on Arctic ice floes show a high degree of regularity.

A clearly defined seasonal trend in concentration is found at all locations in the northern hemisphere. Going from south to north, the annual range of concentration becomes greater and the month of minimum concentration occurs earlier. Separating the samples from aircraft into two groups based upon the temperature of the air at the point of sampling (see table 1), the values for the high latitude, or polar, air are seen to have a greater seasonal range than the comparable values for air of the temperate zone. In contrast, data for the southern hemisphere do not indicate any seasonal variation. Data from Downwind Cruise in November and December, 1957, suggest that the concentrations observed over Antarctica prevail at all southern latitudes of the Pacific ocean.

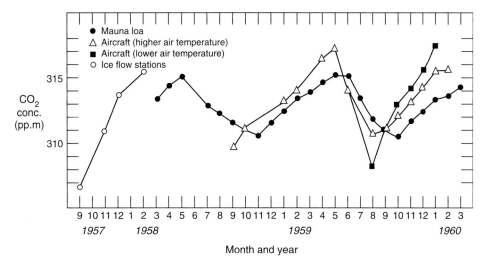

Fig. 1. Variation in concentration of atmospheric carbon dioxide in the Northern Hemisphere.

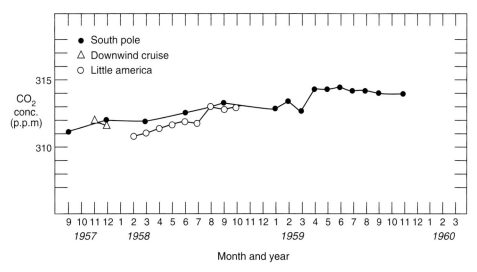

Fig. 2. Variation in concentration of atmospheric carbon dioxide in the Southern Hemisphere.

A seasonal variation in the isotopic abundance of carbon 13 in atmospheric carbon dioxide may also exist (KEELING, 1961). In figure 4 are plotted the observed seasonal variations in isotopic abundance and concentration, based upon samples collected at various stations near the Pacific coast of the United States during 1955 and 1956. Changes in relative abundance of carbon 13 with concentration obey a relationship found for air near plants (KEELING, 1958). The isotopic composition of the carbon dioxide associated with the change in concentration, according to this relationship, is approximately—22 per mil, in good agreement with the value found for air of forests (KEELING, 1958)

and for the carbon of plants growing on land (CRAIG, 1954, WICKMAN, 1952). These data, therefore, indicate that the seasonal trend in concentration observed in the northern hemisphere is the result of the activity of land plants. This interpretation receives further support from the fact that maximum concentrations have been found to occur in spring at the outset of the summer growing season for plants in the temperate zone; that minimum concentrations occur in the fall, approximately at the end of the growing season. The observed absence of a seasonal trend in the southern hemisphere is then to be explained by the smaller area of growing plants found in the southern hemisphere at

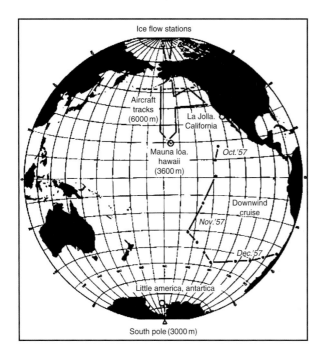

Fig. 3. Location of stations and tracks for sampling of atmospheric carbon dioxide. Circles denote continuous recording stations. Triangles denote flask sampling stations. Elevations in meters are given for locations more than 100 meters above sea level. The approximate mean position of the limiting temperature isotherm, used to separate the aircraft data into groups associated with higher and lower air temperature, is shown by a dotted line.

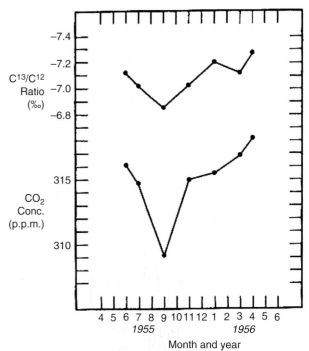

Fig. 4. Variation in concentration and C^{13}/C^{12} ratio of atmospheric carbon dioxide, based on Keeling (1961) for various locations near the Pacific coast of the United States.

temperate and polar latitudes. These conclusions should be considered tentative, however, since they are based on relatively few isotopic measurements.

Where data extend beyond one year, averages for the second year are higher than for the first year. At the South Pole, where the longest record exists, the concentration has increased at the rate of about 1.3 p.p.m. per year. Over the northern Pacific ocean the increase appears to be between 0.5 and 1.2 p.p.m. per year. Since measurements are still in progress, more reliable estimates of annual increase should be available in the future. At the South Pole the observed rate of increase is nearly that to be expected from the combustion of fossil fuel (1.4 p.p.m.), if no removal from the atmosphere takes place (REVELLE and SUESS, 1957). From this agreement, one might be led to conclude that the oceans have been without effect in reducing the annual increase in concentration resulting from the combustion of fossil fuel. Since the seasonal variation in concentration observed in the northern hemisphere is several times larger than the annual increase, it is as reasonable to

suppose, however, that a small change in the factors producing this seasonal variation may also have produced an annual change counteracting an oceanic effect.

ACKNOWLEDGEMENTS

The extensive sampling reported here was made possible through the generous cooperation of the United States Weather Bureau whose personnel carried out the sampling at Little America, the South Pole, Mauna Loa Observatory, and Arctic ice floes "A" and "B"; the 55th Weather Reconnaissance Squadron of the United States Air Force who collected the samples from aircraft; and Dr. Norris W. Rakestraw and Mr. Lee Waterman who collected samples from Downwind Cruise. Special thanks are due Messrs. Charles E. Williams and Jack C. Pales who operated the continuous recording instruments and who reduced the data to preliminary tabular form at Little America and Mauna Loa Observatory, respectively; and to Maj. George T. McClelland, Maj. Bernard M. Rose, Lieut. H. T. Fukuda, and Capt. J. D. Sharp for special effort in planning and executing the sampling from aircraft.

REFERENCES

Bray, J. R., 1959: An analysis of the possible recent change in atmospheric carbon dioxide concentration. *Tellus* II, pp. 220–230.

Callendar, G. S., 1958: On the amount of carbon dioxide in the atmosphere. *Tellus* 10, pp. 243–248.

Craig, H., 1954: Carbon-13 in plants and the relationship between carbon-13 and carbon-14 variations in nature. *J. Geol.* 62, pp. 115–149.

Fonselius, S., Koroleff, F., Wärme, K. E., 1956: Carbon dioxide variations in the atmosphere. *Tellus* 8, pp. 176–183.

Glueckauf, E., 1951: The composition of atmospheric air. *Compendium of Meteorology*, Amer. Meteorol. Soc., Boston, pp. 3–10.

Keeling, C. D., 1958: The concentration and isotopic abundances of atmospheric carbon dioxide in rural areas. *Geochim. et Cosmochim. Acta* 13, pp. 322–334.

Keeling, C. D., 1961: The concentration and isotopic abundances of carbon dioxide in rural and marine air. *Geochim. Cosmochim. Acta*, 24, pp. 277–298.

Revelle, R., Suess, H. E., 1957: Carbon dioxide exchange between atmosphere and ocean and the question of an increase of atmospheric CO_2 during the past decades. *Tellus* 9, pp. 18–27.

Slocum, G., 1955: Has the amount of carbon dioxide in the atmosphere changed significantly since the beginning of the twentieth century? *Month. Weath. Rev.* 83, pp. 225–231.

Smith, V. N., 1953: A recording infrared analyser. *Instruments* 26, pp. 421–427.

Stepanova, N. A., 1952: A selective annotated bibliography on carbon dioxide in the atmosphere. *Meteor. Abs. Bibl.* 3, pp. 137–170.

Wickman, F. E., 1952: Variations in the relative abundance of the carbon isotopes in plants, *Geochim. et Cosmochim. Acta* 2, pp. 243–254.

Is Carbon Dioxide from Fossil Fuel Changing Man's Environment?

CHARLES D. KEELING

Professor of Oceanography, Scripps Institution of Oceanography
(Read April 25, 1969, in the Symposium on Atmosphere Pollution: Its Long-term Implications)

I ORIGINALLY proposed as the title of this talk: "If carbon dioxide from fossil fuels is changing man's environment, what will we do about it?" It was my meaning to inquire into what might be the response of scientists, philosophers, and decision-makers if specialists assert that accelerated use of fossil fuels may be harmful. I was requested to modify the title to read "Is CO_2 from fossil fuel changing man's environment?" either because a shorter title might suggest a shorter, more acceptable talk, or because I obviously cannot answer the first question. I cannot answer the second question either; but I will now give you my views on both questions.

Atmospheric carbon dioxide as a factor in man's environment has been so clearly described by Roger Revelle in a report which he prepared in 1965 for the President's Science Advisory Committee that I shall not attempt to improve on his words (Revelle, 1965: p. 112):

Only about one two-thousandth of the atmosphere and one ten-thousandth of the ocean are carbon dioxide. Yet to living creatures, these small fractions are of vital importance. Carbon is the basic building block of organic compounds, and land plants obtain all of their carbon from atmospheric carbon dioxide....

Over the past several billion years, very large quantities of carbon dioxide have entered the atmosphere from volcanoes. The total amount was at least forty thousand times the quantity of carbon dioxide now present in the air. Most of it...was precipitated on the sea floor as limestone or dolomite. About one-fourth of the total quantity, at least ten thousand times the present atmospheric carbon dioxide, was reduced by plants to organic carbon compounds and became buried as organic matter in the sediments. A small fraction of this organic matter was transformed into the concentrated deposits we call coal, petroleum, oil shales, tar sands, or natural gas. These are the fossil fuels that power the world-wide industrial civilization of our time.

Throughout most of the...[million or so] years of man's existence on earth, his fuels consisted of wood and other remains of plants which had grown only a few years before they were burned. The effect of this burning on the content of atmospheric carbon dioxide was negligible, because it only slightly speeded up the natural decay processes that continually recycle carbon from the biosphere to the atmosphere. During the last few centuries, however, man has begun to burn fossil fuels that were locked in the sedimentary rocks,...and this combustion is measurably increasing the atmospheric carbon dioxide.

The present rate of production from fossil fuel combustion is about a hundred times the...[rate of natural removal by] weathering of silicate rocks....Within a few short centuries, we are returning to the air a significant part of the carbon that was slowly extracted by plants and buried in the sediments during half a billion years.

Not all of this added carbon dioxide will remain in the air. Part of it will become dissolved in the ocean, and part will be taken up by the biosphere, chiefly in trees and other terrestrial plants, and in the dead plant litter called humus. The part that remains in the atmosphere may have a significant effect on climate: carbon dioxide is nearly transparent to visible light, but it is a strong absorber and back-radiator of infrared radiation, particularly in the wave lengths from 12 to 18 microns; consequently, an increase of atmospheric carbon dioxide could act, much like the glass in a greenhouse, to raise the temperature of the lower air.

The annual input to the atmosphere of CO_2 from fossil fuels can be estimated within an accuracy of about 20 per cent from records of the production of coal, petroleum, and the other fossil fuels (Revelle, 1965: p. 116). The rate of input has risen rapidly since 1850 with a doubling time of fifteen to twenty years except for a noticeable slowing during the economic depression, 1930–1940, and during the two world wars. Table 1 shows a sample of the data expressed as the rate of increase in

Keeling, C. D. (1970) Is carbon dioxide from fossil fuel changing man's environment. *Proceedings of the American Philosophical Society* 114 (1): 10–17. Reprinted with permission of American Philosophical Society.

concentration of atmospheric CO_2 in parts per million of dry air (p.p.m.) per year, neglecting any possible removal.

Since 1957 the Scripps Institution of Oceanography, in cooperation with the Environmental Sciences Services Administration of the United States government has monitored the increase in atmospheric CO_2 at two remote stations, Mauna Loa Observatory, Hawaii, and the South Pole, to find out what proportion of the CO_2 from fossil fuel is accumulating in the atmosphere. I shall now briefly describe these measurements (Pales and Keeling, 1965; Brown and Keeling, 1965).

Air is sucked by a diaphragm pump through aluminum tubing from a mast upwind from local sources of CO_2. The air then passes through a trap to remove water vapor and into the cell of an infrared analyzer, where the concentration is compared sequentially with a reference gas to a precision of 0.2 p.p.m. The data, here expressed as 60-minute averages, are for the South Pole monotonously steady (fig. 1) but at Mauna Loa sometimes show locally produced volcanic CO_2 or effects of vegetation 20 kilometers away (fig. 2).

Table 1.

Decade or Year	Rate of input of fossil fuel CO_2 (ppm y^{-1})
1860–69	0.06
1880–89	0.14
1900–09	0.33
1940–49	0.72
1950	0.87
1966	1.77
1970 (estimate)	2.0

Only the steady portions of the record are used in the subsequent analysis.

Over a year both stations show a cyclic pattern owing to the seasonal variation in plant activity. At the South Pole the amplitude of oscillation is less than 2 p.p.m. (fig. 3) while at Mauna Loa it is nearly 6 p.p.m. (fig. 4). At the South Pole the oscillation is so small that it does not mask an increase of about 0.7 p.p.m. owing to the fossil fuel input. From ten years of a nearly uninterrupted record at Mauna Loa we clearly see an increase, superimposed on a remarkably regular annual pulse (fig. 5). We have not been able to separate the cyclic oscillation and secular trend with complete certainty, but by assuming that the cyclic portion repeats exactly each year, a variety of assumed trend functions all yield substantially the pattern shown in the figure (Bainbridge, 1969).

A surprising feature of the Mauna Loa record is the apparent falling off of the slope of the trend during a period when the rate of CO_2 input was increasing (fig. 6). Also surprising is the result of a calculation of the behavior of surface ocean water based on the simple but quite realistic assumption that the CO_2 is entering the oceans at a rate proportional to the CO_2 partial pressure difference produced at the sea surface. A plot of the difference between input and accumulation is nearly linear as though the surface water had maintained a constant CO_2 partial pressure of 311 p.p.m. since 1958.

These features suggest that the surface layer of ocean water has not been the principal agent for removing fossil fuel CO_2 from the air, at least during the past few years. In any case, the surface layer of ocean water is insufficient to hold more than about 10 per cent of the fossil fuel CO_2 which has left the

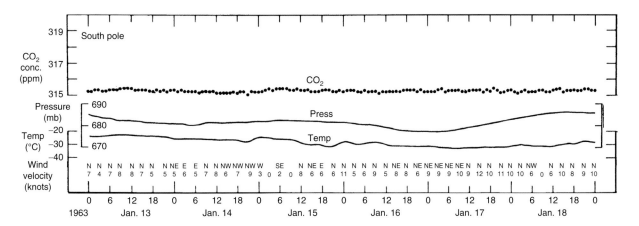

Fig. 1. A typical record of the hourly average concentration of atmospheric CO_2 at the South Pole versus universal (Greenwich) time.

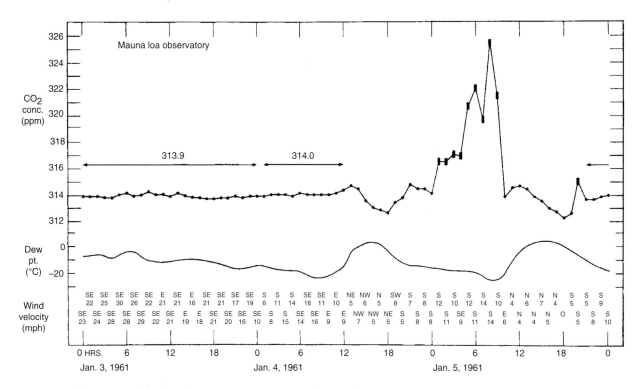

Fig. 2. An illustration of the hourly average concentration of atmospheric CO_2 at Mauna Loa Observatory versus local time. From the beginning of the plot until 12 hr. on January 4 the concentration is steady. During the afternoons of January 4 and January 5 the concentration dips owing to the uptake by vegetation. During the early hours of January 5 the concentration was high and variable owing to volcanic contamination.

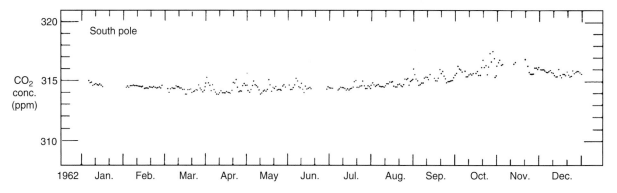

Fig. 3. The daily average concentration of atmospheric CO_2 versus time at the South Pole.

atmosphere; because the depth of the layer which maintains good contact with the atmosphere is shallow, and because the CO_2 pressure of sea water rises nearly ten times as rapidly as the concentration of total carbon when CO_2 is added to the solution (Bolin and Eriksson, 1959: p. 133; their value of 12.5 has been corrected for the influence of boric acid).

It is thus of little consequence to the overall problem of disposing of CO_2 from fossil fuel whether this surface layer has, in fact, taken up CO_2 since 1958 or not—the principal sink for CO_2 must lie in the deep oceans or not be oceanic at all. Our present analysis cannot go much farther: we do not, at the present time, understand how the deep ocean can effectively remove such a large flux of CO_2 because about one thousand years are required for deep water to be renewed by exchange with surface water.

Perhaps land plants, by growing more rapidly by a process plant physiologists call "CO_2 fertilization," are now removing this excess CO_2. But we must put into our balance sheet the large-scale

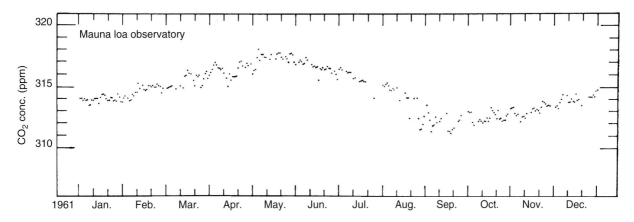

Fig. 4. The daily average concentration of atmospheric CO_2 versus time at Mauna Loa Observatory.

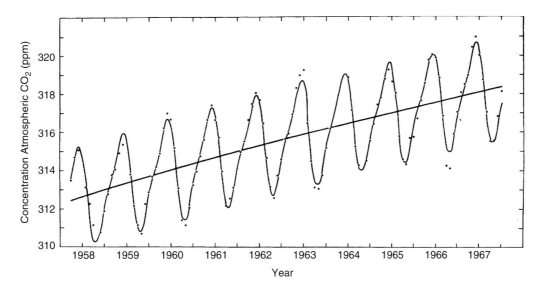

Fig. 5. Long-term variations in the concentration of atmopheric CO_2 at Mauna Loa Observatory. The dots indicate the observed monthly average concentrations. The oscillating curve is a least squares fit of these averages based on an empirical equation containing 6 and 12 month cyclic terms and a trend function. The slowly rising curve is a plot of the trend function, chosen to contain powers of time up to the third.

alteration of virgin lands by cultivation and forest fires, which, since 1900, have released so much CO_2 that a net removal of carbon from the land biosphere seems as likely to have occurred as a net increase (Hutchinson, 1954: p. 390). Thus the true situation is not clearly established, and we are unable, as yet, to predict from our knowledge of the carbon cycle the redistribution of CO_2 from fossil fuel over the coming decades and centuries. From the record so far it appears that about 40 per cent of the CO_2 from fossil fuel is remaining in the atmosphere (fig. 7). Until we obtain better information, I suggest using this figure as a reasonable empirical factor for prediction.

Concerning the apparent anomaly that a larger proportion of the CO_2 from combustion has been removed in 1965–1967 than in 1958–1960 (see fig. 6), our record is still too short to decide whether the anomaly is real.

I will now discuss the possible climatic response to an increase in atmospheric CO_2, passing by some calculations based on static models of the atmosphere (Plass, 1956; Kaplan, 1960; Möller, 1963) to consider the calculations of Manabe and coworkers (Manabe and Strickler, 1964; and Manabe and Wetherald, 1967) which, for the first time, take into account all the major atmospheric factors except circulation. The model

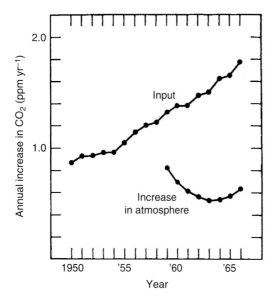

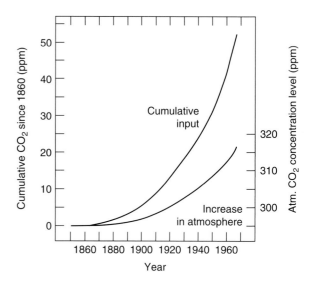

Fig. 6. *Upper curve*: The annual input of fossil fuel CO_2 in the atmosphere versus time expressed as an annual increase in concentration of atmospheric CO_2 neglecting any possible removal. *Lower curve*: The annual increase of atmospheric CO_2 according to the trend function shown in figure 5.

Fig. 7. An estimate of the input of fossil fuel CO_2 into the atmosphere and the increase of atmospheric CO_2 since the mid-nineteenth century.

calculates the state of thermal equilibrium as the asymptotic steady state approached in an initial value problem. Water vapor; ozone; low, middle, and high clouds; and albedo are considered as well as CO_2.

If all the CO_2 were to vanish from the atmosphere, the earth's surface temperature would be reduced substantially: 10 to 20° C, depending on whether the water vapor content remains constant, or decreases to preserve the same relative humidity. If the amount of CO_2 were to rise by a factor of 2, from 300 to 600 p.p.m., the change would be less dramatic but still climatically significant: 2.8° C rise in temperature if the relative humidity is fixed. These calculations are not accurate predictions, however, because they are based on an assumption that other environmental factors do not change when the temperature changes by the predicted amount. The model is especially sensitive to changes in clouds. For example, a 30 p.p.m. increase in CO_2 (approximately what has occurred since 1850) would be compensated for by a 0.3 per cent increase in low clouds or a 0.6 per cent increase in middle clouds. How the amount of clouds would in reality change in response to warming triggered by a CO_2 increase is still a guess, because we lack any dynamic model to predict clouds. Consequently, we are left without a clear prediction. Nevertheless, I believe that

no atmospheric scientist doubts that a sufficiently large change in atmospheric CO_2 would change the climate: we need only compare our atmosphere with the very hot CO_2-laden atmosphere of Venus to guess the consequences of an unrestricted CO_2 increase. We just do not yet know what increase in CO_2 the earth's atmosphere will accept without warming noticeably.

My assessment of the situation is that the increase in CO_2 is of no special concern to our immediate well-being. The rise in CO_2 is proceeding so slowly that most of us today will, very likely, live out our lives without perceiving that a problem may exist.

But CO_2 is just one index of man's rising activity today. We have rising numbers of college degrees, rising steel production, rising costs of television programming and broadcasting, high-rising apartments, rising numbers of marriages, relatively more rapidly rising numbers of divorces, rising employment, and rising unemployment. At the same time we have diminishing natural resources, diminishing distraction-free time, diminishing farm land around cities, diminishing virgin lands in the distant countryside. The event described in the following newspaper quotation could not have occurred fifteen years ago (*Los Angeles Times*, Oct. 27, 1968):

SAN DIEGO—Memorial services will be held Monday for Dr. Robert W. Pidd, ranking authority on thermionics and senior scientist at Gulf General Atomics.

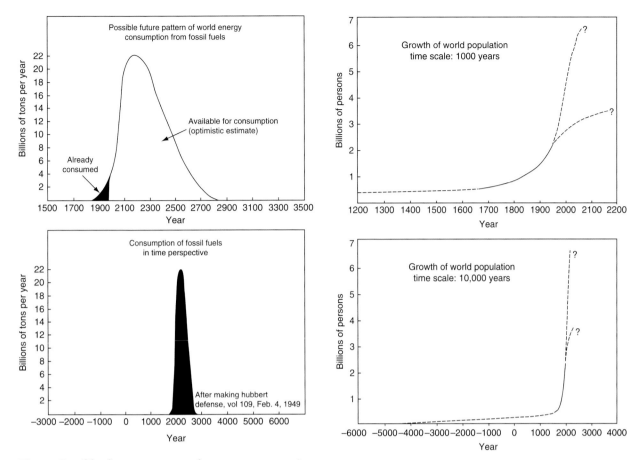

Fig. 8. Possible future patterns of energy consumption from fossil fuels. Two time scales are shown (from Brown, 1954, p. 169).

Fig. 9. Growth of world population on two time scales (from Brown, 1954, p. 49).

Dr. Pidd, 47, was killed Friday in a one-car traffic accident while driving to Borrego Springs. His family said he was going to the desert resort to escape recent heavy smog in San Diego that had aggravated an asthmatic condition.

Now it just happened that the very same day I drove my family to the desert to escape that smog attack. My judgment of the seriousness of the attack was supported by measurements of atmospheric CO_2 at the Scripps Institution of Oceanography which indicate that during the past year La Jolla has sometimes enjoyed levels of contaminated air comparable to severe smog in Los Angeles. Severe air pollution is new to La Jolla. During two years of nearly continuous measurements of CO_2 at Scripps during 1959–1961, the CO_2 level never reached the concentrations observed recently. The smog build-up at La Jolla is the inevitable consequence of more and more people driving automobiles in the still rapidly expanding metropolitan area of Los Angeles, 100 miles to the northwest of La Jolla.

It may be instructive to compare plots of past and predicted increases in CO_2 and population (figs. 8 and 9). I am struck by the the obvious transient nature of the CO_2 rise. The rapid changes in all factors I mentioned, including the rapid rise in world population, are probably also transient; these changes, so familiar to us today, not only were unknown to all but the most recent of our ancestors but will be unknown to all but the most immediate of our descendants.

Thus we are privileged to live in exciting times. For many scientists it has been a period of unprecedented opportunity. Burgeoning populations bring about greater interaction of man and nature. By observing the stimulating and powerful new connections of man with his total environment, indeed even occasionally by manipulating nature deliberately and scientifically, mankind may gain far-reaching insight into nature's secrets.

For example Thomas Jukes has felt this excitement. While a research biochemist at American Cyanamid Company, Jukes (1963: p. 335) wrote:

The story of the world-wide population increase...is a tale of events that have spread through all countries; a story of the changes wrought by science and technology; of the effects of...public health education, farm machinery, vitamins, agricultural genetics, fertilizers, synthetic fibers, and, above all, pesticides. It concerns millions of people whose hopes and aspirations have been increased....

The insecticide DDT...has probably had a greater effect on disease and hunger than has any other man-made chemical substance...it has killed billions of insects.... The sociological effects have been far-reaching, including reclamation of land for agriculture and urban expansion, decreases in absenteeism, higher earnings and improved economic status. The population of Madagascar, which had been stationary for many years, doubled between 1947 and 1959 following the initiation of an antimalaria campaign in 1949.

Jukes (1963: p. 360) states further:

The best hope for coping with the need for food throughout the world lies in extending the superb agricultural technology of the United States into use in other countries...the issue is not one that merely involves 2 per cent of the sales of the chemical industry; at stake is no less than the protection of the the free world from hunger and disease.

Coming closer to my own profession of oceanography is the following quotation from an editorial by Athelstan F. Spilhaus (1964: p. 993):

Man *is* going to colonize the oceans, and it might just as well be *our* men. To compete successfully, we must be able to move faster in the sea, to go deeper, to stay down longer, than anyone else. We must understand more about the sea....

Already phosphorites and diamonds are being taken commercially from the shelves in the sea, and nodules may be an inexhaustible source of other minerals in the future.

This quotation may, indeed, be one of the last times a scientist ever speaks of inexhaustible resources.

In our academic establishments enthusiasm may be tempered by the reflective powers of the scholarly mind. Kenneth Boulding (1964: p. 18) in a book on *The Great Transition*, as he calls it, states:

Attitudes toward the transition can range from rejection, through a grudging acceptance, to a cautious and critical acceptance, and to an enthusiastic and uncritical acceptance... I take my stand somewhere about the third of these positions. I welcome the transition as an event of enormous evolutionary potential in line with the general development of the universe as we know it.

These scientists who, I might say, are intoxicated with the success of twentieth-century science, show a deep faith that science and science-led technology may pull us out of any environmental predicament.

There also exist scientists who are alarmed, especially the ecologists whose natural environment is being so transformed recently that many find their professional work hindered. They are joined by hikers, fishermen, and other throw-backs to Cro-Magnon man who see recreational land disappearing and by scholars like Harrison Brown who, without obvious personal motives, nevertheless register concern.

For example, F. E. Egler (1964: p. 119), a sarcastic ecologist, calls the recent ecological changes such as the CO_2 increase, a Revolution in the Environment. He believes that "in general we have acted with remarkable arrogance to the whole nature of which we are a part" in allowing this revolution to take place without restraint. Now we must direct this revolution so

that we here and after thrive and survive in it. The problem is not one of growing and selling at a suitable profit this year's surplus crop of wheat. The problem is not in destroying the rats in your cellar, the cockroaches in your kitchen, the mosquitoes on your lawn, or even the bats in your belfry. The problem is not in growing timber on a 70-year rotation. The problem is not in saving your child from a lethal disease, or the unreasonable perpetuation of your own life. The problem is not in the continuing growth of our national economy. The problem is not in the balance of payments for a nation, or whether that nation and all nations are to survive under communism or under capitalism. The problem is whether the entire earth, man and the environment he is modifying can persist successfully and happily, generation after generation, meaningfully, without man either destroying himself, or losing himself in a whirling dervish of economic prosperity, linked with individual nonentity made tolerable only by perpetual tranquilization.

Harrison Brown (1954: p. 222) writes:

Our ancestors had available large resources of high-grade ores and fuels that could be processed by the most primitive technology—crystals of copper and pieces of coal that lay on the surface of the earth, easily mined iron, and petroleum in generous pools reached by shallow drilling. Now we must dig huge caverns and follow seams ever further underground, drill oil wells thousands of feet deep, many of them under the bed of the ocean, and find ways of extracting elements from the leanest of ores—procedures that are possible only because of our highly complex modern techniques, and practical only to an intricately mechanized culture which would not have been developed without the high-grade resources that are so rapidly vanishing.

... We are quickly approaching the point where, if a machine civilization should, because of some catastrophe, stop functioning, it would probably never again come into existence.

Finally, we might consider the view of college students. Here is a sample from the Berkeley campus of the University of California (Luten, 1964: p. 45):

The image which comes to my mind as applicable today is that of a Kafka-like toboggan, running down a slope at ever-increasing speed. Most of the passengers are completely unaware that the slope is becoming steeper; in the front, the official drivers are too busy quarreling over possession of the steering-bar to notice anything at all. Here and there among the passengers are a few individuals who recognize the danger. Some of these, convinced that a precipice lies ahead, shrilly exhort the others to "Turn! Turn!"—how, they cannot say, and usually they indicate a direction somewhere along the receding track. Others knowledgeable in toboggan construction, offer wise expedients to hold the vehicle together for a few moments longer, in the hope that the slope will level off.

Whom shall we heed? The sober individuals with the bailing wire, just emerged from conference, speak with authority as they point out that, although the slope is becoming steeper, it cannot yet be considered a precipice. But it looks rather like a precipice to us, and we have just remembered that it was they who a few miles back told us how to grease the runners when we so wanted to feel the rush of crisp, winter air.

I recite these diverging points of view to illustrate that we hold widely divergent views concerning a possible peril. Have you noticed that practically all master plans do not project beyond the year 2000 A.D.? Our college students, however, today expect, or at least nourish the hope, to live beyond that date, and I predict that they will be the first generation to feel such strong concern for man's future that they will discover means of effective action. This action may be less pleasant and rational than the corrective measures that we promote today, but thirty years from now, if present trends are any sign, mankind's world, I judge, will be in greater immediate danger than it is today, and immediate corrective measures, if such exist, will be closer at hand. If the human race survives into the twenty-first century with the vast population increase that now seems inevitable, the people living then, along with their other troubles, may also face the threat of climatic change brought about by an uncontrolled increase in atmospheric CO_2 from fossil fuels.

REFERENCES

Bainbridge, Arnold E. 1970. "The Secular Increase of Atmospheric Carbon Dioxide." To be submitted to *Jour. Geophy. Res.*

Bolin, Bert, and Erik Eriksson. 1958. "Changes in the Carbon Dioxide Content of the Atmosphere and Sea Due to Fossil Fuel Combustion." *Rossby Memorial Volume* (edited by B. Bolin, Rockefeller Institute Press, New York, 1959), pp. 130–142.

Boulding, Kenneth E. 1964. *The Meaning of the Twentieth Century—The Great Transition* (New York, Harper and Row).

Brown, Craig W., and Charles D. Keeling. 1965. "The Concentration of Atmospheric Carbon Dioxide in Antarctica." *Jour. Geophy. Res.* **70**: pp. 6077–6085.

Brown, Harrison. 1954. *The Challenge of Man's Future* (New York, Viking Press).

Egler, Frank E. 1964. "Pesticides—in our Ecosysem." *American Scientist* **52**: pp. 110–136.

Hutchinson, G. Evelyn. 1954. "The Biochemistry of the Terrestrial Atmosphere. *The Earth as a Planet* (chap. 8, edited by G. Kuiper, U. of Chicago Press), pp. 371–433.

Jukes, Thomas H. 1963. "People and Pesticides." *American Scientist* **51**: pp. 355–362.

Kaplan, Lewis D. 1960. "The Influence of Carbon Dioxide Variations on the Atmospheric Heat Balance." *Tellus* **12**: pp. 204–208.

Luten, Daniel B. 1964. "Numbers Against Wilderness." *Sierra Club Bull.* **49**: pp. 43–48.

Manabe, Syukuro, and Robert F. Strickler. 1964. "Thermal Equilibrium of the Atmosphere with a Convective Adjustment." *Jour. Atm. Sci.* **21**: pp. 361–385.

Manabe, Syukuro, and Richard T. Wetherald. 1967. "Thermal Equilibrium of the Atmosphere with a Given Distribution of Relative Humidity." *Jour. Atm. Sci.* **24**: pp. 241–259.

Möller, F. 1963. "On the Influence of Changes in the CO_2 Concentration in Air on the Radiation Balance of the

Earth's Surface and on the Climate." *Jour. Geophy. Res.* **68**: pp. 3877–3886.

Pales, Jack C., and Charles D. Keeling. 1965. "The Concentration of Atmospheric Carbon Dioxide in Hawaii." *Jour. Geophy. Res.* **70**: pp. 6053–6076.

Plass, Gilbert N. 1956. "The Carbon Dioxide Theory of Climatic Change." *Tellus* **8**: pp. 140–154.

Revelle, Roger. 1965. "Atmospheric Carbon Dioxide, Appendix Y4 of Restoring the Quality of Our Environment, a Report of the Environmental Pollution Panel." *President's Science Advisory Committee, The White House*, pp. 111–133.

Spilhaus, Athelstan F. 1964. "Man in the Sea." *Science* **145**: p. 993.

15

One If by Land

Bolin, B. (1977). Changes of land biota and their importance for carbon cycle. *Science*, **196** (4290), 613–615. Times Cited: 174. 3 pages.

Tans, P.P., Fung, I.Y. & Takahashi, T. (1990). Observational constraints on the global atmospheric CO_2 budget. *Science*, **247**, 1431–1438.

Cox, P.M., Betts, R.A., Jones, C.D., *et al.* (2000). Acceleration of global warming due to carbon-cycle feedbacks in a coupled climate model. *Nature*, **408** (6809), 184–187. 3 pages.

The rise in atmospheric CO_2 concentration each year can be accounted for by human emissions of carbon; in fact, we emit more than enough to explain the new CO_2 every year. The problem has been to figure out where the extra CO_2 is going. The two possibilities are into the land surface, the trees and soils and elephants and so on, and the oceans.

Carbon storage on land, both to date and into the future, depends strongly on human land use practices. A climax forest holds a lot of carbon per acre of land. It probably releases carbon in the fall and winter and takes it up in the spring, but averaged over a year the forest will not be a net carbon source or sink if the average density of trees and concentration of carbon in soils have reached steady, climax values. If you cut down a forest to plant crops, there is a net source of carbon to the atmosphere from the decomposition or the burning of the trees and degradation of carbon compounds in the soil. Bolin was among the first to identify deforestation, which today is taking place mainly in the tropics, as a source of CO_2 to the atmosphere, not as much as fossil fuel combustion, but a significant addition to the story.

The land surface can also take up carbon, in places that have not been deforested. The reason for this is not clear; it could be that a longer growing season encourages plants to grow, or the higher CO_2 concentration in the air. It could be nitrate deposition from acid rain acting as a fertilizer, or even suppression of forest fires building up carbon stocks. Balancing the other known sources and sinks of carbon, it appears that the land surface is taking up about 2 billion metric tons of carbon per year, several times more than the mass of the entire human race. One would think it would be easy to find this much stuff accumulating, but it turns out that the best way to track it down is by paying attention to the concentrations of CO_2 in the air as the winds blow around, an approach pioneered by Tans *et al.* The last word has not been written, but most scientists believe that the so-called "missing sink" for carbon is located in the high northern latitudes on land.

The land surface is therefore acting as a natural carbon sink, at least in places where we are not cutting it down, a stabilizing negative feedback to our CO_2 emissions. However, this could change as demonstrated by Cox *et al.* As the Earth warms, the organic carbon in soils decomposes more quickly. Soil carbon concentrations are much lower in tropical soils than they are in temperate and high latitudes, and so if we expand the tropics by warming the planet, it could provoke the land surface to release carbon, amplifying rather than ameliorating the human-driven atmospheric CO_2 rise.

Changes of Land Biota and Their Importance for the Carbon Cycle

The increase of atmospheric carbon dioxide may partly be due to the expansion of forestry and agriculture

BERT BOLIN

It is well established that the carbon dioxide content of the atmosphere is increasing. Observations at the South Pole, on Hawaii (1), and in the upper troposphere of the Northern Hemisphere (2) are available for almost 20 years. The increase has been about 16 parts per million (ppm)—from 311 ppm in 1957 to 327 ppm in 1975—that is, about 5 percent (3). Attempts to estimate the probable increase since the beginning of the industrial revolution have yielded a value between 30 and 35 ppm—more than 10 percent. The major cause of this increase is undoubtedly emission of carbon dioxide into the atmosphere from the combustion of fossil fuels. An accurate compilation of the total output of carbon dioxide from burning fossil fuels and also from the cement industry has been made by Keeling (4). Keeling defines the airborne fraction as the ratio of the annual increase of carbon dioxide in the atmosphere to the annual output, which has varied between 20 and 80 percent during the last 20 years, the average being close to 50 percent. A considerable transfer of carbon dioxide therefore must have occurred into other natural reservoirs, and it seems likely that the oceans have played the most important role in this regard (5). Increased assimilation by land biota cannot be excluded, however, since it is well established that plants grow faster in atmospheres richer in carbon dioxide. Keeling (6) analyzed this question carefully and also concluded that land biota may have served as a sink, but this cannot be proved by direct measurements. It is possible that the slightly more acid seawater now present in the surface layers of the sea due to a net transfer of carbon dioxide into the oceans may have brought some calcium carbonate into solution. The sea may thus have been a more effective sink

for carbon dioxide than previously considered. This problem remains unresolved.

Humans are, however, intervening in the carbon cycle in more ways than by burning fossil fuels. Deforestation and the cultivation of land for agricultural purposes are examples of major changes in the land biota that may well have had significant implications for the global carbon cycle. Data on these changes are few, and it is difficult to assess precisely what may have happened during the last decades or centuries. I will therefore limit myself to some rather rough estimates, which will show, however, that these changes imply a significant decrease of the amount of carbon in organic matter on land during the last century. Thus humans have also been adding carbon dioxide to the atmosphere in this way.

As a background, the present (that is, 1975) carbon dioxide concentration in the atmosphere, 327 ppm, corresponds to a carbon content of about 690×10^9 tons. The corresponding figure at the beginning of the industrial revolution has been estimated as 610 to 620×10^9 tons. Thus, the amount of carbon in the atmosphere has probably increased by 70 to 80×10^9 tons. These values should be compared with an accumulated output due to fossil fuel combustion and cement production of about 140×10^9 tons, of which about half has been emitted during the last two decades.

CHANGES IN THE WORLD'S FORESTS

In recent years several attempts have been made to assess the net primary production and the total mass of land biota. Table 1 is an extract of a summary presented by Whittaker and Likens (7), which is probably the best estimate available so

Bolin, B. (1977) Changes of land biota and their importance for the carbon cycle. *Science* 196 (4290): 613–615.
Reprinted with permission from AAAS.

Table 1. Net primary production and biomass of land biota (7).

Type	Area (10⁶ km²)	Total net production (10⁹ ton/year)	Total biomass (10⁹ tons)
Tropical rain forest	17.0	15.3	340
Tropical seasonal forest	7.5	5.1	120
Temperate forest, evergreen	5.0	2.9	80
Temperate forest, deciduous	7.0	3.8	95
Boreal forest	12.0	4.3	108
Subtotals (forests)	48.5	31.4	743
Woodland and shrubland, savanna	23.0	6.9	49
Temperate grassland	9.0	2.0	6.3
Tundra and alpine	8.0	0.5	2.4
Desert and semidesert	18.0	0.6	5.4
Extreme desert, rock, ice, sand	24.0	0.04	0.2
Cultivated land	14.0	4.1	7.0
Swamp and marsh	2.0	2.2	13.6
Lake and stream	2.5	0.6	0.02
Totals	149.0	48.3	827

far. The figures refer to the year 1950. For comparison, Bazilevich *et al.* (*8*) give about 50 percent larger figures. I will use the values in Table 1, but keep in mind the uncertainty of the estimates.

Even though dense forests cover only about 30 percent of the land surface, about 90 percent of the living biomass is found there. Any significant change in the amount of carbon in land biota would be observed in the atmosphere if it were not masked by the steady increase due to the emission of carbon dioxide from burning fossil fuels. It has long been realized that significant changes of the land biota may have occurred in the past, but because of lack of data no serious attempts have been made to assess quantitatively what may have happened on a global scale during the last century or two. The area of cultivated land has increased considerably during this time. If 25 percent of this area was previously covered by forests, the extension of agriculture has meant a net return of carbon to the atmosphere of 30 to 55 × 10⁹ tons.

Recent changes in forests may be roughly assessed with the aid of studies of world forest resources by

the Food and Agriculture Organization (*9–11*). It is, however, important to realize that estimates of forest resources must not be directly interpreted in terms of the total biomass of the forest or the recorded and projected changes, since these only include amounts of timber and pulp for industrial purposes and, with great uncertainties, production of fuel wood. Thus, for boreal (coniferous) forests the FAO estimate of the gross volume of timber is 110 × 10⁹ cubic meters (*9*), about 50 × 10⁹ tons of carbon, which should be compared with 240 × 10⁹ tons of carbon in the biomass as given in Table 1.

In Europe the area covered by forests has increased slightly from 1950 to 1975, while no significant changes seem to have occurred in the United States during this time. Judging from these studies, the net change of the forests in developed countries (including Canada and the Soviet Union) during the last 20 years is probably less than 1 percent, which is insignificant in the present context. It should be realized, however, that the use of forests in some of these countries has changed considerably during the 20th century. Particularly in Europe, they were previously used extensively for cattle grazing. This was especially the case for deciduous forests, but in Sweden, for example, boreal forests were also used in this way. During the last 50 years this practice has been almost completely abandoned, and the forests have grown more dense. As an example, from the middle of the 1920's until about 1970, the total amount of timber in Swedish forests increased by almost 40 percent (*10*). This does not mean that the total biomass has changed by the same percentage, since with modern forestry more of the primary production in the forests has undoubtedly been timber. It is still likely that a significant increase of the biomass has occurred, but it is impossible to assess it more precisely without a detailed investigation. A 10 percent increase of the biomass of Swedish forests represents an increase of the carbon content by 0.5 × 10⁹ tons. For Europe as a whole this figure might be several times larger. On the other hand, when a virgin forest is subjected to regular exploitation its average biomass decreases, since the forest is harvested before the biomass reaches its maximum. To some extent this effect is counteracted by the prevention of forest fire (which is one mechanism for virgin forest renewal) in exploited forests. I conclude that the changes of forests in developed countries at present do not represent a significant net flux of carbon to or from the atmosphere.

In developing countries major changes are taking place. Forest plantation has increased rapidly.

Persson (9) estimates that during the last three decades an area of 0.3 to 0.6 × 10^6 km^2 has been planted in China and about 0.1 × 10^6 km^2 has been planted in other developing countries. Some of these forests consist of rapidly growing species, particularly *Eucalyptus* and some pine species. Many of these plantations have been on soils that were not recently covered by forests, and not all of them have been successful. I estimate the net amount of carbon being withdrawn from the atmosphere by these forests by assuming an annual net production of 0.5 kilogram per square meter per year, which yields a value of 0.2 to 0.4 × 10^9 tons per year.

Simultaneously with this forestation, clearing of natural forests for other uses has proceeded at an accelerating rate, particularly in tropical countries. The main reason for this is the need for additional land for agriculture. Persson (11) estimates that about 0.02 × 10^6 km^2 is cleared every year in Africa, and the corresponding figure for Latin America is 0.06 × 10^6 km^2 (12). For the developing countries as a whole about 0.12 × 10^6 km^2 of natural forest is cleared and burned every year. It is estimated that the areas of closed forests in the developing countries will have decreased by 20 to 25 percent by the turn of the century if this trend continues. If we assume that all of the biomass in these forests burns (20 kg of carbon per square meter) we arrive at an annual input of 1.5 × 10^9 tons of carbon into the atmosphere in the form of carbon dioxide. This is likely to be an overestimate, partly because not all of the living biomass burns and partly because some of these areas again become covered by vegetation, although usually not forests.

Finally, we should consider the use of wood for industrial purposes and as fuel. Table 2 shows FAO estimates of the production of wood in 1974 (9). Measured in amounts of carbon, this production corresponds to 0.4 × 10^9 tons of carbon in the form of wood, a considerable part (perhaps half) of which goes into long-lasting structures. The paper produced from pulpwood is mostly burned within a year, rapidly returning the carbon to the

atmosphere. The estimated production of fuel wood, equivalent to about 0.5 × 10^9 tons of carbon, is probably considerably too low. In view of the rapidly increasing population in developing countries, it is likely that cutting of wood for fuel has increased appreciably during the last decades and that the present use exceeds the growth of new trees and bushes in these areas. There may be a net input to the atmosphere of 0.2 to 0.4 × 10^9 tons of carbon annually, because of inadequate forest growth for renewal of fuel wood in the developing countries.

CHANGES OF ORGANIC MATTER IN THE SOIL

Estimates of the amount of carbon in the form of organic matter in the soil vary widely. The most recent and perhaps the most accurate is Bohn's (13) value, about 3000 × 10^9 tons, which is considerably higher than most previous estimates. Since this carbon pool is larger (perhaps four times larger) than the carbon pool of living land biota, the turnover time of carbon in the soil is much longer. The average residence time of carbon in land biota is about 15 years. More than half of the organic matter produced annually—that is, most of what is contained in leaves, grass, and so on—has a short residence time and is decomposed rapidly in the soil. The average turnover time for organic matter in the soil in steady state is longer in the same proportion as the soil reservoir is larger than the reservoir of living organic matter. Under special circumstances—for example, in peat bogs—the turnover time may be many hundred years. Changes of the carbon content of the soil therefore take place rather slowly.

A considerable part of agricultural land originally consisted of bogs and marshes, which have been drained. In Sweden such land constitutes almost 10 percent of the total land area used for agriculture. Cultivation of such areas represents a change in the rate of decomposition of organic matter. Ploughing brings oxygen deeper into the soil, which previously may even have been anaerobic because of high water table. Estimates from the cultivation of such areas on the island of Gotland in the Baltic Sea yield values as high as 0.5 kg of carbon per square meter per year. A similar process occurs when land previously covered by forests is cleared and used for agriculture. Paul (14) has shown that the chernozem soils in Canada have lost about 50 percent of their organic matter

Table 2. Production of wood in 1974 (9).

Form	Amount (10^6m^3)
Logs	800
Pulpwood	340
Pitprops, posts, and so on	200
Fuel wood	1170
Total	2510

during 60 to 75 years of cultivation. For the brown soils this corresponds to 2.5 kg/m² and for the black soils 6.5 kg/m². A comparison of land used for cattle grazing with a nearby forest in southern Sweden shows 5 and 25 kg of carbon per square meter, respectively, indicating an even more marked change of the organic content of the soil due to human utilization of the land (15).

If we assume that 25 to 50 percent of the land cultivated today has lost 2.5 to 6 kg of carbon per square meter since the early 19th century, a total of 10 to 40 × 10⁹ tons of carbon has been added to the atmospheric pool during this time. The annual input to the atmosphere today may be similarly assessed as 0.1 to 0.5 × 10⁹ tons per year. This problem warrants further consideration.

DISCUSSION

The data compilation presented above is incomplete and a much more detailed review of available information should be made. Nevertheless, it is possible to give some overall limits on the way in which humans are modifying the exchange of carbon dioxide between land biota, soil, and the atmosphere. Table 3 is a summary of the estimates, including rough uncertainties. The accumulated input of carbon to the atmosphere is 50 ± 25 percent of the amount transferred to the atmosphere from fossil fuel combustion, and the present annual input is 10 to 35 percent of the present emission from the use of fossil fuels. Possible compensation through more rapid growth of land biota in other parts of the world due to increasing amounts of carbon dioxide in the atmosphere might reduce these figures somewhat, but it seems unlikely that such compensation would be complete. If nutrients are growth-limiting increased assimilation would be unlikely, but if water supply is the important factor assimilation would proceed more rapidly during periods of moist weather, as the flux of carbon dioxide through the open stomata would be enhanced while evapotranspiration remained unchanged.

Even if our understanding of the magnitude of man's effect on the exchange of carbon between the plants, the soil, and the atmosphere is incomplete, it is interesting to carry the discussion one step further. First, during the last decade the average input of carbon to the atmosphere from fossil fuel combustion has been 4.0 × 10⁹ ton/year, of which 2.0 × 10⁹ ton/year has remained airborne. The total input due to human activities

Table 3. Summary of present net average annual input of carbon (in the form of carbon dioxide) into the atmosphere and accumulated input since the early 19th century due to human modifications of land biota and soils. The basis for the estimates is presented in the text.

	Input (10⁹ tons)	
Source	Present average annual	Accumulated
Reduction of forests		
Developed countries	0 ± 0.1	
Developing countries		45 ± 15
Forestation	−0.3 ± 0.1	
Deforestation	0.8 ± 0.4	
Use of fuel wood	0.3 ± 0.2	
Changes of organic matter in soil	0.3 ± 0.2	24 ± 15
Totals	1.0 ± 0.6	70 ± 30

has thus been 5.0 ± 0.6 × 10⁹ ton/year. The airborne fraction has been 40 ± 5 percent, rather than 50 percent as estimated previously. Furthermore, the total accumulated input since the early 19th century has been 210 ± 30 × 10⁹ tons. Assuming that the airborne fraction has remained unchanged throughout this period, the amount of carbon in the atmosphere has increased by 90 ± 15 × 10⁹ tons; that is, the carbon dioxide concentration has increased by 35 to 50 ppm. The concentration may have been as low as 275 ppm during the early 19th century.

It follows that the oceans must have served as a more effective sink for carbon dioxide than previously considered. This result is difficult to reconcile with present models of the role of the oceans in the carbon cycle, and the problem warrants further study [see also (16)].

Attempts to predict the future increase in the carbon dioxide content of the atmosphere have been based on the assumption that the airborne fraction of the net output due to human activities has been about 50 percent during the last 20 years. If, instead, it is 40 ± 5 percent, the future increase might be slower and possible secondary effects such as climatic changes might be delayed, provided depletion of the world's forests is stopped. There are other and more immediate reasons why we need to take great care in dealing with the global ecosystem, but the modifications of the global carbon cycle described here are another and in the long run also an important reason (17).

REFERENCES AND NOTES

1   C. Ekdahl and C. D. Keeling, in *Carbon and the Biosphere* (Atomic Energy Commission, Washington, D.C., 1973), p. 51.

2   W. Bischof, Report AC-36, Department of Meteorology, University of Stockholm (1976).

3   The concentrations cited have not been adjusted in accordance with the recent absolute calibration of Keeling (*4*); however, this is of no consequence in assessing the changes that have occurred.

4   C. D. Keeling, *Tellus* **25**, 179 (1973).

5   B. Bolin, *Sci. Am.* **223**, 124 (September 1970).

6   C. D. Keeling, in *Chemistry of the Lower Atmosphere*, S. I. Rasool, Ed. (Plenum, New York, 1973), p. 251.

7   R. Whittaker and G. Likens, in *Primary Productivity of the Biosphere*, H. Lieth and R. Whittaker, Eds. (Springer-Verlag, New York, 1973).

8   N. Bazilevich, L. Rodin, N. Rozov, *Sov. Geogr.* **12**, 293 (1971).

9   R. Persson, in preparation.

10 A. Eriksson, and K. Janz, *R. Coll. For. Dep. For. Surv. Res. Notes No. 21* (1975).

11 R. Persson, *Forest Resources in Africa*. part II, *Regional Analysis* (Department of Forest Survey, Royal College of Forestry, Stockholm, 1976).

12 Food and Agriculture Organization, paper presented at the 12th session of the Latin American Forestry Commission, Havana, 3 to 7 February 1976.

13 H. L. Bohn, *Soil Sci. Soc. Am. Proc.* **40**, 468 (1976).

14 E. A. Paul, in *Environmental Biogeochemistry*, J. O. Nrigu, Ed. (Ann Arbor Science Publishers, Ann Arbor, Mich., 1976), vol. 1, p. 225.

15 N. Nykvist, *Skånes Nat. No. 53* (1966).

16 C. D. Keeling, in *Energy and Climate: Outer Limits to Growth* (National Academy of Sciences, Washington, D.C., 1976).

17 The essence of this article was presented at the Dahlem Workshop on Global Chemical Cycles and Their Alteration by Man, Berlin, November 1976. Similar conclusions were advanced by M. Woodwell and R. A. Houghton and by W. A. Reiners and H. Wright, Jr. The discussions at the workshop revealed that a detailed and thorough study of all possible changes of carbon in organic form, not least in the oceans, is urgently needed to obtain more accurate estimates of fluxes in the carbon cycle. Even though the available data are meager it was generally concluded that man's direct interference with land vegetation, particularly forests, most likely decreases the amount of carbon in land plants and the soil.

Observational Constraints on the Global Atmospheric CO$_2$ Budget

PIETER P. TANS, INEZ Y. FUNG, TARO TAKAHASHI

Observed atmospheric concentrations of CO$_2$ and data on the partial pressures of CO$_2$ in surface ocean waters are combined to identify globally significant sources and sinks of CO$_2$. The atmospheric data are compared with boundary layer concentrations calculated with the transport fields generated by a general circulation model (GCM) for specified source-sink distributions. In the model the observed north-south atmospheric concentration gradient can be maintained only if sinks for CO$_2$ are greater in the Northern than in the Southern Hemisphere. The observed differences between the partial pressure of CO$_2$ in the surface waters of the Northern Hemisphere and the atmosphere are too small for the oceans to be the major sink of fossil fuel CO$_2$. Therefore, a large amount of the CO$_2$ is apparently absorbed on the continents by terrestrial ecosystems.

Rising atmospheric CO$_2$ concentrations are expected to lead to significant global climatic changes during the coming decades (1). After 30 years of measurements in the atmosphere and the oceans, the global atmospheric CO$_2$ budget is still surprisingly uncertain. An improved understanding of the CO$_2$ cycle is essential to predict the future rate of atmospheric CO$_2$ increase and to plan eventually for an international CO$_2$ management strategy.

Combustion of fossil fuels, the amount of which is well documented (2), is a major contributor to the observed concentration increase of CO$_2$ in the atmosphere. The measured rise was about 57% of the fossil fuel input from 1981 to 1987. Other sources may have also contributed to the rise, but the amount of CO$_2$ released by changes in land use remains uncertain (3, 4), as is the response of terrestrial ecosystems to higher CO$_2$ levels and to other climatic and environmental perturbations (5). Estimates of the uptake of CO$_2$ by the oceans have been based entirely on computational schemes of varying complexity (6), from "box" models to three-dimensional ocean circulation models (7). The "consensus" among these studies is that the oceans might be absorbing between 26 and 44% of the fossil CO$_2$. This would leave no room for any significant net loss of C from terrestrial ecosystems, but instead would require net C uptake on the land (except for the highest ocean uptake estimates) to balance the atmospheric CO$_2$ budget (6, 7).

The inorganic carbon chemistry that describes the ultimate uptake capacity of the oceans is well understood; however, the capacity of the oceans for uptake of CO$_2$ also depends sensitively on their circulation dynamics and the biological processes in them. The atmosphere exchanges CO$_2$ with the ocean surface layer, in which biological processes keep the partial pressure of CO$_2$ (pCO$_2$) much lower than in deeper waters. High-latitude areas, where deep water outcrops at the sea surface during winter, are an exception. The high pCO$_2$ in waters below about 300 m depth is attributed mainly to the downward transport of C, via gravitational settling of biogenic debris produced in the photic zone, and the slow vertical mixing rate of deep water. The models that have been used to estimate the uptake of CO$_2$ by the oceans incorporate these oceanic features in varying degrees and have been validated with observed distributions of tracers such as ^{222}Rn, ^{14}C, ^{3}H, chlorofluorocarbons (CFCs), nutrient salts, and O$_2$. However, none of these tracers behaves exactly like CO$_2$. Furthermore, in all models the circulation is assumed to be in steady state, and in many of them changes in biological processes and the seasonal nature of C uptake are not included.

P. P. Tans is with the Cooperative Institute for Research in Environmental Sciences, University of Colorado/National Oceanic and Atmospheric Administration, Campus Box 216, Boulder, CO 80309. I. Y. Fung is with the National Aeronautics and Space Administration, Goddard Space Flight Center, Institute for Space Studies, 2880 Broadway, New York, NY 10025. T. Takahashi is with the Lamont-Doherty Geological Observatory, Columbia University, Palisades, NY 10964.

Tans, P. P., Fung, I. Y, Takahashi, T. (1990) Observational constraints on the global atmospheric CO$_2$ budget. *Science* 247: 1431–1438. Reprinted with permission from AAAS and the author.

Table 1. Annual average concentrations of CO_2 above 300 ppmv (by volume) in dry air. Years for which the data quality was deemed insufficient have been omitted (dashes), and the lack of an ongoing program is indicated by blanks. For the calculation of the 1981 to 1987 average, all years were first normalized to 1987 by adding the globally averaged difference between 1987 and that year. In order to avoid biasing the global averages by the addition or omission of stations, the averages were calculated from third-degree polynomial curve fits to the available yearly data. The reported SD is a measure of the variability of the annual averages at each station after normalization to 1987.

Name	Code	Location	1981	1982	1983	1984	1985	1986	1987	Average	SD
South Pole	SPO	90°S	38.5	39.3	40.7	42.2	43.6	44.6	46.8	46.59	.17
Halley Bay	HBA	76°S, 26°W			41.2	—	—	45.0	47.2	47.11	.23
Palmer Station	PSA	65°S, 64°W		39.5	40.9	42.7	43.9	—	47.0	46.91	.13
Cape Grim	CGO	41°S, 145°E				42.5	43.7	44.6	46.5	46.54	.11
Amsterdam Island	AMS	38°S, 78°E		39.3	41.1	42.4	43.9	45.0	—	46.82	.20
Samoa	SMO	14°S, 171°W	39.3	40.3	41.4	43.5	44.7	45.2	47.1	47.44	.27
Ascension Island	ASC	8°S, 14°W	39.8	40.7	42.6	43.9	45.0	45.8	48.1	48.07	.33
Seychelles	SEY	5°S, 55°E	40.2	40.5	41.1	44.1	45.2	46.1	—	47.93	.41
Christmas Island	CHR	2°N, 157°W				44.7	45.9	46.3	48.5	48.56	.32
Guam	GMI	13°N, 145°E		41.0	42.7	44.4	46.0	—	—	48.64	.19
Virgin Island	AVI	18°N, 65°W	40.3	40.9	42.0	43.4	45.4	46.4	48.2	48.13	.28
Cape Kumukahi	KUM	20°N, 155°W	40.6	41.2	42.6	44.3	45.6	46.5	48.5	48.52	.14
Key Biscayne	KEY	26°N, 80°W				45.2	46.7	47.6	49.5	49.47	.06
Midway	MID	28°N, 177°W						47.6	49.7	49.61	.21
Azores	AZR	39°N, 27°W		41.2	43.0	44.5	—	—	—	48.77	.21
Shemya Island	SHM	53°N, 174°E						48.9	50.0	50.39	.52
Cold Bay	CBA	55°N, 163°W	41.0	41.8	43.3	45.5	47.2	48.1	49.7	49.58	.34
Station "M"	STM	66°N, 2°E	41.8	42.1	43.1	45.5	46.5	48.2	48.8	49.49	.42
Point Barrow	BRW	71°N, 157°W	41.4	42.6	43.7	45.4	46.4	48.6	49.5	49.73	.39
Mould Bay	MBC	76°N, 119°W	41.8	42.4	43.6	45.6	46.7	48.6	49.8	49.85	.28
Alert	ALT	83°N, 62°W						48.0	49.5	49.68	.25
Global average			40.00	40.65	42.03	43.91	45.27	46.26	48.10	48.10	

Measurements of pCO_2 in the surface waters and of total inorganic carbon (TCO_2) dissolved in the oceans have not yet led to a direct confirmation of the amount of fossil CO_2 removed from the atmosphere by the oceans (8), in part because the expected increases are small compared to the natural variation. For example, if half of the cumulative fossil fuel CO_2 emitted since 1850 were distributed uniformly in the upper 1000 m of the oceans, TCO_2 would have increased by only 1%.

Any geographical distribution of CO_2 sources and sinks is reflected in the spatial and temporal variations of CO_2 concentration patterns in the atmosphere. Numerical models of atmospheric transport can simulate these patterns; they thereby allow us to test hypotheses of the atmospheric CO_2 budget (9, 10). With the use of two-dimensional models (latitude, height) the observed concentration gradients in the atmospheric boundary layer can be inverted directly to yield the net surface source as a function of latitude and time (11). In this article, we use three-dimensional (3-D) transport fields to simulate the global distribution of CO_2 in response to specific assumptions about the strength and location of surface fluxes of CO_2. Global CO_2 budgets are constructed as linear com-

binations of separate sources and sinks, including new estimates for the oceanic fluxes. The mean annual meridional gradient observed from 1981 to 1987 is then compared with the model values, calculated as the corresponding linear combinations of the distributions generated separately for each source or sink, and thus used to select acceptable CO_2 source-sink scenarios.

ATMOSPHERIC OBSERVATIONS

The Geophysical Monitoring for Climatic Change (GMCC) division of the National Oceanic and Atmospheric Administration (NOAA) has been collecting air samples in flasks for CO_2 analysis from more than 20 sites since 1980 (Table 1) (12). All flasks have been analyzed on the same nondispersive infrared analyzer in Boulder, Colorado, and referenced to the international manometric mole fraction scale (13) adopted for CO_2 monitoring. The seasonal cycles of CO_2 concentration observed at these sites have been used to estimate the seasonal net ecosystem production of the major terrestrial biomes of the world (10, 14). In this study we have used the average of the annual

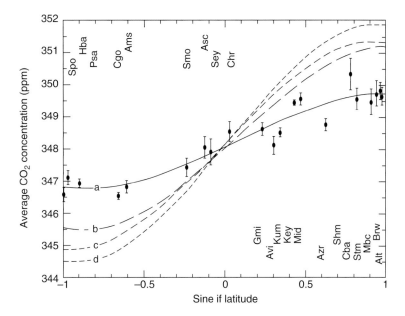

Fig. 1. Observed atmospheric CO_2 concentrations at the sites of the NOAA/GMCC flask network. The three-letter station codes are explained in Table 1. The error bars represent 1 SD of the annual averages at each site after adjustment to 1987. Curve (a) is a least-squares cubic polynomial fit to the data. The residual SD of the points with respect to the curve is 0.39 ppm. The concentration distributions at the NOAA/GMCC sites have also been calculated with the NASA/GISS GCM transport fields. Other curves are polynomial fits to the calculated CO_2 distributions (not shown) with fossil fuel emissions, seasonal vegetation (no net annual source or sink), tropical deforestation of 0.3 Gt of C per year, and three different cases of ocean uptake: (c), the compilation of CO_2 uptake based on the ΔpCO_2 data (Table 2) and our empirical transfer coefficients; (b), CO_2 uptake based on the same ΔpCO_2 map, but calculated with the Liss-Merlivat (*22*) relation for air-sea exchange; (d), an earlier estimate of ocean uptake (*21*) totaling 2.6 Gt of C per year.

mean concentrations for 1981 to 1987 (Table 1 and Fig. 1). We have not used the data from all of the GMCC sites. Records from Niwot Ridge, Colorado, as well as Mauna Loa Observatory, Hawaii, were not used because mountainous terrain is not resolved well in the transport model. Specifically, we do not know what effective model height to assign to these sites. At some other sites, such as Cape Meares, Oregon, the data are too noisy to extract annual averages with sufficient confidence. The data yield a large-scale meridional gradient that corresponds closely to those obtained by other atmospheric CO_2 monitoring programs (*14, 15*).

OCEANIC OBSERVATIONS AND CO₂ FLUX ESTIMATES

The observed pCO_2 difference (ΔpCO_2) between the surface ocean and the atmosphere represents the thermodynamic driving potential for transfer of CO_2 gas across the sea surface and includes implicitly the combined effects of all the processes that influence the CO_2 distribution in the oceans and atmosphere. We have analyzed measurements

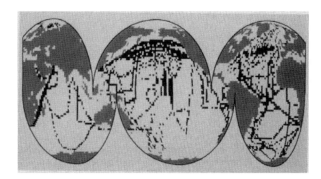

Fig. 2. The distribution of measurements of ΔpCO_2 since 1972. Where observations were made quasi-continuously, the values have been averaged over 2° intervals in longitude and latitude, and each of these intervals is represented by a single dot on the map.

of ΔpCO_2 obtained from 1972 to 1989 (*16*) (Fig. 2). El Niño events, occurring irregularly every few years, reduce the CO_2 flux from the Eastern and Central Equatorial Pacific waters to virtually zero (*17*), but the equatorial measurements during the 1982–1983 and 1986–1987 events have been excluded. The oceans were divided into 2° by 2° "pixels", and the mean ΔpCO_2 value for each pixel

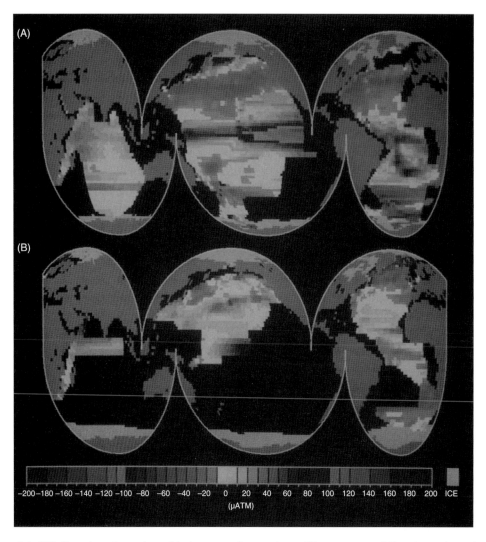

Fig. 3. Observed $\Delta p\mathrm{CO}_2$ (in microatmospheres) between surface waters of the oceans and the atmosphere during two seasonal periods, (A) January through April and (B) July through October. These maps have been compiled from direct observations made since 1972 (Fig. 2) and represent the mean distributions during the past 16 years, excluding the El Niño conditions in the equatorial Pacific. Areas of ice cover are indicated in light gray.

was computed separately for two seasonal periods, January through April and July through October (*18*) (Fig. 3). To estimate the global distribution of $\Delta p\mathrm{CO}_2$ during each of the two seasonal periods, we extrapolated the observed values into regions where observations were lacking using relations between water temperature and surface water $p\mathrm{CO}_2$ observed in various oceanographic regimes (*19*).

The net CO_2 flux (*F*) across the air-sea interface was computed from

$$F = E\Delta p\mathrm{CO}_2 = V_p S\Delta p\mathrm{CO}_2 \qquad (1)$$

where E is the gas transfer coefficient, V_p is the gas transfer piston velocity, and S is the solubility of CO_2 in seawater; V_p depends on turbulence in both

media and hence on the wind speed, W. Because the effects of temperature on V_p and S nearly cancel each other, E is mainly a function of wind speed alone. Measurements of V_p made under various wind regimes in the field and in wind tunnels show that V_p is nearly zero for $W < 3\,\mathrm{m\,s^{-1}}$. They also show a wide range of variation (about a factor of 2) in V_p for $W > 3\,\mathrm{m\,s^{-1}}$, the cause of which is not clearly understood. For $W > 3\,\mathrm{m\,s^{-1}}$ (the wind speed at 10 m above the sea surface), we adopted the relation (*20*)

$$E(\text{moles of }\mathrm{CO}_2\ \mathrm{m^{-2}\ year^{-1}\ \mu atm^{-1}})$$
$$= 0.016[W(\mathrm{m s^{-1}})^{-3}] \qquad (2)$$

whereas E is taken to be zero for $W < 3\,\mathrm{m\,s^{-1}}$. This relation yields V_p values slightly lower than the

Table 2. Estimates of sea-to-air CO$_2$ flux (Gt of C per year) based on the compilation of ΔpCO$_2$ in microatmospheres in various oceans (Fig. 3) and transfer coefficients depending on wind speeds (see text). The seasonality of the ΔpCO$_2$ and the winds has been taken into account. The north Indian Ocean is included in the equatorial oceans. Extrapolation of ΔpCO$_2$ into ocean areas with no measurements is based on water temperatures (*19*).

Ocean	Location	January to April		July to October		Annual mean	
		ΔpCO$_2$	Flux	ΔpCO$_2$	Flux	ΔpCO$_2$	Flux
Atlantic subarctic	>50°N; 90°W to 20°E	−22	−0.15	−53	−0.31	−37	−0.23
Atlantic gyre	15°N to 50°N; 90°W to 20°E	−29	−0.58	−1	−0.02	−15	−0.30
North Pacific	>15°N; 110°E to 90°W	−11	−0.44	14	0.33	2	−0.06
Equatorial	15°S to 15°N; 180°W to 180°E	37	1.56	28	1.69	33	1.62
Southern gyres	50°S to 15°S; 180°W to 180°E	−9	−1.46	−25	−3.31	−17	−2.39
Antarctic	>50°S	−23	−0.38	−10	−0.03	−17	−0.20
Global		3	−1.5	−1	−1.7	1	−1.6

upper limit of the wind-tunnel data (*21*). For comparison, Liss and Merlivat [(*22*), see also (*23*)], using results of experiments in wind tunnels and in the field (*24*), chose values about one half of our values. If their values are adopted, the resulting CO$_2$ transfer flux would be halved for a given value of ΔpCO$_2$.

We calculated monthly values of E for every 2° by 2° pixel using Eq. 2 and monthly climatological wind speeds compiled by Esbensen and Kushnir (*25*). The resulting annual mean global value for E is 0.067 mol of CO$_2$ m^{-2} year^{-1} µatm^{-1}, which is consistent with the global mean CO$_2$ gas exchange rate of 20 ± 3 mol of CO$_2$ m^{-2} year^{-1}, based on the distribution of ^{14}CO$_2$ in the atmosphere and oceans (*21*) (hence Eq. 2 is "empirical"). The ocean fluxes were calculated from the seasonal ΔpCO$_2$ maps (Fig. 3), Eq. 2, and the monthly climatological winds (*25*) (Table 2). This analysis gave a net CO$_2$ uptake of 1.6 Gt of C per year (1 Gt equals 10^{15} g), which corresponds to about 30% of the current rate of fossil fuel emissions.

A rigorous error analysis for this estimate cannot be made at this time, but most of the uncertainty is attributed to the sparsity of data in the South Pacific and South Indian oceans. In the North Pacific Ocean, where 26 trans-Pacific transects have been made during various seasons from 1984 to 1989, the uncertainty in ΔpCO$_2$ due to the finite number of samples can be estimated. We removed an east-west transect data set (about 40 values) and computed pixel values using the remaining data (about 260 values for a seasonal map), after which we compared the values on the computed map with the removed transect. This comparison was made for three separate data sets, representing transects across the northern high-latitude areas in summer and winter, respectively, and one across the mid-

latitudes during the winter. The root-mean-square difference between individual computed and measured values was 8 µatm, whereas the mean difference was about 1 µatm. This result suggests good consistency between the transects and only minor statistical sampling errors in this ocean basin, but does not address possible systematic errors. A systematic error of 1 µatm in the annual average ΔpCO$_2$ would lead to a total flux error of about 0.07 Gt of C per year for the Arctic, North Atlantic, and North Pacific oceans combined. On the other hand, the same error in ΔpCO$_2$ for the Southern Hemisphere oceans (south of 10°S) would cause an error in the net flux of about 0.15 Gt of C per year, mainly because of the greater area.

TRANSPORT MODELING WITH SURFACE SOURCES AND SINKS

We used a global 3-D atmospheric tracer model derived from the general circulation model (GCM) developed at Goddard Institute for Space Studies (GISS) of the National Aeronautics and Space Administration (*26*) to model the distribution of CO$_2$ in the atmosphere. The 3-D model is fully seasonal in terms of its transport and mixing characteristics (including parameterized diffusion) as well as in the sources and sinks of CO$_2$. The parent GCM has diurnal and seasonal cycles, and four hourly mass fluxes, as well as monthly averaged convective frequencies, were saved for the tracer transport model. In addition to producing realistic simulations of the largescale features of the general circulation of the atmosphere, the GCM transport fields have been validated by the simulation of inert tracers (*27*). For tracers with Northern Hemisphere mid-latitude sources, the interhemispheric exchange time has been adjusted via a subgrid diffusion parameterization to 1.0 year, intermediate between what is needed to match the observed north-south distributions of CFCs and ^{85}Kr (exchange times of 0.9 and 1.1 years, respectively).

Two-dimensional models based on transport coefficients derived (28) from the GCM developed at the Geophysical Fluid Dynamics Laboratory (GFDL) have an interhemispheric exchange time for ^{85}Kr (11) nearly identical to that in the GISS model. An independent 3-D transport model based on analyzed winds, as obtained by the European Center for Medium Range Weather Forecasting, together with a convective vertical mixing scheme, gives an interhemispheric transport time for ^{85}Kr of 1.39 years (29). The calculated vertically and hemispherically averaged difference between the hemispheres for the fossil fuel source by the GISS model is the same as that derived for a simple atmospheric two-box model with an interhemispheric exchange time of 1 year. Also, our two-dimensional model based on the transport derived from the GFDL GCM gave a virtually identical result.

The GISS transport model has been used to simulate the effects of seasonal CO_2 exchange with the terrestrial biosphere (30). The modeled annual oscillations are similar to those observed at the surface sampling sites, as well as to those found in aircraft data from Scandinavia, Japan, Australia (Fig. 4), and from various latitudes in the Northern Hemisphere at 500 and 700 mbar (31).

The covariance of seasonal transport and seasonal CO_2 sources and sinks may lead to annually averaged concentration differences between different sites, both in the model and in the atmosphere, even in the absence of net annual sources: If transport is less vigorous during the season when a surface region is a source rather than when it is a sink, a positive net annual concentration anomaly will result. With purely seasonal annually balanced sources, the GISS 3-D model calculates annual mean concentrations for the GMCC sites in the Northern Hemisphere that are on average 0.25 ppm higher than for the sites in the Southern Hemisphere, whereas a 2-D model (11) gives a difference of only 0.05 ppm. There are no independent tracers to validate this aspect of the models. The most important reason for the difference is the summer-to-winter variability of vertical convective mixing at high latitudes. The greater vertical stability in winter would tend to keep the respired CO_2 closer to the ground, which would result in higher annual average surface CO_2 concentrations in the Northern Hemisphere.

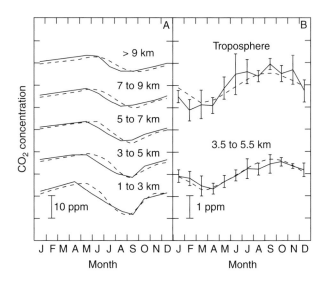

Fig. 4. Comparison of the observed (31) (solid line) and GISS-model calculated (30) (dashed line) annual cycle of CO_2 at different altitudes in the troposphere over (**A**) Scandinavia (67°N, 20°E) and (**B**) Bass Strait (40°S, 150°E).

We used the 3-D model to test hypotheses about global CO_2 budgets, constructed as linear combinations of separate source-sink patterns. We first calculated the CO_2 distribution for each source separately by running the model with that source for 3 years, during which the annual average concentration gradients became stabilized. The CO_2 distributions computed for the last year of the simulations were used in our analyses. As a sign convention, fluxes into the atmosphere (sources) are positive, fluxes from the atmosphere (sinks) negative. For any hypothesized global budget to be acceptable, it must satisfy two observational criteria: first, the total atmospheric inventory must increase by 3.0 Gt of C per year (corresponding to 1.4 ppm per year), and second, the corresponding linear combination of the modeled response distributions must reasonably resemble the observed atmospheric concentration differences at the stations.

The residuals, departures of the modeled annual average CO_2 concentrations from those observed at the GMCC sites, were fit with a third-order polynomial and with a straight line. In this way we were looking for consistent patterns of disagreement between the model and the data, because we did not want to adjust sources solely on the basis of discrepancies at single points. A source scenario is considered not plausible if the slope of the linear fit or any structure in the polynomial fit is statistically significant. The linear slope constraint requires that the strength of extratropical sources and sinks in the Northern relative to those in the Southern Hemisphere be determined to within about 0.2 Gt of C per year.

A TEST OF SOME CURRENT VIEWS OF THE CO$_2$ BUDGET

The geographical distribution of fossil fuel combustion (*32*) was combined with several global compilations of CO$_2$ exchange with the oceans and the terrestrial biosphere. The fossil fuel source was 5.3 Gt of C per year, typical of that from 1980 to 1987, when the global fossil fuel consumption remained fairly constant. Seasonal exchange with the terrestrial biosphere (*30*) was included although it does not affect the global budget. Tropical deforestation was assumed to be a source of 0.3 Gt of C per year, at the low end of the release estimates. Three ocean estimates were tested. In the first, our ocean data analysis presented above (Table 2) was used, and in this case an additional CO$_2$ sink of 1 Gt of C per year is required to balance the budget because the observed atmospheric increase is 3.0 Gt of C per year. In the second, the ΔpCO$_2$ values were combined with the air-sea transfer coefficients proposed by Liss and Merlivat (*22*); this scenario results in a total ocean uptake of only 0.8 Gt of C per year, in which case an extra sink of 1.8 Gt of C per year is required. In the third, we set the global net ocean sink to 2.6 Gt of C per year (*21*), thus balancing the budget.

The simulated difference in atmospheric CO$_2$ between the north and south poles resulting exclusively from fossil fuel combustion without any CO$_2$ sinks was 4.4 ppm. The uncertainty in the CO$_2$ production from fossil fuel combustion is estimated to be between 6 and 10% (*33*), and about 5% (*34*) of the fuel carbon is only partially oxidized to CO during combustion. This CO is oxidized in the atmosphere by reaction with OH radicals, which are concentrated at lower latitudes. This effect is neglected in the scenario, so that the calculated pole-to-pole gradient for fossil fuel combustion alone could be between 3.8 to 4.6 ppm. The seasonal terrestrial CO$_2$ exchange and tropical deforestation together are calculated to add another 0.6 ppm to the pole-to-pole gradient. The inclusion of the oceanic sink, acting strongly in the Southern Hemisphere, resulted in a meridional gradient between both poles of 5.7 to 7.3 ppm, depending on the ocean scenario. These values are contradicted in all cases by the atmospheric data (Fig. 1) which exhibit a difference of only 3.0 ppm. What

is wrong? In order to decrease the modeled gradient due to the fossil fuel source alone, either the extratropical net sink in the Northern Hemisphere must be larger than in the Southern Hemisphere, or there is a serious problem with the simulations of atmospheric transport in the GCM.

The annually averaged interhemispheric transport in the GCM is constrained by the [85]Kr and CFC calibrations, and we estimate that this part of the uncertainty in the calculated pole-to-pole concentration gradient is 10% or less. The behavior of the seasonal cycle of CO$_2$ as a function of altitude is well represented by the model (*30, 31*) in the few places where data are available. Inverse calculations with two-dimensional transport models (*11*) have similarly shown that the sink of CO$_2$ needs to be substantially larger in the Northern Hemisphere than in the Southern Hemisphere. As the peak-to-trough amplitude of the mean Northern Hemisphere CO$_2$ annual cycle is about 8 ppm, it is unlikely that covariation of this seasonal source and seasonal transport could produce a north-south counter-gradient as large as 3 to 4 ppm to allow the southern oceans to be the dominant sink of fossil fuel CO$_2$. Therefore, the presence of a large sink of C in the Northern Hemisphere is a more likely cause for the discrepancy than problems with the model transport.

CO$_2$ PATTERNS FROM SINGLE SOURCE REGIONS

Before we discuss CO$_2$ source-sink scenarios, that is, linear combinations of sources and sinks that satisfy the two constraints, we describe the series of "basis" sources and simulations of the corresponding CO$_2$ response distributions made with the 3-D model. Atmospheric CO$_2$ patterns were calculated separately for eight oceanic source regions: the equatorial oceans between 15°N and 15°S, the North Pacific gyre north of 15°N, the North Atlantic north of 50°N, the North Atlantic gyre between 15°N and 50°N, the South Atlantic, South Pacific and Indian ocean gyres each between 15°S and 50°S, and the Antarctic Ocean south of 50°S. In each of these cases the source was assumed, as a first approximation, to be constant in time and uniformly distributed in its respective area. The resulting concentration patterns were as expected: for example, if there is a CO$_2$ source of 1 Gt of C per year in the North Atlantic gyre, the CO$_2$ concentrations at AZR, KEY, and AVI (Table 1) stand out from values at Pacific stations at similar latitudes

Table 3. Four modeled scenarios of the global atmospheric cycle. Fluxes are in units of gigatons of C per year and $\Delta p CO_2$ is in microatmospheres. The terrestrial sources and sinks correspond to the basis functions: (i) tropical deforestation, (ii) carbon sequestering by temperate ecosystems, and (iii) CO_2 fertilization (see text). Fossil fuel combustion and the seasonality of the terrestrial biosphere is included in all cases. After fluxes to and from the terrestrial biosphere have been postulated, uptake by the oceans is adjusted to minimize the SD (in parts per million, last line) of the residual differences between the observed and calculated atmospheric CO_2 concentrations. The required annual average $\Delta p CO_2$ is estimated for ocean basins with empirical air-sea gas transfer coefficients.

Source or sink	Scenario 1, flux $\Delta p CO_2$		Scenario 2, flux $\Delta p CO_2$		Scenario 3, flux $\Delta p CO_2$		Scenario 4, flux $\Delta p CO_2$	
Tropical deforestation	0.3		0.3		2.0		2.0	
Temperate ecosystem uptake	0.0		0.0		0.0		−1.0	
CO_2 fertilization	0.0		−1.0		0.0		0.0	
Total terrestrial	0.3		−0.7		2.0		1.0	
North Atlantic (>50°N)	−0.7	−72	−0.5	−52	−0.7	−72	−0.5	−52
North Atlantic gyre (15° to 50°N)	−1.0	−52	−0.8	−42	−1.4	−73	−1.0	−52
North Pacific gyre (>15°N)	−1.0	−24	−0.7	−17	−1.4	−34	−1.0	−24
Equatorial (15°S to 15°N)	1.0	22	1.0	22	1.0	22	1.0	22
Combined southern gyres (15° to 50°S)	−1.4	−14	−1.1	−11	−2.3	−23	−2.3	−23
Antarctic (>50°S)	0.5	9	0.5	9	0.5	9	0.5	9
Total oceans	−2.6		−1.6		−4.3		−3.3	
SD of residuals	0.25		0.24		0.26		0.25	

by about 0.6 ppm. To reduce the number of independent variables, we assumed that the fluxes were proportional to area in the three southern ocean gyres, and we held the equatorial ocean source fixed at 1 Gt of C per year (*21*). We then had five ocean areas left as variables, the North Atlantic, the two north temperate gyres, the combined southern gyres, and the waters around Antarctica.

We considered four "basis functions" of net annual CO_2 exchange with the terrestrial biosphere: (i) net release due to deforestation in the tropics (*3*); (ii) C sequestering by temperate ecosystems; (iii) storage of C by high latitude boreal ecosystems; (iv) and a hypothetical sink due to enhanced net photosynthesis, which is referred to as CO_2 fertilization. For the second basis function, the C sink was uniformly distributed among locations associated with cold-deciduous forests (13×10^6 km²); similarly for the third, the sink was distributed among evergreen needle-leaved forests and woodlands (12×10^6 km²) and tundra (7×10^6 km²). Carbon sequestering in these regions may be through processes such as reforestation (*35*) or accumulation of organic matter in soils. For the fourth sink, we assumed that the net fertilization is proportional to net primary productivity (NPP); thus, this sink is intense in tropical regions because of their high NPP (*36*). A global fertilization effect of 1 Gt of C per year, for example, would represent an increase of only 2% of NPP if ecosystem respiration remained the same. This amount is easily within the uncertainties of global NPP estimates (*36*).

In the simulations we took into account the covariance of the annually balanced seasonal CO_2 exchange (*30*) with terrestrial plants (no net annual flux) and the seasonality of the transport as a separate "basis" source scenario. The inclusion of this scenario significantly improved the comparison between the modeled and the observed concentrations with respect to the longitudinal variability.

ADJUSTMENT OF OCEANIC UPTAKE TO TERRESTRIAL SCENARIOS

After we specified a priori certain combinations of gain and loss of C on the continents, uptake by the oceans was adjusted in each case until satisfactory agreement with the atmospheric observations was obtained. The four scenarios (Table 3 and Fig. 5) all fit the atmospheric observations equally well; these data by themselves do not permit us to determine whether any one is more likely. In fitting the data, we could, to a limited extent, trade off uptake of C by terrestrial ecosystems against uptake by the oceans, for example, boreal forest and tundra ecosystems against the North Atlantic. Monitoring techniques need to be developed for and extended to the continental interiors to preclude such freedom in modeling and to pinpoint the source-sink distributions much more definitively.

The disagreement with Table 2 for the uptake of CO_2 by the southern oceans stems mainly from the limited number of $\Delta p CO_2$ observations in the

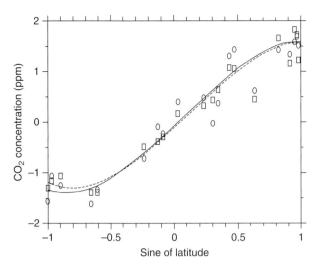

Fig. 5. Results of model calculations (scenario 1, Table 3) of the atmospheric CO_2 concentrations at the GMCC sites (squares and dashed curve) are compared with the observed concentrations (circles and solid curve). All values are relative to the global mean. The curves are least-squares cubic polynomial fits; the differences between the curves are not statistically significant.

high-latitude waters near Antarctica. The atmospheric data seem to indicate that there is a CO_2 source in the waters around Antarctica. This estimate for the Antarctic waters rests on the concentration difference between HBA and PSA on the one hand and SPO, CGO, and AMS on the other hand (Fig. 1). Recent oceanographic measurements (*37*) appear to have provided some confirmation for the presence of a CO_2 source (Fig. 3B).

Atmospheric CO_2 concentrations at AVI, KEY, and AZR (Table 1) in the Atlantic relative to KUM and MID in the Pacific suggest that the average pCO_2 of the North Pacific should be higher than the North Atlantic. The ΔpCO_2 observations confirm this (Fig. 3).

All of the scenarios (Table 3), however, are equally unrealistic because the mean annual ΔpCO_2 required for the Northern Hemisphere oceans is much greater than observed (Fig. 3). The discrepancy is much larger than can be explained by the uncertainty in the ΔpCO_2 data. Use of the gas exchange rates of Liss and Merlivat (*22*) would double this discrepancy.

ADJUSTMENT OF TERRESTRIAL EXCHANGE TO OBSERVED ΔpCO_2

Because of the conflict of the ΔpCO_2 required by the foregoing scenarios and the observed ΔpCO_2 of the northern oceans, we constructed several sce-

narios in which CO_2 fluxes in better known oceanic regions were kept fixed (with linear interpolation for the intervening months), namely uptake by the northern oceans and CO_2 outgassing from the equatorial oceans (Table 2). Exchange with the terrestrial ecosystems and with the southern oceans was varied to produce agreement with the atmospheric data.

Several types of scenarios (four are presented in Table 4) all agreed about equally well with the atmospheric data. The constraint of the observed north-south gradient imposes two important common features. First, a large terrestrial sink at northern temperate latitudes is necessary, and second, total CO_2 uptake by the oceans is considerably less than uptake by terrestrial systems. The total terrestrial sink at high northern and temperate latitudes (including its share of the global CO_2 fertilization) varies between 2.0 and 2.7 Gt of C per year in the four scenarios. The sum of the temperate and high-latitude sources and sinks is tightly constrained, but the two can be traded off against one another to some extent. However, a large temperate sink requires a smaller high-latitude source to prevent the modeled CO_2 concentration at arctic sites from becoming too low.

The following scenarios were unsuccessful: The additional absorption of more than a few tenths of a gigaton of C by high latitude ecosystems or the Arctic Ocean resulted in predicted concentrations for the five northernmost stations that were too low. Balancing the global budget by uptake via CO_2 fertilization proportional to NPP (and no tropical deforestation) left the concentrations at equatorial latitudes too low; half of the NPP takes place in the tropics, so that the area would in that case act as a net sink for CO_2.

The modeled CO_2 gradients are not very sensitive to the magnitude of tropical deforestation because the GMCC sites are remote from deforestation activities and the released CO_2 is dispersed rapidly via vigorous vertical mixing. If the release of CO_2 from tropical forest destruction is balanced by the fertilization effect, half of the extra CO_2 is taken up in the tropics themselves, and thus smaller amounts of carbon uptake are required at temperate latitudes in both hemispheres. A large amount of tropical deforestation (scenario 8, Table 4) can only be accommodated if CO_2 fertilization is a strong sink, so that the modeled tropical CO_2 concentrations do not become significantly larger than those observed.

We have not included in the simulations the atmospheric oxidation of CO, which produces a

Table 4. Four modeled scenarios of the global atmospheric C cycle in which uptake by the northern and equatorial oceans is held fixed. Fluxes are in units of Gt of C per year. Equatorial ocean outgassing is lower than in Table 2 by 0.32 Gt of C per year to take into account El Niño episodes occurring about once every 4 years. After the rate of tropical deforestation has been postulated, CO_2 exchange with terrestrial ecosystems and the southern oceans is varied (indicated by asterisk) to produce agreement with the atmospheric observations. The estimates of uptake by the oceans are based on observed seasonal ΔpCO_2 values, monthly climatological winds and two sets of air-sea gas transfer coefficients, our empirical relation (Emp), and the Liss-Merlivat (22) relation (LM). In the latter case the equatorial oceanic source is smaller, so that less uptake is required at temperate latitudes in both hemispheres to balance the budget.

Source or sink	Scenario 5		Scenario 6		Scenario 7		Scenario 8	
	Emp	LM	Emp	LM	Emp	LM	Emp	LM
Tropical deforestation	0.0	0.0	1.0	1.0	1.0	1.0	2.5	2.5
CO_2 fertilization*	0.0	0.0	0.0	0.0	−1.0	−1.0	−3.0	−3.0
Temperate uptake*	−2.0	−2.0	−3.0	−2.9	−2.3	−2.0	−1.9	−1.9
Boreal source*	0.0	0.0	0.4	0.4	0.4	0.2	0.7	0.7
Total terrestrial	−2.0	−2.0	−1.6	−1.5	−1.9	−1.8	−1.7	−1.7
Arctic and sub-arctic (>50°N)	−0.23	−0.12	−0.23	−0.12	−0.23	−0.12	−0.23	−0.12
Combined northern gyres (15°N to 50°N)	−0.36	−0.18	−0.36	−0.18	−0.36	−0.18	−0.36	−0.18
Equatorial (15°S to 15°N)	1.30	0.65	1.30	0.65	1.30	0.65	1.30	0.65
Combined southern gyres* (50°S to 50°S)	−1.5	−1.1	−1.9	−1.6	−1.6	−1.3	−1.8	−1.4
Antarctic (>50°S)	0.5	0.5	0.5	0.5	0.5	0.5	0.5	0.5
Total oceans	−0.3	−0.3	−0.7	−0.8	−0.4	−0.5	−0.6	−0.6
SD of residuals (ppm)	0.26	0.28	0.27	0.29	0.27	0.28	0.28	0.29

total of 0.85 ± 0.25 Gt of C per year of CO_2 (34). Simulations with a two-dimensional model of a latitudinal and seasonal distribution of CO oxidation totaling 1 Gt of C per year globally (38) suggest that a broad maximum in CO_2 concentrations forms at 30°N that decreases by 0.3 ppm toward the South Pole and by 0.15 ppm toward the North Pole. The inclusion of this process would have reinforced the need for a northern mid-latitude sink. As a related problem, a small part of the terrestrial sink for CO_2 that we infer will not contribute to C storage on the land because C is recycled by the biosphere into reduced volatile compounds that are oxidized, often via CO, to CO_2 in the atmosphere.

CONCLUSIONS

From 1981 to 1987 atmospheric CO_2 increased at an average rate of 3.0 Gt of C per year. The release of CO_2 from fossil fuel burning (5.3 Gt of C per year) and land use modification (0.4 to 2.6 Gt of C per year) is being partially balanced by the uptake of CO_2 by the oceans and by terrestrial ecosystems. Observations and simulations of the meridional gradient of CO_2 in the atmosphere suggest that these sinks are larger in the Northern Hemisphere than in the Southern Hemisphere.

The atmospheric gradient constrains the combined uptake by the southern ocean gyres and

Antarctic waters to be from 0.6 to 1.4 Gt of C per year. In consideration of the large data base of seasonal ΔpCO_2 measurements in the surface waters of the Northern Hemisphere, the uncertainties in ΔpCO_2 are most likely not large enough to accommodate the values of C removal required without a large terrestrial sink. We infer that the global ocean sink is at most 1 Gt of C per year. Our analysis thus suggests that there must be a terrestrial sink at temperate latitudes to balance the carbon budget and to match the north-south gradient of atmospheric CO_2. The mechanism of this C sink is unknown; its magnitude appears to be as large as 2.0 to 3.4 Gt of C per year, depending on the sources in the tropical and the boreal and tundra regions.

The global C cycle is not well understood. Unraveling the contemporary CO_2 cycle and the development of future mitigation strategies requires a concerted measurement program to determine the seasonal fluxes of CO_2 between the atmosphere, land, and oceans. Our hypothesis suggests that annually averaged ΔpCO_2 values in the combined southern oceans are small negative values. Collection of data on air-sea exchange of CO_2 in these areas in all seasons should be given high priority. Understanding the role of the land in the C budget must include a reanalysis of the contribution of mid-latitude reforestation as well as studies of the feedbacks between ecosystem functioning, climate, and atmospheric composition.

The atmosphere integrates the fluxes from all sources and sinks. It thus contains the large-scale signatures of CO$_2$ source areas that are often highly variable, and therefore hard to measure, on smaller scales. Data from the present international network of CO$_2$ monitoring sites, located almost exclusively in oceanic areas, cannot be used to resolve longitudinal gradients, and thus identification of the important source-sink areas is currently difficult. In addition, high-precision measurements of the large-scale variations of ^{13}C/^{12}C ratios in CO$_2$ and the concentration of atmospheric O$_2$ are needed to untangle the contributions of the land and oceans.

REFERENCES AND NOTES

1   R. E. Dickinson and R. J. Cicerone, *Nature* **319**, 109 (1986); V. Ramanathan, *Science* **240**, 293 (1988); J. Hansen *et al.*, *J. Geophys. Res.* **93**, 9341 (1988).

2   G. Marland *et al.*, *Estimates of CO$_2$ Emissions from Fossil Fuel Burning and Cement Manufacturing* (*Oak Ridge Natl. Lab. Rep. ORNL/CDIAC-25*, National Technical Information Service, Springfield, VA, 1989).

3   B. Bolin, *Science* **196**, 613 (1977); G. M. Woodwell *et al.*, *ibid.* **199**, 141 (1978); R. P. Detwiler and C. A. S. Hall, *ibid.* **239**, 42 (1988).

4   R. A. Houghton *et al.*, *Tellus* **39B**, 122 (1987).

5   E. R. Lemon, Ed., *CO$_2$ and Plants* (*Am. Assoc. Adv. Sci. Select. Symp. Ser. 84*, Westview, Boulder, 1983); B. R. Strain and J. D. Cure, Eds., *Direct Effects of Increasing Carbon Dioxide on Vegetation* (*U.S. Dept. Energy Rep. DOE/ER-0238*, National Technical Information Service, Springfield, VA, 1985); W. R. Emanuel, G. E. G. Killough, J. S. Olson, in *Carbon Cycle Modelling*, B. Bolin, Ed. (Wiley, New York, 1981), pp. 335–354; L. H. Allen *et al.*, *Glob. Biogeochem. Cyc.* **1**, 1 (1987).

6   W. S. Broecker, T. Takahashi, H. J. Simpson, T.-H. Peng, *Science* **206**, 409 (1979).

7   C. D. Keeling, in *Chemistry of the Lower Atmosphere*, S. I. Rasool, Ed. (Plenum, New York, 1973), pp. 251–329; H. Oeschger, U. Siegenthaler, U. Schotterer, A. Gugelmann, *Tellus* **27**, 168 (1975); R. Bacastow and A. Bjorkstrom, in *Carbon Cycle Modelling*, B. Bolin, Ed. (Wiley, New York, 1981) pp. 29–80; T.-H. Peng, *Radiocarbon* **28**, 363 (1986); E. Maier-Reimer and K. Hasselmann, *Climate Dyn.* **2**, 63 (1987); J. R. Toggweiler, K. Dixon, K. Bryan, *J. Geophys. Res.* **94**, 8217 (1989); *ibid.*, p. 8243 (1989).

8   J. R. Trabalka, Ed., *Atmospheric Carbon Dioxide and the Global Carbon Cycle* (*U.S. Dept Energy Rep. DOE/ER-0239*, National Technical Information Service, Spring-field, VA, 1985).

9   G. I. Pearman and P. Hyson, *J. Geophys. Res.* **85**, 4468 (1980).

10 I. Fung *et al.*, *ibid.* **88**, 1281 (1983).

11 I. G. Enting and J. V. Mansbridge, *Tellus* **41B**, 111 (1989); P. P. Tans, T. J. Conway, T. Nakasawa, *J. Geophys. Res.* **94**, 5151 (1989).

12 W. D. Komhyr *et al.*, *J. Geophys. Res.* **90**, 5567 (1985); T. J. Conway *et al.*, *Tellus* **40B**, 81 (1988). Since 1983 the flask samples have also been analyzed for methane [L. P. Steele *et al.*, *J. Atmos. Chem.* **5**, 125 (1987)].

13 All values are relative to the World Meteorological Organization X85 scale for CO$_2$ [P. R. Guenther and C. D. Keeling, *Scripps Reference Gas Calibration System for CO$_2$-in-Air Standards: Revision of 1985* (Scripps Institute of Oceanography, La Jolla, CA, 1985)].

14 G. I. Pearman and P. Hyson, *J. Atmos. Chem.* **4**, 81 (1986).

15 C. D. Keeling and M. Heimann, *J Geophys. Res.* **91**, 7782 (1986).

16 The data base used consists mainly of the measurements obtained by the Lamont-Doherty Geological Observatory. This has been supplemented by data in the North Atlantic by M. Roos and G. Gravenhorst [*J. Geophys. Res.* **89**, 8181 (1984)]; in the equatorial Atlantic by C. Andrie, C. Oudot, C. Genthon, and L. Merlivat [*ibid.* **91**, 11,741 (1984)]; and in the North and South Pacific oceans and the eastern Indian Ocean by R. Gammon [personal communication]. When measurements were made continuously with a flow-through equilibrator, a mean value for each 2° longitude or latitude interval was computed and used as a data point (Fig. 2). The ΔpCO$_2$ values were obtained by subtracting the atmospheric pCO$_2$ values at nearby locations from the oceanic values. We computed the atmospheric values by using the mole fraction concentration in dry air (measured with the same instrument as that used for pCO$_2$ measurements in sea water), the barometric pressure, and the saturated water vapor pressure at sea surface temperature.

17 R. A. Feely *et al.*, *J. Geophys. Res.* **92**, 6545 (1987).

18 The measured values were weighted inversely proportional to the square of the distance from the center of the pixel, and those obtained in different years were weighted equally; ΔpCO$_2$ values in pixels with no measurements, but surrounded by pixels with measured ΔpCO$_2$ values, were estimated in gyre areas by linear interpolation in both latitude and longitude. In the equatorial zone, where currents are dominated by zonal flows, the value interpolated along the same latitude was used.

19 To extrapolate ΔpCO$_2$ values into areas where measurements were not available (black areas in Fig. 3), the seawater pCO$_2$ was assumed to be a function of temperature alone. The following temperature coefficients were determined on the basis of the measurements made during various seasons and are assumed to be independent of seasons: 1.6% °C^{-1} in the western North Atlantic (10°N to 40°N) and the south Indian Oceans (10°S to 34°S); 2.3% °C^{-1} in the South Atlantic (10°S to 34°S) and South Pacific (10°S to 34°S); 4.3% °C^{-1} in the eastern North Pacific (10°N to 34°N, 84°W to 154°W); 1.2% °C^{-1} in the Southern Ocean (34°S to 62°S). The climatological sea surface temperature data compiled by S. Levitus [Climatological Atlas of the World Ocean, *NOAA Prof. Pap. 13*, pp. 173 (1982)] were used. In the Pacific coastal areas along the Central and South Americas, where high ΔpCO$_2$ values occur because of upwelling of deep water, the ΔpCO$_2$ data obtained outside the two seasonal periods have been used with the assumption that the values do not change seasonally.

20 T.-H. Peng and T. Takahashi, in *Biogeochemistry of CO$_2$ and the Greenhouse Effect*, M. P. Farrell, Ed. (*Am Chem. Soc. Symp.*, CRC/Lewis, Boca Raton, FL, in press).

21 W. S. Broecker *et al., J. Geophys. Res.* **91**, 10517 (1986); T. Takahashi *et al., Seasonal and Geographic Variability of Carbon Dioxide Sink/Source in the Oceanic Areas* (*Tech. Rep. for Contr. MRETTA 19X-89675C*, Lamont-Doherty Geological Observatory, Palisades, NY, 1986); H.-C. Broecker, J. Petermann, W. Siems, *J. Mar. Res.* **36**, 595 (1978).

22 P. Liss and L. Merlivat, in *The Role of Air-Sea Exchange in Geochemical Cycling*, P. Buat-Menard, Ed. (*Adv. Sci. Inst. Ser. 185*, Reidel, Hingham, 1986), pp. 113–127. Their formulation of the wind-speed-dependent gas exchange is

$E = 0.00048W$ for $0 \leq W \leq 3.6 \text{ms}^{-1}$

$E = 0.00083(W - 3.39)$ for $3.6 \leq W \leq \text{ms}^{-1}$

$E = 0.017(W - 8.36)$ for $W\ 13 \geq \text{ms}^{-1}$

23 J. Etcheto and L. Merlivat, *J. Geophys. Res.* **93**, 15669 (1988). If high-frequency wind speed data are used with the Liss-Merlivat relation, the seasonal mean gas transfer rate would increase at high latitudes by about 25% in the northern oceans and by 50% in the southern oceans because of the nonlinear character of the relation [J. Etcheto and L. Merlivat, *Adv. Space Res.* **9**, 141 (1989)].

24 R. Wanninkhof, J. R. Ledwell, W. S. Broecker, *Science* **227**, 1224 (1985); R. Wanninkhof, J. R. Ledwell, W. S. Broecker, M. Hamilton, *J. Geophys. Res.* **92**, 14,567 (1987).

25 S. K. Esbensen and Y. Kushnir, *The Heat Budget of the Global Oceans: An Atlas Based on Estimates from the Surface Marine Observations* (*Clim. Res. Inst. Rep. 29*, Oregon State University, Corvallis, OR, 1981).

26 G. L. Russell and J. A. Lerner, *J. Appl. Meteor.* **20**, 1483 (1981); J. G. Hansen *et al., Mon. Weather Rev.* **111**, 609 (1983). The version used in this study and in (*30*) has 4° by 5° resolution and has improved simulation of the higher moment statistics of the general circulation.

27 M. Prather, M. McElroy, S. Wofsy, G. Russell, D. Rind, *J. Geophys. Res.* **92**, 6579 (1987); D. J. Jacob, M. J. Prather, S. C. Wofsy, M. B. McElroy, *ibid.*, p. 6614.

28 R. A. Plumb and J. D. Mahlman, *J. Atmos. Sci.* **44**, 298 (1987).

29 M. Heimann and C. D. Keeling, in *Aspects of Climate Variability in the Pacific and the Western Americas*, D. H. Peterson, Ed. (*Geophysical Monograph 55*, American Geophysical Union, Washington, DC, 1989).

30 I. Y. Fung, C. J. Tucker, K. C. Prentice, *J. Geophys. Res.* **92**, 2999 (1987).

31 We have made comparisons with aircraft data from C. D. Keeling, T. B. Harris, E. M. Wilkins, *J. Geophys. Res.* **73**, 4511 (1968); B. Bolin and W. Bischof, *Tellus* **22**, 431 (1970); G. I. Pearman and D. J. Beardsmore, *ibid.* **36B**, 1 (1984); M. Tanaka, T. Nakasawa, S. Aoki, *ibid* **39B**, 3 (1987).

32 G. Marland, R. M. Rotty, R. L. Treat, *Tellus* **37B**, 243 (1985).

33 G. Marland and R. M. Rotty, *ibid.* **36B**, 232 (1984).

34 W. Seiler and R. Conrad, in *The Geophysiology of Amazonia*, R. E. Dickinson, Ed. (Wiley, New York, 1987), pp. 133–160.

35 T. V. Armentano and C. W. Ralston, *Can. J. Forest. Res.* **10**, 53 (1980); W. C. Johnson and D. M. Sharpe, *ibid.* **13**, 372 (1983).

36 H. Lieth, in *Primary Productivity of the Biosphere*, H. Lieth and R. Whittaker, Eds. (*Ecolog. Stud. 14*, Springer, New York, 1975), pp 203–213.

37 H. Inoue and Y. Sugimura, *Tellus* **40B**, 308 (1988).

38 P. Logan, personal communication.

39 We thank T. Conway, K. Masarie, K. Thoning, and L. Waterman for obtaining the atmospheric data of the NOAA/GMCC flask network, and many people at the field sites for collecting the air samples over the years. J. John executed 3-D model runs and, together with J. Jonas and P. Palmer, provided support for the color graphics. Assistance by D. Chipman, J. Goddard, S. Sutherland, and E. A. Takahashi is appreciated. R. Gammon and E. Garvey contributed their data to this study. This work has been supported by the Geophysical Monitoring for Climatic Change division of NOAA, the National Science Foundation, Martin Marietta's Carbon Dioxide Information Analysis and Research program for the U.S. Department of Energy under contract DE-AC05-84OR21400, and the EXXON Research and Engineering Company.

Acceleration of Global Warming Due to Carbon-Cycle Feedbacks in a Coupled Climate Model

PETER M. COX*, RICHARD A. BETTS*, CHRIS D. JONES*, STEVEN A. SPALL*
& IAN J. TOTTERDELL[†]

*Hadley Centre, The Met Office, Bracknell, Berkshire RG12 2SY, UK
[†]Southampton Oceanography Centre, European Way, Southampton SO14 3ZH, UK

The continued increase in the atmospheric concentration of carbon dioxide due to anthropogenic emissions is predicted to lead to significant changes in climate[1]. About half of the current emissions are being absorbed by the ocean and by land ecosystems[2], but this absorption is sensitive to climate[3,4] as well as to atmospheric carbon dioxide concentrations[5], creating a feedback loop. General circulation models have generally excluded the feedback between climate and the biosphere, using static vegetation distributions and CO_2 concentrations from simple carbon-cycle models that do not include climate change[6]. Here we present results from a fully coupled, three-dimensional carbon–climate model, indicating that carbon-cycle feedbacks could significantly accelerate climate change over the twenty-first century. We find that under a 'business as usual' scenario, the terrestrial biosphere acts as an overall carbon sink until about 2050, but turns into a source thereafter. By 2100, the ocean uptake rate of 5 Gt C yr^{-1} is balanced by the terrestrial carbon source, and atmospheric CO_2 concentrations are 250 p.p.m.v. higher in our fully coupled simulation than in uncoupled carbon models[2], resulting in a global-mean warming of 5.5 K, as compared to 4 K without the carbon-cycle feedback.

The general circulation model (GCM) that we used is based on the third Hadley Centre coupled ocean–atmosphere model, HadCM3[7], which we have coupled to an ocean carbon-cycle model (HadOCC) and a dynamic global vegetation model (TRIF-FID). The atmospheric physics and dynamics of our GCM are identical to those used in HadCM3, but the additional computational expense of including an interactive carbon cycle made it necessary to reduce the ocean resolution to 2.5° × 3.75°, necessitating the use of flux adjustments in the ocean component to counteract climate drift. HadOCC accounts for the atmosphere–ocean exchange of CO_2, and the transfer of CO_2 to depth through both the solubility pump and the biological pump[8]. TRIFFID models the state of the biosphere in terms of the soil carbon, and the structure and coverage of five functional types of plant within each model gridbox (broadleaf tree, needleleaf tree, C_3 grass, C_4 grass and shrub). Further details on HadOCC and TRIFFID are given in Methods.

The coupled climate/carbon-cycle model was brought to equilibrium with a 'pre-industrial' atmospheric CO_2 concentration of 290 p.p.m.v., starting from an observed landcover data set[9]. The resulting state was stable, with negligible net land–atmosphere and ocean–atmosphere carbon fluxes in the long-term mean, and no discernible drift in atmospheric CO_2 concentration. This simulation produces the locations of the main land biomes, and estimates of ocean carbon (38,100 Gt C), vegetation carbon (493 Gt C), soil carbon (1,180 Gt C) and terrestrial net primary productivity (60 Gt C yr^{-1}) that are within the range of other estimates[2,10–12]. Ocean primary productivity is also compatible with results derived from remote sensing[13,14], producing a global-mean total of 53 Gt C yr^{-1}, and realistic seasonal and latitudinal variations[15].

The simulated carbon cycle displays significant interannual variability, which is driven by the model-generated El Niño/Southern Oscillation (ENSO). A realistic response to internal climate variability is an important prerequisite for any carbon-cycle model to be used in climate change predictions. Fluctuations in annual-mean atmospheric CO_2 are correlated with the phase of ENSO, as indicated by the Nino3 index (Fig. 1). During El Niño conditions (positive Nino3), the model simulates an increase in atmospheric CO_2; this

Cox, P. M. (2000) Acceleration of global warming due to carbon-cycle feedbacks in a coupled climate model. *Nature* 408: 184–187.
Reprinted by permission from Macmillan Publishers Limited.

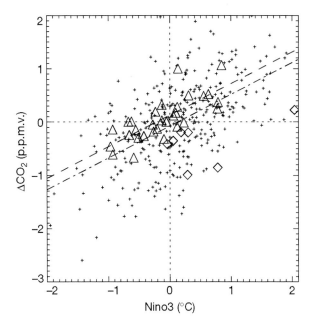

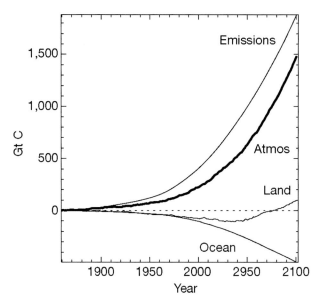

Fig. 1. Modelled and observed interannual variability in the atmospheric CO_2 concentration. The figure shows the anomaly in the growth rate of atmospheric CO_2 versus the Nino3 index, taken from our pre-industrial control simulation (crosses) and the Mauna Loa observations (triangles). (The Nino3 index is the annual mean sea surface temperature anomaly in the tropical Pacific, 150° W–90° W, 5° S–5° N.) The gradients of the dashed and dot-dashed lines represent the sensitivity of the carbon cycle to ENSO, as given by the observations and the model, respectively. We have excluded observations that immediately follow major volcanic events (data points shown by diamonds), since during these years the carbon cycle may have been significantly perturbed by the induced tropospheric cooling.

Fig. 2. Budgets of carbon during the fully coupled simulation. The thick line shows the simulated change in atmospheric CO_2. The thinner lines show the integrated impact of the emissions, and of land and ocean fluxes, on the atmospheric CO_2 increase, with negative values implying net uptake of CO_2. We note that the terrestrial biosphere takes up CO_2 at a decreasing rate from about 2010 onwards, becoming a net source at around 2050. By 2100 this source from the land almost balances the oceanic sink, so that atmospheric carbon content is increasing at about the same rate as the integrated emissions (that is, the airborne fraction is ~1).

increase results from the terrestrial biosphere acting as a large source (especially in Amazonia[16]), which is only partially offset by a reduced outgassing from the tropical Pacific Ocean. The opposite is true during the La Niña phase. The overall sensitivity of the modelled carbon cycle to ENSO variability is consistent with the observational record[17], demonstrating that the coupled system responds realistically to climate anomalies.

Transient simulations were carried out for 1860–2100, using CO_2 emissions as given by the IS92a scenario[18]. Other greenhouse gases were also prescribed from IS92a, but the radiative effects of sulphate aerosols were omitted. Three separate runs were completed to isolate the effect of climate/carbon-cycle feedbacks; an experiment with prescribed IS92a CO_2 and fixed vegetation (that is, a 'standard' GCM climate change simulation), an experiment with interactive CO_2 and dynamic vegetation but no direct effects of CO_2 on

climate (akin to 'offline' carbon-cycle projections that neglect climate change[6]), and a fully coupled climate/carbon-cycle simulation.

Figure 2 shows results from the fully coupled run. From 1860 to 2000, the simulated stores of carbon in the ocean and on land increase by about 100 Gt C and 75 Gt C, respectively. However, the atmospheric CO_2 is 15–20 p.p.m.v. too high by the present day (corresponding to a timing error of about 10 years). Possible reasons for this include an overestimate of the prescribed net land-use emissions and the absence of other important climate forcing factors. The modelled global mean temperature increase from 1860 to 2000 is about 1.4 K (Fig. 3b), which is higher than observed[19], probably due to the absence of cooling from anthropogenic aerosols[20]. Offline tests suggest that such a relative warming can suppress the terrestrial carbon sink by enhancing soil and plant respiration[11]. Nevertheless, the rate of increase of CO_2 from 1950 to 2000 closely follows the recent observational record, which implies that the airborne fraction is being well simulated over this period.

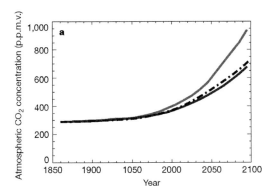

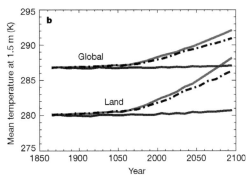

Fig. 3. Effect of climate/carbon-cycle feedbacks on CO_2 increase and global warming. a, Global-mean CO_2 concentration, and b, global-mean and land-mean temperature, versus year. Three simulations are shown; the fully coupled simulation with interactive CO_2 and dynamic vegetation (red lines), a standard GCM climate change simulation with prescribed (IS92a) CO_2 concentration and fixed vegetation (dot-dashed lines), and the simulation which neglects direct CO_2-induced climate change (blue lines). The slight warming in the latter is due to CO_2-induced changes in stomatal conductance and vegetation distribution.

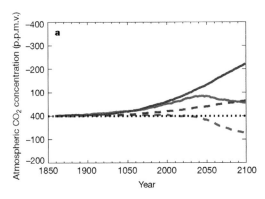

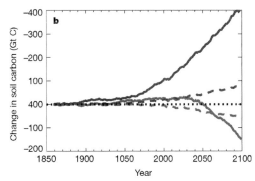

Fig. 4. Effect of global warming on changes in land carbon storage. The red lines represent the fully coupled climate/carbon-cycle simulation, and the blue lines are from the 'offline' simulation which neglects direct CO_2-induced climate change. The figure shows simulated changes in vegetation carbon (a) and soil carbon (b) for the global land area (continuous lines) and South America alone (dashed lines).

For the 20 years centred on 1985, the mean land and ocean uptake of carbon are 1.5 and 1.6 Gt yr^{-1}, respectively (compare best estimates for the 1980s of 1.8 ± 1.8 and 2.0 ± 0.8 Gt yr^{-1})[2]. The model therefore captures the most important characteristics of the present-day carbon cycle.

The simulated atmospheric CO_2 diverges much more rapidly from the standard IS92a concentration scenario in the future. First, vegetation carbon in South America begins to decline, as a drying and warming of Amazonia initiates loss of forest (Fig. 4a). This is driven purely by climate change, as can be seen by comparing the fully coupled run (red lines) to the run without global warming (blue lines). The effects of anthropogenic deforestation on land-cover are neglected in both cases. A second critical point is reached at about 2050, when the land biosphere as a whole switches from being a weak sink for CO_2 to being a strong source (Fig. 2). The reduction in terrestrial carbon from around 2050 onward is associated with a widespread climate-driven loss of soil carbon (Fig. 4b). An increase in the concentration of atmospheric CO_2 alone tends to increase the rate of photosynthesis and thus terrestrial carbon storage, provided that other resources are not limiting[4]. However, plant maintenance and soil respiration rates both increase with temperature. As a consequence, climate warming (the indirect effect of a CO_2 increase) tends to reduce terrestrial carbon storage[11], especially in the warmer regions where an increase in temperature is not beneficial for photosynthesis. At low CO_2 concentrations the direct effect of CO_2 dominates, and both vegetation and soil carbon increase with atmospheric CO_2. But as CO_2 increases further, terrestrial carbon begins to decrease, because the direct effect of CO_2 on photosynthesis saturates but the specific soil respiration rate continues to increase with temperature. The transition between these two regimes occurs abruptly at around 2050 in this experiment (Fig. 4b). The carbon stored on land decreases by about 170 Gt C from 2000 to 2100, accelerating the rate of atmospheric CO_2 increase over this period.

The ocean takes up about 400 Gt C over the same period, but at a rate which is asymptotically approaching 5 Gt Cyr^{-1} by 2100. This reduced efficiency of oceanic uptake is partly a consequence of the nonlinear dependence of the partial pressure of dissolved CO_2 on the total ocean carbon concentration, but may also have contributions from climate change[3]. Although the thermohaline circulation of the ocean weakens[21] by about 25% from 2000 to 2100, this is much less of a reduction than seen in some previous simulations[22], and the corresponding effect on ocean carbon uptake is less significant. In our experiment, increased thermal stratification due to warming of the sea surface suppresses upwelling, which reduces nutrient availability and lowers primary production by about 5%. However, ocean-only tests suggest a small effect of climate change on oceanic carbon uptake, as this reduction in the biological pump is compensated by reduced upwelling of deep waters which have high concentrations of total carbon.

By 2100 the modelled CO_2 concentration is about 980 p.p.m.v. in the coupled experiment, which is more than 250 p.p.m.v. higher than the standard IS92a scenario or that simulated in the 'offline' experiment (Fig. 3a). As a result, the global-mean land temperatures increase from 1860 to 2100 by about 8 K, rather than the 5.5 K of the standard scenario (Fig. 3b).

These numerical experiments demonstrate the potential importance of climate/carbon-cycle feedbacks, but the magnitude of these in the real Earth system is still highly uncertain. Terrestrial carbon models differ in their responses to climate change[11,12], owing to gaps in basic understanding of processes. For example, the potential conversion of the global terrestrial carbon sink to a source is critically dependent upon the long-term sensitivity of soil respiration to global warming, which is still a subject of debate[23]. The experiments presented here exclude the potentially large direct human influences on terrestrial carbon uptake through changes in landcover and land management. Local effects, such as the possible climate-driven loss of the Amazon rainforest, rest upon uncertain aspects of regional climate change, and may be 'short-circuited' by direct human deforestation. A full assessment of the uncertainties must await further coupled experiments utilizing alternative representations of processes and including a more complete set of natural and anthropogenic forcing factors (for example, land-use change, forest fires, sulphate aerosol concentrations and nitrogen deposition). However, our results indicate that it will be essential to accurately represent previously neglected carbon-cycle feedbacks if we are to successfully predict climate change over the next 100 years.

METHODS

Ocean carbon-cycle model

The inorganic component of HadOCC has been extensively tested as part of the Ocean Carbon Cycle Intercomparison Project; it was found to reproduce tracer distributions to an accuracy consistent with other ocean GCMs[24]. The biological component treats four additional ocean fields: nutrient, phytoplankton, zooplankton and detritus[8]. The phytoplankton population changes as a result of the balance between growth, which is controlled by light level and the local concentration of nutrient, and mortality, which is mostly as a result of grazing by zooplankton. Detritus, which is formed by zooplankton excretion and by phyto- and zooplankton mortality, sinks at a fixed rate and slowly remineralizes to reform nutrient and dissolved inorganic carbon. Thus both nutrient and carbon are absorbed by phytoplankton near the ocean surface, pass up the food chain to zooplankton, and are eventually remineralized from detritus in the deeper ocean. The model also includes the formation of calcium carbonate and its dissolution at depth (below the lysocline).

Terrestrial carbon-cycle model

TRIFFID (top-down representation of interactive foliage and flora including dynamics) has been used offline in a comparison of dynamic global vegetation models[11]. Carbon fluxes for each vegetation type are calculated every 30 minutes as a function of climate and atmospheric CO_2 concentration, from a coupled photosynthesis/

stomatal-conductance scheme[25,26], which utilizes existing models of leaf-level photosynthesis in C_3 and C_4 plants[27,28]. The accumulated fluxes are used to update the vegetation and soil carbon every 10 days. The natural landcover evolves dynamically based on competition between the vegetation types, which is modelled using a Lotka–Volterra approach and a tree–shrub–grass dominance hierarchy. We also prescribe some agricultural regions, in which grasslands are assumed to be dominant. Carbon lost from the vegetation as a result of local litterfall or large-scale disturbance is transferred into a soil carbon pool, where it is broken down by microorganisms that return CO_2 to the atmosphere. The soil respiration rate is assumed to double for every $10 K$ of warming[29], and is also dependent on the soil moisture content[30]. Changes in the biophysical properties of the land surface[5], as well as changes in terrestrial carbon, feed back onto the atmosphere.

ACKNOWLEDGEMENTS

We thank J. Mitchell and G. Jenkins for comments on earlier versions of the manuscript. This work was supported by the UK Department of the Environment, Transport and the Regions.

Correspondence and requests for materials should be addressed to P.M.C. (e-mail: pmcox@meto.gov.uk).

RECEIVED 6 JANUARY; ACCEPTED 26 SEPTEMBER 2000

1 Houghton, J. T. *et al.* (eds) *Climate Change 1995: The Science of Climate Change* (Cambridge Univ. Press, Cambridge, 1996).

2 Schimel, D. *et al.* in *Climate Change 1995: The Science of Climate Change* Ch. 2 (eds Houghton, J. T. *et al.*) 65–131 (Cambridge Univ. Press, Cambridge, 1995).

3 Sarmiento, J., Hughes, T., Stouffer, R. & Manabe, S. Simulated response of the ocean carbon cycle to anthropogenic climate warming. *Nature* **393**, 245–249 (1998).

4 Cao, M. & Woodward, F. I. Dynamic responses of terrestrial ecosystem carbon cycling to global climate change. *Nature* **393**, 249–252 (1998).

5 Betts, R. A., Cox, P. M., Lee, S. E. & Woodward, F. I. Contrasting physiological and structural vegetation feedbacks in climate change simulations. *Nature* **387**, 796–799 (1997).

6 Enting, I., Wigley, T. & Heimann, M. *Future Emissions and Concentrations of Carbon Dioxide; Key Ocean/ Atmosphere/Land Analyses* (Technical Paper 31, Division of Atmospheric Research, CSIRO, Melbourne, 1994).

7 Gordon, C. *et al.* The simulation of SST, sea ice extents and ocean heat transports in a version of the Hadley Centre coupled model without flux adjustments. *Clim. Dyn.* **16**, 147–168 (2000).

8 Palmer, J. R. & Totterdell, I. J. Production and export in a global ocean ecosystem model. *Deep-Sea Res.* **48**, 1169–1198, (2001).

9 Wilson, M. F. & Henderson-Sellers, A. A global archive of land cover and soils data for use in general circulation climate models. *J. Clim.* **5**, 119–143 (1985).

10 Zinke, P. J., Stangenberger, A. G., Post, W. M., Emanuel, W. R. & Olson, J. S. *Worldwide Organic Soil Carbon and Nitrogen Data* (NDP-018, Carbon Dioxide Information Center, Oak Ridge National Laboratory, Oak Ridge, Tennessee, 1986).

11 Cramer, W. *et al.* Global response of terrestrial ecosystem structure and function to CO_2 and climate change: results from six dynamic global vegetation models. *Glob. Change Biol.* **7**, 357–373 (2001).

12 VEMAP Members. Vegetation/ecosystem modelling and analysis project: comparing biogeography and biogeochemistry models in a continental-scale study of terrestrial responses to climate change and CO_2 doubling. *Glob. Biogeochem. Cycles* **9**, 407–437 (1995).

13 Longhurst, A., Sathyendranath, S., Platt, T. & Caverhill, C. An estimate of global primary production in the ocean from satellite radiometer data. *J. Plank. Res.* **17**, 1245–1271 (1995).

14 Field, C., Behrenfeld, M., Randerson, J. & Falkowski, P. Primary production of the biosphere: integrating terrestrial and oceanic components. *Science* **281**, 237–240 (1998).

15 Antoine, D., Andre, J.-M. & Morel, A. Oceanic primary production 2. Estimation at global scale from satellite (Coastal Zone Color Scanner) chlorophyll. *Glob. Biogeochem. Cycles* **10**, 57–69 (1996).

16 Tian, H. *et al.* Effects of interannual climate variability on carbon storage in Amazonian ecosystems. *Nature* **396**, 664–667 (1998).

17 Keeling, C. D., Whorf, T., Whalen, M. & der Plicht, J. V. Interannual extremes in the rate of rise of atmospheric carbon dioxide since 1980. *Nature* **375**, 666–670 (1995).

18 Houghton, J. T., Callander, B. A. & Varney, S. K. (eds) *Climate Change 1992: The Supplementary Report to the IPCC Scientific Assessment* (Cambridge Univ. Press, Cambridge, 1992).

19 Nicholls, N. *et al.* in *Climate Change 1995: The Science of Climate Change* Ch. 3 (eds Houghton, J. T. *et al.*) (Cambridge Univ. Press, Cambridge, 1996).

20 Mitchell, J. F. B., Johns, T. C., Gregory, J. M. & Tett, S. F. B. Climate response to increasing levels of greenhouse gases and sulphate aerosols. *Nature* **376**, 501–504 (1995).

21 Wood, R. A., Keen, A. B., Mitchell, J. F. B. & Gregory, J. M. Changing spatial structure of the thermohaline circulation in response to atmospheric CO_2 forcing in a climate model. *Nature* **399**, 572–575 (1999).

22 Sarmiento, J. & Quere, C. L. Oceanic carbon dioxide uptake in a model of century-scale global warming. *Nature* **274**, 1346–1350 (1996).

23 Giardina, C. & Ryan, M. Evidence that decomposition rates of organic carbon in mineral soil do not vary with temperature. *Nature* **404**, 858–861 (2000).

24 Orr, J. C. in *Ocean Storage of Carbon Dioxide, Workshop 3: International Links and Concerns* (ed. Ormerod, W.) 33–52 (IEA R&D Programme, CRE Group Ltd, Cheltenham, UK, 1996).

25 Cox, P. M., Huntingford, C. & Harding, R. J. A canopy conductance and photosynthesis model for use in a GCM land surface scheme. *J. Hydrol.* **212–213**, 79–94 (1998).

26 Cox, P. M. *et al.* The impact of new land surface physics on the GCM simulation of climate and climate sensitivity. *Clim. Dyn.* **15**, 183–203 (1999).

27 Collatz, G. J., Ball, J. T., Grivet, C. & Berry, J. A. Physiological and environmental regulation of stomatal conductance, photosynthesis and transpiration: a model that includes a laminar boundary layer. *Agric. Forest Meteorol.* **54**, 107–136 (1991).

28 Collatz, G. J., Ribas-Carbo, M. & Berry, J. A. A coupled photosynthesis-stomatal conductance model for leaves of C4 plants. *Aust. J. Plant Physiol.* **19**, 519–538 (1992).

29 Raich, J. & Schlesinger, W. The global carbon dioxide flux in soil respiration and its relationship to vegetation and climate. *Tellus B* **44**, 81–99 (1992).

30 McGuire, A. *et al.* Interactions between carbon and nitrogen dynamics in estimating net primary productivity for potential vegetation in North America. *Glob. Biogeochem. Cycles* **6**, 101–124 (1992).

16

Two If by Sea

Broecker, W.S. & Takahashi, T. (1978). Neutralization of fossil fuel CO_2 by marine calcium carbonate. In *The Fate of Fossil Fuel CO_2 in the Oceans* (eds. N.R. Andersen & A. Malahoff), pp. 213–248. Plenum Press, New York,. 35 pages but they're small format.

Walker, J.C.G. & Kasting, J.F. (1992). Effects of fuel and forest conservation on future levels of atmospheric carbon dioxide. *Paleogeography, Paleoclimatology Paleoecology (Global and Planetary Change Section)*, **97**, 151–189. 39 pages.

Kennett, J.P. & Stott, L.D. (1991). Abrupt deep sea warming, paleoceanographic changes and benthic extinctions at the end of the paleocene. *Nature*, **353**, 319–322.

Two papers, Broecker and Takahashi (1978) and Walker and Kasting (1991) take on the question of what happens to the fossil fuel CO_2 after it is released to the atmosphere, where it goes, and how long it takes to get there. Both sets of authors had worked primarily on understanding of natural carbon cycle in the geologic past, and here applied their understanding to the problem of the anthropogenic perturbation.

Broecker and Takahashi explore the mechanics of the pH neutralizing reaction between CO_2, which dissolves in water to form H_2CO_3 carbonic acid, and $CaCO_3$, a solid form of a chemical base. Reaction of CO_2 with carbonate rocks was mentioned as a potentially important process by Revelle and Suess (1957), and Bolin (1958). Broecker and Takahashi take the question a step further by considering the distribution of $CaCO_3$ on the sea floor and the mechanics of how it could dissolve. They discuss the implication of recent measurements of chemical tracers such as tritium produced by nuclear weapons testing, which suggested that fossil fuel CO_2 should enter the ocean in the Atlantic, where most of the $CaCO_3$ is. On the sea floor, only $CaCO_3$ on the very surface of the sediment will feel the change in ocean chemistry. Ultimately, any $CaCO_3$ within reach of the mixing of surface sediment by burrowing organisms could be in play. Broecker and Takahashi find that the oceans have access to about as much $CaCO_3$ as it would take to neutralize the potential amount of fossil fuel CO_2.

Walker and Kasting (1991) consider the impact $CaCO_3$ neutralization among a variety of other processes that can affect atmospheric CO_2, including uptake or release from the terrestrial biosphere, and the ultimate uptake of the carbon back to the solid Earth by reaction with the CaO component of igneous rocks to form sedimentary $CaCO_3$. The land surface contains enough carbon to be significant, especially when soil carbon is considered, but eventually it could be swamped by the amount of fossil carbon available. Land-use decision affect carbon fluxes, as well as a response of the natural landscape to changes in CO_2 concentration ("CO_2 fertilization") and climate. The idea of a "silicate weathering feedback" mechanism that stabilizes the climate of the Earth dates back to Urey (1957), and was developed in a quantitative way by Walker, Hayes, and Kasting in 1981, and Berner, Lasaga, and Garrells in 1983. Walker's thermostat was faster than Berner's, but both predicted that the silicate thermostat ought to take hundreds of thousands of years to control the atmospheric CO_2 concentration.

Broecker and Takahashi predicted that it should take thousands of years for $CaCO_3$ to neutralize the fossil fuel CO_2. They did not explicitly consider the impact of the $CaCO_3$ dissolution on the atmospheric CO_2 concentration, but Walker and Kasting find that a significant amount of CO_2 remains in the atmosphere even after neutralization by $CaCO_3$, leaving behind a million-year long tail to the elevated CO_2 concentrations in the atmosphere (their Figure 29).

The Paleocene Eocene Thermal Maximum (PETM) is an excursion of carbon and oxygen isotopes preserved in $CaCO_3$ sediments in the deep sea from 55 million years ago. The carbon isotopic composition tells us of a sudden release of CO_2 to the atmosphere. The source of the carbon is not yet pinned down, nor is the amount of the carbon. The oxygen isotopic composition of the $CaCO_3$ tells us of a global warming of about 5°C. Both the CO_2 and the warming decayed back toward their normal values on a time scale of about 100 000 years, consistent with the lifetime of fossil fuel CO_2 projected by Walker and Kasting (this volume). Although the murkiness of the details surrounding the PETM event make it impossible to really test the details of the climate models, such as the climate sensitivity to doubling CO_2 concentration, the PETM provides the clearest analogue in the geologic record to the potential severity and longevity of the global warming climate event in our future.

Neutralization of Fossil Fuel CO_2 by Marine Calcium Carbonate

W. S. BROECKER[1] AND T. TAKAHASHI[2]*

[1]*Lamont-Doherty Geological Observatory* and [2]*Queens College*

ABSTRACT

The $CaCO_3$ stored in marine sediments will ultimately neutralize the CO_2 generated by fossil fuel combustion. Details of this process are explored and a model of the early phases of this process in the western basin of the deep Atlantic Ocean is presented. The amount of $CaCO_3$ available for dissolution is derived from the calcite content of marine sediments and the extent to which these sediments are stirred by benthic organisms. The conclusion is that the available calcite is about equivalent to the CO_2 which would be released if our known resources of natural gas, oil and coal were burned. The rate at which this dissolution will proceed is estimated from the rate at which natural dissolution has proceeded during the Holocene. The conclusion is that if linear kinetics are followed, the time constant will be on the order of 1500 years. On the other hand, if the exponential kinetics found in the laboratory by Berner and Morse (1972) apply, then the time constant will be much shorter. It is possible that the dissolution process will become limited by the rate stirring of the mixed layer in deep sea sediments. The insoluble residue of clay minerals and opal built up in the upper few millimeters of the sediment as the result of dissolution must be mixed down into the underlying sediment if more than the upper few millimeters of sediment is to be attacked.

In order to model the change in the carbonate ion content of the waters in the deep Atlantic Ocean over the next few hundred years it is necessary to know:

1) the rate of chemical fuel use over this period;
2) the distribution of this CO_2 between the terrestrial biosphere, upper ocean and atmosphere;
3) the relationship between the $CO_3^=$ content of newly formed deep water and the pCO_2 in the atmosphere;
4) the ventilation time of the northern component of the deep water in the Atlantic Ocean and the distribution of this component in the deep mixing zone;
5) the relationship between the in situ $CO_3^=$ content of deep water and the $CO_3^=$ content necessary to trigger dissolution;
6) the relationship between the rate of dissolution of deep sea sediments and the degree of undersaturation of the bottom water.

Preliminary estimates have been made of these factors and they have been combined into a dissolution model.

INTRODUCTION

CO_2 generated by the combustion of fossil fuels will ultimately be neutralized through combination with sedimentary calcium carbonate via the reaction

$$CO_2 + CaCO_3 + H_2O \rightleftarrows 2HCO_3^- + Ca^{++} \quad (1)$$

Regardless of whether this dissolution takes place on land or in the sea, the calcium and bicarbonate ions generated will end up as part of the ocean's

---

* Present address: Lamont–Doherty Geological Observatory

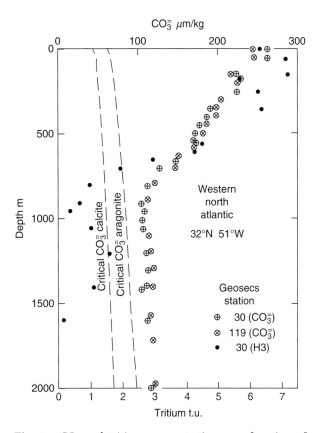

Fig. 1a. $CO_3^=$ and tritium concentration as a function of water depth in the upper water column at 32°N and 51°W in the western North Atlantic Ocean. Curves for the critical carbonate ion content for calcite and for aragonite are given for reference. They are based on those given by *Broecker and Takahashi* (in press) with a temperature correction for the decrease in solubility with increasing water temperature. The carbonate ion results were calculated from the titration data given in the GEOSECS leg reports using the constants adopted by *Takahashi et al.* (1976). At station 30 the leg report ΣCO_2 results were decreased by *15 μm/kg* to bring the leg 3 results into agreement with the *p*CO₂ data and with the results from the other GEOSECS Atlantic Ocean legs. The tritium results are those of *Ostlund et al.* (1974) on samples collected in Sept. 1972.

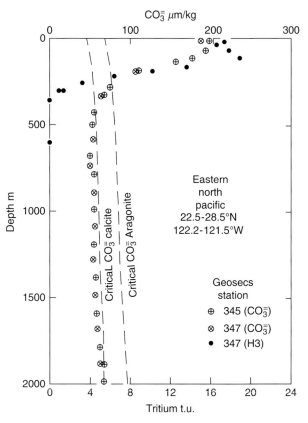

Fig. 1b. Similar results for the North Pacific Ocean (ST 345 23°N; 122°W; ST 347 29°N; 122°W). In accord with the study by *Broecker and Takahashi* (in press) the Pacific Ocean leg report ΣCO_2 results were reduced by *15 μm/kg* to bring the Pacific Ocean results into agreement with Pacific Ocean *p*CO₂ results and with the Atlantic Antarctic results. This correction increases the $CO_3^=$ concentrations by about *8 μm/kg*. The tritium results are those of *Roether* (1974) for samples collected in November 1971.

dissolved salt load. In this paper some of the details of this process are considered.

The primary contributor to this neutralization will be deep sea sediments. The deep sea will be gradually "acidified" by the downward mixing of surface water enriched in fossil fuel CO_2 and its sediments will become subject to enhanced calcium carbonate dissolution. Shallow marine deposits are less promising because of their relatively small area and their generally low $CaCO_3$ content, and because the water in which they are bathed is highly saturated with respect to both calcite and aragonite (see Figure 1). It is unlikely

that the CO_2 content of the atmosphere will ever become great enough to bring these waters to a state of undersaturation. Although enhanced continental weathering will contribute to the neutralization, the fact that the decay of organic material maintains the CO_2 content of soil and ground waters far higher than the atmospheric equilibrium content suggests that the higher atmospheric CO_2 contents will not have a very large impact.

For the purposes of this discussion, the sediments of the sea will be divided into five categories.

1) Those lying beneath the calcite compensation depth. As these sediments contain little calcium carbonate they will not contribute significantly to the neutralization process.
2) Those lying between the calcite compensation depth and the lysocline horizon (see Figure 2).

These sediments lie in the transition zone between depths where natural dissolution by bottom water is small and depths where natural

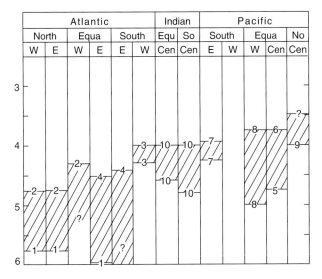

Fig. 2. Summary of lysocline and carbonate compensation depths for various oceanic regions as given by *Broecker and Takahashi* (in press). The numbers refer to the source of the information as given in the original paper.

dissolution by bottom water removes virtually all of the calcite from the accumulating sediment (see Figure 3). These sediments still have calcite available for the neutralization of fossil fuel CO_2. As the bottom water with which they are in contact is already understaturated, any further decrease in $CO_3^=$ content will produce an immediate effect.

3) Those lying above the lysocline and below a water depth of 2800 meters. These sediments are located mainly on the flanks of the oceanic ridges and rises and are on the average very rich in calcium carbonate (i.e., *>80%*). Although they currently lie in non-corrosive water, as the $CO_3^=$ content of the deep sea falls they will eventually come under attack. The time elapsed before the onset of dissolution in any given area will increase with elevation above the lysocline. (The shallower the sediment the more the $CO_3^=$ content will have to fall before dissolution is triggered.)

4) Those sediments lying above the ridge crests but beneath the 100 fathom contour. These sediments lie primarily along the continental rises and in marginal seas. They are in general low in calcite content (e.g., ~25%) because of dilution

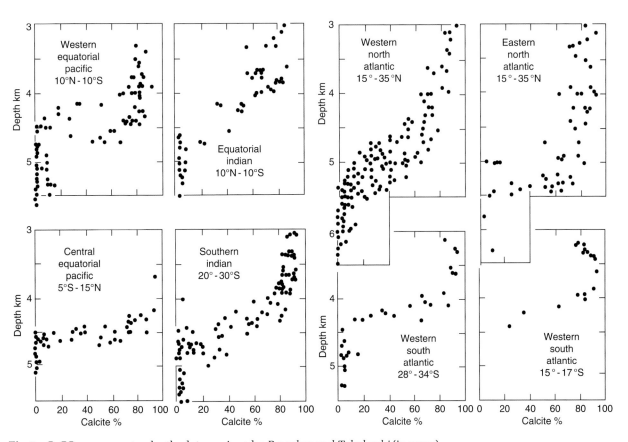

Fig. 3. *CaCO₃* versus water depth plots as given by *Broecker and Takahashi* (in press).

Table 1. Sediment Area Distribution*

(Units 10^{14} m^2)

Region	Depth Range	Atlantic	Indian	Pacific	Total
Above Shelf break	<200	0.14	0.03	0.10	0.27
Shelf break to Ridge crest	200–2800	0.22	0.11	0.21	0.54
Ridge crest to Lysocline	–	0.35	0.18	0.18	0.71
Lysocline to Compensation	–	0.21	0.14	0.14	0.49
Below Compensation	–	0.14	0.30	1.15	1.59
TOTAL	–	1.06	0.76	1.78	3.60

* Based on the area versus water depth data given in Sverdrup (1942) and the lysocline and compensation depths given in Fig. 2.

Table 2. Available $CaCO_3$ in Deep Sea Sediment*

(Units 10^{16} moles)

Region	Atlantic	Indian	Pacific	Total
Shelf break to Ridge crest	0.7	0.3	0.6	1.6
Ridge crest to Lysocline	18.0	9.2	9.2	36.4
Lysocline to Compensation	1.3	0.8	0.9	3.0
TOTAL	20.0	10.3	10.7	41

$$* = \text{Dry density (i.e., } \sim 1\,\text{gm/cm}^3) \times \text{Area} \times \frac{\text{Mixing Depth (i.e., 9 cm)}}{\text{Frac. non} - \text{CaCO}_3} \times \text{Frac. CaCO}_3$$

with continental detritus. Since the acidification of the waters which they contact will occur more rapidly than that of the subridge crest waters, their contribution could be significant during the *early* stages of the neutralization process.

5) Shallow water sediments (e.g., <200 meter depths). The waters in this depth range will respond very quickly (i.e., a few years) to changes in atmospheric CO_2 content. Those sediments in contact with mixed layer water (i.e., upper 30–150 meters) will not, however, be subject to attack because the mixed layer is too highly supersaturated with respect to calcite and aragonite to ever become corrosive. However, as the $CO_3^=$ content drops sharply with depth below the mixed layer in some regions of the ocean, dissolution of sediments in contact with upper thermocline waters could contribute to the neutralization of fossil fuel CO_2. Assessment of this potential will prove complex. Region by region studies will be necessary. Examples are given in Figure 1. Certainly on the long time scale these sediments will prove far less important than deep sea sediments in neutralizing CO_2. Their area and $CaCO_3$ content are just too low.

The depth of the lysocline varies from place to place in the world ocean ranging from 4.7 km in

the North Atlantic Ocean to less than 3.0 km in the North Pacific Ocean. The thickness of the transition zone beneath the lysocline varies from 200 to 1000 meters (see Figure 2). Using this information, the area of each of these sediment types has been estimated (see Table 1) using the area versus water depth data given by *Sverdrup et al.* (1942).

As of today about 1.4×10^{16} moles of CO_2 have been generated through the burning of gas, oil and coal (*Broecker et al.*, 1971). If the production rate is increased by 2% per year, then in the year 2100, 18×10^{16} moles of CO_2 will have been produced. Our total reserves of coal, oil and gas, if burned, would produce about 50×10^{16} moles of CO_2 (*Hubbert*, 1972). How much sediment $CaCO_3$ is available for the neutralization of this CO_2? An estimate of this amount can be obtained from the following expression:

$$\textit{Sediment density} \times \textit{frac.}\ CaCO_3 \times \frac{\textit{burrowing depth}}{\textit{frac. non} - CaCO_3} \quad (2)$$

The assumption is that once a carbonate-free layer one burrowing depth thick mantles the sea floor, the underlying calcium carbonate will be immune to dissolution regardless of the acidity of the overlying water. From radiocarbon data on deep sea cores it appears that the mean depth of burrowing is about $9\,cm$ (see *Peng et al.*, 1977). The following average $CaCO_3$ contents are adopted for the sediment provinces mentioned above:

shelf break – ridge crest	25%
ridge crest – lysocline	85%
lysocline – compensation	40%

The corresponding amounts of available $CaCO_3$ are shown in Table 2. Two points stand out. First, the amount of "available" $CaCO_3$ is about equivalent to the amount of CO_2 locked up in recoverable fuel. Second, about half of this calcium carbonate lies in the Atlantic Ocean.

BASIS FOR FUTURE DISSOLUTION RATE ESTIMATES

In order to calculate the rate at which the deep sea $CaCO_3$ will dissolve, four basic pieces of information are needed:

1) projected atmospheric CO_2 contents;
2) the carbonate ion content of newly formed deep water;
3) the rate of fossil fuel CO_2 neutralization by sediment per unit drop in the carbonate ion content of the ocean water in contact with the sediment; and
4) the ventilation times for various deep water masses.

For the decade prior to the sharp increase in oil prices, the use of chemical fuels was increasing by about 4.5% per year. Since the petroleum price hike this pace has slowed. As comprehensive energy policies have yet to be developed, any projection of chemical fuel use into the future is bound to be little more than a guess. For the calculations presented here, a 2% per year increase in the growth rate of the atmospheric CO_2 content until the year 2100 (see Figure 4) has been adopted and it has been further assumed that no further increase will occur during the following century. This choice allows a contrast of the situation during a period of exponential growth to that during a period of atmospheric stability.

As about half of the deep water in the ocean is generated at the northern end of the Atlantic

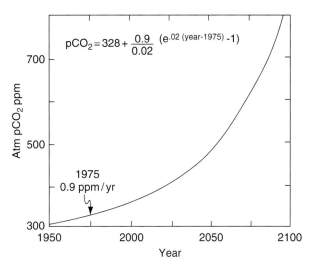

Fig. 4. Partial pressure of CO_2 versus time calculated assuming that the $0.9\,ppm/yr$ increase rate measured by Dr. C. Keeling and his coworkers of Scripps Institution of Oceanography over the last several years will increase by 2% per year through the next century.

Ocean, it is appropriate to take the waters in this region as an example. Table 3 summarizes the composition of a number of surface waters in the northern Atlantic Ocean. These waters are more saline and, of course, much warmer than those descending to form North Atlantic Deep Water (NADW). If these waters are cooled and freshened at equilibrium with the atmosphere, then, as shown in Table 4, they must change significantly in their ΣCO_2 content. The question is whether these waters maintain equilibrium with the atmosphere as they are cooled. The best way to answer this question is to look at newly produced deep water. As shown in Figure 5 there are three major contributors to NADW; water produced at the surface of the Labrador Sea (LSW), water produced in the Norwegian Sea which enters the Atlantic Ocean by spillage over the Denmark Straits (DSW) and water produced in the Norwegian Sea which enters the Atlantic Ocean by spillage over the Iceland-Scotland rise followed by passage through the Gibbs Fracture Zone (GFZW). As shown in Table 5 these three sources have similar alkalinities and ΣCO_2. All are deficient in O_2, presumably because of respiration at depth in their source regions. If the composition of these waters prior to alteration by *in situ* respiration is reconstructed by bringing the O_2 content to the atmospheric contact value (i.e., 5% supersaturated), then the CO_2 partial pressure drops to roughly the atmospheric value and the ΣCO_2 content and $CO_3^=$ contents

Table 3. Composition of Some North Atlantic Warm Surface Water Samples*

GEOSECS STATION NO.	24	25	26	27	28	29	30	119
T °C	11.2	20.3	21.8	22.4	25.0	25.2	26.6	21.1
S ‰	34.7	35.6	35.8	36.1	36.4	36.0	36.6	36.7
O_2 μm/kg	291	229	220	218	210	209	205	230
SiO_2 μm/kg	1.9	–	0.5	0.8	0.6	1.3	0.9	0.8
PO_4 μm/kg	0.41	–	0.09	0.03	0.02	0.05	0.04	0.04
NO_3 μm/kg	5.0	–	0.0	0.1	0.0	0.0	0.0	0.0
ALK μeq/km	2278	2308	2308	2359	2377	2357	2384	2389
ΣCO_2 μm/kg	2088	2073	2064	2085	2039	2007	2020	2040
$H_2BO_3^-$	67	79	82	90	110	112	116	113
$2CO_3^= + HCO_3^-$	2211	2229	2226	2269	2267	2245	2268	2276
$HCO_3^{-2}/CO_3^= CO_2$	1555	1345	1310	1285	1233	1233	1187	1306
HCO_3^- μm/kg	1931	1885	1872	1873	1789	1749	1752	1784
$CO_3^=$ μm/kg	140	172	177	198	239	248	258	246
CO_2 μm/kg	17.0	15.3	15.1	13.8	10.8	10.0	10.0	9.9
pCO_2 10^{-6} atm	407	480	490	455	383	355	368	315

* Based on the titration alkalinity and total dissolved inorganic carbon data, nutrient element and dissolved oxygen and hydrographic data given in the GEOSECS leg reports. The constants used are those adopted by Takahashi *et al.* (in press).

Table 4. Station 119 Surface Water Freshened and Cooled at Equilibrium with the Atmosphere*

	21.1°C 36.7%	2.25°C 34.93‰
$HCO_3^{-2}/CO_3^= \cdot CO_2$	1306	1725
ALK μeq/km	2389	2273
$H_2BO_3^-$ μm/kg	113	65
$2CO_3^= + HCO_3^-$	2275	2209
HCO_3^-	1785	1953
$CO_3^=$	245	128
CO_2	9.9	17.5
pCO_2 10^{-6} atm	315	329
σCO_2 μm/kg	2040	2098

* The constants used are those adopted by Takahashi *et al.* (in press).

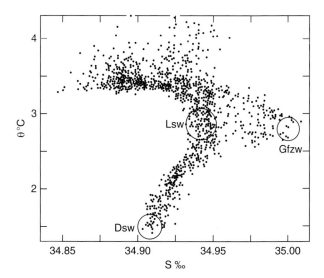

Fig. 5. Potential temperature versus salinity diagram for deep waters found at the northern end of the western basin of the North Atlantic Ocean. The end members are DSW, water spilling over the Denmark Straits; GFZW, water entering the western basin through the Gibbs Fracture Zone; and LSW, water generated at the surface of the Labrador Sea. A mixture of these three end members forms the northern component to basal North Atlantic Deep Water (NADW).

become similar to that for the cooled and freshened northern Atlantic Ocean surface water (see Table 4). Hence, the contact time between the atmosphere and the cooling water appears to be long enough to bring the ΣCO_2 content to equilibrium with the atmosphere. Thus, an estimate as to how the $CO_3^=$ content of downwelling deep water will change with atmospheric CO_2 content can be obtained by considering either "freshened and cooled" northern Atlantic Ocean surface water (Table 4) *or* "oxygenated" NADW (Table 5). An example using the latter is shown in Table 6. For a 50% increase in atmospheric CO_2 content, the $CO_3^=$ content drops by 27%. As shown in Table 7

this result is not dependent on the exact composition of the water. When the same calculation is made for warm surface water, the carbonate ion content drops by 24% for a 50% increase in atmospheric CO_2 content. In Figure 6 a plot of the

Table 5. North Atlantic Deep Water Components (GFZW, DSW, LSW); Their Average and Their "Oxygenated" Average+

Water Type	GFZW	DSW	LSW	NADW	NADW*
Station	24	25	26	–	–
Depth (m)	2470	4550	1990	–	–
θ °C	2.81	1.78	3.62	2.25	2.25
S ‰	35.00	34.91	34.95	34.93	34.93
O_2 μm/kg	276	288	274	280	345
SiO_2 μm/kg	13.9	–	12.0	13	–
NO_3^- μm/kg	16.1	–	16.8	17	10
$PO_4^{\equiv}$ μm/kg	1.06	–	1.16	1.14	0.66
ALK μeq/kg	2292	2290	2288	2290	2290
ΣCO_2 μm/kg	2168	2172	2168	2168	2119
$H_2BO_3^-$ μm/kg	50	48	48	49	63
$2CO_3^{\equiv}+HCO_3^-$	2242	2242	2240	2241	2227
HCO_3^-	2044	2051	2046	2046	1975
$CO_3^{\equiv}$	99	95	97	98	126
CO_2	25	25	25	24	18
pCO_2 10^{-6} atm	436	433	463	430	310

* to 5% O_2 supersaturation assuming $\Delta O_2/\Delta \Sigma CO_2 = -1.33$
+ The hydrographic, nutrient element and dissolved oxygen data are from the GEOSECS leg reports. The constants used are those adopted by Takahashi *et al.* (in press).

Table 6. Change in Composition of Sinking NADW with Increasing Atm. CO_2 Content*

Year	1975	2050	Δ	$\dfrac{2050}{1975}$
$HCO_3^{-2}/CO_3^{\equiv} \cdot CO_2$	1725	1725	–	–
ALK μeq/kg	2290	2290	0	–
$H_2BO_3^-$ μm/kg	63	46	–17	–
$2CO_3^{\equiv} + HCO_3^-$	2227	2244	+17	–
HCO_3^- μm/kg	1975	2062	+85	–
$CO_3^{\equiv}$ μm/kg	126	91	–35	0.73
CO_2 μm/kg	18	27	+9	1.50
pCO_2 10^{-6} atm	310	468	–	1.50
ΣCO_2 μm/kg	2119	2180	+61	1.029

$$\frac{\%\,\text{inc atm}\,CO_2}{\%\,\text{inc ocean}\,CO_2} = \frac{50}{2.9} = 17.2$$

* The constants used are those adopted by Takahashi *et al.* (in press).

fractional change in $CO_3^{\equiv}$ content of "oxygenated" NADW against atmospheric CO_2 content is given.

The most reliable way to assess the rate of attack caused by a given lowering of the $CO_3^{\equiv}$ content of the water in contact with the sediment is to use the Holocene record for calibration. This record is interpreted to indicate that no measurable calcite dissolution will occur until the $CO_3^{\equiv}$ content drops below what is herein called the critical carbonate ion content given by the following equation (*Broecker and Takahashi*, in press).

Table 7. Change in Composition of Warm Surface Water with Increasing Atm. CO_2 Content*

Year	1975	2050	Δ	$\dfrac{2050}{1975}$
$HCO_3^{-2}/CO_3^{\equiv} \cdot CO_2$	1306	1306	–	–
ALK μeq/kg	2389	2389	0	–
$H_2BO_3^-$ μm/kg	114	87	–27	–
$2CO_3^{\equiv} + HCO_3^-$	2275	2302	+27	–
HCO_3^- μm/kg	1785	1924	143	–
$CO_3^{\equiv}$ μm/kg	245	187	–58	0.76
CO_2 μm/kg	10	15	+5	1.50
pCO_2 10^{-6} atm	317	484	–	1.50
ΣCO_2 μm/kg	2040	2130	90	1.044

$$\frac{\%\,\text{inc. atm}\,CO_2}{\%\,\text{inc. ocean}\,CO_2} = \frac{50}{4.4} = 11.4$$

* The constants used in these calculations are those adopted by Takahashi *et al.* (in press).

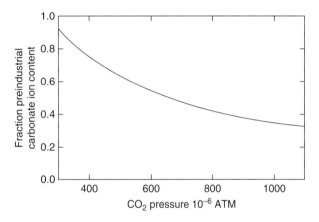

Fig. 6. $CO_3^{\equiv}$ content of input northern component water (Table 6) relative to its pre-industrial value as a function of atmospheric CO_2 content. The pre-industrial $CO_3^{\equiv}$ ion content is calculated to be *135 μm/kg* (compared to *126 μm/kg* in 1975).

$$\left[CO_3^{\equiv}\right]_{\text{CRIT}} = 93\,e^{0.14(z-4)} \;\; \mu m/kg \qquad (3)$$

The shape of the transition zone calcite content profile can be satisfactorily explained if the dissolution rate, R, is assumed to follow the equation

$$R = \sqrt{f}\; 0.025\left(\left[CO_3^{\equiv}\right]_{\text{CRIT}} - \left[CO_3^{\equiv}\right]\right)$$
$$\text{moles}/m^2\,yr \qquad (4)$$

Where f is the fraction of calcite in the sediment (*Broecker and Takahashi*, in prep.)[1]. For small drops in $CO_3^=$ content (i.e., up to $15\,\mu m/kg$) this relationship should be quite valid. For larger drops (20–$50\,\mu m/kg$), its validity depends on whether dissolution follows linear kinetics or Morse–Berner exponential kinetics (*Morse and Berner*, 1972). As it is not yet known what constitutes the rate limiting step for dissolution, it is not possible to say for sure which type of kinetics applies. Therefore, both will be considered.

Through radiocarbon dating, a general idea of the ventilation times for most parts of the deep sea has been developed. They range up to about 1500 years for the deep water in the North Pacific Ocean (*Bien et al.*, 1965). The deep Atlantic Ocean, which is of considerable interest because of its large calcite reserves, is ventilated on the time scale of several hundred years (*Broecker et al.*, 1960; *Stuiver*, 1976). The presence of bomb-produced tritium and radiocarbon in upper thermocline waters demonstrates that this region of the sea is ventilated on the time scale of a few years to a few tens of years (see Figure 1).

THE TIME CONSTANT FOR SEDIMENT DISSOLUTION

A rough idea of the rate at which deep sea sediments will dissolve can be obtained. Assume that 23×10^{16} moles of CO_2 are generated through fossil fuel burning and allowed to come to equilibrium with the ocean. For simplicity, the ocean will be taken to be entirely NADW. In such a case the pCO_2 of the atmosphere would be 660×10^{-6} atm. The deep ocean ΣCO_2 content would then be $120\,\mu m/kg$ higher than during pre-industrial time and the $CO_3^=$ content of the deep water about half its pre-industrial value (i.e., down by about $55\,\mu m/kg$). Of the 23×10^{16} moles of CO_2 generated, 6.7×10^{16} would reside in the atmosphere and 16.3×10^{16} in the ocean. The amount of excess CO_2 per m^2 of ocean surface would be 640 moles.

Under such circumstances, deep sea sediments would be subjected to waters with carbonate ion deficiencies (relative to the critical carbonate ion content) ranging from about $55\,\mu m/kg$ for the transition zone to between 15 and $25\,\mu m/kg$ at the ridge crests. The average for all the $CaCO_3$-bearing sediments would be about $40\,\mu m/kg$. Using the dissolution coefficient of 0.025 moles/m^2 yr per $\mu m/kg$ undersaturation obtained for Holocene dissolution in the ocean (*Broecker and Takahashi*, in prep.) and linear kinetics, yields an average of 1 mole/m^2 yr excess dissolution from the carbonate-

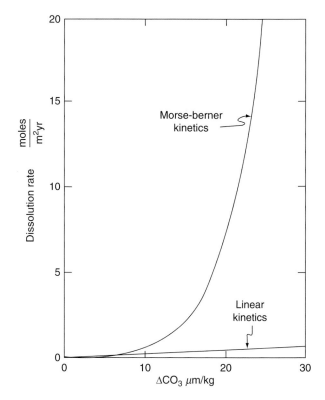

Fig. 7. Comparison of the dissolution rates for the linear kinetic model as calibrated using the Holocene $CaCO_3$ distribution with that for the exponential kinetic model as calibrated using the Holocene $CaCO_3$ distribution. The assumptions of these models are outlined by *Takahashi and Broecker* (1977).

bearing sediments of the deep ocean. Since the area of deep floor bearing $CaCO_3$-rich sediment is about half the total area (see Table 1), this corresponds to 0.5 mole/m^2 yr dissolution rate as referred to the *entire* ocean floor. Thus, the initial rate of dissolution in this well mixed ocean would be an amount of $CaCO_3$ equivalent to one part in 1280 of the excess CO_2 in the ocean-atmosphere system. This yields a time constant (i.e., about 1280 years) for neutralization *comparable* to the time scale for deep ocean ventilation. Thus, any realistic model must consider dissolution and deep mixing together.

If Morse-Berner exponential kinetics (*Morse and Berner*, 1972) are used, then as shown in Figure 7, dissolution rates many times higher would be expected. They would be so high in fact, that the sediments would neutralize the excess CO_2 roughly as fast as it arrived in the deep sea.

Of course, as the dissolution proceeded, dilution of the sedimentary calcite by the buildup of a noncalcite residue would slow down the dissolution process. Eventually dissolution would become limited by the rate at which the residue was stirred into the sediment column by benthic burrowers. Thus, any realistic model of the fate of CO_2 will have to simultaneously deal with oceanic mixing, interface dissolution *and* sediment stirring.

A MODEL FOR THE DISSOLUTION OF SEDIMENT IN THE WESTERN BASIN OF THE ATLANTIC OCEAN

Since the NADW ventilation time is small compared to that for the deep Pacific Ocean, events in the Atlantic Ocean will dominate during the next hundred or so years. The deep mixing model selected here is designed with this fact in mind.

The radiocarbon data from the GEOSECS Program show that the water in the western basin of the central deep Atlantic Ocean is replaced on a time scale of about 200 years (*Stuiver*, 1976). Furthermore the "age" of the water does not show any strong dependence on depth or latitude. Thus, in modeling it is appropriate to assume that throughout the western basin the NADW component is being replaced at the rate of one part in 200 each year. The water exiting the western basin of the central Atlantic Ocean is assumed to pass in part directly to the Antarctic and in part to the eastern basin of the Atlantic Ocean.

The water below the ridge crests in the Atlantic Ocean is a mixture of NADW and Antarctic Bottom Water (AABW). The salinity section (Figure 8) through the western basin shows the geographic and depth dependence of the composition of these mixtures. The end members have salinities of 34.93‰ (NADW) and 34.67‰ (AABW). The 34.80

isohaline thus marks a 50–50 mixture of the two end members.

As AABW contains a large component of recirculated Deep Pacific Ocean Water which has *not* reequilibrated with the atmosphere, its $CO_3^=$ content will be slow to change. The time scale for its alteration will be many hundreds of years. Thus, for the calculations to be carried out here, the $CO_3^=$ content of AABW will be assumed to remain unchanged. Therefore, as the mixture is renewed, only the NADW component will change in $CO_3^=$ content (at least during the next hundred or so years).

Plots of $CO_3^=$ content for deep waters in the northern and southern parts of the central Atlantic Ocean are shown in Figure 9 along with the critical carbonate ion curve (i.e., the $CO_3^=$ content at which the dissolution rate becomes geologically significant). In the northern western basin where NADW dominates the deep water column, the present day crossover depth (i.e., depth where the *in situ* $CO_3^=$ content becomes less than the critical $CO_3^=$ content) is about 4700 meters. In the western basin of the South Atlantic Ocean, where AABW underrides and mixes upward into the NADW, the crossover is at a shallower depth (~4000 meters).

Figure 10 illustrates what the situation will be when the ion content of NADW has dropped by *17 μm/kg*. The crossover depth at this time will have risen to about 3700 meters throughout the Atlantic Ocean. The sediments in the transition zone between the pre-industrial lysocline and the compensation depths will be subjected at that time to much more vigorous dissolution than during the pre-industrial era. Superlysocline sediments will also become subject to attack by the acidified deep water.

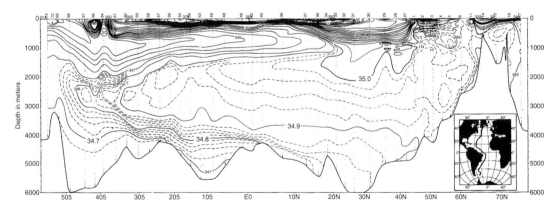

Fig. 8. Salinity section along the GEOSECS track in the western basin of the Atlantic Ocean. The mixing zone of interest to the $CaCO_3$ dissolution problem is that below 3 kilometers where AABW water ($S = 34.67$‰) mixes up into NADW water ($S > 34.90$‰).

(a)

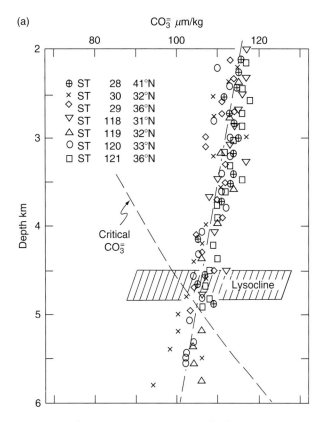

Fig. 9a. Carbonate ion content versus depth in the western North Atlantic Ocean as calculated from the GEOSECS titration data using the constants adopted by *Takahashi et al.* (in press). The lysocline depth is based on the data of *Kipp* (1976). The critical carbonate ion content curve is that of *Broecker and Takahashi* (in press).

(b)

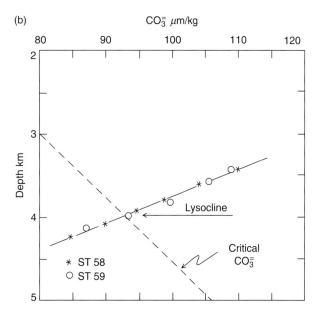

Fig. 9b. Carbonate ion content in the basal NADW-AABW mixing zone of the western South Atlantic as calculated from potential temperature using the relationship given by *Broecker and Takahashi* (in press). The critical carbonate ion content curve is given for reference.

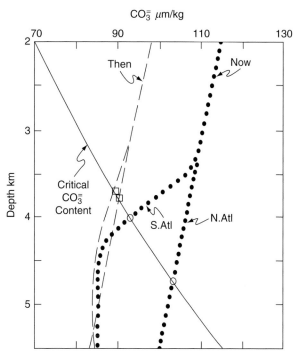

Fig. 10. $CO_3^=$ versus depth curves for the western basin of the Atlantic Ocean (dotted curves) prior to the introduction of fossil fuel CO_2 and after fossil fuel CO_2 introduction has reduced carbonate ion content of the NADW component of the mixing zone water by $17 \mu m/kg$ (dashed curves). The circles represent the present depths at which the *in situ* $CO_3^=$ content currently cross the critical value and the squares the depths at which these crossover will lie after the NADW has been "acidified" to the level where its $CO_3^=$ ion content has dropped $17 \mu m/kg$.

In the sections that follow, an attempt will be made to model the attack of the sediments in the western basin of the Atlantic Ocean by the progressively acidified NADW. A simple box model approach is used. The NADW component of waters below 3000 meters in the western basin of the Atlantic Ocean is assumed to be well mixed. Three processes tend to change the ΣCO_2 content and $CO_3^=$ content of this water.

1) The input of new NADW progressively enriched in total dissolved inorganic carbon: The volume of water in the western basin of the Atlantic Ocean beneath 3000 meters between 30°S and 40°N is $32 \times 10^{15} m^3$. Of this about 75% is NADW. If this water is to be replaced once every 200 years, the flux of new NADW must be $1.2 \times 10^{14} m^3/yr$ (i.e., about 4 Sverdrups). The $CO_3^=$ content of this input is calculated as follows.

$$\left[CO_3^=\right]_{INPUT}^t = \frac{[CO_3^=]_{INPUT\ NADW}^t}{[CO_3^=]_{INPUT\ NADW}^{1850}}[CO_3^{=\ 1850}_{NADW}] \quad (5)$$

The ratio $\left[CO_3^=\right]^t_{INPUT\ NADW} \Big/ \left[CO_3^=\right]^{1850}_{INPUT\ NADW}$ is that given in Figure 6. In order to eliminate the subsurface processes taking place in the natural (i.e., pre-industrial) system, the actual input value is taken to be the pre-industrial mean $CO_3^=$ content of *in situ* NADW (i.e., *115 μm/kg*) times this ratio. This assumes that the natural respiration processes and the natural dissolution processes continue at the same rate as in pre-industrial time. Since these processes are limited by the availability of NO_3^- and $PO_4^\equiv$, rather than by CO_2, this assumption is likely valid. As the alkalinity of the incoming water should remain unchanged, the carbonate system is uniquely defined by assigning the carbonate ion content of the input water.

2) Loss of excess industrial CO_2 by outflow from the western basin to the eastern basin and to the Antarctic: The excess CO_2 concentration in the exiting water is assumed to be equal to the mean excess for the water mass itself.

3) Addition of Ca^{++} and $CO_3^=$ to the deep water through excess dissolution of deep sea sediments (the natural dissolution has been taken care of through our handling of the input concentrations): For sediments below the lysocline the rate of excess dissolution at any point on the sea floor depends on the decrease in the $CO_3^=$ content of the water from its pre-industrial value. In a separate paper (*Broecker and Takahashi,* in prep.) it will be shown that for sediments moderately rich in $CaCO_3$ (i.e., >40%), dissolution proceeds at a rate of .025 moles/m^2 yr per μm/kg decrease in the $CO_3^=$ content below the critical carbonate ion content. It will be assumed that for each additional decrease of *1 μm/kg* in $CO_3^=$ content the rate of dissolution of these carbonate-rich sediments will rise by 0.025 moles/m^2 year (i.e., linear kinetics). For sediments above the natural lysocline, no dissolution will take place until the $CO_3^=$ content of the water reaches the critical carbonate ion value. Once it drops below this value the same dissolution rate versus $\Delta CO_3^=$ relationship is used as for the sub-natural lysocline sediment. In the North Atlantic Ocean the lysocline will rise 100 meters per *1.7 μm/kg* drop in the $CO_3^=$ content of NADW. Thus as the $CO_3^=$ content drops, both the area of sediment exposed to excess dissolution *and* the rate of the excess dissolution at any given depth will increase.

Before launching out on the model calculation, one basic assumption needs justification. Are the products of the dissolution well distributed in the deep water column? As the volume of water per unit of area of sediment contacted decreases with each depth slice down the water column, if vertical mixing does *not* occur then the impact of the dissolution process will increase strongly with water depth.

Whether or not the products of calcite dissolution are mixed throughout the water column can be best tested by considering the natural steady state. In the northern western basin nearly all the carbonate falling below about 5100 meters dissolves. In the southern western basin the corresponding depth is 4400 meters. The area of sediment nearly free of $CaCO_3$ in the western basin is about $10 \times 10^{12}\ m^2$. If the mean rain rate of $CaCO_3$ is taken to be *0.8 gm/cm² 10³ yrs* (i.e., 8×10^{-6} moles/cm² yr) and if the products of this dissolution are assumed to have been confined to the volume below the top of the dissolution zone (volume of 6×10^{15} m³) then during its residence time of 200 years the water would increase in alkalinity by *55 μeq/kg*. There would, of course, then be no alkalinity increase for waters above the dissolution zone. The observed increase is on the order of only *10 μeq/kg* (see Figure 11) and shows no strong correlation with depth or geographic location. On the other hand, if the dissolution products are mixed upward to a depth of 3000 meters (volume of 32×10^{15} m³), then the expected alkalinity increase would be only *10 μeq/kg* – a value roughly consistent with the GEOSECS results. The near uniformity of the alkalinity increase throughout the mixing zone beneath the Two Degree Discontinuity (TDD) suggests that the products of dissolution are not confined to the level from which they were released, but rather that they are mixed well up into the water column. Mixing along isopycnal surfaces (i.e., along the isohalines in Figure 8) could produce the observed distribution.

The model calculation is begun in the year 1900. Since the amount of CO_2 added to the NADW mass up to this time was quite small, it will be assumed that the NADW was still at its steady state $CO_3^=$ content (i.e., *115 m/kg*) and at its mean alkalinity (*2330 μeq/kg*). The calculation is carried out in 10 year increments. The CO_2 pressure in the atmosphere and the composition of the input water are taken to be those for the mid-point of the decade over which the calculation is being carried out. The $CO_3^=$ content of the NADW reservoir used in the dissolution and outflow calculations is taken to be that calculated for the end of the preceding decade. At the end of each decade

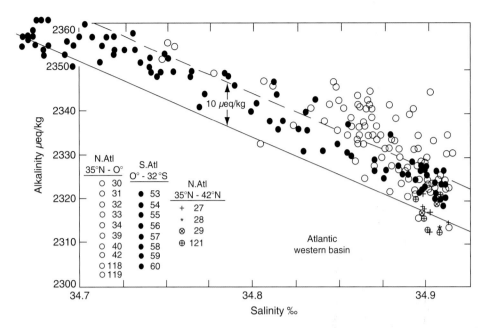

Fig. 11. Alkalinity versus salinity in the mixing zone between AABW and basal NADW. The lower solid line represents the values expected if there were no $CaCO_3$ dissolution in this mixing zone. The upper solid line represents the alkalinity expected if the products of natural $CaCO_3$ dissolution were spread uniformly through this mixing zone.

the net change in the ΣCO_2 content and in the alkalinity of the NADW is recomputed. The $CO_3^=$ content obtained from these values is used for the next decade.

The area of sediment in the *western* basin of the Atlantic Ocean lying between the lysocline and the compensation depth is about 7×10^{12} m². These sediments are assumed to yield 0.025 moles of excess $CaCO_3$ dissolution per square meter year for each $\mu m/kg$ drop in the $CO_3^=$ ion content of NADW. The sediments beneath the compensation depth are assumed to contribute nothing. The area of sediment above the natural lysocline contributing to the excess dissolution is taken to be 0.4×10^{12} m² per $\mu m/kg$ drop in $CO_3^=$ content of the NADW reservoir. As the difference between the critical carbonate ion content and the *in situ* carbonate ion content for this depth interval *averages* one half the drop in carbonate content of the NADW, a factor of 1/2 must be introduced to account for this gradient in $\Delta CO_3^=$ in the super-lysocline region. Thus the dissolution rate, R, is calculated as follows:

$$R = \Delta CO_3^= \left(A_{L-C} + \frac{1}{2} A_{<L} \right) b \qquad (6)$$

where

$$\Delta CO_3^= = CO_3^{= \, INITIAL}_{NADW} - CO_3^{= \, t}_{s \, NADW}$$

A_{L-C} = area of sediments between the natural lysocline and natural compensation depth

$A_{<L}$ = area of sediment above the natural lysocline which has come under attack (i.e., $0.4 \times 10^{12} \times \Delta CO_3^=$)

b = coefficient of dissolution (i.e., 0.025 moles/m² yr per $\mu m/kg$)

The results of this calculation for the period between 1900 and 2105 are given in Table 8. This is the period where the rate of the CO_2 content increase for the atmosphere is assumed to rise by 2% per year.* The yearly increase rate goes from 0.22 ppm/yr in 1905 to 0.90 ppm per year in 1975 to 12.1 ppm in 2105. The results for the period 2105 to 2205, during which the atmospheric CO_2 content is held constant, are shown in Table 9. At the end of this period the dissolution rate of $CaCO_3$ from deep Atlantic Ocean sediment becomes about one third the input rate of excess CO_2. At steady state (i.e., when the NADW has come to equilibrium with this new atmospheric CO_2 content) about half of the incoming excess will be neutralized within the Atlantic Ocean. Table 10 summarizes the change in chemical composition of the input water to the NADW and of the mean NADW itself over this period. Table 11 compares the fluxes of excess CO_2 into the NADW and of $CaCO_3$ dissolution of sediments in contact with NADW with the flux of CO_2 into the atmosphere.

Table 8. Western Basin Atlantic Model Calculation Carried Out Over the Period During Which the Atmospheric CO_2 Rise Rate is Assumed to be Increasing by 2% per Year

	pCO_2 10^{-6} atm	$[CO_3^=]$ Input to NADW µm/kg	$\Delta\Sigma CO_2$ Input to NADW µm/kg	$[CO_3^=]$ NADW µm/kg	$\Delta\Sigma CO_2$ NADW µm/kg	$\Delta CO_3^=$ NADW µm/kg	Increase ΣCO_2 via inflow 10^{11} moles/yr	Increase ΣCO_2 via dissolution 10^{11} moles/yr
1900–1910	294	109	5	115	0.0	0.0	6.0	0.0
1910–1920	297	108	6	115	0.2	0.1	7.2	0.2
1920–1930	300	107	8	115	0.4	0.2	9.6	0.4
1930–1940	303	106	10	115	0.7	0.3	12.0	0.5
1940–1950	308	105	13	115	1.1	0.5	15.6	0.9
1950–1960	313	104	16	114	1.6	0.7	19.2	1.3
1960–1970	320	102	19	114	2.1	1.0	22.8	1.8
1970–1980	328	101	23	114	2.8	1.3	27.6	2.3
1980–1990	338	99	27	113	3.6	1.6	32.4	2.9
1990–2000	350	97	32	113	4.6	2.0	38.4	4.0
2000–2010	365	94	38	113	5.8	2.4	45.6	4.8
2010–2020	383	91	46	112	7.1	3.0	55.2	6.0
2020–2030	405	86	55	111	8.8	3.6	66.0	7.2
2030–2040	432	82	65	111	10.8	4.4	78.0	8.8
2040–2050	465	77	75	110	13.1	5.3	90.0	10.6
2050–2060	506	72	86	109	15.8	6.3	103.2	12.6
2060–2070	555	67	97	108	18.8	7.4	116.4	16.3
2070–2080	615	61	110	106	21.9	8.5	132.0	18.7
2080–2090	689	55	125	105	25.8	9.8	150.0	21.6
2090–2100	779	49	139	104	30.2	11.3	166.8	24.9
2100–2110	888	44	157	101	35.1	13.0	188.4	28.6

Table 9. Western Basin Atlantic Model for Period During Which the Atmospheric CO_2 Content is Assumed to Remain Constant

	$CO_3^=$	$\Delta\Sigma CO_2$	$\Delta CO_3^=$	ΣCO_2 In	ΣCO_2 Out	ΣCO_2 Dis moles/	ΣCO_2 Net
	⟵	µm/kg	⟶	⟵	10^{11}	yr	⟶
2100–2110	101	35	13.0	188	−42	29	175
2110–2120	100	41	14.8	188	−49	33	172
2120–2130	99	46	16.4	188	−55	36	169
2130–2140	97	51	18.0	188	−61	40	167
2140–2150	96	56	19.4	188	−67	46	167
2150–2160	94	62	20.6	188	−74	50	164
2160–2170	93	67	21.6	188	−80	52	160
2170–2180	93	72	22.4	188	−86	54	156
2180–2190	92	77	23.2	188	−92	56	152
2190–2200	91	82	23.8	188	−98	58	147
2200–2210	91	86	24.3	188	−102	60	146

$pCO_2^{atm} = 888$ ppm

$\Sigma CO_{2\ input} = 2282\,µm/kg$

$ALK_{input} = 2290\,µeq/kg$

$CO_{3\ input}^= = 44\,µm/kg$

$\Sigma CO_{2\ NADW}^{pre\text{-}ind.} = 2125\,µm/kg$

$ALK_{NADW}^{pre\text{-}ind.} = 2290\,µm/kg$

$CO_{3\ NADW}^{=\ pre\text{-}ind.} = 115\,µm/kg$

How long can this go on? At a $CO_3^=$ depletion of $20\,µm/kg$, the rate of dissolution is $0.5\,m/m^2$ yr (i.e., $0.5\,mm/decade$). For sediment averaging 70% $CaCO_3$, the amount of available $CaCO_3$ is about 2000 moles/m^2. Hence, the time constant for depletion is on the order of 4000 years. This is about threefold longer than the time constant for the replacement of deep Indian Ocean and Pacific Ocean waters.

Table 10. Summary of Composition of Input Water to the NADW and of NADW Itself Over Period Covered by the Model Calculations

		← Input →		← NADW →		
	pCO_2	$CO_3^=$	ΣCO_2	$CO_3^=$	ΣCO_2	ALK
1850	283	115	2125	115	2125	2290
1975	328	101	2148	114	2128	2290
2105	888	44	2282	101	2160	2300
2205	888	44	2282	91	2211	2325

To be more realistic this model would have to include:

1) The dependence of dissolution rate on the $CaCO_3$ content of the sediment.
   a) If the rate limiting step is a "stagnant boundary film" at the sediment-water interface then the dependence will be quite small until very low calcite contents are achieved.
   b) If the rate limiting step is the resaturation time of the sediment pore waters then the rate of dissolution will vary with the *square root* of the fraction calcite.

2) The dependence of the dissolution rate on $\Delta CO_3^=$. If the rate limiting step for dissolution is the release of ions from the crystal surfaces, then as shown by the experiments of *Morse and Berner* (1972) an exponential rather than linear dependence on $\Delta CO_3^=$ would have to be used. If so, dissolution will occur about an order of magnitude faster.

3) The relative rates of burrowing and dissolution. At a dissolution rate of 0.5 m/m^2 yr, in the absence of bioturbation an insoluble residue one characteristic diffusion length in thickness would be built up in a decade or so. Therefore, if burrowing occurs on the time scale of centuries, rather than years, it will quite soon become the rate limiting step.

CONCLUSIONS

The rate of deep sea sediment dissolution is certainly fast enough that this process will take place concurrently with the transfer of CO_2 from the atmosphere-shallow ocean reservoir to the deep sea reservoir. The amount of calcite "kinetically" available for dissolution is comparable to the amount of carbon locked up in recoverable fossil fuels. The major uncertainties remaining to be resolved before adequate modeling can be done are:

Table 11. Summary of Fossil Fuel CO_2 Fluxes Contrasting the Total Flux into the System with that Leaving the Surface Ocean for the Deep North Atlantic and with that of Fossil Fuel Induced Dissolution of Western Basin Sediments

	10^{12} moles/yr		
	Total CO$_2$ Influx	Flux to NADW	Flux to NADW Sediments
1850	0	0	0.0
1975	177	3	0.2
2105	2140	19	2.9
2205	–	19	6.0

1) The identification of the rate limiting step for dissolution on the sea floor (i.e., a distinction between linear and exponential kinetics must be made).
2) The quantification of mixing rates of sediments on the sea floor (the mechanical eddy diffusivity as a function of depth in the sediment column must be determined).

ACKNOWLEDGEMENTS

Much of the $CO_3^=$, ΣCO_2, pCO_2 and alkalinity data used in this paper was generated by the GEOSECS Program. Financial support for this work was provided by a grant to Lamont-Doherty Geological Observatory from the Energy Research and Development Administration (E(11–1)2185) and by a grant to CUNY from IDOE (OCE72-06419). Lamong-Doherty Geological Observatory Contribution No. 2513.

REFERENCES

Bien, G. S., N. W. Rakestraw and H. E. Suess. 1965. Radiocarbon in the Pacific and Indian Oceans and its relation to deep-water movements. *Limnol. Oceanogr.* 10, Supplement R25–R37.

Broecker, W. S., R. Gerard, M. Ewing and B. C. Heezen. 1960. Natural radiocarbon in the Atlantic Ocean. *J. Geophys. Res.* 65. 2903–2931.

Broecker, W. S., Y.-H. Li and T.-H. Peng. 1971. Carbon dioxide – Man's unseen artifact, In: *Impingement of Man on the Oceans*, D. W. Hood (Ed.), John Wiley and Sons, Inc., N.Y. 297–324.

Broecker, W. S. and T. Takahashi. In press. The relationship between lysocline depth and *in situ* carbonate ion concentration. *Deep-Sea Res.*

Broecker, W. S. and T. Takahashi. In prep. A model for the sea floor dissolution of calcite.

Hubbert, M. King. 1972. Man's conquest of energy: Its ecological and human consequences, In: *The Environmental and Ecological Forum 1970–1971*, USAEC Report TID-25857.

Kipp, N. G. 1976. New transfer function for estimating past sea surface conditions from sea bed distribution of planktonic foraminiferal assemblages in the North Atlantic. *Geol. Soc. Am. Memoir No. 145*, (R. Cline and J. Hays, Eds.).

Morse, J. W. and R. A. Berner. 1972. Dissolution kinetics of calcium carbonate in sea water: II. A kinetic origin for the lysocline. *Amer. Jour. Sci. 272*. 840–851.

Ostlund, H. G., H. G. Dorsey and C. G. Rooth. 1974. Geosecs North Atlantic radiocarbon and tritium results. *Earth and Plan. Sci. Letters, 23*. 69–86.

Peng, T.-H., W. S. Broecker, G. Kipphut and N. Shackleton. 1977. Benthic mixing in deep sea cores as determined by 14_C dating and its implications regarding climate stratigraphy and the fate of fossil fuel CO$_2$. In: *The Fate of Fossil Fuel CO$_2$*, N. R. Andersen and A. Malahoff (Eds.), Plenum, N.Y., (this volume).

Roether, W. 1974. The tritium and carbon-14 profiles at the GEOSECS I (1969) and GOGOI (1971) North Pacific Stations. *Earth and Planet. Sci. Letters, 23*. 108–115.

Stuiver, M. 1976. The ^{14}C distribution in west Atlantic abyssal waters. *Earth Planet. Sci. Letters 32*. 322–330.

Sverdrup, H. U., M. W. Johnson and R. H. Fleming. 1942. *The Oceans: Their Physics, Chemistry and General Biology*. Prentice-Hall, New York. 1087 pp.

Takahashi, T. and W. S. Broecker. 1977. Mechanisms for calcite dissolution on the sea floor. In: *The Fate of Fossil Fuel CO$_2$*, N. R. Andersen and A. Malahoff (Eds.), Plenum, N.Y., (this volume).

Takahashi, T., P. Kaiteris, W. S. Broecker and A. E. Bainbridge. 1976. An evaluation of the apparent dissociation constants of carbonic acid in seawater. *Earth and Planet. Sci. Letters 32*. 458–467.

EDITOR NOTES

1 Since this paper was written, the authors have found the coefficient of dissolution, 0.025, to be a factor of two lower than that used in this contribution.

2 If, as well may be the case, the acceleration of fossil fuel use is stemmed before the end of the next century, then the carbon ion content of NADW will not drop as fast as indicated by these calculations.

Effects of Fuel and Forest Conservation on Future Levels of Atmospheric Carbon Dioxide

JAMES C.G. WALKER[a] AND JAMES F. KASTING[b]

[a]Department of Atmospheric, Oceanic, and Space Sciences, University of Michigan, Ann Arbor, MI 48109, USA

[b]Department of Geosciences, The Pennsylvania State University, University Park, PA 16802, USA

ABSTRACT

Walker, J.C.G. and Kasting, J.F., 1991. Effects of fuel and forest conservation on future levels of atmospheric carbon dioxide. Palaeogeogr., Palaeoclimatol., Palaeoecol. (Global Planet. Change Sect.), 97: 151–189.

We develop a numerical simulation of the global biogeochemical cycles of carbon that works over time scales extending from years to millions of years. The ocean is represented by warm and cold shallow water reservoirs, a thermocline reservoir, and deep Atlantic, Indian, and Pacific reservoirs. The atmosphere is characterized by a single carbon reservoir and the global biota by a single biomass reservoir. The simulation includes the rock cycle, distinguishing between shelf carbonate and pelagic carbonate precipitation, with distinct lysocline depths in the three deep ocean reservoirs. Dissolution of pelagic carbonates in response to decrease in lysocline depth is included.

The simulation is tuned to reproduce the observed radiocarbon record resulting from atomic weapon testing. It is tuned also to reproduce the distribution of dissolved phosphate and total dissolved carbon between the ocean reservoirs as well as the carbon isotope ratios for both ^{13}C and ^{14}C in ocean and atmosphere. The simulation reproduces reasonably well the historical record of carbon dioxide partial pressure as well as the atmospheric isotope ratios for ^{13}C and ^{14}C over the last 200 yr as these have changed in response to fossil fuel burning and land use changes, principally forest clearance. The agreements between observation and calculation involves the assumption of a carbon dioxide fertilization effect in which the rate of production of biomass increases with increasing carbon dioxide partial pressure. At present the fertilization effect of increased carbon dioxide outweighs the effects of forest clear-ance, so the biota comprises an overall sink of atmospheric carbon dioxide sufficiently large to bring the budget approximately into balance.

This simulation is used to examine the future evolution of carbon dioxide and its sensitivity to assumptions about the rate of fossil fuel burning and of forest clearance. Over times extending up to thousands of years, the results are insensitive to the formulation of the rock cycle and to the dissolution of deep sea carbonate sediments. Atmospheric carbon dioxide continues to increase as long fossil fuel is burned at a significant rate, because the rate of fossil fuel production of carbon dioxide far exceeds the rates at which geochemical processes can remove carbon dioxide from the atmosphere. The maximum concentration of carbon dioxide achieved in the atmosphere depends on the total amount of fossil fuel burned, but only weakly on the rate of burning. The future course of atmospheric carbon dioxide is, however, very sensitive to the fate of the forests in this simulation because of the important role assigned to carbon dioxide fertilization of plant growth rate. Forest clearance drives up atmospheric carbon dioxide not only by converting biomass into atmospheric carbon dioxide but more importantly by reducing the capacity of the biota to sequester fossil fuel carbon dioxide. In this simulation, atmospheric carbon dioxide levels could be sustained indefinitely below 500 parts per million (ppm) if fossil fuel combustion rates were immediately cut from their present value of 5×10^{14} m/y to 0.2×10^{14} m/y (a factor of 25 reduction) and if further

Walker, J. C. G. (1992) Effects of fuel and forest conservation on future levels of atmospheric carbon dioxide. *Global and Planetary Change* 5(3): 151–189. Reproduced with permission of Elsevier.

forest clearance were halted. If neither of these conditions is met and if we consume most of the world's fossil fuel reserves, peak carbon dioxide concentrations of 1000–2000 ppm are probable within the next few centuries.

1. INTRODUCTION AND OVERVIEW

It is by now widely recognized that man has the capability of warming Earth's climate by causing a buildup of greenhouse gases in the atmosphere. The principal culprit is carbon dioxide released by the burning of fossil fuels and by changes in land use patterns, especially deforestation. Other trace gases (methane, nitrous oxide, freons) are also contributing to the increase in the greenhouse effect, but none of these gases has the same long lifetime as CO_2 and, hence, the same capability of producing long-lasting climatic change. In this paper we consider how CO_2 and climate may vary on time scales of hundreds to hundreds of thousands of years in response to man's activities.

1.1 Previous studies (brief summary)

Projections of how atmospheric CO_2 may vary during the next few centuries in response to fossil fuel burning have made by several different investigators (Keeling and Bacastrow, 1977; Revelle and Munk, 1977; Bolin et al., 1979; Bacastrow and Bjorkstrom, 1981; Broecker and Peng, 1982; Sundquist, 1986). The greatest uncertainty in any such calculation is the future rate of burning and the total amount of fossil fuel that will eventually be consumed. A typical assumption is that the burning rate will increase from its present value of 5×10^{14} moles C yr^{-1} (Rotty and Marland, 1986), henceforth abbreviated as 5×10^{14} m/y to four or five times that value by the year 2100 and then decline to nearly zero over the next 300 yr as the fossil fuel reserves are exhausted (Fig. 1a). The total amount of recoverable fossil fuels (most of which is coal) is considered to be in the range of 4000 to 6000 Gtons carbon, or $(3.5–5.0) \times 10^{17}$ moles C (Bacastrow and Bjorkstrom, 1981; Broecker and Peng, 1982). This is 7–10 times the amount of carbon dioxide that was present in the atmosphere prior to the onset of the industrial age, when the CO_2 concentration was about 280 ppm (Neftel et al., 1985).

Most carbon cycle modelers, including us, assume that the principal sink for carbon dioxide during the next few centuries will be the ocean. The rate at which the ocean can take up CO_2 is

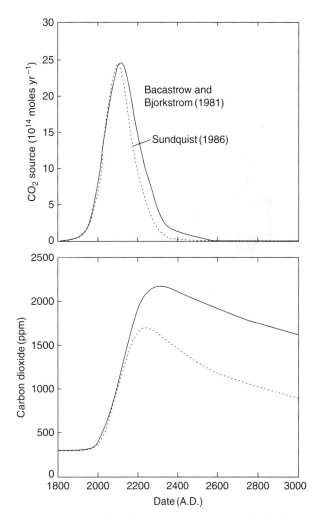

Fig. 1. Assumed fossil burning rates (top) and calculated atmospheric CO_2 concentrations (bottom) in two previous models of the global carbon cycle. The total amount of fossil fuel consumed in 5×10^{17} moles C for Bacastrow and Bjorkstrom (1981) and 4.2×10^{17} moles C for Sundquist (1986). The Bacastrow and Bjorkstrom calculation is their "slow burning" case (n = 0.25).

estimated by determining how fast it absorbs tracers such as ^{14}C and tritium. By combining an ocean uptake model with a postulated fossil fuel burning scenario, one can estimate how atmospheric CO_2 concentrations will vary during the next few centuries. The results of two such model calculations are shown in Fig. 1b. The models predict a five- to seven-fold increase in atmospheric CO_2 levels over the next 300 yr, followed by a long, slow decline. Of the two calculations shown here, Sundquist predicts a smaller increase and a faster decline because he assumes a smaller net consumption of fossil fuels (5000 Gtons C versus 6156 Gtons C) and because he includes dissolution of seafloor carbonates. More will be said about this in the next section. For the present, we simply note that the predictions of these models are roughly comparable, as are our own predictions when we make similar assumptions (Section 5).

1.2 This study

Our model differs from the two shown in Fig. 1 in three main respects. First, it includes a simple, and admittedly speculative, representation of the exchange of carbon between the atmosphere, forests, and soils. This terrestrial carbon cycle is needed to allow the model to reproduce the past histories of atmospheric CO_2 and atmospheric and oceanic carbon isotopes. It also allows us to bound the extent to which the terrestrial biota may influence future atmospheric CO_2 levels. Second, we also include the long term carbonate-silicate cycle, in which carbon is exchanged between the atmosphere/ocean system and carbonate rocks. This allows us to carry our computations into the distant future and to calculate how long it will take for the atmospheric CO_2 concentration to decay to its original preindustrial value. No previous model of fossil fuel CO_2 uptake has done this explicitly, although Sundquist (1986) has discussed the problem qualitatively. Finally, we consider a wider range of burning scenarios than has been looked at in most studies. In particular, we examine the extent to which future atmospheric CO_2 levels could be reduced by implementing strict energy conservation measures in the immediate future. We do this by performing simulations in which the recoverable fossil fuels are consumed at much slower rates than those shown in Fig. 1.

We begin with a general review of the global carbon cycle in which we explore the carbon reservoirs and the processes most likely to influence atmospheric carbon dioxide on a time scale of decades to millennia. This review guides the development of the mathematical model, which is presented in Section 2. We have sought to keep our model as simple as possible, consistent with a plausible representation of the most important processes, so that underlying realities shall not be obscured by computational complexities. The fluxes of water and biogenic particles included in the oceanic component of the model are tuned to reproduce observed values of carbon isotope ratios, nutrient concentrations, and total dissolved carbon. The predictive capabilities of the model are then tested against the observed record of atmospheric carbon dioxide and the carbon isotopes using the history of release of fossil fuel carbon

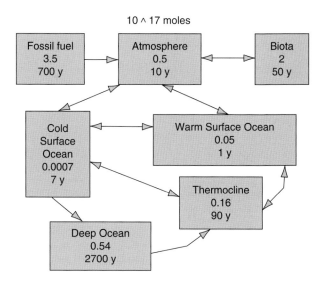

Fig. 2. Various reservoirs that affect atmospheric carbon dioxide. Reservoir capacities are quoted in units of 10^{17} moles. The capacities of oceanic reservoirs are given by the carbonate ion concentration, CO_3^{2-}, not the total dissolved carbon concentration, because absorption of carbon dioxide without change in alkalinity converts carbonate ions to bicarbonate ions. The times associated with the oceanic reservoirs are the times to respond to a change in atmospheric carbon dioxide. For fossil fuel, atmosphere, and biota reservoirs, the times are the carbon residence times at current rates of exchange.

dioxide and the history of land use changes affecting biomass. Our results show reasonable agreement both with the present distribution of tracers in the sea and with the history of carbon dioxide and its isotopes. However, our model is a very simple representation of the real world and incorporates a number of quite arbitrary assumptions. The agreement between calculation and observation does not necessarily mean that the calculations are correct, therefore. All we can say at this point is that the model is not obviously wrong.

1.3 Constraints on carbon exchange

The carbon dioxide content of the atmosphere is affected by release of carbon dioxide in the burning of fossil fuels, exchange with the biota, and exchange with the ocean. Figure 2 shows the carbon exchange rates and the carbon absorption capacities of the reservoirs that are most important on the short time scale. The present rate of fossil fuel burning is large enough to double the carbon dioxide content of the atmosphere in a time of about 100 yr. At the present rate of burning it would take 700 yr to exhaust the estimated reserve of recoverable fossil fuel. The large

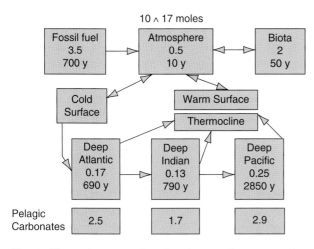

Fig. 3. The carbon capacity of various surface reservoirs, with more detailed description of the deep ocean. The ability of the deep ocean reservoirs to absorb carbon dioxide is greatly enhanced by the dissolution of pelagic carbonates. The indicated reservoir capacities correspond to the carbonate content in the top 30 cm of pelagic sediments integrated over the area of the sea floor above the lysocline.

reservoir of organic carbon as biomass includes soil organic carbon, but not dissolved organic carbon in sea water. Exchange between the atmosphere and this large biomass reservoir is rapid.

Because of the constraints of charge balance, the capacity of the ocean to absorb carbon dioxide is limited. As long as carbon dioxide is the only constituent being exchanged with the ocean the dissolution reaction is, in effect

$$CO_3^{2-} + H_2O + CO_2 \rightarrow 2HCO_3^-$$

The capacity of sea water to absorb carbon dioxide is therefore limited by the concentration of dissolved carbonate ions, a concentration only one-tenth as large as the concentration of total dissolved carbon. The absorption capacities of the oceanic reservoirs in Fig. 2 are calculated by multiplying the reservoir volume by the corresponding concentration of carbonate ions. The surface ocean reservoirs exchange carbon rapidly with the atmosphere, but their capacities are too small for them to exert a controlling influence on atmospheric carbon dioxide. Because of its much larger volume, the capacity of the deep ocean reservoir is significant, but the time scale associated with carbon transfer into the deep ocean is 1000 yr or more, so the behavior of atmospheric carbon dioxide on a time scale of decades to centuries is little affected by exchange with the deep ocean. It is for this reason that it is not necessary to describe the ocean circulation in great detail in order to

predict atmospheric concentrations of carbon dioxide over times of decades to centuries. The oceanic reservoirs that exchange rapidly with the atmosphere are small; the large, deep ocean reservoir exchanges slowly with the atmosphere.

The deep ocean reservoirs are shown in more detail in Fig. 3. Deep sea water ages as it moves from the Atlantic to the Indian to the Pacific Ocean because deep water forms principally in the Atlantic. The times in Figs. 2 and 3 are the times to respond to an atmospheric perturbation. Because of the pressure and temperature dependence of the solubility of calcium carbonate, the deepest portions of the oceans are undersaturated with respect to calcium carbonate while the shallower portions are supersaturated. Addition of carbon dioxide to the deep ocean renders the waters more corrosive, making possible the dissolution of deep sea pelagic carbonate sediments. Carbonate dissolution affects the charge balance in such a way as to permit more carbon dioxide to dissolve. The effective reaction is

$$CaCO_3 + H_2O + CO_2 \rightarrow Ca^{2+} + 2HCO_3^-$$

The capacity of the deep ocean reservoirs to absorb carbon dioxide is therefore very significantly affected by the dissolution of pelagic carbonates. Figure 3 shows an estimate of the carbon absorption capacity associated with the dissolution of pelagic carbonates. The values have been calculated to include the carbonate minerals in the top 30 cm of pelagic sediments, 6000 moles/m², over the area of sea floor above the lysocline (Broecker and Peng, 1982). With allowance for the dissolution of pelagic carbonates, the capacity of deep sea water to absorb carbon dioxide is large, but this process still does not begin to affect the atmosphere on a time scale shorter than the ventilation time of the deep sea, about 1000 yr. The dissolution of shallow water carbonates is not expected to occur because shallow waters are strongly supersaturated with respect to carbonate minerals. The dissolution of terrestrial carbonate deposits in weathering reactions contributes a flux of calcium and dissolved carbon carried to the oceans by rivers that is much smaller than any of the fluxes in Figs. 2 and 3.

On a time scale of decades to centuries, atmospheric carbon dioxide is affected by rapid release from the relatively large fossil fuel reserve, rapid exchange with a large reservoir of terrestrial biomass, and rapid exchange with surface ocean reservoirs with little capacity to absorb carbon dioxide. Exchange with the large deep sea reservoir, for which the absorption capacity is enhanced by the solution of pelagic carbonates, takes millennia.

2. MODEL DEVELOPMENT

2.1 Reservoirs

For computational purposes the surface environment is partitioned into a small number of homogeneous reservoirs. The simulation calculates the evolution of the carbon content of each of these reservoirs in response to specified carbon sources and sinks and the exchange of material between reservoirs. The reservoirs and their interconnections are shown in Fig. 4. The atmosphere exchanges carbon with the biomass reservoir, with warm surface water, and with cold surface water. The pre-industrial concentration of carbon dioxide in the atmosphere was 280 parts per million (Neftel et al., 1985; Friedli et al., 1986), which is defined as 1 PAL (present atmospheric level), equal to an amount of 0.495×10^{17} moles of carbon dioxide. The carbon content of the biomass reservoir is taken initially to be 2×10^{17} moles, of which 1.25×10^{17} moles is in the form of soil humus (Schlesinger, 1986) and 0.75×10^{17} moles is living biomass and litter (Woodwell et al., 1983; Esser, 1987). The division of the ocean into separate reservoirs is based largely on Broecker and Peng

(1982), with volumes and areas from Menard and Smith (1966). The warm surface water reservoir has an area of 3.61×10^{14} m^2 and a thickness of 75 m. The cold surface water reservoir has an area of only 0.009×10^{14} m^2 and a thickness of 1 km. This area has no geographical significance. Its value is a tuning parameter that affects both the carbon cycle and the global temperature distribution. Both of these surface water reservoirs exchange water with the thermocline in mixing processes. The thermocline has a volume of 3.1×10^{17} m^3 and a thickness of 925 m. The mixing processes in the ocean denoted by double-headed arrows are ones in which the fluxes of water in each direction are equal. In addition to mixing with the thermocline water is exchanged between the surface reservoirs in a mixing process.

Circulation through the deep ocean is represented not as a mixing process but as advection, in which the flow is in one direction only. These flows are designated by the large single-headed arrows in the figure. Water flows from the cold surface water reservoir into the deep Atlantic and from there partly into the deep Indian reservoir and partly into the thermocline reservoir. There is an upwelling flux from the deep Indian reservoir into the thermocline and a flow also on into the deep Pacific reservoir. The thermohaline circulation is closed by upwelling from the deep Pacific into the thermocline reservoir and by a flow from the thermocline reservoir back into the cold surface water reservoir (Gordon, 1986). The volumes and areas of the deep ocean reservoirs are designated by v and a. The upwelling velocity is assumed to be the same in each ocean basin, so the flux from deep ocean to thermocline is proportional to the area of the deep ocean reservoir. The other fluxes are calculated to conserve water. A better representation of the thermohaline circulation would include a second source of bottom water flowing directly into each ocean basin (Broecker and Peng, 1982, p. 371).

The choice of how to partition the surface environment into a small number of homogeneous reservoirs is to some extent arbitrary. The trick is to develop a system that can reasonably simulate the real world while maintaining computational and conceptual simplicity. In the rest of this paper we will not change the volumes of the oceanic reservoirs or the nature of the exchange processes, whether mixing or advective. The exchange coefficients and fluxes will be tuned so that the simulation reproduces various observed properties of the system, in a manner described in Section 3.

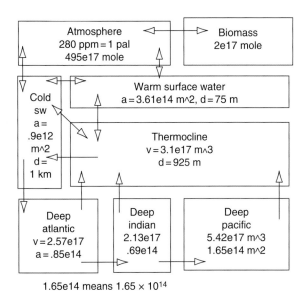

1.65e14 means 1.65×10^{14}

Fig. 4. Dimensions of the carbon reservoirs included in the numerical simulation. The areas of oceanic reservoirs are denoted by a, their vertical dimensions by d, and their volumes by v. Exchange processes are indicated by doubled-headed arrows and advective processes are indicated by single-headed arrows.

2.2   Species and processes

The simulation incorporates the rock cycle, but the processes of weathering and precipitation of chemical sediments occur too slowly to have significant impact on the short-term response of the ocean–atmosphere system. We shall therefore postpone discussion of the parameters of the rock cycle until later in the paper, describing first the processes that influence the short-term response.

2.2.1   Phosphorus and photosynthesis

Dissolved phosphorus is added to and removed from the ocean through the warm surface reservoir only. Addition is by rivers and removal is by incorporation into organic matter deposited in shelf sediments (Holland, 1978), the removal rate is assumed to be proportional to biological productivity in warm surface water which is in turn assumed to be proportional to the dissolved phosphorus concentration, so these interactions effectively fix the concentration of dissolved phosphorus in warm surface water. Dissolved phosphorus is transported from one oceanic reservoir to another by the thermohaline circulation and by the mixing processes already described. In addition, phosphorus is transported downwards from the warm surface water reservoir into the underlying thermocline and deep ocean reservoirs as a constituent of particulate organic matter produced by plankton. We neglect the transport of phosphorus associated with non-refractory dissolved organic matter.

We calculate the downward flux of particulate organic matter, loosely described as productivity, by assuming that the phosphorus budget of the warm surface water reservoir is in instantaneous balance and that the concentration of dissolved phosphorus in this reservoir is a fixed fraction of the dissolved phosphorus concentration in the deep Pacific reservoir. This fraction, designated f_{PD}, we take equal to 0.04. Then, with all of the water fluxes specified, the amount by which the flow of dissolved phosphorus into the warm surface water reservoir exceeds the flow of dissolved phosphorus out of this reservoir is readily calculated. The difference is equal to the flux out of the reservoir in the form of particulate organic matter. The parameters have been set so that the concentration of dissolved phosphorus in river water is 1 millimole/m³, the concentration in warm surface water is 0.1 millimole/m³, and the concentration in deep Pacific water is 2.5 millimole/m³ (Broecker and Peng, 1982, p. 309). The concentrations in the other reservoirs depend on the biogeochemical

Table 1. Selected model parameters

Parameter	Code name	Value
R_{CP}	CTPR	120
R_{CO}	CORAT	0.09
f_{PT}	TCPFRAC	0.925
f_{PD}	PFP	0.04
z_1	LYCON1	5.8 km
z_2	LYCON2	50 km/(mmol/l)
τ_{OA}	DISTIME	10 years
C_{Bio}	BIOMZ	2×10^{17} moles
M_{atm}	MATMCO2	4.95×10^{16} moles

circulation of the ocean. Other formulations can be imagined for the relationship between productivity and phosphate circulation (Volk, 1989a). In this simulation, productivity is constant in time and space, so the formulation makes no difference to the results. Biological production in the cold surface water reservoir is neglected because this reservoir has small area and large depth.

Some of the particulate phosphorus is converted back into dissolved form in the thermocline reservoir; the rest settles into the underlying deep ocean reservoirs, at rates proportional to the areas of these reservoirs. We assume a constant value for the fraction of phosphorus returned to dissolved form in the thermocline, a constant designated f_{PT}. For simplicity, we assume that all particulate organic matter settling into the deep ocean reservoirs is converted back into dissolved form in these reservoirs, releasing dissolved phosphorus and total dissolved carbon. We ignore the small fraction of the particulate flux that is preserved to become a constituent of pelagic sediments.

Some of the particulate phosphorus is converted back into dissolved form in the thermocline reservoir; the rest settles into the underlying deep ocean reservoirs, at rates proportional to the areas of these reservoirs. We assume a constant value for the fraction of phosphorus returned to dissolved form in the thermocline, a constant designated f_{PT}. For simplicity, we assume that all particulate organic settling into the deep ocean reservoirs is converted back into dissolved form in these reservoirs, releasing dissolved phosphorus and total dissolved carbon. We ignore the small fraction of the particulate flux that is preserved to become a constituent of pelagic sediments.

2.2.2   Carbon

Carbon dioxide is added to the atmosphere by volcanic and metamorphic processes at a rate that is held constant in this simulation. Additional sources of atmospheric carbon dioxide are the burning of fossil fuel, changes in the terrestrial biomass, and the oxidative weathering of sedimentary organic carbon. Carbon dioxide is removed from the atmosphere by exchange with the terrestrial biomass and by dissolution in cold warm surface water reservoirs. The carbon dioxide flux between ocean and atmosphere is proportional to the area of the surface water reservoir and to the difference between the partial pressure of carbon dioxide in the atmosphere and the partial pressure of dissolved carbon dioxide in equilibrium with dissolved species in surface sea water. The calculation of equilibrium partial pressure is described below. This simulation is not intended for the study of carbon transfer through the atmosphere between warm and cold surface reservoirs, and it does not represent this process well.

Dissolved carbon is added to the warm surface water reservoir by rivers. Carbon is removed from the ocean by incorporation into shelf sediments as organic carbon at a rate that just balances the rate of kerogen oxidation. A larger sink for dissolved carbon is the precipitation of shelf carbonate sediments at a rate proportional to the concentration of dissolved carbonate ions in warm surface water. The largest sink for oceanic carbon is the accumulation of pelagic carbonate sediments. In the long-term steady state, the rock cycle fixes the partial pressure of carbon dioxide in the atmosphere at a level that ensures that silicate weathering reactions can consume carbon dioxide as fast as it is added to the atmosphere by volcanic and metamorphic processes (Holland, 1978; Walker et al., 1981; Berner et al., 1983). On shorter time scales, however, atmospheric carbon dioxide is affected by redistribution of carbon species between atmosphere, ocean, and terrestrial biomass.

The evolution of biomass is described by a single differential equation: Rate of change of biomass equals source minus sink. The source is initially assumed to be constant at 4×10^{15} m/y (Woodwell et al., 1983; Esser, 1987; Keeling et al., 1989a). The sink is first order in biomass with a time constant of 50 yr.

The precipitation of carbonate minerals on the continental shelves and the preservation of carbonate sediments in the deep ocean depend on the partitioning of total dissolved carbon amongst the various carbon species: dissolved CO_2, carbonate ions, and bicarbonate ions. This partitioning depends on the concentration of total dissolved carbon and the net charge concentration of all of the dissolved species other than carbon, which we shall call alkalinity. In calculating carbonate equilibria

we use effective equilibrium constants that are linear functions of temperature, T_s, in degrees K

$$K_{carb} = 5.75 \times 10^{-4} + 6 \times 10^{-6}(T_s - 278)$$
$$K_{co_2} = 0.035 + 0.0019(T_s - 278) \, \text{PAL m}^3 \, \text{mole}^{-1}$$

where $K_{carb} = K_2/K_1$ and $K_{CO2} = K_2/(\alpha K_1 \times 280 \times 10^{-6} \text{atm})$ from Broecker and Peng (1982, p. 151) and PAL represents the preindustrial atmospheric level of CO_2 (= 280×10^{-6} atm). The expressions for dissolved carbon species as functions of total dissolved carbon and alkalinity are

$$\left[HCO_3^- \right] = \frac{\Sigma - \sqrt{\Sigma^2 - A(2\Sigma - A)(1 - 4K_{carb})}}{1 - 4K_{carb}}$$
$$\left[CO_3^{2-} \right] = \frac{A - \left[HCO_3^- \right]}{2}$$
$$\left(pCO_2 \right)_s = \frac{K_{CO_2} \left[HCO_3^- \right]^2}{\left[CO_3^{2-} \right]}$$

where Σ is total dissolved inorganic carbon and A is alkalinity. $(pCO_2)_s$ is expressed in PAL; concentrations of dissolved species are in moles/m^3. These expressions were derived by writing Σ equals the sum of carbonate ions, bicarbonate ions, and dissolved CO_2, while A equals bicarbonate ions plus twice carbonate ions. One then expresses carbonate ion and dissolved CO_2 in terms of A and bicarbonate ion and solves the resulting quadratic equation for the bicarbonate concentration. In order to keep the solution general, we have not here assumed that the concentration of dissolved carbon dioxide is negligibly small.

Total dissolved carbon, Σ, is transported between oceanic reservoirs by the thermohaline circulation and the mixing processes that exchange water between reservoirs. In addition, total dissolved carbon is carried out of the warm surface water reservoir into the underlying thermocline and deep ocean reservoirs in the form of biogenic particles of both organic carbon and calcium carbonate minerals. This transport is taken to be proportional to organic productivity, discussed above in connection with the phosphorus budget. The ratio of organic carbon to phosphorus in the biogenic particulate rain, R_{CP}, is assigned a value of 120. The ratio of particulate carbonate minerals to particulate organic carbon, R_{CO}, is set equal to 0.09. This ratio, which refers just to calcite, controls the lysocline depth and the rate of accumulation of pelagic carbonate sediments. We could have used a larger value if we had included a flux of aragonite, which dissolves completely in the deep sea. The organic carbon is converted back into total dissolved carbon by respiration and decay processes partly in the thermocline and partly in the deep ocean reservoirs. The partitioning of organic carbon between these reservoirs is the same as the partitioning of phosphorus. A fraction $f_{PT} = 0.925$ is restored in the thermocline; the remainder is partitioned between deep ocean reservoirs in proportion to their areas. Both R_{CO} and f_{PT} are determined by tuning the model to reproduce observed distributions of dissolved inorganic carbon. The value of f_{PT} is a tuning parameter that determines the vertical gradient of $\delta^{13}C$.

Sea water at shallow depths is supersaturated with respect to carbonate minerals while deep water is undersaturated, partly because of the temperature dependence of the carbonate solubility but largely because of the pressure dependence. Carbonate particles therefore do not dissolve in the thermocline reservoir but continue on into the deep ocean reservoirs. The fraction of the particulate carbonate flux into the deep ocean reservoirs that dissolves depends on the carbonate saturation state of these reservoirs. We characterize this saturation state in terms of the depth of the lysocline, taken in this simulation to be a depth below which dissolution of carbonate particles is complete and above which no dissolution occurs. The carbonate saturation state of a deep ocean reservoir depends on the concentration of carbonate ions in that reservoir. The concentration of calcium ions varies negligibly between ocean reservoirs on the time scales of this study. Calcium ion concentrations are therefore assumed to be constant. The depth of the lysocline, z_{ly}, is taken to be a linear function of the carbonate ion concentration (in moles/m³)

$$z_{1y} = z_1 + z_2 \left(\left[CO_3^{2-} \right] - 0.1 \right)$$

The coefficient $z_1 = 5.8\,km$ determines the overall depth of the lysoclines in the three deep ocean reservoirs and therefore the overall rate of accumulation of pelagic carbonates. It is adjusted in procedures described below. The parameter z_2 describes how rapidly lysocline depth changes with changing carbonate ion concentration. Its value of $50\,km/(mole/m^3)$ was deduced from results presented by Broecker and Peng (1982, p. 283). The lysocline depth for each ocean basin is then combined with average total ocean hypsometry presented by Menard and Smith (1966) to calculate the fraction of the area of each basin that

lies below the lysocline. This is the fractional area of each ocean bottom exposed to undersaturated water. We assume that productivity and the rain of carbonate particles is spatially homogeneous, so the proportion of this rain that dissolves in each deep ocean reservoir is proportional to the area of the basin below the lysocline. The remaining flux of carbonate particles settles to the bottom in water supersaturated with respect to calcium carbonate and accumulates as pelagic carbonate sediments. This carbon is removed at least temporarily from the ocean–atmosphere system. The carbonate particles that dissolve in the deep ocean reservoirs add total dissolved carbon and alkalinity to these reservoirs. The rate of precipitation of shelf carbonate sediments, which is small in this simulation, is taken to be proportional to the concentration of carbonate ions in warm surface water. The expression we use for this rate is presented in Section 5.1.

2.2.3 Temperature

A single differential equation describes the evolution of global average surface temperature, T_s (degrees K)

$$\frac{dT_S}{dt} = \frac{Q - F_{IR}(T_S)}{C_{surf}}$$

The source term in this equation is global average insolation

$$Q = \frac{1365}{4}(1 - \alpha)Wm^{-2}$$

where the albedo, α, equals 0.3 (Ramanathan et al., 1989). The sink term is the outgoing flux of longwave infrared radiation, a linear function of global average surface temperature

$$F_{IR}(T_S) = A_{IR} + B_{IR}T_S \ Wm^{-2}$$

(Marshall et al., 1988). The coefficients in this expression depends on the partial pressure of carbon dioxide as follows

$$A_{IR} = -352.08 + 9.56\ln(pCO_2)$$
$$B_{IR} = 2.053 - 0.0514\ln(pCO_2)$$

with pCO_2 expressed in PAL. The effective heat capacity, C_{surf}, used in the calculation of surface temperature evolution is 50 joule s/m²/deg/yr. The time units (s/yr) enter in because time in this calculation is measured in years while the energy fluxes are in W/m^2. This heat capacity is equivalent to a column of water 377 m thick, a thickness appropriate to global warming on a time scale of decades.

The temperature of cold surface water and the temperature of the deep ocean reservoirs are fixed at 2°C. The temperature of the warm surface water reservoir is calculated from global average surface temperature and the fraction of the globe covered by cold surface water at a temperature of 2°C. This fraction is taken to be 0.0025. In this calculation the temperature plays a role only in determining the values of carbonate equilibrium coefficients. These equilibrium coefficients are different for warm surface water and for the cold ocean reservoirs. The difference in the equilibrium constants and thus in the carbon speciation between warm and cold surface water depends on atmospheric carbon dioxide partial pressure through the greenhouse effect.

In a future study we might calculate also the temperature of the various water masses, providing thermal constraints on the assumed circulation and permitting possible explanation of the effects on climate and carbon dioxide of assumed changes in circulation.

2.2.4   Alkalinity

Alkalinity is defined in this simulation as the sum of the biocarbonate ion concentration and twice the carbonate ion concentration. It is the net concentration of charge on the dissolved carbon species. Its importance is that the requirement of overall charge neutrality implies that alkalinity must be equal to the net positive charge on all dissolved species other than the carbon species. Alkalinity therefore depends on the concentration of other dissolved charged species, and it controls the speciation of total dissolved carbon.

Alkalinity is provided to the warm surface water reservoir by rivers, being supplied by dissolution of terrestrial carbonate rocks and silicate rocks. It is removed from the ocean by the precipitation of carbonate sediments on the shelves and on the deep sea floor. The way in which pelagic carbonate accumulation rate depends on the rain of carbonate particles of biological origin and on the saturation state of the deep ocean reservoirs has been described above in the sub-section on carbon.

Alkalinity is transported between ocean reservoirs by fluxes of water associated with the thermohaline circulation and mixing processes. It is extracted from the warm surface water reservoir by the formation of calcium carbonate particles at a rate proportional to productivity, and it is restored to deep ocean reservoirs by the dissolution of these calcium carbonate particles at a rate proportional to the area of the deep sea floor overlain by undersaturated water. In addition, there is a small transport of alkalinity associated with the formation of organic nitrogen from dissolved nitrate in the warm surface water reservoir and the restoration of nitrate in the thermocline and deep sea reservoirs. This nitrogen transport is directly proportional to the transport of organic carbon. Because of the negative charge on dissolved nitrate the transport of alkalinity by descending particles of organic matter is upward. The ratio of alkalinity transport to organic carbon transport is 0.15 (Broecker and Peng, 1982).

2.2.5   Carbon isotopes

The conservation equations for carbon isotopes are, of course, directly related to the conservation equations for carbon. There is isotopic fractionation when carbon dioxide dissolves in sea water with a value that depends on temperature according to

$$\Delta C_S^{13}\,(‰) = 9.5 - (T_s - 298)/10$$

where T_s is expressed in degrees K (Mook et al., 1974; Broecker and Peng, p. 310). There is a different isotopic fractionation for carbon dioxide coming out of the solution. Its value is $\Delta C_A^{13} = 16‰$ (Siegenthaler and Munnich, 1981). These fractionation coefficients refer to ^{13}C relative to ^{12}C. In all cases, the fractionation coefficients for ^{14}C are twice as large. Terrestrial photosynthesis produces isotopically light organic carbon, with a fractionation relative to atmospheric carbon dioxide of 20‰ for ^{13}C and 40‰ for ^{14}C. Photosynthesis in the warm surface water reservoir also fractionates, by 25‰ for ^{13}C and by twice as much for ^{14}C (Anderson and Arthur, 1983). This fractionation is with respect to the carbon isotopic composition of warm surface water. Carbonate minerals are assumed to be slightly fractionated with respect to dissolved carbon, by 1.7‰ for ^{13}C and by twice as much for ^{14}C (Anderson and Arthur, 1983). The overall level of the ^{13}C ratio in the ocean is determined by the rock cycle and the rates of carbonate dissolution, kerogen oxidation, volcanic release of carbon dioxide, and accumulation of organic carbon in shelf sediments.

In addition to the fractionation between reservoirs already described, radiocarbon is affected by radioactive decay with a decay time equal to 8267 yr (Broecker and Peng, 1982). In the steady state, this decay is balanced by an atmospheric source provided by the interaction of cosmic rays with atmospheric nitrogen. This is the only source for radiocarbon, but radioactive decay occurs in all of the carbon reservoirs, atmosphere, terrestrial biomass, and ocean.

3.   TUNING THE MODEL

The coefficients that control the rates of exchange of carbon between reservoirs are adjusted to yield reasonable agreement between the simulation and observations. The process is known as tuning. We shall approach the tuning process in a systematic manner, seeking values for the various coefficients that can be left unchanged for the rest of the calculations described in this paper. We seek to arrive at a single simulation that can describe the response of the biogeochemical cycles of carbon on a variety of time scales, yielding reasonable agreement with such observations as are available.

We begin with observations of radiocarbon produced by atmospheric testing of nuclear weapons during the middle decades of the 20th century. These observations constrain the rates of exchange of carbon between the atmosphere, biomass, and surface ocean reservoirs. We consider next the long-term steady state distribution of carbon and its isotopes in atmosphere and oceanic reservoirs as well as the distribution of dissolved phosphorus in the ocean and the saturation state of the deep ocean. These observations constrain the biogeochemical circulation of the ocean. With the ocean simulation calibrated in this manner we shall in the next section test the simulation by applying it to the historical record of change brought about by the industrial revolution.

3.1   Bomb radiocarbon

Important information about the exchange of carbon between the atmosphere and adjacent reservoirs on a time scale of years is provided by the response of atmospheric radiocarbon to the very large source resulting from atmospheric testing of nuclear weapons in the 50's and 60's (Broecker and Peng, 1982). The markers in Fig. 5 show the observed history of radiocarbon in the atmosphere and in warm surface water, and the solid lines

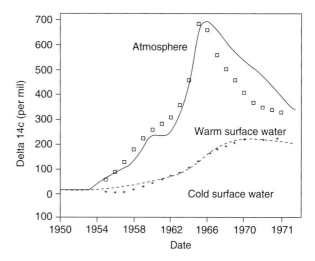

Fig. 5. Calculated response of the atmosphere and surface ocean reservoirs to the input of bomb radiocarbon. Observations of radiocarbon response are indicated by the markers.

show the simulated response of these reservoirs and the cold surface water reservoir during the period 1950 to 1976. The data are from Broecker and Peng (1982, p. 389). In calculating this history we assumed for the rate of production of radiocarbon in the atmosphere the values shown in Fig. 6. The history of this source is based on observations of the rate of fall-out of tritium and radioactive strontium (Broecker and Peng, 1982, pp. 386–387). We assumed that the radiocarbon source was proportional to the fallout rate. The amplitude of the source, with a peak value approaching 45,000 m/y, was adjusted to yield the correspondence between theory and observation shown in Fig. 5.

Because of the short time involved in the response to bomb radiocarbon, the computed results are sensitive mainly to the rate of exchange of carbon between atmosphere and biomass and between atmosphere and surface water reservoirs. The rate of mixing of the surface water reservoirs with the thermocline also has an effect, but the response of the system on this time scale is not sensitive to the thermohaline circulation. The evolution of the biomass reservoir will be discussed in more detail below. As already noted it had an initial mass of 2×10^{17} moles of carbon. By 1950, in this simulation, this mass had hardly changed. Exchange of carbon between atmosphere and biomass was assumed to depend on atmospheric carbon dioxide partial pressure and to be first order in biomass, with a residence time of carbon in the biomass reservoir equal to 50 yr. These parameters were not adjusted to reproduce

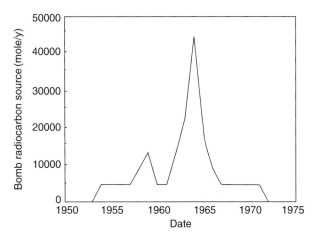

Fig. 6. The history of the bomb radiocarbon source. The source is assumed proportional to the concentration of strontium in fallout. The proportionality constant is adjusted to achieve the agreement between theory and observation illustrated in Fig. 5.

the radiocarbon response shown in Fig. 5, but were adjusted to reproduce the 100-yr history of carbon isotopes and atmospheric carbon dioxide in a tuning process discussed below.

The parameters that were adjusted to reproduce the radiocarbon response are the rate of exchange of carbon between atmosphere and surface water and the rate of exchange between the warm surface water reservoir and the thermocline. The carbon flux between atmosphere and surface water is proportional to the difference between atmospheric partial pressure and the partial pressure in equilibrium with the concentrations of dissolved carbon species in the surface water. The expression for the carbon flux is

$$F_{OA} = \frac{\left[(pCO_2)_s - pCO_2\right]A_s M_{atm}}{\tau_{OA}}$$

where $(pCO_2)_s$ is the equilibrium partial pressure and pCO_2 the atmospheric value, both in PAL. M_{atm} is the amount of carbon dioxide in the preindustrial atmosphere, 4.95×10^{16} moles. The total exchange between the atmosphere and each of the surface water reservoirs is proportional to the fractional area, A_s, of that reservoir. A value for the proportionality constant of $\tau_{OA} = 10$ years was deduced from the comparison between calculation and observation shown in Fig. 5 and was used in all of the calculations presented in this paper. The tuned value of this parameter depends on the volumes and carbon contents of the surface ocean reservoirs.

As pointed out by Broecker and Peng, the bomb radiocarbon data provide an additional constraint on the system, related to how quickly the added radiocarbon is mixed down into the thermocline. This constraint takes the form of the total amount of bomb radiocarbon in the ocean in 1973. The measured value, reported by Broecker and Peng (1982, p. 548), was 3.14×10^{28} atoms. The calculated value in this simulation was 3.09×10^{28} atoms. This calculated value depends on the rate of mixing between the warm surface water reservoir and the thermocline. This rate was adjusted to yield the agreement just reported. The deduced value of the mixing parameter was a vertical mixing velocity of 10.5 m/y. Thus the total amount of water per year exchanged between the warm surface water reservoir and the thermocline is the area of the reservoir times 10.5 m. Alternatively, the residence time of warm surface water before it is mixed into the thermocline is a thickness of 75 m divided by the vertical mixing velocity of 10.5 m/y, approximately 7 yr.

In adjusting the simulation to reproduce the observations shown in Fig. 5, then, we determine the values of the three parameters: amplitude of the bomb radiocarbon source, exchange coefficient between atmosphere and surface water, and mixing coefficient between warm surface water and thermocline. The three observations that constrain the values of these parameters were: the history of radiocarbon in the atmosphere, the history of radiocarbon in warm surface water, and the total amount of radiocarbon in the ocean in 1973. The mixing and exchange parameters just specified were not changed for any of the remaining calculations presented in this paper. The model used for the calculations shown in Fig. 5 was the fully developed and tuned model that we are in the course of describing. Because we make no further use of bomb radiocarbon results, for the rest of this paper the bomb radiocarbon source will be set to zero.

3.2 Pre-industrial steady state

In this section we tune the oceanic component of the model by adjusting exchange fluxes between oceanic reservoirs to reproduce with reasonable fidelity the distribution of tracers in the sea. This process involves a fair amount of iteration, but in order to keep the development clear we shall describe the tuning strategy in a way that associates observable parameters of the model with the adjustable parameters that influence them most

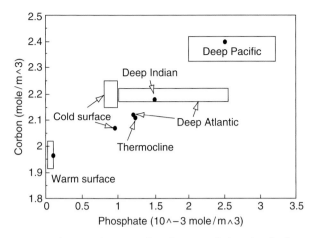

Fig. 7. The steady state distribution of dissolved phosphate and total dissolved carbon in the ocean. The points are calculated values, while observations are indicated by rectangles (Broecker and Peng, 1982, p. 70).

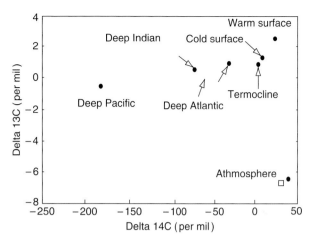

Fig. 8. The steady state distribution of carbon isotopes in ocean and atmosphere. The calculated values are indicated by points while observations are represented by rectangles (Broecker and Peng, 1982, pp. 249–252, 309).

Table 2. Preindustrial steady state concentrations

Reservoir	Phosphate*	Σ	A	HCO_3^-	CO_3^{2-}	$(pCO_2)_{eq}$**
Warm surface	0.10	1.96	2.12	1.78	0.171	0.996
Cold surface	0.96	2.07	2.12	1.96	0.078	1.45
Deep Atlantic	1.22	2.12	2.15	2.02	0.067	1.79
Deep Indian	1.51	2.18	2.20	2.08	0.059	2.16
Deep Pacific	2.50	2.40	2.38	2.29	0.046	3.34
Thermocline	1.25	2.11	2.12	2.01	0.053	2.26

* Phosphate concentrations in millimoles m^{-3}; ** pCO_2 in PAL (1 PAL = 280 × 10^{-6} atm); all other concentrations in moles m^{-3}.

strongly. The targets of this tuning process are the distribution of total dissolved carbon and dissolved phosphate in the various oceanic reservoirs, shown in Fig. 7 and the distribution of ^{13}C and ^{14}C in the oceanic reservoirs and in the atmosphere, shown in Fig. 8. Numerical values of dissolved carbon and phosphate and other carbonate species are listed in Table 2.

In this section we tune the oceanic component of the model by adjusting exchanges fluxes between oceanic reservoirs to reproduce with reasonable fidelity the distribution of tracers in the sea. This process involves a fair amount of iteration, but in order to keep the development clear we shall describe the tuning strategy in a way that associates observable parameters of the model with the adjustable parameters that influence them most strongly. The targets of this tuning process are the distribution of total dissolved carbon and dissolved phosphate in the various oceanic reservoirs, shown in Fig. 7 and the distribution of ^{13}C and ^{14}C in the oceanic reservoirs and in the

atmosphere, shown in Fig. 8. Numerical values of dissolved carbon and phosphate and other carbonate species are listed in Table 2.

The spread in radiocarbon ratios between cold surface water and the deep Pacific water depends mainly on how long it takes water to travel in the thermohaline circulation from the cold surface reservoir through the deep Atlantic and deep Indian reservoirs to the deep Pacific reservoir. This deep sea ventilation time is controlled by the vertical upwelling velocity that carries water from the deep ocean reservoirs into the thermocline at fluxes proportional to the areas of the deep sea reservoirs. This upwelling velocity is adjusted to reproduce as closely as possible the observed radiocarbon age of deep Pacific water. The adjusted value is 1.15 m/y. The absolute values of radiocarbon depend on the magnitude of the atmospheric source of radiocarbon and also on the total dissolved carbon content of ocean and atmosphere. The deduced value of the radiocarbon source is 506.22 m/y, within the range of values reported by O'Brien (1979) and Stuiver and Quay (1980). The difference in radiocarbon ratios between atmosphere and surface water is controlled by the exchange parameters already determined in the tuning to bomb radiocarbon observations.

The next tuning target is the distribution of phosphate in the deep ocean reservoirs. In this simulation the phosphate content of warm surface water is determined by the rock cycle, because the marine sink of dissolved phosphate is proportional to the phosphate concentration in warm surface water and the river source of phosphate is fixed. Because we have assumed that the phosphate concentration of warm surface water is a fixed fraction, $f_{PD} = 0.04$, of the phosphate concentration in deep Pacific Water, the concentration in deep Pacific water is also fixed. The parameters are such that the concentration in warm surface water is 0.1×10^{-3} moles/m³ and the concentration in deep Pacific water is 2.5×10^{-3} moles/m³. The goal of the tuning exercise is to yield a spread between these limiting values of concentrations in the deep Atlantic and deep Indian ocean reservoirs that is reasonably close to observations. With the ocean simulation as structured here, the tuning parameter is the fraction of particulate phosphate that dissolves in the thermocline as opposed to the fraction that dissolves in the deep ocean reservoirs. In this simulation the thermocline reservoir is uniform across all ocean basins and so does not contribute to the difference between Atlantic and Pacific Oceans. If little phosphorus survives dissolution in the thermocline reservoir to settle into the deep ocean there will be little difference between the phosphate concentrations in the different deep sea reservoirs. Therefore an increase in the fraction of phosphorus dissolving in the thermocline results in an increase in the phosphate concentration in the deep Atlantic reservoir. For the results shown in Fig. 7, it was assumed that a fraction $f_{PT} = 0.925$ of particulate phosphate dissolves in the thermocline reservoir.

The tuning exercises already carried out have fixed the phosphate concentration in the various oceanic reservoirs as well as the fluxes of water between these reservoirs. These quantities in turn determine the biological productivity of the warm surface water reservoir, which depends on the flux into the reservoir of dissolved phosphate. The difference in the ¹³C ratios in the surface water reservoir and the deep ocean reservoirs then depends on the carbon to phosphorus ratio

in the descending particles of organic matter. This same ratio controls also the gradient in carbon isotope ratio between the deep Atlantic reservoir, which accumulates relatively little organic debris, and the deep Pacific reservoir, which accumulates a lot. An increase in the carbon to phosphorus ratio in biogenic particles reduces the value of the ¹³C ratio in the deep Pacific reservoir. The results in Fig. 8 correspond to a carbon to phosphorus ratio $R_{CP} = 120$ atoms of carbon per atom of phosphorus. This value is close to the Redfield ratio, with some allowance for preferential dissolution of phosphorus. The difference between carbon isotopes in the atmosphere and in surface water is not subject to adjustment, but depends on the equilibrium isotope fractionation factor. The absolute values of the isotope ratios in the ocean–atmosphere system are controlled by the rock cycle. The most important controlling factors are the isotope ratio in the carbon supplied by carbonate dissolution, for which we use a value of +1.32‰, the isotope ratio in volcanic and metamorphic carbon dioxide, for which we use a value of −0.7‰, and the fractionation involved in the precipitation of carbonate minerals, for which we use a value of −1.7‰ relative to warm surface sea water. These values result in a steady state isotopic composition for the pre-industrial atmosphere of −6.35‰, close to that derived from measurements on bubbles in polar ice (Friedli et al., 1986) or from the changes deduced from studies of tree rings (Freyer, 1986). The difference in lysocline depths between Atlantic, Indian, and Pacific Oceans depends on the carbonate ion concentrations in these reservoirs and therefore on the relative additions of biogenic carbon and calcium carbonate. An increase in the ratio of calcium carbonate to organic carbon in the biogenic particles causes an increase in the Pacific lysocline depth relative to the Atlantic lysocline depth. We use a value of $R_{CO} = 0.09$ for this ratio, which results in lysocline depths of 4.15, 3.74, and 3.10 km in the Atlantic, Indian, and Pacific Oceans. The corresponding fractional areas of these ocean basins bathed in supersaturated water are 0.50, 0.41, and 0.29.

The overall level of total dissolved carbon and alkalinity in the ocean is constrained by the requirement that carbonates be deposited as rapidly as they are being added to the ocean by carbonate dissolution, weathering, and volcanism. In the steady state, the atmospheric pressure of carbon dioxide is fixed by the rock cycle, as already

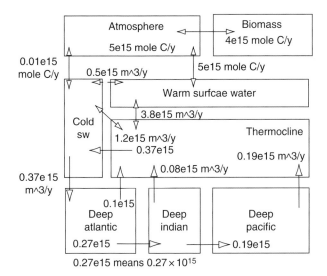

Fig. 9. Exchange fluxes in the simulation tuned to yield the steady state values of Figs. 7 and 8, and the bomb radiocarbon history of Fig. 5. Oceanic fluxes are stated in terms of volumes of water exchanged per year. The fluxes connecting to the atmosphere are quoted in terms of moles carbon per year.

noted, so only one parameter is subject to further adjustment. This parameter is the constant z_1 in the expression for lysocline depth as a function of carbonate ion concentration. We use a value of 5.8 km for this constant to derive the steady state values of total dissolved carbon shown in Fig. 7. This value is not strictly in accord with the solubility constant for calcite. Our simulation does not deal with additional factors affecting lysocline depth such as the oxidation of organic carbon within the bioturbation zone and the variation with depth of carbonate ion concentration in the deep sea.

The tuning process is iterative because adjustments to some of the parameters described later in this section cause changes in some of the observable quantities referred to earlier. The values quoted in this discussion and illustrated in Fig. 9 are the final results of this iterative tuning process. Some of the parameters of the simulation have little effect on the results and so have values that can not be determined in the tuning process. These include the mixing processes that exchange water between the surface water reservoirs and between the thermocline and the cold surface water reservoir. These mixing fluxes are given by the volume of the warm surface water reservoir divided by a mixing time of 50 yr for exchange between the surface water reservoirs and by the volume of the thermocline

divided by a mixing time of 250 yr for exchange between the thermocline and the cold surface water reservoir.

4. INDUSTRIAL REVOLUTION

The computer simulation has been described in previous sections. We have shown how the parameters of the simulation can be adjusted, in a process known as tuning, to reproduce observations. The particular observations we have used to tune this simulation are the record of bomb radiocarbon in the atmosphere and ocean and the pre-industrial distribution in the ocean of carbon isotopes, phosphorus, and total dissolved carbon. In this section we apply the simulation to the observed record of atmospheric carbon dioxide and carbon isotopes during the industrial revolution. This comparison of theory and observation will reveal some inadequacies in the simulation. We will introduce additional features into the simulation that improve the agreement of the theory with observations.

4.1 Fossil fuel

For the first numerical simulation of the industrial revolution we add a source of atmospheric carbon dioxide corresponding to the burning of fossil fuel. All other parameters of the simulation are left unchanged. Figure 10 shows the fossil fuel source, the observed atmospheric partial pressure of carbon dioxide, and the calculated partial pressure. The fossil fuel source is based on a tabulation by Keeling et al. (1989a). The observations of atmospheric carbon dioxide partial pressure are from Keeling et al. (1989a) since 1958 and from Neftel et al. (1985) and Friedli et al. (1986) for the older record. The pre-1958 data are derived from air bubbles in polar ice.

Both calculated and observed carbon dioxide mixing ratios equal about 280 parts per million prior to 1800. Indeed, the simulation was tuned in the previous section to reproduce this pressure. The observed carbon dioxide values show a steady increase during the 19th century and the first half of the 20th century and a more rapid increase in the second half of the 20th century. This failure is a direct consequence of the very small values of fossil fuel release rate during most of the 19th century. In this simulation the only perturbation is carbon dioxide from fossil fuel burning. The rate of this process was negligibly

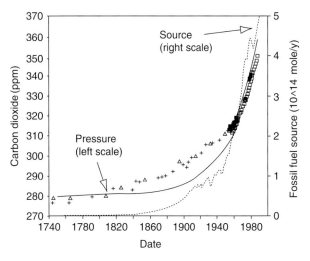

Fig. 10. History of carbon dioxide during the last few centuries. Observations are indicated by markers. The calculated values are indicated by solid lines. The response of the system is driven by fossil fuel combustion only, with the fossil fuel source as shown by the dashed line.

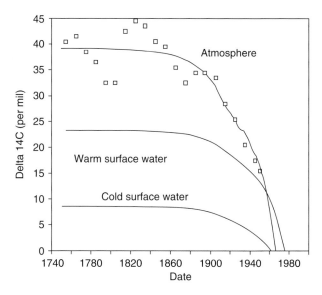

Fig. 11. Response of radiocarbon in atmosphere and surface ocean reservoirs to combustion of fossil fuel over the last few centuries. Theoretical results are plotted as solid lines. The markers are observations.

small before about 1850. Therefore the simulation shows no response before about 1850. The observed increase in carbon dioxide during the 19th century is not due principally to fossil fuel burning but almost certainly due to forest clearance in the so called pioneer effect (Woodwell et al., 1983; Hobbie et al., 1984). We will take up the pioneer effect in the next section, but first let us examine the isotope results.

Figure 11 compares the calculated radiocarbon ratio in the atmosphere with observations based on measurements of carbon preserved in tree rings. The observations are from Stuiver (1982, 1986). Also shown in the figure are calculated values of the radiocarbon ratio in the surface ocean. The fluctuations in measured radiocarbon values in the earlier part of the record are associated with changes in solar activity and in the cosmic ray source of radiocarbon in the atmosphere (Stuiver and Quay, 1980). We have made no attempt to simulate these changes. The point of interest in this study is the rapid decrease in radiocarbon in the 20th century caused by the release of fossil fuel carbon dioxide. Because of the great age of fossil fuels, they contain no radiocarbon.

The agreement between theory and observation for atmospheric radiocarbon is good. Since the source, fossil fuel carbon dioxide, is fairly well determined, and the radiocarbon content of this source is well known, namely zero, this agreement

relates mainly to the size of the reservoirs exchanging carbon with the atmosphere on a time scale of decades. The agreement indicates that our simulation has about the right rate of exchange of carbon between surface ocean and atmosphere and about the right rate of exchange of carbon between the biomass and the atmosphere. Since the simulation has already been tuned to reproduce the history of bomb radiocarbon in ocean and atmosphere, a history with more or less the same time scale, it is perhaps not surprising that the agreement between theory and observation for atmospheric radiocarbon is good.

Theory and observation are compared for stable carbon, ^{13}C in Fig. 12. The calculations use values for the isotope ratio in fossil fuel carbon as a function of time tabulated by Keeling et al. (1989a) and plotted in Fig. 13. The ratio has changed with time as a result of changing proportions of coal, oil, and natural gas consumed. The observations of atmospheric isotope ratio are based on direct mass spectrometer measurements of atmospheric carbon dioxide reported by Keeling et al. (1989a) for the period since 1958. The older data are based on tree ring measurements summarized by Freyer (1986) and on ice core measurements by Friedli et al. (1986). Fossil fuel carbon is isotopically light, so the burning of fossil fuels has caused a reduction in the isotope ratio in the atmosphere. Agreement between calculation and observation is reasonably good.

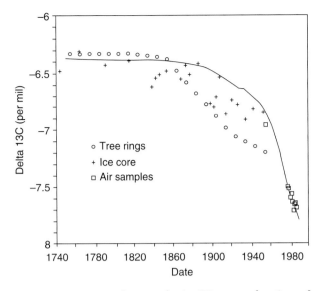

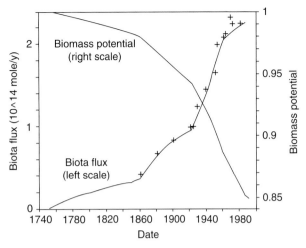

Fig. 12. Response of atmospheric ¹³C to combustion of fossil fuel. Observations are indicated by markers, the calculations by a solid line.

Fig. 14. The history of the flux of carbon from biota to atmosphere caused by changes in land use (forest clearance). Values deduced from land use studies appear as markers. The calculated fit to these values with the assumption of no change with time in carbon storage unit per area is plotted as a solid line, referring to the left-hand scale. The history of biomass potential has been tuned to yield the illustrated values of biota flux. The derived values of biomass potential are plotted by a solid line referring to the right-hand scale.

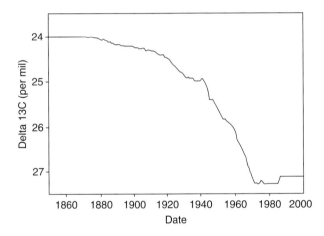

Fig. 13. The history of the average isotopic composition of fossil fuel carbon.

4.2 Pioneer effect

The problems with atmospheric carbon dioxide identified here have been encountered in previous studies of the carbon cycle (e.g., Enting and Pearman, 1986; Post et al., 1990). It is clear that there has been an important contribution to the carbon cycle associated with the clearance of forests as part of the spread of population, particularly in the temperate mid-latitudes of the Northern Hemisphere in the 19th century and in the tropics in the latter half of the 20th century. The process is called the pioneer effect in reference to the activities of colonial pioneers.

Careful studies of land use over the globe have been carried out by Woodwell et al., (1983; Houghton et al., 1983; Houghton, 1986) to assess the history of the flux of carbon dioxide to the atmosphere caused by forest clearance. These studies assemble data on changes in the area of different kinds of agricultural and natural biomes. These estimates of areal change are combined with estimates of carbon storage per unit area (Schlesinger, 1986), assumed to be constant for a given land use class, to derive a rate of release of carbon dioxide. Integration of these rates of release over all land use classes yields a global flux of carbon dioxide to the atmosphere associated with changing land use patterns (Houghton et al., 1983; Detwiler and Hall, 1988). This history of biomass flux is plotted as points in Fig. 14. The solid line that goes through these points is the flux that actually enters into the simulation. For reasons that will be explained below, we have chosen not to use the biomass flux as a fixed and tabulated value but instead as the solution to a differential equation. Our differential equation calculates the evolution of total biomass in response to changing land use patterns and possibly other perturbations (there are related equations for the carbon isotopic composition of biomass).

The differential equation is

$$\frac{dM_{Bio}}{dt} = \frac{F_{Bio} - M_{Bio}}{50}$$

where the first term represents the growth of new biomass in units of moles per year and the second term represents the decay of old biomass, described here as a first order decay process with an effective decay time of 50 yr. The real situation is of course much more complicated. Some biomass decays rapidly. Some survives for long periods of time (Goudrian, 1989). The 50 yr residence time represents an average that might be appropriate to the more massive components of total biomass. The initial value of F_{Bio} is 2×10^{17} moles, equal to the initial value of M_{Bio}.

The actual record of changing land use involves a mosaic of different changes between forest and agriculture, between forest and different kinds of forest, between different agricultural uses, and between agriculture and desert. In order to greatly simplify the system without destroying its essentially dynamic features we represent the source of biomass as proportional to a quantity we call biomass potential, B_p. Biomass potential is loosely defined as an integral of area weighted by carbon storage per unit area, normalized to 1 at the beginning of the industrial revolution. Forest clearance, then, reduces biomass potential from its initial value of 1. In this way, we use a single number to represent a whole multiplicity of changes in land use. We then adjust the history of biomass potential in such a way as to reproduce the biomass changes that have been deduced from studies of the history of land use. The components of this calculation are shown in Fig. 14. The points mark the biomass flux deduced from land use studies. This flux is equal to the negative of the rate of change of biomass and is calculated under the assumption that the carbon storage per unit area does not change with time in given land use classes. The solid line through the points represents the biomass flux derived from our solution of the differential equation for biomass evolution. The illustrated fit has been achieved by adjusting the history of biomass potential to the values plotted in the figure. Reduced biomass potential leads to reduced biomass and therefore the release of carbon dioxide from biomass. The response of biomass and the biomass flux to change in biomass potential is delayed because of the 50 yr residence time of carbon in the biomass. This delay is incorporated into

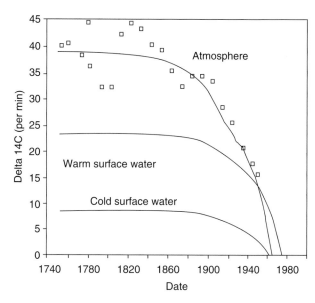

Fig. 15. Response of radiocarbon in the atmosphere and surface ocean to the release of carbon dioxide by combustion of fossil fuels and by forest clearance. Observations appear as markers, calculated values as solid lines.

the solution of the differential equation. The results illustrated in the figure suggest that forest clearance and changing patterns of land use from uses that store more carbon per unit area to uses that store less carbon per unit area have reduced biomass potential by about 15%. If all biomass were in a single kind of forest, then the area of that forest would have been reduced by 15%. Alternatively if there were no further changes in land use and if the system were allowed to come to steady state, the final biomass would be 15% smaller than the initial biomass. In this calculation the initial biomass was taken to be 2×10^{17} moles (Woodwell et al., 1983; Schlesinger, 1986). The value and the residence time of carbon in biomass are to some extent constrained by the results for radiocarbon, both the response of the system to bomb radiocarbon and to the release of fossil fuel carbon.

Inclusion of the pioneer effect has no significant impact on the history of radiocarbon in the atmosphere. Theory and observation are compared in Fig. 15. The simulation in this case includes both fossil fuel carbon dioxide and the flux of carbon dioxide from biomass as shown in Fig. 14. Forest clearance has little effect on atmospheric radiocarbon because the difference between radiocarbon ratios in biomass and radiocarbon ratios in the atmosphere is small compared with the difference between atmosphere and fossil fuel. The difference between these two reservoirs is not great because the residence time of carbon in

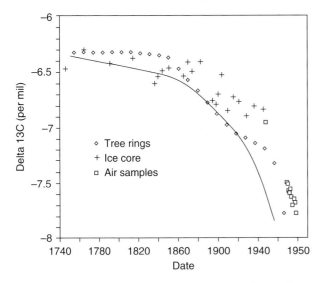

Fig. 16. Response of ^{13}C in the atmosphere to release of carbon dioxide by fossil fuel combustion and forest clearance. Observations appear as markers, theoretical values as a solid line.

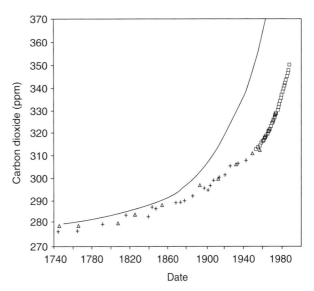

Fig. 17. Response of the carbon dioxide concentration in the atmosphere to the source provided by fossil fuel combustion and forest clearance. Observations appear as markers and the theoretical results as a solid line.

biomass is short compared with the decay time of radiocarbon.

The history of ^{13}C in the atmosphere is, on the other hand, significantly changed by the pioneer effect, as shown in Fig. 16. Biomass carbon is significantly lighter with respect to ^{13}C that atmospheric carbon because photosynthesis fractionates in favor of the lighter carbon isotope. Release of biomass carbon reduces the isotope ratio in the atmosphere. The agreement between theory and observation for ^{13}C is bad. The absolute values of calculated isotope ratio could be adjusted, but the amplitude of the calculated perturbation is too large.

The agreement between calculated and observed carbon dioxide mixing ratio, shown in Fig. 17, is also bad. The simulation now includes both fossil fuel release and the pioneer effect. The agreement between theory and observation is fair until the middle of the 19th century, but after that time the calculated carbon dioxide amounts are too large compared with observation. This result is well known in carbon cycle modelling (Enting and Pearman, 1986; Post et al., 1990). There is a missing sink for carbon dioxide.

4.3 Fertilization

There has been in this field a tendency for oceanographers to say that the missing sink must involve biomass and for biologists to say that the missing sink must be in the ocean. To us it appears that the data on changing land use are good and

our understanding of the capacity of the ocean to adsorb carbon dioxide is good also, as evidenced by the success of this simulation in reproducing the history of bomb radiocarbon. In order to make progress it is necessary to guess the nature of the missing sink so that it can be incorporated into the simulation. At this point we cannot claim that our identification of the sink is better than a guess. We shall make some assumptions about the sink and proceed to work out what the consequences of those assumptions might be. The exercise provides an indication of where more research is needed. It seems to us a more profitable approach than simply pointing the finger at other people's data, measurements, and methodologies.

Specifically, we assume that the studies of the history of land use are essentially correct in the identification of changes in areas of different land use classes and so, in the usage of this simulation, in their calculation of the history of biomass potential. What we suggest is that the assumption of constant carbon storage per unit area and land use class is incorrect because of the fertilization effect of carbon dioxide. For purpose of this simulation we shall assume that the growth rate of biomass is proportional not only to biomass potential but also increases with the partial pressure of carbon dioxide. In this way, even as the area of more carbon-rich biomes decreases, the carbon storage per unit area increases with increasing atmospheric carbon dioxide, so total global biomass does not decline as fast. Keeling et al. (1989b) and Esser (1987) have also ascribed the missing sink to carbon dioxide fertilization, which may be the cause of the increasing amplitude of seasonal change in carbon dioxide (Bacastrow et al., 1985; Houghton, 1987).

For the functional dependence of growth rate on carbon dioxide we adapt the expression of Esser (1987), derived from ecological and physiological data

$$F_{\text{Bio}} = C_{\text{Bio}} B_{\text{P}} F(\text{CO}_2)$$

where $C_{\text{Bio}} = 2 \times 10^{17}$ moles is the equilibrium value of biomass for carbon dioxide mixing ratio = 280 ppm and biomass potential $B_{\text{p}} = 1$. Here

$$F(\text{CO}_2) = 2.22\left(1 - \exp\left[-0.003\left(p\text{CO}_2 - 80\right)\right]\right)$$

with $p\text{CO}_2$ expressed in ppm. The fertilization function is plotted in Fig. 18. This fertilization function yields reasonable agreement between calculated and observed carbon dioxide mixing ratios, as shown in Fig. 19.

It is, moreover, consistent with the characterization of the missing sink by Tans et al. (1990). From careful observational and theoretical studies of the meridional gradient in carbon dioxide amount these authors were able to establish the approximate magnitude of the missing sink in the early years of the 1980's as between 2 and 3×10^{14} m/y, and its location as being in middle to high latitudes of the Northern Hemisphere. This Northern Hemisphere location is consistent with our identification of the missing sink as associated with biomass because the Northern Hemisphere is where most middle to high latitude biomass is located. Moreover, the magnitude of the missing sink deduced by Tans et al. is reasonably consistent with the deductions of our simulation with the fertilization assumption. Fig. 20 shows the biomass flux calculated in the simulation with biomass potential as shown in Fig. 14, but with carbon dioxide fertilization. With these assumptions there was little change in biomass until about 1950. Increase in biomass due to the fertilization effect of increasing carbon dioxide has just about cancelled decrease caused by forest clearance, represented here by decreasing biomass potential. Since about 1950, the fertilization effect has dominated over the clearance effect, global biomass has been increasing, and the biota has been absorbing carbon dioxide.

Because carbon dioxide has increased by about 25% in the last century our assumption of a ferti-

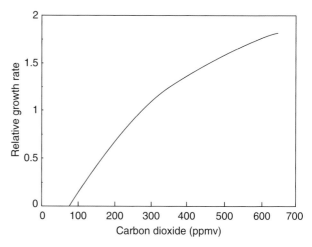

Fig. 18. Assumed fertilization function. The figure shows how the global rate of photosynthesis is assumed to depend on the carbon dioxide concentration in the atmosphere.

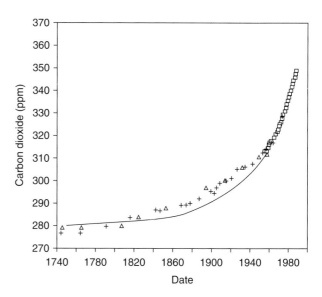

Fig. 19. Response of atmospheric carbon dioxide concentration to the release of carbon dioxide by fossil fuel combustion and forest clearance with the assumption of carbon dioxide fertilization of the global rate of photosynthesis. Observations appear as markers and the theoretical results as a solid line.

lization effect almost proportional to carbon dioxide implies an increase in stored carbon per unit area of a comparable 25%. We are not aware of any observations that could rule out such a change. Carbon concentrations quoted for each land use class generally vary by as much as a factor of 2 or 3, reflecting variability within the class. There is no way to say that these numbers have not increased by 25% in the last century. Looked at in another way, if we suppose that most of the

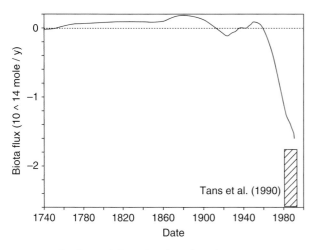

Fig. 20. The derived flux of carbon from biomass to atmosphere plotted as a solid line, and an estimate of this flux derived from observations of latitudinal gradients in carbon dioxide concentration, plotted as a rectangle. The calculated flux represents the combined effects of land use change and carbon dioxide fertilization of the rate of photosynthesis.

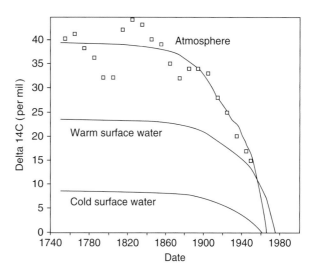

Fig. 21. Response of radiocarbon in atmosphere and surface ocean to the release of carbon dioxide from fossil fuel combustion and forest clearance, with allowance for carbon dioxide fertilization of the global rate of photosynthesis. Observations appear as markers and the calculated values as solid lines.

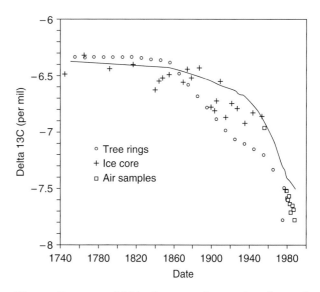

Fig. 22. Response of ^{13}C in the atmosphere to the release of carbon dioxide by fossil fuel combustion and forest clearance, with allowance for carbon dioxide fertilization of the global rate of photosynthesis. Observations appear as markers and the theoretical results as a solid line.

biomass is in the form of trees, and that the mass of carbon in the tree varies as the cube of the height of a tree we are looking at an increase in the height of the average mature tree by just 8% in a century. [Actually, the diameter of a tree must increase more rapidly than the height to prevent buckling, so the height increase is less than 8% (T. Volk, private commun., 1991). On the other hand it may be that increased height would lead to increased spacing between trees (K. Caldeira, private commun., 1991). Such a change can not be ruled out on the basis of observation.

As already noted, biomass changes have little effect on the radiocarbon content of the atmosphere so the agreement between theory and observation shown in Fig. 21 is good in this case as in the previous cases. Results for the case carbon isotopes are shown in Fig. 22. The agreement in this case is worse than in the case of fossil fuel only (Fig. 12), particularly in the comparison with direct measurements in the last decade or so. The biota is isotopically light. As it increases in mass it drives the atmosphere isotopically heavy. The calculated ^{13}C amounts for the recent past are larger than the observed.

4.4 Agriculture and isotopes

We have now developed a simulation that reasonably reproduces bomb radiocarbon data, the pre-industrial state in the ocean, atmospheric carbon dioxide, and radiocarbon during the industrial

revolution. There is a remaining problem of substantial disagreement between calculated and observed ratios of atmospheric ^{13}C. For the sake of neatness we make a further speculative suggestion about a possible resolution of the ^{13}C discrepancy.

Some plants, the C3 type, fractionate carbon by 20‰ during the course of photosynthesis. Other

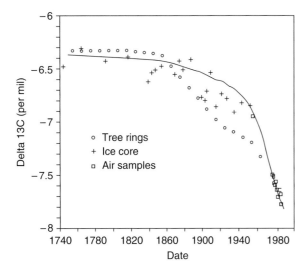

Fig. 23. Response of atmospheric ¹³C to the release of carbon dioxide by fossil fuel combustion and forest clearance, with allowance for carbon dioxide fertilization of the global rate of photosynthesis and a small change in the global average carbon fractionation in photosynthesis associated with an agricultural shift from C3 plants to C4 plants.

plants, the C4 type, fractionate by only 6‰ (Van der Merwe, 1982). Among the C4 plants are such important agricultural types as maize and sugar cane. Is it possible that during the course of changing patterns of agricultural and land use there has been a change in the global ratio of C3 plants to C4 plants that has resulted in a change in biomass fractionation with respect to atmospheric carbon isotopes?

In order to explore this possibility we have repeated the calculation that includes fossil fuel burning, land use change, and carbon dioxide fertilization with the introduction of a change in carbon isotope fractionation during the course of photosynthesis. Specifically the expression we have used for the fractionation factor in photosynthesis is

$$\Delta_{phot}\left(‰\right) = 20 - f(t)$$

where $f(t) = 0$ before 1960 and increases linearly from 0 in 1960 to 0.84 in 1990. The amplitude of the change is tuned to reproduce the observed carbon isotope history. Theory and observations are compared in Fig. 23. With this further assumption, the agreement between theory and observation is as good for ¹³C as for ¹⁴C and carbon dioxide partial pressure. The remaining discrepancy in the late 1800's and early 1900's for carbon dioxide could be removed by adjusting the fertilization function. Adjustments to the assumed history of photosyn-

thetic fractionation could alter the fit for δ¹³C. Both of these functions are otherwise unconstrained.

Is this a reasonable resolution of the ¹³C problem? It is hard to say. The required change in the fractionation factor associated with global productivity is only 4‰, corresponding to a replacement of only 6‰ of C3 photosynthesis by C4 photosynthesis. We are not talking about a very large replacement of C3 plants by C4 plants. Perhaps such a change has occurred.

The rest of this paper does not depend on our suggestion concerning the influence of agriculture on carbon isotopes. The calculation just presented serves mainly to call attention to a factor that may have affected the observation record of atmospheric ¹³C and to show the limited diagnostic power of the carbon isotope records.

5. EXTRAPOLATION

The simulation has now been tuned to the steady state distributions of dissolved phosphorus, total dissolved carbon, and carbon isotopes in the oceanic reservoirs and to the history of bomb radiocarbon. The tuned simulation has been tested and further modified to reproduce the industrial age history of atmospheric carbon dioxide, radiocarbon, and stable carbon isotopes. In this section we shall use the simulation without further modification to explore possible extrapolations of carbon dioxide evolution into the future. The approach is to assume scenarios for fossil fuel burning and for forest clearance and to calculate the evolution of atmospheric carbon dioxide in response to these various scenarios. This process is not correctly described as prediction, because the calculated carbon dioxide levels depend entirely on the assumed scenarios, and we make no effort to justify the scenarios. Ideally, different components of a prediction system would be used to develop scenarios for carbon dioxide release in response to changes in global population, climate, and technology. The use of assumed scenarios enables us to explore some options for the future and to find out what control measures might most effectively influence future levels of carbon dioxide. The simulation is a tool for exploring possible consequences of future courses of action, not a tool for predicting the future.

We consider limiting scenarios for fossil fuel combustion and for forest clearance. These scenarios are illustrated in Fig. 24. The two fossil fuel scenarios both consume the same total reserve of

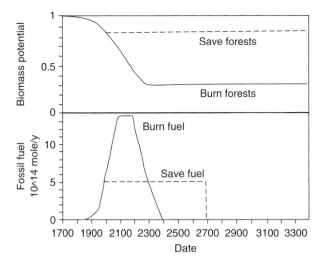

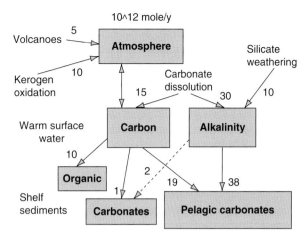

Fig. 24. Scenarios for the consumption of a fixed fossil fuel reserve and for the destruction of global biomass.

Fig. 25. The rock cycle, showing values of fluxes used in this simulation. Fluxes are expressed in units of 10^{12} moles per year.

recoverable fossil fuel, 35×10^{16} moles (Broecker and Peng, 1982). The profligate scenario assumes that the rate of consumption of fossil fuel continues to increase until the reserves begin to run out in a hundred years or so. There follows a turnaround and a rapid decrease in the rate of consumption of fossil fuel, with the reserve exhausted before 2400 A.D. This scenario is labeled "burn fuel" in Fig. 24. It is similar to the scenarios shown in Fig. 1 except that the peak burning rate is only three times the present value, as compared to five times this value, and the total amount of carbon consumed is slightly smaller. For the austere scenario, labeled "save fuel" in Fig. 24, we assume that the rate of combustion of fossil fuel is held constant at its present day value (5×10^{14} m/y) until the reserve is exhausted in 700 yr time. The fossil fuel combustion rate drops to zero shortly before 2700 A.D. These scenarios seem to us to more or less encompass the range of possibilities, subject to a fixed total reserve of fossil fuel. For forest clearance, the austere scenario, labelled "save forests" in Fig. 24, assumes no further reduction of biomass potential from that of the present day. This scenario implies an immediate end to the clearance of forest and the conversion of land from uses sustaining high carbon content per unit area to uses sustaining lower carbon content. The profligate scenario, labelled "burn forests" in Fig. 24 assumes that land use changes and forest clearance continue at about the present rate until biomass potential has been reduced to 0.35 of its value in 2300 A.D., remaining constant at that value into the future. Again, we view these sce-

narios as more or less bounding the range of possible future changes in land use.

We are particularly interested in extrapolations on a time scale of centuries to millennia, a time scale that has received little recent attention. Exploration of the possible consequences of global change on a time scale of several generations is most definitely needed. On the longer time scale, however, the rock cycle begins to influence the results. We shall therefore first describe our formulation of the rock cycle and test the sensitivity of simulated carbon dioxide levels to the description of the rock cycle. We shall then explore the possible responses of atmospheric carbon dioxide to the various scenarios for fossil fuel burning and forest clearance.

5.1 Rock cycle

Our formulation of the rock cycle is derived fairly directly from the work of Walker et al. (1981), Berner et al. (1983), Kasting et al. (1986), and other references. The components are illustrated in Fig. 25. Carbon dioxide is released to the atmosphere by volcanic and metamorphic processes at a constant rate of 5×10^{12} m/y (1/100 of the present rate of CO_2 release from fossil fuel burning). This carbon dioxide is neutralized by the weathering of silicate minerals, which releases calcium and magnesium ions to the warm surface water reservoir at a rate that we take to depend on the carbon dioxide partial pressure, a dependence that includes possible dependences on temperature and runoff (Walker et al., 1981; Berner et al., 1983). Possible dependences on vegetation (Volk, 1989b) are not included in this study. The calcium and magnesium ions and the volcanic carbon dioxide are ultimately removed from the system by precipitation as carbonate minerals. Further weathering reactions dissolve terrestrial carbonate rocks, adding alkalinity and total dissolved carbon to the warm surface water reservoir at a rate that depends on the carbon dioxide partial pressure. In this formulation there is an additional source of atmospheric carbon dioxide provided by the oxidation of sedimentary organic carbon, called kerogen. The rate of this oxidation process is 10^{13} m/y. It is exactly balanced in our formulation by the precipitation of fresh organic carbon in shelf sediments (Berner, 1982), although short-term imbalances are possible in the real world.

Most accumulation of carbonate sediments occurs in the deep sea, according to the formulation described above. The rain of carbonate particles into the deep sea is proportional to planktonic productivity, which depends on the flux of dissolved phosphate into the warm surface water reservoir. Carbonate particles which fall below the lysocline dissolve while those that fall into shallower waters are preserved. The rain of particulate carbonate is assumed to be spatially uniform. Lysocline depth is proportional to the carbonate ion concentration in each deep ocean reservoir. The fraction of area of each ocean above the lysocline is based on observed average ocean hypsometry. We include a small flux of carbonate to the shallow shelves at a rate proportional to the carbonate ion concentration in warm surface sea water. This flux is given by $10^{12}[CO_3^{2-}]/0.26$ mole/y, with $[CO_3^{2-}]$ expressed in moles/m^3 (Wilkinson and Walker, 1989). In the steady state, the total carbonate flux is 20 times larger.

As previous work has shown, the key element in the rock cycle is how the carbonate and silicate dissolution rates depend on carbon dioxide partial pressure. In the very long term it is this dependence that determines the steady state partial pressure of carbon dioxide in the atmosphere. In the shorter term it is this dependence that determines how quickly the system of ocean and atmosphere will recover from the impulsive input of fossil fuel carbon dioxide. Various formulations of the weathering rate laws have been proposed by Walker et al. (1981), Berner et al. (1983), and Volk (1987). In our view, all of these rate laws are subject to very serious uncertainty. On the other hand, the weathering rate law controls the carbon dioxide response only on a time scale of thousands to millions of years and so is not at the heart of the

current study. For simplicity, we shall not include in our formulation an explicit dependence of weathering rates on the hydrologic cycle or on global average temperature, expressing the weathering rates simply as functions of carbon dioxide partial pressure. The expressions we use are $W_{carb} = 15 \times 10^{12}$ pCO_2 and $W_{sil} = 5 \times 10^{12}$ $(pCO_2)^{0.3}$ moles/y, with pCO_2 expressed in PAL. The silicate weathering law is based on laboratory data cited by Walker et al. (1981). The rate at $pCO_2 = 1$ PAL is set equal to the assumed volcanic outgassing rate. (These two rates must be in balance at steady state.) Its dependence on pCO_2 is intermediate between that of Walker et al. (1981) and Berner et al. (1983). Carbonate weathering is assumed to have a stronger dependence on pCO_2; this choice is made so as to maximize potential CO_2 uptake by this process. In the next section we show how modification of these expressions affects the calculated levels of carbon dioxide.

The steady state values of carbon isotope ratios in ocean and atmosphere depend on the isotope ratios associated with the sources and sinks of carbon in the rock cycle. The values we use are as follows: volcanoes, −0.7; kerogen oxidation, −22.3; carbonate dissolution, 1.7; new organic carbon fractionated by −25‰; new carbonates fractionated by +1.7‰. These values have been tuned, with only modest observational constraints, to yield the steady state isotopic composition of ocean and atmosphere that gave the results of Fig. 8.

5.2 Dissolution of pelagic carbonate sediments

As fossil fuel carbon dioxide moves into the deep ocean the waters become more corrosive, the lysocline moves to shallower depths, and previously deposited pelagic carbonate sediments become susceptible to dissolution. Dissolution of these sediments adds alkalinity and total dissolved carbon to the deep sea reservoirs, increasing their capacity to absorb the added carbon dioxide. In this section we describe how this process is represented in our simulation. In developing this representation we have sought an upper limit on the role of pelagic carbonate dissolution by assuming rapid reaction of a large mass. We can test the sensitivity of our results to pelagic carbonate dissolution by suppressing the process completely.

We treat our calculated lysocline depth as marking a sharp transition between shallower levels in which no carbonate dissolution occurs and deeper levels in which no carbonate accumulation

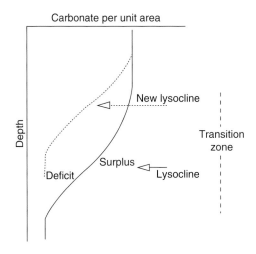

Fig. 26. The carbonate transition zone in pelagic sediments is represented in this calculation by a sudden change in dissolution rate, with no carbonate dissolution occurring above the lysocline and complete dissolution occurring below the lysocline.

takes place. The actual transition in the real ocean is more gradual as shown schematically in Fig. 26. In terms of the total carbonate accumulation rate, our step function representation incurs a deficit by ignoring accumulation below the calculated lysocline depth, but incurs a balancing surplus by neglecting dissolution above the calculated lysocline depth.

Suppose now that the lysocline moves to a shallower depth, as indicated by the dashed lines in Fig. 26. Carbonate sediments at depths between the new transition and the old transition become subject to dissolution. The total mass of carbonate that might be dissolved is given by the area of sea floor between the new lysocline depth and the old lysocline depth multiplied by the mass of soluble carbonate sediment per unit area. This total mass can be represented with little error by the step function distribution of carbonate accumulation rate because the areas of accumulation surplus just balance the areas of accumulation deficit. Following Broecker and Peng (1982) we assume that all of the carbonate minerals in the top 30 cm of sediment are susceptible to dissolution. If the carbonate sediments consist of one-third insoluble clay minerals, on average, dissolution of 30 cm of sediment will build up a 10 cm thick layer of clay which should halt further dissolution. The carbonate content of the top 30 cm of sediment is taken to be 6000 moles/m². We regard this as a generous estimate of the amount of calcium carbonate that is likely to dissolve.

From the rate of change of total dissolved carbon and alkalinity in the deep ocean reservoirs we calculate the rate of change of the carbonate ion concentration. From the rate of change of carbonate ion concentration we calculate the rate of change of lysocline depth, using the linear relationship between lysocline depth and carbonate ion concentration. From the rate of change of the lysocline depth we calculate the rate of change of ocean area above the lysocline, using observed average ocean hypsometry. This rate of change of ocean area above the lysocline is multiplied by 6000 moles/m² to get the rate at which pelagic carbonate minerals are added to a reservoir of dissolving pelagic carbonate sediments. There is one such reservoir of dissolving pelagic carbonate sediments for each deep ocean basin. These sediments dissolve according to a first order rate law, with a dissolution time of 10 yr. On a time scale of decades to centuries, therefore, the dissolution of pelagic carbonates is effectively instantaneous. Rapid reaction is assumed in order to maximize the potential impact of this process. The dissolution of these pelagic carbonates adds alkalinity and total dissolved carbon to the appropriate deep ocean reservoirs.

When the lysocline moves to greater depths there is a corresponding extraction of alkalinity and total dissolved carbon from the deep ocean reservoirs. To see this, consider the transition shown in Fig. 26 between the shallow lysocline indicated by dashed lines and the deeper lysocline indicated by solid lines. When the lysocline moves to deeper levels the transition region where carbonate dissolution is partial has to be filled with the appropriate stock of carbonate minerals before the step function representation of carbonate accumulation rate can be considered a reasonable approximation. The filling of this transition zone to bring the deficit accumulation region back into balance with the surplus accumulation region extracts total dissolved carbon and alkalinity from the deep ocean reservoir. The rate of extraction is the 6000 moles/m² of available carbonate times the rate of increase of area of the sea floor bathed by saturated sea water. The process is symmetrical. Movement of the lysocline to shallower depths exposes previously deposited carbonate sediments in the transition zone to dissolution, releasing carbon and alkalinity to the sea water. Movement of the lysocline to greater depths causes a reduction in the dissolution of carbonate particles that fall into the transition zone, effectively extracting carbon and alkalinity from the sea water.

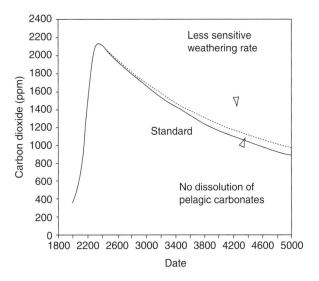

Fig. 27. Sensitivity of carbon dioxide predictions to the formulation of the rock cycle. These results correspond to a profligate scenario in which both fossil fuel and forests are rapidly burned. If dissolution of pelagic carbonates is suppressed in the simulation, there is a modest increase in carbon dioxide concentrations a few thousand years in the future. Results are more sensitive to assumptions concerning the dependence of carbonate weathering rate on carbon dioxide partial pressure, but even this process has little effect on carbon dioxide prior to the maximum in about the year 2400.

Our formulation deals with changes in lysocline depth either upward or downward. We have sought an upper limit on the role of pelagic carbonate dissolution. We now show the effect of suppressing this process completely.

5.3   Sensitivity to the rock cycle

For the sensitivity test we adopt as standard a profligate scenario in which fossil fuel and forests are both burned. The evolution of carbon dioxide partial pressure to the year 5000 A.D. is shown by the solid line in Fig. 27. In this scenario carbon dioxide increases to almost 2200 parts per million before fossil fuel reserves are exhausted in about the year 2400 A.D. Thereafter, the carbon dioxide partial pressure decreases as the added carbon dioxide is transferred into the deep ocean and is gradually neutralized by weathering reactions.

We test the sensitivity of the result to pelagic carbonate dissolution by suppressing the process altogether. The results are shown by the dashed line in Fig. 27. They are indistinguishable from the standard run until after the maximum in carbon dioxide partial pressure at about 2400 A.D. Thereafter, the suppression of pelagic carbonate

dissolution results in higher carbon dioxide partial pressures, as would be expected. Pelagic carbonate dissolution has virtually no influence on the result in the first few hundred years of the simulation because the thermohaline circulation takes hundreds of years to carry the extra carbon dioxide into the deep ocean reservoirs. In this formulation we deliberately separated the deep ocean reservoir into separate Atlantic, Indian, and Pacific Ocean basins in order to see whether more rapid ventilation of the Atlantic Ocean might have a more immediate effect on carbon dioxide levels. However, the residence time of water in our deep Atlantic reservoir, calculated by dividing the volume of the reservoir by the influx of water from the cold surface water reservoir is still 690 yr, as shown in Fig. 3. It takes this long for the saturation state of this reservoir to change significantly in response to changes in the atmosphere. The process could be carried further, by sub-dividing the deep Atlantic reservoir into a north Atlantic reservoir with still more rapid ventilation and a south Atlantic reservoir with a longer residence time, but the carbonate sediments available for dissolution in such a north Atlantic deep reservoir would not be very abundant. We conclude that dissolution of pelagic carbonates is not likely to influence the levels of atmospheric carbon dioxide before several hundred years have elapsed. The main conclusions of this paper, which concern maximum carbon dioxide concentrations, are not sensitive to our formulation of the pelagic dissolution process.

On the other hand, the influence of dissolution of pelagic carbonates on lysocline depth is profound, as would be expected. The evolution of lysocline depths in the standard run and in the run pelagic dissolution suppressed is shown in Fig. 28. The figure shows how the carbon dioxide perturbation is felt first in the Atlantic Ocean and last in the Pacific Ocean and how the response of lysocline depth is very much larger in the simulation in which pelagic carbonate dissolution is suppressed. By releasing alkalinity to the deep ocean reservoirs, pelagic carbonate dissolution serves to neutralize the effect of fossil fuel carbon dioxide, limiting the amplitude of excursions in lysocline depth. By overestimating the dissolution of pelagic carbonates, our simulation underestimates changes in both lysocline depth and carbon dioxide partial pressure. We regard this as a conservative treatment of unavoidable uncertainties.

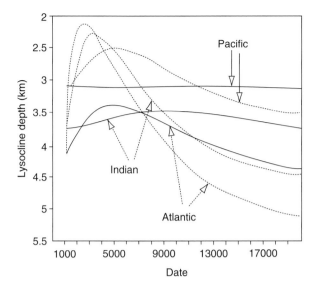

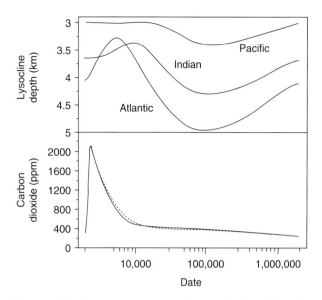

Fig. 28. The effect on lysocline depths of pelagic carbon dissolution. These results are for the same profligate scenario as Fig. 27. The solid lines show the evolution of lysocline depths in the different ocean basins in the standard run. If pelagic carbonate dissolution is suppressed, changes in lysocline depth are much more pronounced, as shown by the dashed lines.

Fig. 29. The long term recovery of atmospheric carbon dioxide after the injected of fossil fuel and forest carbon dioxide according to the profligate scenario. The different lines in the carbon dioxide plot correspond to the standard scenario, plotted as a solid line, the case with pelagic carbonate dissolution suppressed, plotted as a dashed line, and the case with reduced sensitivity to carbon dioxide pressure of the carbonate weathering rate, plotted as a dotted line. The evolution of lysocline depths in the standard case is plotted at the top of the figure. The time scale is logarithmic. Complete recovery takes more than a million years.

Also shown in Fig. 27 are results that illustrate the sensitivity of carbon dioxide partial pressure to the weathering rate law. For purposes of this comparison, the carbonate weathering rate was taken to be proportional to the square root of carbon dioxide partial pressure instead of the first power, and the silicate weathering rate was left unchanged, proportional to the carbon dioxide partial pressure raised to the power of 0.3. This change in the weathering rate law has no significant impact on calculated carbon dioxide levels prior to about 2200 A.D. Thereafter, reduced weathering fluxes result in larger carbon dioxide pressures.

The relative buffering capacities of terrestrial carbonate weathering and pelagic carbonate dissolution depend on poorly quantified weathering and dissolution rate laws. Because we have assumed that the rate of weathering of terrestrial carbonates is directly proportional to pCO_2, this rate is increased by a factor of six in our standard run at the time of the carbon dioxide maximum. This corresponds to a very large flux of alkalinity to the ocean, some 2×10^{14} equivalents/year. Moreover, the rate of terrestrial weathering responds immediately to carbon dioxide changes, but pelagic carbonate dissolution is delayed by the ventilation time of the deep sea. Our calcula-

tion suggests that terrestrial carbonate weathering may be more important than previously thought, but is also suggests that more work is needed to determine the probable response of this process to future carbon dioxide increases.

The long-term recovery of this system is controlled by the rock cycle. Figure 29 shows the long-term evolution of atmospheric carbon dioxide, comparing the cases with modified weathering rate and without pelagic carbonate dissolution, and shows also the evolution of lysocline depths in the standard case. The time scale here is logarithmic because such a scale best shows the changes that occur on short time scales initially and then on the very much longer time scales associated with the rock cycle. What we see here is that the carbon dioxide content of the atmosphere continues to increase as long as fossil fuel is being burned. The accessible reservoirs, biomass, and surface ocean, are too small to absorb the added carbon dioxide, which mostly accumulates in the atmosphere. The large deep ocean reservoirs are not accessible on the time scale of a few hundred years associated with exhaustion of the fossil fuel reserve. Atmospheric carbon dioxide

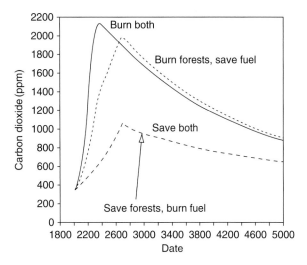

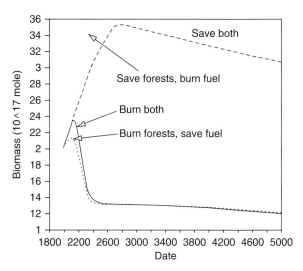

Fig. 30. Responses of atmospheric carbon dioxide to the various scenarios for fossil fuel combustion and forest clearance. These results correspond to the assumption of carbon dioxide fertilization of the global rate of photosynthesis.

Fig. 31. The evolution of biomass in response to the various scenarios for fossil fuel combustion and forest clearance. The results correspond to the carbon dioxide results of Fig. 30.

begins to decrease as soon as fossil fuel combustion has ended. This decrease is fast at first as the deep ocean comes to equilibrium with surface ocean and atmospheric reservoirs. There follows a much slower recovery associated with the gradual neutralization of the added carbon dioxide by rock weathering. It has not generally been realized how long the fossil fuel perturbation may last. The duration, of course, depends on the weathering rate law, but an indication of the time for recovery can be obtained by dividing the total fossil fuel reserve, 3.5×10^{17} moles by the expected enhancement in silicate weathering rate. If silicate weathering varies as the carbon dioxide partial pressure raised to the power of 0.3 then a doubling of carbon dioxide partial pressure will increase the silicate weathering rate by a factor of $2^{0.3}$ for an augmentation in the rate of about 1×10^{12} moles/y. Such an augmented rate would require 350,000 yr to neutralize the total fossil fuel release. The recovery will be more rapid if the increase in weathering rate with carbon dioxide partial pressure is more marked.

5.4 Sensitivity of carbon dioxide to fuel and forest scenarios

All of the calculations of this section are begun in 1990 A.D. with the values calculated in the section on the industrial revolution and run into the future for a few million years. We consider four

scenarios for fuel and forest burning, as described above, and illustrated in Fig. 24. Results for the carbon dioxide partial pressure over a few thousand years are shown in Fig. 30. The solid line shows the results already presented in which both fuel and forests are burned rapidly. Carbon dioxide partial pressure peaks at almost 2200 parts per million a little before the year 2400 A.D. If fuel is conserved while forests are burned, the maximum carbon dioxide concentration is reduced only to about 2000 parts per million, but the maximum is postponed to about the year 2700 A.D. It is at this time that the fossil fuel reserve is exhausted under the "save fuel" scenario.

In this calculation, saving the forests leads to a marked reduction in the maximum carbon dioxide partial pressure, to about 1100 parts per million or just 3 times the present level. This reduction results from the fertilization effect and the absorption of fossil fuel carbon dioxide by the preserved forests. The point can be seen most clearly in Fig. 31, which illustrates the evolution of biomass for these various scenarios. In the "save forest" scenarios biomass increases from its present value by almost a factor of 2. When forests are burned, an initial increase in biomass corresponding to the fertilization effect is soon overwhelmed by forest clearance, so both fuel burning and forest burning contribute to the calculated increase in atmospheric carbon dioxide.

In both the forest burn and the forest save cases, the maximum carbon dioxide partial pressure occurs at the time of exhaustion of the fossil fuel reserve. This is because fossil fuel combustion rates are much greater than any of the geochemical rates in this system. Oceans and rocks cannot absorb the carbon dioxide as fast as it is being added. It simply accumulates in the atmosphere and biota until the fossil fuel reserve is exhausted. For the same reason, the maximum value of the carbon dioxide partial pressure is relatively insensitive to the rate at which fossil fuels are burned. In the time scale for exhaustion of the fossil fuel reserve, 700 yr in our austere scenario, very little of the added carbon dioxide can be absorbed by the deep ocean. The maximum value of carbon dioxide partial pressure depends mainly on the total amount of fossil fuel carbon dioxide released and only weakly on the rate of release. Energy conservation can delay the occurrence of large carbon dioxide partial pressures, but those large pressures will finally arrive unless the fossil fuel is left in the ground.

On the other hand, this simulation suggests that we might be able to substantially limit future increases in atmospheric carbon dioxide by preserving global biomass. This conclusion depends directly on our assumption of a significant enhancement in plant growth rate caused by carbon dioxide increases—an idea that has been suggested numerous times before (e.g. Revelle and Munk, 1977; Freyer, 1979, refs. on p. 90). This assumption has been criticized (Bolin et al., 1979; Houghton and Woodwell, 1989) because it does not include any possibly unfavorable impact of climate change on biomass or, more specifically, on carbon storage in soils. In this sense the "save forest" calculations shown in Fig. 30 may be wildly optimistic. They do serve, however, to call attention to the potential importance of biomass as a sink for fossil fuel carbon dioxide and suggest that conservation of biomass may be as important to the limitation of global change as conservation of fossil fuel.

5.5 Sensitivity to biomass growth and decay rates

In order to explore the sensitivity of the results to assumptions made about rates of growth and decay of the biomass we have conducted some further calculations. For purposes of the sensitivity study we use the scenario in which forests are preserved at their present biomass potential but

fossil fuel is rapidly consumed. This is the "save forests, burn fuel" scenario of Fig. 24. The standard simulation, for which results have already been presented, appears in Figs. 32 and 33 as the "enhanced growth, constant decay" case. Enhanced growth refers to the fertilization of biomass growth rate by carbon dioxide. Constant decay refers to the residence time of carbon in biomass and soils, in this case assumed constant at 50 yr.

In order to explore the sensitivity of the results to assumptions made about rates of growth and decay of the biomass we have conducted some further calculations. For purposes of the sensitivity study we use the scenario in which forests are preserved at their present biomass potential but fossil fuel is rapidly consumed. This is the "save forests, burn fuel" scenario of Fig. 24. The standard simulation, for which results have already been presented, appears in Figs. 32 and 33 as the "enhanced growth, constant decay" case. Enhanced growth refers to the fertilization of biomass growth rate by carbon dioxide. Constant decay refers to the residence time of carbon in biomass and soils, in this case assumed constant at 50 yr.

The results presented for comparison in Figs. 32 and 33 are for cases in which biomass growth rate is held constant at its 1990 value. There is no further carbon dioxide fertilization and no further change in biomass potential. This is referred to as the "constant growth" case. The "enhanced decay" case is one in which the residence time of carbon in biomass is reduced by a factor of 2 for an increase of global average temperature by 10°C. Our energy balance climate model yields rather modest increases in global average temperature up to a maximum of about 6°C. We assume that the biomass decay rate is a linearly increasing function of global average temperature. The "constant growth, enhanced decay" case yields maximum carbon dioxide concentrations about twice those of the "enhanced growth, constant decay" case (see Fig. 32). The other cases are intermediate. Corresponding changes in biomass appear in Fig. 33.

The lesson to be drawn here is that there is no guarantee that forest conservation will indeed lead to dramatic reductions in future atmospheric carbon dioxide levels. The terrestrial biota appear to have been absorbing CO_2 during the past half century, but may cease to do so as the climate warms. From a policy standpoint, we should attempt to preserve the forests but we should not

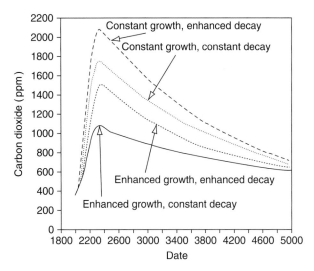

Fig. 32. Sensitivity of future levels of carbon dioxide to assumptions about biomass changes. These results correspond to the scenario in which fossil fuel is rapidly burned, but forests are conserved. The standard run includes dioxide fertilization, but no increase in biomass decay rate in response to increases in global average temperature. Comparative runs show the effect of increasing rates of biomass decay and the suppression of the carbon dioxide fertilization effect.

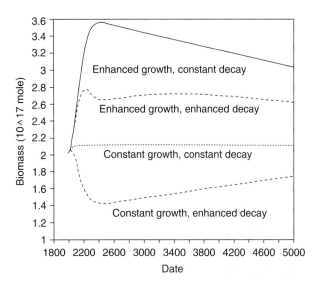

Fig. 33. Sensitivity of biomass changes in response to various assumptions concerning growth rates and decay rates. These results for biomass correspond to the results for carbon dioxide concentration in Fig. 32.

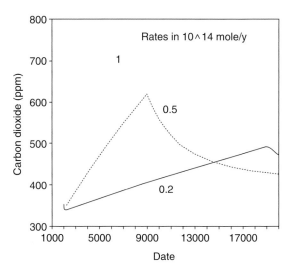

Fig. 34. Response of atmospheric carbon dioxide concentrations to various constant rates of fossil fuel combustion. These results assume that forest clearance is immediately halted and that the total fossil fuel reserve is consumed at the constant rate indicated on the figure. The present rate of fossil fuel combustion is 5×10^{14} moles per year. These results indicate that an immediate reduction in fossil fuel combustion by a factor of 25 would hold carbon dioxide concentrations below 500 parts per million.

5.6 Towards a sustainable economy

As a tentative exploration of a possible goal we consider in this section whether it might be possible by rigorous energy conservation measures to hold the partial pressure of atmospheric carbon dioxide permanently below 500 parts per million while still exhausting our fossil fuel reserves. For purposes of this calculation we assume an immediate end to forest clearance, holding biomass potential constant at its present value of 0.85, and we use the (most favorable) "enhanced growth, constant decay" model. We assume an immediate reduction in the rate of fossil fuel combustion to various levels smaller than the present level, holding this rate constant until the reserve is exhausted. The results appear in Fig. 34. They show that sustainability in this sense could be achieved if the fossil fuel combustion rate were permanently reduced from its present level of 5×10^{14} m/y to 0.2×10^{14} m/y— a reduction by a factor of 25, not the factor of 2 that is sometimes quoted in the press. Since the assumptions made in doing this calculation are optimistic to begin with, we conclude that keeping atmospheric CO_2 levels below 500 ppm would probably require switching to alternative energy sources long before our fossil fuel reserves are depleted.

assume that doing so will protect us from the extremely high atmospheric CO_2 levels predicted in our worst case scenarios or by studies like those shown in Fig. 1. The important role of biomass in our standard simulation is a direct consequence of our assumption of a large fertilization effect.

6. CONCLUSION

In this study we have put together a simple numerical simulation of the global biogeochemical cycles affecting atmospheric carbon dioxide. We have worked with the smallest number of reservoirs and the simplest formulation that seemed able to reproduce reasonably well the observations that relate to this problem and also to represent the processes likely to be important on the time scales of decades to millenia of principal interest to us. We have adjusted the parameters of our simulation to reproduce such observations as the distribution of phosphate and total dissolved carbon between ocean reservoirs, the isotope ratios for both ^{13}C and ^{14}C in ocean and atmosphere, and the histories of carbon dioxide partial pressure and isotope ratios through the 19th and 20th centuries, including the response of radiocarbon to its production by atomic weapons tests. With this tuned simulation we have calculated future levels of atmospheric carbon dioxide for various assumed scenarios concerning the future course of fossil fuel burning and land use change, including forest clearance. This does not constitute a prediction of future levels of carbon dioxide but, rather, a tentative look at how those levels might respond to various changes in human practice. The goal of our study has been to provide an indication of what possible courses of action might be fruitful and what issues merit further study. The major conclusions are not particularly sensitive to the details of our simulation.

REFERENCES

Anderson, T.F. and Arthur, M.A., 1983. Stable isotopes of oxygen and carbon and their application to sedimentological and paleoenvironmental problems. In: M.A. Arthur (Editor), Stable Isotopes in Sedimentary Geology. SEPM Course No. 10, Soc. Econ. Paleontol. Mineral., Tulsa, Okla., pp. 1.1–1.151.

Bacastrow, R.B. and Bjorkstrom, A., 1981. Comparison of ocean models for the carbon cycle. In: B. Bolin (Editor), Carbon Cycle Modelling, SCOPE 16 Wiley, New York, pp. 29–79.

Bacastrow, R.B., Keeling, C.D. and Whorf, T.P., 1985. Seasonal amplitude increase in atmospheric CO_2 concentration at Mauna Loa, Hawaii, 1959–1982. J. Geophys. Res., 90: 10529–10540.

Berner, R.A., 1982. Burial of organic carbon and pyrite in the modern ocean: Its geochemical and environmental significance. Am. J. Sci., 282: 451–473.

Berner, R.A., Lasaga, A.C. and Garrels, R.M., 1983. The carbonate-silicate geochemical cycle and its effect on atmospheric carbon dioxide over the past 100 million years. Am. J. Sci., 283: 641–683.

Bolin, B., Degens, E.T., Duvigneaud, P. and Kempe, S., 1979. Carbon Cycle Modelling. In: B. Bolin, E.T. Degens, S. Kempe and P. Ketner (Editors), The Global Carbon Cycle, SCOPE Rept. 13. Wiley, New York, pp. 1–28.

Broecker, W.S. and Peng, T.-H., 1982. Tracers in the Sea. Lamont-Doherty Geological Observatory, Palisades, New York, 690 pp.

Correspondence to: J.F. Kasting, Department of Geosciences, The Pennsylvania State University, University Park, PA 16802, USA

Detwiler, R.P. and Hall, C.A.S., 1988. Tropical forests and the global carbon cycle. Science, 239: 42–47.

Enting, I.G. and Pearman, G.I., 1986. The use of observations in calibrating and validating carbon cycle models. In: J.R. Trabalka and D.E. Reichle (Editors), The Changing Carbon Cycle: A Global Analysis. Springer, New York, pp. 425–458.

Esser, G., 1987. Sensitivity of global carbon cycle pools and fluxes to human and potential climatic impacts. Tellus, 39B: 245–260.

Freyer, H.-D., 1979. Variations in the atmospheric CO_2 content. In: B. Bolin, E.T. Degens, S. Kempe and P. Ketner (Editors), The Global Carbon Cycle: A Global Analysis. Scope 13. Wiley, New York, pp. 79–100.

Freyer, H.D., 1986. Interpretation of the Northern Hemispheric record of $^{13}C/^{12}C$ trends of atmospheric CO_2 in tree rings. In: J.R. Trabalka and D.E. Reichle (Editors). The Changing Carbon Cycle: A Global Analysis. Springer, New York, pp. 125–150.

Friedli, H., Lotscher, H., Oeschger, H., Siegenthaler, U. and Stauffer, B., 1986. Ice core record of the $^{13}C/^{12}C$ ratio of atmospheric CO_2 in the past two centuries. Nature, 324: 237–238.

Gordon, A.L., 1986. Interocean exchange of thermocline water. J. Geophys. Res., 91: 5037–5046.

Goudrian, J., 1989. Modelling biospheric control of carbon fluxes between atmosphere, ocean, and land in view of climatic change. In: A. Berger, S. Schneider and J.C. Duplessy (Editors), Climate and Geo-Sciences. Kluwer Academic Publishers, Boston, pp. 481–499.

Hobbie, J., Cole, J., Dungan, J., Houghton, R.A. and Peterson. B., 1984. Role of the biota in global CO_2 balance: The controversy. BioScience, 34: 492–498.

Holland, H.D., 1978. The Chemistry of the Atmosphere and Oceans. Wiley, New York, 351 pp.

Houghton, R.A., 1986. Estimating changes in the carbon content of terrestrial ecosystems from historical data. In: J.R. Trabalka and D.E. Reichle (Editors), The Changing Carbon Cycle: A Global Analysis. Springer, New York, pp. 173–193.

Houghton, R.A., 1987. Biotic changes consistent with the increased seasonal amplitude of atmospheric CO_2 concentrations. J. Geophys. Res., 92: 4223–4230.

Houghton, R.A. and Woodwell, G.M., 1989. Global climatic change. Sci. Am., 260 (4): 36–44 [April.]

Houghton, R.A., Hobbie, J.E., Melillo, J.M., Moore, B., Peterson, B.J., Shaver G.R. and Woodwell, G.M., 1983. Changes in the carbon content of terrestrial biota and soils between 1860 and 1980: A net release of CO_2 to the atmosphere. Ecol. Monogr., 53: 235–262.

Kasting, J.F., Richardson, S.M., Pollack, J.B. and Toon, O.B., 1986. A hybrid model of the CO_2 geochemical cycle and its application to large impact events. Am. J. Sci., 286: 361–389.

Keeling, C.D. and Bacastrow, R.B., 1977. Impact of industrial gases on climate. In: Energy and Climate: Studies in Geophysics. National Academy of Sci., Washington, D.C., pp. 72–95.

Keeling, C.D., Bacastrow, R.B., Carter, A.F., Piper, S.C., Whorf, T.P., Heimann, M., Mook, W.G. and Roeloffzen, H., 1989a. A three-dimensional model of atmospheric CO_2 transport based on observed winds. 1. Analysis of observational data. In: D.H. Peterson (Editor), Aspects of Climate Variability in the Pacific and the Western Americas. Am. Geophys. Union, Washington, D.C., pp. 165–236.

Keeling, C.D., Piper, S.C. and Heimann, M., 1989b. A three-dimensional model of atmospheric CO_2 transport based on observed winds. 4. Mean annual gradients and interannual variations. In: D.H. Peterson (Editor), Aspects of Climate Variability in the Pacific and the Western Americas. Am. Geophys. Union, Washington, D.C., pp. 305–363.

Lasaga, A.C., Berner, R.A. and Garrels, R.M., 1985. An improved model of atmospheric CO_2 fluctuations of the past 100 million years. In: E.T. Sundquist and W.S. Broecker (Editors), The Carbon Cycle and Atmospheric CO_2: Natural Variations Archean to Present. Am. Geophys. Union, Washington, D.C., pp. 397–411.

Marshall, H.G., Walker, J.C.G. and Kuhn, W.R., 1988. Long-term climate change and the geochemical cycle of carbon. J. Geophys. Res., 93: 791–802.

Menard, H.W. and Smith. S.M., 1966. Hypsometry of ocean basin provinces. J. Geophys. Res., 71: 4305–4325.

Mook, W.G., Bommerson, J.C. and Staverman, W.H., 1974. Carbon isotope fractionation between dissolved bicarbonate and gaseous carbon dioxide. Earth Planet. Sci. Lett., 22: 169–176.

Neftel, A., Moor, E., Oeschger, H. and Stauffer, B., 1985. Evidence from polar ice cores for the increase in atmospheric CO_2 in the past two centuries. Nature, 315: 45–47.

O'Brien, K., 1979. Secular variations in the production of cosmogenic isotopes in the Earth's atmosphere. J. Geophys. Res., 78: 423–431.

Post, W.M., Peng, T.-H., Emanuel, W.R., King, A.W., Dale, V.H. and DeAngelis, D.L., 1990. The global carbon cycle. Am. Sci., 78: 310–326.

Ramanathan, V., Cess, R.D., Harrison, E.F., Minnis, P. and Barkstrom, B.R., 1989. Cloud-radiative forcing and climate: Results from the Earth Radiation Budget Experiment. Science, 243: 57–63.

Revelle, R. and Munk, W., 1977. The carbon dioxide cycle and the biosphere. In: Energy and Climate: Studies in Geophysics. National Academy of Sci., Washington, D.C., pp. 140–158.

Rotty, R.M. and Marland. G., 1986. Fossil fuel combustion: Recent amounts, patterns, and trends of CO_2. In: J.R. Trabalka and D.E. Reichle (Editors), The Changing Carbon Cycle: A Global Analysis. Springer, New York, pp. 474–490.

Schlesinger, W.H., 1986. Changes in soil carbon storage and associated properties with disturbance and recovery. In: J.R. Trabalka and D.E. Reichle (Editors), The Changing Carbon Cycle: A Global Analysis, Springer, New York, pp. 194–220.

Siegenthaler, U. and Munnich, K.O., 1981. $^{13}C/^{12}C$ fractionation during CO_2 transfer from air to sea. In: B. Bolin (Editor), Carbon Cycle Modelling, SCOPE 16. Wiley, New York, pp. 249–257.

Stuiver, M., 1982. The history of the atmosphere as recorded by carbon isotopes. In: E.D. Goldberg (Editor) Atmospheric Chemistry. Springer, Berlin, pp. 159–179.

Stuiver, M., 1986. Ancient carbon cycle changes derived from tree-ring ^{13}C and ^{14}C. In: J.R. Trabalka and D.E. Reichle (Editors), The Changing Carbon Cycle: A Global Analysis. Springer, New York, pp. 109–124.

Stuiver, M. and Quay, P.D., 1980. Changes in atmospheric carbon- 14 attributed to a variable sun. Science, 207: 11–19.

Sundquist, E.T., 1986. Geologic analogs: Their value and limitations in carbon dioxide research. In: J.R. Trabalka, J.A. Edmonds, J.M. Reilly, R.H. Gardner and D.E. Reichle (Editors), The Changing Carbon Cycle: A Global Analysis. Springer, New York, pp. 371–402.

Tans, P.P., Fung, I.Y. and Takahashi, T., 1990. Observational constraints on the global atmospheric CO_2 budget. Science, 247: 1431–1438.

Van der Merwe, N.J., 1982. Carbon isotopes, photosynthesis, and archaeology. Am. Sci., 70: 596–606.

Volk, T., 1987. Feedbacks between weathering and atmospheric CO_2 over the last 100 million years. Am. J. Sci., 287: 763–779.

Volk, T., 1989a. Effect of the equatorial Pacific upwelling on atmospheric CO_2 during the 1982–1983 El Nino. Global Biogeochem. Cyc., 3: 267–279.

Volk, T., 1989b. Rise of angiosperms as a factor in long-term climatic cooling. Geology, 17: 107–110.

Walker, J.C.G., Hays, P.B. and Kasting, J.F., 1981. A negative feedback mechanism for the long-term stabilization of Earth's surface temperature. J. Geophys. Res., 86: 9776–9782.

Wilkinson, B.H. and Walker, J.C.G., 1989. Phanerozoic cycling of sedimentary carbonate. Am. J. Sci., 289: 525–548.

Woodwell, G.M., Hobbie, J.E., Houghton, R.A., Melillo, J.M., Moore, B., Peterson, B.J. and Shaver, G.R., 1983. Global deforestation: Contribution to atmospheric carbon dioxide. Science, 222: 1081–1086.

Abrupt Deep-Sea Warming, Palaeoceanographic Changes and Benthic Extinctions at the End of the Palaeocene

J. P. KENNETT* & L. D. STOTT[†]

* Marine Science Institute and Department of Geological Sciences, University of California,
Santa Barbara, California 93106-6150, USA
† Department of Geological Sciences, University of Southern California, Los Angeles,
California 90089-0740, USA

A remarkable oxygen and carbon isotope excursion occurred in Antarctic waters near the end of the Palaeocene (~57.33 Myr ago), indicating rapid global warming and oceanographic changes that caused one of the largest deep-sea benthic extinctions of the past 90 million years. In contrast, the oceanic plankton were largely unaffected, implying a decoupling of the deep and shallow ecosystems. The data suggest that for a few thousand years, ocean circulation underwent fundamental changes producing a transient state that, although brief, had long-term effects on environmental and biotic evolution.

WE describe foraminiferal oxygen and carbon isotope changes over the Palaeocene/Eocene transition at high stratigraphic resolution in an Antarctic sedimentary sequence. We infer palaeoenvironmental changes that caused major deep-sea benthic faunal extinctions at the end of the Palaeocene, perhaps the largest during the last 90 Myr (refs 1,2). This event profoundly affected oceanic benthic communities deeper than the continental shelf (>100 m; neritic zone)[1–4] resulting in a 35–50% species reduction of benthic foraminiferal taxa[4,5]. The extinction level pre-dates the last appearance of *Morozovella velascoensis* (at the tropical P6a/P6b boundary[1,6,7]) and is close to other biostratigraphic levels at or close to the Palaeocene/Eocene boundary[1,3]. It is also located within one of the largest negative $\delta^{13}C$ changes of the Cenozoic, beginning in the late Palaeocene at ~60 Myr and continuing into the early Eocene[6,9–11]. The $\delta^{13}C$ values immediately preceding the event are the highest for the entire Cenozoic[9]. This event is also located within a long-term negative $\delta^{18}O$ trend of >1‰ (refs 6, 9, 12–15), interpreted[16] to reflect a warming of Antarctic surface waters by ~5 °C.

What could have caused such extensive changes in the ocean deeper than the continental shelf, an ecosystem that forms >90% by volume of the Earth's total habitable environments[17]? Before now, the extinction event has been studied at insufficient stratigraphic resolution to determine its speed and potential causes. During our earlier stable isotope investigations[12,16] of Antarctic Palaeogene sequences, we discovered a sudden, brief, striking negative isotope excursion in planktonic and benthic $\delta^{13}C$ and $\delta^{18}O$ coinciding with the major benthic foraminiferal extinctions[5] close to the Palaeocene/Eocene boundary. Our subsequent detailed studies reveal correlations in the timing of a number of isotopic changes and the extinctions (Figs 1, 2), thus providing a stronger basis for evaluation of the possible environmental changes that caused them.

The Palaeocene/Eocene transition has been examined in Ocean Drilling Project (ODP) hole 690B (65° 09′ S, water depth 2,914 m) on the flank of Maud Rise, Weddell Sea, Antarctica. Palaeodepths during the Palaeocene/Eocene transition were ~2,100 m (ref. 12). The Palaeogene sequence[18,19] consists of a nannofossil ooze with well-preserved planktonic and benthic foraminifera. There is a minor terrigenous sedimentary fraction (5–15% clay and ~10–15% mica)[18]. Colour photographs of the cores indicate a lack of bioturbation for several centimetres (core 19–3, 75–65 cm) encompassing the extinction level, whereas underlying and overlying sediments were clearly bioturbated. The cores show no drilling disturbance.

Kennett, J. P. (1991) Abrupt deep-sea warming, palaeoceanographic changes and benthic extinctions at the end of the Palaeoscene. *Nature* 353: 319–322. Reprinted by permission from Macmillan Publishers Limited.

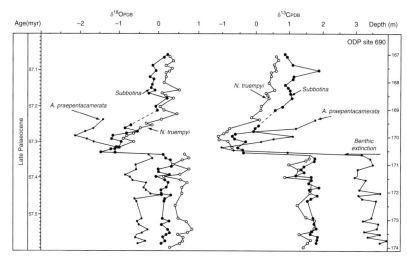

Fig. 1. Changes in $\delta^{18}O$ and $\delta^{13}C$ ([$(^{18}O/^{16}O)_{sample}$ / $(^{18}O/^{16}O)_{standard}$ − 1] × 1.000; [$(^{13}C/^{12}C)_{sample}$/$(^{13}C/^{12}C)_{standard}$ − 1] × 1.000) of the planktonic foraminifers *Acarinina praepentacamerata* and *Subbotina* and the benthic foraminifer *Nuttallides truempyi* in the latest Palaeocene (57.0 to 57.6 Myr) of Ocean Drilling Program (ODP) hole 690B, Maud Rise, Antarctica. Note the sharp negative shifts at 57.33 Myr coinciding with the main level of extinction in benthic foraminifera. Oxygen isotope values have been adjusted to account for nonequilibrium effects[6,12]. The gap in isotope records of *Subbotina* is between the last appearance of *S. patagonica* and the first appearance of *S. varianta*. The upper extent of isotope records of *A praepenta-camerata* marks the disappearance of this species. Age framework is after ref. 23. The age assignments will change, however, because the age of the Palaeocene/Eocene boundary is being revised to 2 Myr younger[56]. Depth scale at right represents metres in section below sea floor. Samples were analysed with a Finnigan MAT 251 mass spectrometer linked to a Carousel-48 automatic carbonate preparation device. The laboratory gas standard is calibrated to PDB by cross-calibration using the standard NBS 20. Precision is 0.1‰ or better for both $\delta^{18}O$ and $\delta^{13}C$. We applied the standard $\delta^{18}O$ correction factor of +0.5‰ to *Nuttallides truempyi*[6,12]. No correction factor was made to $\delta^{13}C$ for this taxon.

CHRONOLOGY AND APPROACHES

The Palaeocene/Eocene boundary in hole 690B is located in a long reversed-polarity zone identified[20,21] as Magnetochron 24R. As before[20], we place the boundary magnetostratigraphically at 57 Myr within Chron 24R[21,22] at 166 m below sea floor (mbsf). This is close to the first appearance of *Planorotalites australiformis* (which defines the AP4/AP5 boundary at 170.05 mbsf; core 19–3, 15–18 cm) and below the first appearance of *Acarinina wilcoxensis berggreni* (159.8 mbsf). The principal benthic extinction level (core 19–3, 70 cm) is ~40 kyr older, and is assigned an age of 57.33 Myr (Fig. 1). Our chronology is based on an age for the top of Chron 25N (185.2 mbsf) of 58.64 Myr and the bottom of Chron 24N (154.62 mbsf) of 56.14 Myr (ref. 21). This provides an average sedimentation rate of 1.23 cm per 1,000 yr (1 cm in ~800 yr). In low-latitude sequences, the Palaeocene/Eocene boundary has been correlated to the last appearance of *M. velascoensis*, marking the boundary between the P6a and P6b zones with an assigned age of 57.8 Myr (refs 6, 7, 22). More recent correlations[23] place the

younger boundary at 57 Myr and within zone P6b, postdating the extinction of *M. velascoensis*. This species is not present in site 690 and the AP4/AP5 boundary remains to be accurately correlated with low-latitude biostratigraphy including the P6a/P6b boundary. We believe that the chief deep-sea extinction level near the end of the Palaeocene is synchronous in marine sequences, and that offsets between our age estimates (Fig. 1) and others[6,22] result either from biostratigraphic miscorrelations or the necessarily broad age interpolations between magnetochrons at site 690.

Foraminiferal samples were analysed for oxygen and carbon isotope composition between 174.00 mbsf (57.59 Myr) and 166.95 mbsf (57.06 Myr) at a maximum sampling interval of 10 cm, a stratigraphic resolution of ~4–7.5 kyr. We located the extinction horizon within a 4-cm interval between 170.62 mbsf (core 19–3, 72 cm) and 170.58 mbsf (core 19–3, 68 cm), estimated to represent ~3 kyr (57.334 and 57.331 Myr) (Fig. 2). This we sampled at 1-cm intervals, providing an age resolution of ~800 yr (Fig. 2).

Isotope analyses were conducted on monospecific samples of the planktonic foraminifers *Acarinina praepentacamerata*, *Subbotina patagonica* and

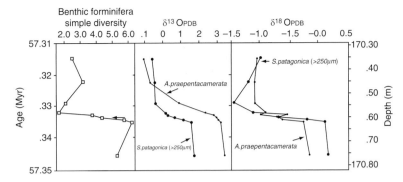

Fig. 2. Changes in oxygen and carbon isotope composition of the planktonic foraminifers *A. praepentacamerata* (>63 μm fraction) and *Subbotina patagonica* (>250 μm fraction) and in simple diversity of benthic foraminiferal assemblages at high stratigraphic resolution during the negative isotope excursion in the latest Palaeocene shown at 57.33 in Fig. 1 (ODP hole 690B). Note the large, rapid decrease in diversity coinciding with the negative oxygen isotope shift, followed by a return to higher diversities. The large latest-Palaeocene extinction in benthic foraminifera coincides with the drop in diversity. Arrow at left indicates the last appearance of a distinctive Palaeocene benthic foraminifer *Stensioina beccariiformis*. All benthic foraminifera (>150 μm) were picked from samples in the early part of the excursion to determine simple diversity. Because specimens were abundant before the excursion, simple diversity was determined by partial picking and thorough scanning of sample residues (>150 μm). No relationship is apparent between the original dried weight of samples and the simple diversity and abundances of benthic foraminifera. Depth scale at right represents metres in section below sea floor.

Subbotina varianta, and the benthic foraminifer *Nuttallides truempyi*. This benthic species occurs throughout the record, except for a brief gap close to the extinction event. We used standard approaches to extract and prepare the foraminifera for faunal and isotopic analyses[12]. Oxygen isotopic palaeotemperatures were determined assuming an ice-free world (mean ocean water $\delta^{18}O$ = −1.2‰ relative to present day[13]).

THE TERMINAL PALAEOCENE ISOTOPIC EXCURSION

The planktonic foraminifer *A. praepentacamerata* records the lowest oxygen and the highest carbon isotope values (Fig. 1). Such values are consistent with a *near-surface habitat*, possibly during the austral spring/summer months[12,16]. The species of *Subbotina* record higher $\delta^{18}O$ and lower $\delta^{13}C$ values (Fig. 1) than *A. praepentacamerata*, indicating a *deeper-water planktonic* habit and/or a preference for cooler months of the year[16]. As expected in a thermally stratified ocean with deep waters richer in nutrients, the highest $\delta^{18}O$ and lowest $\delta^{13}C$ values are shown by the benthic foraminifer *N. truempyi*.

Between 57.6 and 57.33 Myr, isotope values showed relatively little change (Fig. 1). Surface water isotopic temperatures are estimated to have been 13–14 °C and deep water temperatures ~10 °C. The oxygen and carbon isotopic composition of

plank-tonic and benthic foraminifera (Fig. 1) show a considerable excursion between ~57.33 and 57.22 Myr. This excursion (the 'terminal Palaeocene isotopic excursion') began abruptly at ~57.33 Myr with a conspicuous decrease in $\delta^{18}O$ and $\delta^{13}C$ values, followed by a return to values only slightly lower than before the excursion (Fig. 1).

Oxygen isotopic changes

Between ~57.33 and 57.32 Myr, the marked decrease in $\delta^{18}O$ occurred in all three foraminiferal groups. The change was largest (2.0‰) in the benthic forms, intermediate (1.5‰) in deeper-dwelling planktonics and smallest (1.0‰) in shallow-dwelling planktonics. *Surface water* temperatures increased to ~18 °C. This isotopic shift coincided with the benthic extinctions (Figs 1 and 2). Isotope values in *shallow* surface water further decreased to ~−2.2‰, suggesting temperatures of ~21 °C, between 57.31 and 57.28 Myr, probably the warmest of the Cenozoic[16]. In contrast, the isotope temperatures of *Subbotina* decreased after 57.32 Myr. The rapid warming is associated with an increase in warm subtropical planktonic microfossils in Antarctic waters, including discoasters[24] and morozovellids. Also, an associated peak in *kaolinite* suggests warm, humid climate on Antarctica, the inferred source of the clays[25].

A brief gap (20 kyr) exists in the range of *N. truempyi* during the excursion (Fig. 1). When *N. truempyi* reappeared at ~57.31 Myr, its oxygen isotope values were similar (−1.1‰) to the planktonic forms (Fig. 1). Thus, deep waters warmed to temperatures close to those at the ocean surface, temporarily eliminating the vertical temperature gradient at this location. This gradient was re-established ~30 kyr after the initial isotope decrease (Fig. 1). Following the warming peak, temperatures decreased, but remained ~1–2 °C higher than before the beginning of the excursion (Fig. 1).

Carbon isotopic changes

The pattern of $\delta^{13}C$ change during the excursion (Fig. 1) was similar to that of $\delta^{18}O$: a marked initial decrease in values at ~57.33 Myr followed by a more gradual increase. The change was very large (>4‰) in the surface-dwelling planktonic *A. praepentacamerata*. *Subbotina* showed a 2.0‰ decrease. The absence of benthic values during the shift prevents a detailed comparison, but values rapidly decreased by at least 2.6‰, from 1.5‰ at 57.33 Myr to −1.1‰ by 57.29 Myr. From ~57.33 to 57.31 Myr, $\delta^{13}C$ values of surface and deep planktonics are similar; *Subbotina* values are between ~−0.5‰ and −0.8‰. Benthic $\delta^{13}C$ values at ~57.32 Myr are almost identical to surface-water planktonic values. The large $\delta^{13}C$ gradient (~2.0‰) that had existed between surface and deep waters before the excursion was virtually eliminated at ~57.32 Myr. After 57.30 Myr, surface-water $\delta^{13}C$ values began to increase again and a vertical gradient was re-established.

A decrease in benthic foraminiferal diversity (Fig. 2) coincided with the $\delta^{18}O$ and $\delta^{13}C$ changes in the deeper-dwelling planktonic *Subbotina* and the $\delta^{18}O$ changes in the shallow-dwelling planktonic *A. praepentacamerata*. Most of the reduction in benthic diversity had already occurred, however, before the initiation (~3 kyr later) of the principal $\delta^{13}C$ decrease in the surface-dwelling *A. praepentacamerata* (Fig. 2). Thus, the interval of rapid isotope change is not the result of a hiatus. Within ~3 kyr, surface water temperatures at hole 690B are estimated to have increased by 3–4 °C, with a total warming in subsurface waters of 6 °C in 6 kyr. The $\delta^{13}C$ shift (2.0‰) in *Subbotina* lasted 6 kyr. Most (3.5‰) of the $\delta^{13}C$ shift in *A. praepentacamerata*, which began slightly later (Fig. 2), occurred within 11 kyr, although the trend continued for more than 20 kyr. After the shift, the vertical $\delta^{18}O$ and $\delta^{13}C$ gradients were temporarily eliminated or even reversed.

Changes in carbon isotope gradients ($\Delta\delta^{13}C$) between shallow and deep forms (Fig. 2) indicate

that characteristics of the vertical water column differed before, during and after the isotope excursion. A distinct $\delta^{13}C$ gradient between surface and deep waters existed before 57.33 Myr. Furthermore, at that time the similarity between the $\delta^{13}C$ composition of the benthic foraminifera and the subbotinids suggests the existence of a midwater zone with a strong minimum in oxygen concentration. During the $\delta^{13}C$ shift beginning at ~57.33 Myr, the gradient between all groups was eliminated (Fig. 2). After 57.29 Myr, the vertical $\delta^{13}C$ gradient was gradually re-established (Fig. 1). But after re-establishment, by ~57.28 Myr, a distinct isotope offset existed between benthic foraminifera and the subbotinids, suggesting a reduction in the strength of the oxygen minimum zone.

BIOTIC CHANGES

Large changes in the taxonomic composition of benthic foraminifera clearly define the extinction event and the pattern of decreasing diversity[5]. Before the extinction event, the simple diversity of benthic foraminiferal assemblages (>150 μm) averaged ~55–60 species (Fig. 2) or higher[5]. Assemblages containing abundant benthic foraminifera include a wide range of morphotypes inferred to be of both infaunal and epifaunal habit[5,26], including trochospiral and other coiled forms. A considerable decrease in diversity (>150 μm) coincides with the $\delta^{18}O$ shift (Fig. 2). Simple diversity dropped by 72% from ~60 to 17 species (>150 μm) in <4 cm (<3 kyr). Many distinctive taxa such as *Stensioina beccariiformis* and *Neoflabellina* disappeared early and rapidly (<1.5 kyr) during the isotope shift marking the extinction event (Fig. 2). Most trochospiral forms such as *N. truempyi* had disappeared by the midpoint of the shift. Most lost taxa are not observed again even in the immediately overlying 1-cm sample interval, supporting sedimentary and isotope evidence for a lack of bioturbation over this interval. Benthic foraminifera (>150 μm) also became uncommon to rare (as low as ~30 specimens per sample) during the extinction event. Small specimens (<150 μm), however, remained abundant throughout this interval. Ostracods decreased markedly in diversity and abundance during the shift, leaving a rare assemblage composed almost exclusively of small, smooth, thin-walled forms. Both planktonic and benthic foraminiferal assemblages are well preserved throughout, showing no apparent increase in dissolution. During this episode, planktonic foraminifera and calcareous nannofossils, which remained abundant, increased in diversity in hole 690B (refs 20, 24) and elsewhere in the oceans[27], possibly as a result of ocean surface warming.

The 5-kyr interval immediately following the isotope shift (Fig. 2) is marked by low diversities and abundances of benthic foraminifera (>150 μm fraction). Coiled forms are almost completely absent, leaving an assemblage dominated by relatively small and thin-walled uniserial, triserial and other forms typical of an infaunal habitat[5,26]. The relative increase in abundances of *small specimens of benthic foraminifera* associated with a decrease in diversity suggests conditions low in oxygen and high in nutrients[5,28]. The diversity of benthic foraminiferal assemblages (>150 μm) increased thereafter (Fig. 2) to an average of ~30 species (>150 μm). This increase resulted, in part, from the reappearance of forms previously present, but a high proportion of species (~ 35%) in hole 690 B became truly extinct at the beginning of the excursion[5].

Major, roughly coeval extinctions of benthic foraminifera have been recorded in other deep-sea sequences[1,7,29,30] and in upper bathyal marine sequences in Trinidad[3], the Reichenhall and Salzburg Basins, Austria[31] and northern Italy[32]. In contrast, benthic foraminiferal assemblages on the continental shelves were little affected. The loss of so many characteristic Late Cretaceous and Palaeocene deep-sea taxa produced a deep-sea benthic foraminiferal assemblage of Tertiary character[1]. The biotic crisis at the K/T boundary was opposite, in that elements of the oceanic plankton and shallow benthic communities were considerably reduced, whereas deep-sea benthic foraminiferal assemblages experienced little change[4,33–36].

INTERPRETATIONS OF BIOTIC CHANGES

An important question relates to the large-scale character of the oceanographic changes that caused these rapid biotic changes. The lack of major extinctions in oceanic plankton and in shallow-water benthic communities eliminates any possibility that the changes were caused by a bolide impact with the Earth, as has been suggested for the terminal Cretaceous extinctions[34]. A more plausible explanation would involve an oceanic cause. The extinctions were clearly driven by processes that preferentially affected the deep-

sea biota below the thermocline or continental shelf. The process that caused the extinctions, although not instantaneous, was remarkably rapid (< 3 kyr), and any proposed mechanism must have had the capacity to affect vast volumes of the deep ocean rapidly. The extinctions occurred at about the rate of the replacement time of the oceans (currently ~1 kyr (ref. 37), although possibly slower in the early Palaeogene). Also, the speed and magnitude of the associated temperature increase implies global, not solely oceanic, warming involving strong positive feedback mechanisms. The superposition of the rapid, negative $\delta^{13}C$ and $\delta^{18}O$ excursions upon similar, more gradual trends during the late Palaeocene to early Eocene[13,4] suggests a threshold event similar to the oxygen isotope shift near the Eocene/Oligocene boundary[28], although in the opposite sense. Our data indicate that an essentially isothermal water column developed, with deep waters in the Antarctic warming more than shallow waters. The rapidity of the initial changes and the gradual return (over ~100 kyr) to isotope values similar to those before the event suggest an ocean-climate system temporarily in disequilibrium[37]. As deep waters warmed, most benthic foraminiferal species (>150 μm) were eliminated at this location. The disappearance of many more inferred epifaunal than infaunal benthic species (>150 μm) indicates that species living in sediments were more likely to survive. Nevertheless, the disappearance of many infaunal forms indicates that the changes also affected the infaunal environment. Lack of bioturbation early in the excursion suggests that burrowing organisms were also severely affected.

Three main hypotheses have been proposed to account for the deep-sea benthic crisis at the end of the Palaeocene. They are: (1) a rapid warming of the deep ocean and a change in bottom water sources[6]; (2) a deep-sea oxygen deficiency due to the sudden warming and change in circulation of deep water[5,11,12,19,36]; and (3) a sharp drop in surface ocean productivity that reduced trophic resources available for deep-sea benthic organisms[10,39,40].

The benthic extinctions occurred at the same time as rapid warming of deep waters and before the shift in $\delta^{13}C$ in the surface planktonic foraminifera. They therefore occurred before any associated large change in surface water processes that might have affected primary production. Although no large changes are evident in calcareous nannofossils at this location (A. R. Edwards, personal communication), elsewhere an earliest Eocene peak in taxonomic turnover was interpreted as a response to decreasing primary production[41]. Substantial productivity reduction is generally accompanied by a decrease in biogenic calcium carbonate accumulation, which is not observed at the Palaeocene/Eocene boundary at site 690. Considerable reduction in delivery of organic matter to the deep sea during the faunal crisis is unlikely[4,11,12,36], although it could have occurred later during the isotope excursion.

It is more likely that the benthic extinctions were caused by the rapid temperature increase of deep waters and associated reduction of oxygen concentrations[4,16]. In the modern ocean, cold (<6 °C) deep waters contain sufficient oxygen to oxidize organic matter fully and to maintain diverse benthic communities in most areas. At the same nutrient levels and Redfield ratios, however, deep waters of 10 °C would not contain sufficient preformed oxygen to avoid anoxia unless atmospheric oxygen levels were considerably higher or primary production lower than at present. Modelling studies suggest that atmospheric oxygen levels were possibly 10% higher in the early Palaeogene[42], in which case deep waters of 10 °C could have remained aerobic if overall nutrient levels were no higher than in the modern ocean. But at temperatures higher than 15 °C, as at the Palaeocene/Eocene boundary, deep waters would almost certainly have been depleted in oxygen. Deep waters are unlikely to have become completely anoxic during the excursion because there is no increase in organic carbon accumulation at hole 690B and possibly elsewhere[43]. Therefore a balance must have been maintained between deep-sea oxygen levels and the supply of organic carbon from surface waters and of preformed nutrients in deep waters[44]. The decrease in $\delta^{13}C$ gradients suggests reduced primary production, contrary to benthic foraminiferal trends.

INTERPRETATIONS OF OCEANOGRAPHIC CHANGES

What process could have triggered the rapid oceanic warming and inferred reduction in deep-sea oxygen concentrations that produced the extinctions? It has been proposed[4–6,11,12] that the rapid warming resulted from the concurrence of an almost total dominance in the oceans of warm saline deep waters produced at mid-latitudes, and a reduction of high-latitude deep water sources, in an extreme example of the Proteus ocean model[12]. Modelling studies[45] suggest that deep-ocean oxygen concentrations are primarily controlled by deep-water formation processes. The relative importance of high- and mid-latitude sources of deep ocean waters during the early to middle Palaeogene is controversial (G. Mead and D. Hodell, manuscript in preparation; and refs 5, 6, 12). In the modern ocean, almost all dense, oxygen-rich waters are produced at high latitudes as a result of cold temperatures in combination with moderately high salinities[37]. Saline, dense waters also form in the Mediterranean and Red Seas because of high net evaporation. *Because of low buoyancy fluxes*, however, they sink only to thermocline depths[37].

During the early Palaeogene, different global geography and climate[46–48] combined to make the ocean circulation very different from that of the modern ocean[12,46,49]. Significant Antarctic warmth[6,19,20,22] coupled with inferred higher precipitation[25,50–52] may have caused a large reduction in deep waters derived from the high latitudes[6,12]. At the same time, the mid-latitude Tethys Seaway would have produced warm, saline deep waters[12], probably in sufficient volume[45] to alter the density distributions in the oceans considerably. We suggest that during the excursion these factors combined to form, through positive feed-back responses, an extreme case of such an ocean, because a global climate threshold was temporarily penetrated. The nature of the triggering remains unknown.

The character of the $\delta^{13}C$ changes constrains hypotheses for the cause of the extinctions. Of particular importance is that the negative carbon isotope shift occurred throughout the water column (Fig. 2) and was smaller (~2.5‰) in deep waters than in surface waters (~4.0‰). This trend is opposite to that of the $\delta^{18}O$ record. Furthermore, the shift in $\delta^{13}C$ in waters at thermoclinal and greater depths coincided with warming of these waters, but most of the $\delta^{13}C$ shift occurred later in surface waters. A negative $\delta^{13}C$ shift of this magnitude[39] might be interpreted as indicating a transfer of light $\delta^{13}C$ organic carbon from the continents to the oceans[43,53], but this is unlikely because there is no evidence of any significant and rapid drop in global sea level at that time[40,54]. Moreover, the diachronism shown in the carbon isotope changes between surface and deeper-dwelling planktonic forms does not support this hypothesis. It has been suggested[40] that plate-tectonic reorganization at this time increased oceanic volcanism and deep-sea hydrothermal activity, triggering warming of deep waters and global greenhouse warming by release of carbon dioxide. The speed of the palaeoenvironmental changes and brevity of the excursion lend little support to this hypothesis.

SUMMARY

The rapid (<6,000 yr) negative oxygen and carbon isotope shifts in Antarctic waters near the end of the Palaeocene reflect oceanographic changes associated with global warming that caused considerable deep-sea extinctions. These observations show that for at least the early Cenozoic, short (1,000 yr) events can strongly affect the course of environmental and biotic evolution. Furthermore, such change can be effectively decoupled between different parts of the biosphere, in this case the deep and shallow marine ecosystems. The oceanographic changes associated with the excursion were clearly broad and complex, and seem to reflect a transient state between fundamentally different modes of ocean circulation. Brief elimination of the vertical $\delta^{18}O$ and $\delta^{13}C$ gradients indicates vertical ocean mixing and homogenization of nutrient distributions, presumably related to vertical instability of the Antarctic Ocean. The slightly delayed shift, during the excursion, to highly negative $\delta^{13}C$ values in Antarctic surface waters is difficult to explain but suggests higher concentrations of nutrients and total carbon dioxide, partial pressure of carbon dioxide[55]. This would have contributed to greenhouse warming. Detailed examination of the excursion is required elsewhere, especially at low latitudes, for better definition of the character and magnitude of the oceanographic and biotic changes associated with this remarkable event.

ACKNOWLEDGEMENTS

We thank D. Lea, B. Flower, and the reviewers. E. Thomas, K. Miller and J. Zachos, for constructive criticism of the manuscript, and M. Arthur for discussions. We also thank A. Puddicombe and H. Berg for technical assistance. This work was supported by the NSF Division of Polar Programs (J.K.) and Submarine Geology and Geophysics (J.K. and L.S.).

NOTES

Received 13 February; accepted 11 August 1991.

1. Tjalsma, R. C. & Lohmann, G. P. *Micropaleont. spec. Publ.* 4 (Micropaleontology Press, New York, 1983).
2. Douglas, R. & Woodruff, F. in *The Oceanic Lithosphere, The Sea.* Vol. 7 (ed. Emiliani, C.). 1233–1327 (Wiley, New York, 1981).
3. Beckmann, J. P. *Reports Int. Geol. Cong. 21st Session, Norden.* Part V 57–69 (1960).
4. Thomas, E. in *Origins and Evolution of the Antarctic Biota* (ed. Crame. J. A.) *Spec. Publ.* No. 47, 283–296 (Geological Society, London, 1989).
5. Thomas, E. in *Proc. ODP. Scient. Results* **113** (eds Barker. P. F. *et al.*) 571–594 (Ocean Drilling Program, College Station, Texas. 1990).
6. Miller, K. G. *et al. Paleoceanography* **2**, 741–761 (1987).
7. Berggren, W. A. & Miller, K. G. *Micropaleontology* **34**, 362–380 (1988).
8. Boltovskoy, E. & Boltovskoy, D. *Rev. Micropaleont.* **31**(2), 67–84 (1988).
9. Shackleton, N. J. *et al.* in *Init. Reports DSDP* **74** (eds Moore. T. C. *et al.*) 599–612 (US Government Printing Office, Washington DC, 1984).
10. Shackleton, N. J. *Palaeogr. Palaeoclimat. Palaeoecol.* **57**, 91–102 (1986).
11. Katz, M. E. & Miller, K. G. in *Proc. ODP. Scient. Results* **144** (eds Ciesielski, P. F. *et al.*) (Ocean Drilling Program, College Station, Texas.
12. Kennett, J. P. & Stott, L. D. in *Proc. ODP. Scient. Results* **113** (eds Barker, P. F. *et al.*) 865–880 (Ocean Drilling Program, College Station, Texas, 1990).
13. Shackleton, N. J. & Kennett, J. P. in *Init. Reports DSDP* **29** (eds Kennett. J. P. *et al.*) 743–755 (US Government Printing Office. Washington DC, 1975).
14. Savin, S. M., Douglas, R. G. & Stehli, F. G. *Geol., Soc. Am. Bull.* **96**, 1499–1510 (1975).
15. Kennett, J. P. *Mar. Micropaleontol.* **3**, 301–345 (1978).
16. Stott, L. D., Kennett, J. P., Shackleton, N. J. & Corfield, R. M. in *Proc. ODP. Scient Results* 113 (eds Barker, P. F. *et al.*) 849–863 (Ocean Drilling Program, College Station, Texas, 1990).
17. Childress, J. J. in *Oceanography: The Present and Future* (ed. Brewer, P.) 127–135 (Springer, New York, 1983).
18. Barker, P. F. *et al. Proc. ODP. Init. Reports* **113** (Ocean Drilling Program, College Station, Texas, 1988).
19. Kennett, J. P. & Barker, P. F. in *Proc. ODP. Scient. Results* **113** (eds Barker, P. F. *et al.*) 937–960 (Ocean Drilling Program, College Station, Texas, 1990).
20. Stott, L. D. & Kennett, J. P. in *Proc. ODP. Scient Results* **113** (eds Barker, P. F. *et al.*) 549–569 (Ocean Drilling Program, College Station, Texas, 1990).
21. Spiess, V. in *Proc. ODP. Scient Results* **113** (eds Barker, P. F. *et al.*) 261–315 (Ocean Drilling Program, College Station, Texas, 1990).
22. Berggren, W. A., Kent, D. V. & Flynn, J. J. *Mem. Geol. Soc. London* **10**, 141–195 (1985).
23. Aubry, M.-P. *et al. Paleoceanography* **3**, 707–742 (1988).
24. Pospichal, J. J. & Wise, S. W. *Jr Proc. ODP. Scient. Results* **113** (eds Barker, P. F. *et al.*) 613–638 (Ocean Drilling Program, College Station, Texas, 1990).
25. Robert, C. & Maillot, H. *Proc. ODP. Scient. Results* **113** (eds Barker, P. F. *et al.*) 51–70 (Ocean Drilling Program, College Station, Texas, 1990).
26. Corliss, B. H. & Chen, C. *Geology* **16**, 716–719 (1988).
27. Aubry, M.-P., Gradstein, F. M. & Jansa, L. F. *Micropaleontology* **36**(2), 164–172 (1990).

28. Bernerd, J. M. *J. Foram. Res.* **16**, 207–215 (1986).

29. Schnitker, D. in *Init. Reports DSDP* **48** (eds Montadert, L. *et al.*) 377–414 (US Government Printing Office. Washington DC. 1979).

30. Nomura, R. in *Proc. ODP. Scient. Results* **121** (eds Peirce. J. W. *et al.*) (Ocean Drilling Program, College Station, Texas.

31. von Hildebranot, A. *Akad Wiss. (Wien). Math.-Naturw. Klasse, Abh. n. ser.* **106**, 1–182 (1962)

32. Braga, G., De Biase, R., Grunig, A. & Proto Decima, F. in *Monogr. micropaleontol. sul Paleocene e l'Eocene di Possagno. Prov. di Treviso. Italia. Abh. Mem. Suisse Paleontol.* (ed. Botli, H. M.) **97**, 85–111 (1975).

33. Russell, D. A. *Ann. Rev. Earth planet Sci.* **7**, 163–182 (1979).

34. Alvarez, L. W., Alvarez, W., Asaro, F. & Michel, H. V. *Science* **206**, 1095–1108 (1980).

35. Stott, L. D. & Kennett, J. P. *Proc. ODP. Scient Results* **113** (eds Barker, P. F. *et al.*) 829–846 (Ocean Drilling Program, College Station, Texas, 1990).

36. Thomas, E. in *Geol. Soc. Am. spec. Publ.* **247**.

37. Broecker, W. S. & Peng, T.-H. *Tracers in the Sea* (Eldigio, Palisades, 1982).

38. Kennett, J. P. & Shackleton, N. J. *Nature* **260**, 513–515 (1976).

39. Shackleton, N. J. Corfield, R. M. & Hall, M. A. *J. foram. Res.* **15**, 321–336 (1985).

40. Rea, D. K., Zachos, J. C., Owen, R. M. & Gingerich, P. D. *Palaeogr. Palaeoclimat. Palaeoecol.* **79**, 117–128 (1990).

41. Corfield, R. M. & Shackleton, N. J. *Historical Biology* **1**, 323–343 (1988).

42. Berner, R. A. & Canfield, D. E. *Am. J. Science* **289**, 333–361 (1989).

43. Shackleton, N. J. in *Geol. Soc. Spec. Publ.* **26**, 423–434 (1987).

44. Herbert, T. D. & Sarmiento, J. L. *Geology* **19**, 702–705 (1991).

45. Brass, G. W., Southam, J. R. & Peterson, W. H. *Nature* **296**, 620–623 (1982).

46. Kennett, J. P. *J. geophys. Res.* **82**, 3843–3860 (1977).

47. Haq, B. U. *Oceanologica Acta* **4** (Suppl. to Vol. 4, Proc. 26th Int. Geol. Congress) 71–82 (1981).

48. Hay, W. W. *Geol. Soc. Am. Bull.* **100**, 1934–1956 (1988).

49. Benson, R. H. *Historical Biogeography, Plate Tectonics, and the Changing Environment* (eds Gray, J. A. & Boucot, A. J.) 379–389 (Oregon State University Press, 1979).

50. Truswell, E. M. *New Zealand Dept. Sci. ind. Res. Bull.* **237**, 131–134 (1986).

51. Mohr, B. A. R. *Proc. ODP. Scient. Results* **113** (eds Barker, P. F. *et al.*) (Ocean Drilling Program, College Station, Texas, 1990).

52. Case, J. A. *Geol. Soc. Am. Mem.* **169**, 523–530 (1988).

53. Miller, K. G. & Fairbanks, R. G. *The Carbon Cycle and Atmospheric CO_2 Natural Variations. Archean to Present* (eds Sundquist, E. & Broecker, W. S.) *Am. geophys. Un. Monogr. Series* **32**, 469–486 (1985).

54. Heq, B. U. Herdenbol, J. & Vail, P. R. *Science* **235**, 1156–1166 (1987).

55. Sarmiento, J. L. & Toggweiler, J. R. *Nature* **306**, 621–624 (1984).

56. Berggren, W. A., Kent, D. V., Obradovich, J. D. & Swisher, C. C. in *Eocene-Oligocene Climatic and Biotic Evolution* (eds Prothero. D. R. & Berggren, W. A.) (Princeton University Press.

17

On Ocean pH

Caldeira, K. & Wickett, M.E. (2003). Anthropogenic carbon and ocean pH. *Nature*, **425**, 365.

Riebesell, U., Zondervan, I., Rost, B., Tortell, P.D., Zeebe, R.E. & Morel, F.M.M. (2000). Reduced calcification of marine plankton in response to increased atmospheric CO_2. *Nature*, **407**, 364–367.

Carbon dioxide, when dissolved in seawater in contact with $CaCO_3$, produces carbonic acid (H_2CO_3), bicarbonate ion (HCO_3^-), and carbonate ion (CO_3^{2-}). Together, these compounds control the acidity, or pH, or seawater, the same as they do in blood and cell plasma fluids. Adding CO_2 to the ocean from fossil fuel will tend to make the oceans more acidic, as mentioned by many of the early carbon-cycle papers in this volume (Revelle, Bolin, Broecker, and Walker). The excess acidity will ultimately be neutralized by dissolution of $CaCO_3$ (see Broecker and Walker), allowing further CO_2 uptake than would have been possible without the $CaCO_3$ "antacid." The time scale for the $CaCO_3$ cycle to respond to changes in ocean acidity is thousands of years.

Caldeira and Wickett were the first to realize that our present CO_2 rise differs from past natural changes in atmospheric CO_2 in that it is faster than the $CaCO_3$ cycle can keep up with. The ice core CO_2 measurements presented by Barnola *et al.* in this volume showed that atmospheric concentration of CO_2 rose from 200 ppm during the last ice age to about 280 ppm in the preindustrial atmosphere, but this took about 10 000 years. In contrast, CO_2 today is rising much faster, on a time scale of about a century. For this reason, the acidification of the ocean from fossil fuel CO_2 will be much more intense than it generally is from natural carbon cycle changes that the biota were usually subjected to in the past.

The impact of changing the pH of seawater is felt most intensely by organisms that create shells or infrastructure out of the mineral $CaCO_3$. $CaCO_3$ tends to dissolve in the deep ocean, and in the naturally more acidic cold surface waters in high latitudes, but most of the $CaCO_3$ that is formed biologically comes from warm surface waters in the tropics and subtropics. Here the chemistry is such that $CaCO_3$ is thermodynamically more than stable, it is what chemists call supersaturated. Early researchers, Broecker, for example, assumed that a change in pH would have little impact on biological $CaCO_3$ formation as long as the chemistry did not cross the line into undersaturation. The paper by Riebesell *et al.* showed instead that changes in saturation state can affect calcification rates, even in supersaturated surface waters. This effect has now been documented in corals and other types of $CaCO_3$-secreting organisms. These papers demonstrate the seriousness of ocean acidification, the "other problem" with CO_2 emissions.

Anthropogenic Carbon and Ocean pH

KEN CALDEIRA*, MICHAEL E. WICKETT[†]

*Energy and Environment Directorate and
†Center for Applied Scientific Computing, Lawrence Livermore National Laboratory, 7000 East Avenue,
Livermore, California 94550, USA e-mail: kenc@llnl.gov

THE COMING CENTURIES MAY SEE MORE OCEAN ACIDIFICATION THAN THE PAST 300 MILLION YEARS.

Most carbon dioxide released into the atmosphere as a result of the burning of fossil fuels will eventually be absorbed by the ocean[1], with potentially adverse consequences for marine biota[2-4]. Here we quantify the changes in ocean pH that may result from this continued release of CO_2 and compare these with pH changes estimated from geological and historical records. We find that oceanic absorption of CO_2 from fossil fuels may result in larger pH changes over the next several centuries than any inferred from the geological record of the past 300 million years, with the possible exception of those resulting from rare, extreme events such as bolide impacts or catastrophic methane hydrate degassing.

When carbon dioxide dissolves in the ocean it lowers the pH, making the ocean more acidic. Owing to a paucity of relevant observations, we have a limited understanding of the effects of pH reduction on marine biota. Coral reefs[2], calcareous plankton[3] and other organisms whose skeletons or shells contain calcium carbonate may be particularly affected. Most biota reside near the surface, where the greatest pH change would be expected to occur, but deep-ocean biota may be more sensitive to pH changes[4].

To investigate the effects of CO_2 emissions on ocean pH, we forced the Lawrence Livermore National Laboratory ocean general-circulation model[5] (Fig. 1a) with the pressure of atmospheric CO_2 (CO_2) observed from 1975 to 2000, and with CO_2 emissions from the Intergovernmental Panel on Climate Change's IS92a scenario[1] for 2000–2100. Beyond 2100, emissions follow a logistic function for the burning of the remaining fossil-fuel resources (assuming 5,270 gigatonnes of carbon (GtC) in 1750; refs 6, 7). Simulated atmospheric CO_2 exceeds 1,900 parts per million (p.p.m.) at around the year 2300. The maximum pH reduction at the ocean surface is 0.77; we estimate, using a geochemical model[8,9], that changes in temperature, weathering and sedimentation would reduce this maximum reduction by less than 10%.

A review[10] of estimates of palaeo-atmospheric CO_2 levels from geochemical models, palaeosols, algae and forams, plant stomata and boron isotopes concluded that there is no evidence that concentrations were ever more than 7,500 p.p.m. or less than 100 p.p.m. during the past 300 million years (Myr). Moreover, the highest concentrations inferred from the geological record were thought to have developed over many millions of years owing to slow processes involving tectonics and biological evolution.

We estimated the effect of past changes in atmospheric CO_2 levels on ocean pH by using a four-box ocean/atmosphere model[8,9]. Modelled processes include weathering of carbonate and silicate minerals on land, production of shallow-water carbonate minerals, production and oxidation of biogenic organic carbon, production and dissolution of biogenic carbonate minerals in the ocean, air–sea gas exchange of carbon, and transport by advection, mixing and biological processes.

In a series of simulations, atmospheric CO_2 was varied linearly from the pre-industrial value (about 280 p.p.m.) to stabilization values from 100–10,000 p.p.m. over time intervals of 10–10^7 yr. For each simulation, we recorded the maximum predicted perturbation in pH in the surface-ocean boxes (Fig. 1b). When a CO_2 change occurs over a short time interval (that is, less than about 10^4 yr), ocean pH is relatively sensitive to added CO_2. However, when a CO_2 change occurs over a long time interval (longer than about 10^5 yr), ocean chemistry is buffered by interactions with carbonate minerals, thereby reducing sensitivity to pH changes[11].

Caldeira, K. (2003) Anthropogenic carbon and ocean pH. *Nature* 425: 365. Reprinted by permission from Macmillan Publishers Limited.

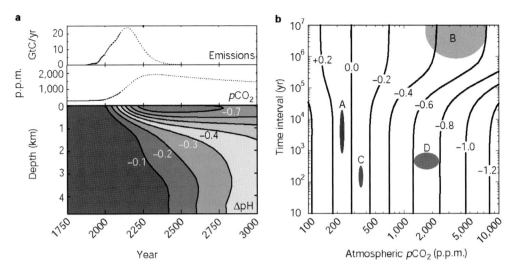

Fig. 1. Atmospheric release of CO_2 from the burning of fossil fuels may give rise to a marked increase in ocean acidity. **a**, Atmospheric CO_2 emissions, historical atmospheric CO_2 levels and predicted CO_2 concentrations from this emissions scenario, together with changes in ocean pH based on horizontally averaged chemistry. **b**, Estimated maximum change in surface ocean pH as a function of final atmospheric CO_2 pressure, and the transition time over which this CO_2 pressure is linearly approached from 280 p.p.m. A, glacial–interglacial CO_2 changes[13]; B, slow changes over the past 300 Myr; C, historical changes[1] in ocean surface waters; D, unabated fossil-fuel burning over the next few centuries.

Based on the record[10] of atmospheric CO_2 levels over the past 300 Myr and our geochemical model[8,9], there is no evidence that ocean pH was more than 0.6 units lower than today. Our general circulation model results indicate that continued release of fossil-fuel CO_2 into the atmosphere could lead to a pH reduction of 0.7 units. We conclude that unabated CO_2 emissions over the coming centuries may produce changes in ocean pH that are greater than any experienced in the past 300 Myr, with the possible exception of those resulting from rare, catastrophic events in Earth's history[8,12].

NOTES

1   Houghton, J. T. *et al.* (eds) *Climate Change 2001: The Scientific Basis* (Contribution of Working Group I to the Third Assessment Report of the IPCC, Cambridge Univ. Press, Cambridge, 2001).

2   Kleypas, J. A. *et al. Science* **284**, 118–120 (1999).

3   Riebesell, U. *et al. Nature* **407**, 364–367 (2000).

4   Seibel, B. A. & Walsh, P. J. *Science* **294**, 319–320 (2001).

5   Caldeira, K. & Duffy, P. B. *Science* **287**, 620–622 (2000).

6   Metz, B., Davidson, O., Swart, R. & Pan, J. (eds) *Climate Change 2001: Mitigation* (Contribution of Working Group III to the Third Assessment Report of the IPCC, Cambridge Univ. Press, Cambridge, 2001).

7   Marland, G., Boden, T. A. & Andres, R. J. Global, regional, and national CO_2 emissions. *Trends: A Compendium of Data on Global Change* http://cdiac.esd.ornl.gov/trends/emis/tre_glob.htm (2002).

8   Caldeira, K. & Rampino, M. R. *Paleoceanography* **8**, 515–525 (1993).

9   Caldeira, K. & Rau, G. H. *Geophys. Res. Lett.* **27**, 225–228 (2000).

10 Crowley, T. J. & Berner, R. A. *Science* **292**, 780–781 (2001).

11 Caldeira, K., Berner, R., Sundquist, E. T., Pearson, P. N. & Palmer, M. R. *Science* **286**, 2043 (1999).

12 Beerling, D. J. & Berner, R. A. *Global Biogeochem. Cycles* **16**, 101–113 (2002).

13 Sanyal, A., Hemming, N. G., Hanson, G. N. & Broecker, W. S. *Nature* **373**, 234–236 (1995).

Competing financial interests: declared none.

Reduced Calcification of Marine Plankton in Response to Increased Atmospheric CO$_2$

ULF RIEBESELL*, INGRID ZONDERVAN*, BJÖRN ROST*, PHILIPPE D. TORTELL[†],
RICHARD E. ZEEBE*[‡] & FRANÇOIS M. M. MOREL[†]

Alfred Wegener Institute for Polar and Marine Research, P.O. Box 120161, D-27515 Bremerhaven, Germany
[†]*Department of Geosciences & Department of Ecology and Evolutionary Biology, Princeton University, Princeton, New Jersey 08544, USA*
[‡]*Lamont-Doherty Earth Observatory, Columbia University, Palisades, New York 10964, USA*

The formation of calcareous skeletons by marine planktonic organisms and their subsequent sinking to depth generates a continuous rain of calcium carbonate to the deep ocean and underlying sediments[1]. This is important in regulating marine carbon cycling and ocean–atmosphere CO$_2$ exchange[2]. The present rise in atmospheric CO$_2$ levels[3] causes significant changes in surface ocean pH and carbonate chemistry[4]. Such changes have been shown to slow down calcification in corals and coralline macroalgae[5,6], but the majority of marine calcification occurs in planktonic organisms. Here we report reduced calcite production at increased CO$_2$ concentrations in monospecific cultures of two dominant marine calcifying phytoplankton species, the coccolithophorids *Emiliania huxleyi* and *Gephyrocapsa oceanica*. This was accompanied by an increased proportion of malformed coccoliths and incomplete coccospheres. Diminished calcification led to a reduction in the ratio of calcite precipitation to organic matter production. Similar results were obtained in incubations of natural plankton assemblages from the north Pacific ocean when exposed to experimentally elevated CO$_2$ levels. We suggest that the progressive increase in atmospheric CO$_2$ concentrations may therefore slow down the production of calcium carbonate in the surface ocean. As the process of calcification releases CO$_2$ to the atmosphere, the response observed here could potentially act as a negative feedback on atmospheric CO$_2$ levels.

By the end of the next century, the expected increase in atmospheric CO$_2$ (ref. 3) will give rise to an almost threefold increase in surface ocean CO$_2$ concentrations relative to pre-industrial values. This will cause CO$_3^{2-}$ concentrations and surface water pH to drop by about 50% and 0.35 units, respectively[4]. Changes of this magnitude have been shown to significantly slow down calcification of temperate and tropical corals and coralline macroalgae[5,6]. Although coral reefs are the most conspicuous life-supporting calcareous structures, the majority of biogenic carbonate precipitation (>80%) is carried out by planktonic microorganisms[1], particularly coccolithophorids[7]. These unicellular microalgae are major contributors to marine primary production and an important component of open ocean and coastal marine ecosystems[8]. Two prominent representatives of the coccolithophorids, *Emiliania huxleyi* and

Gephyrocapsa oceanica, are both bloom-forming and have a world-wide distribution. *G. oceanica* is the dominant coccolithophorid in neritic environments of tropical waters[9], whereas *E. huxleyi*, one of the most prominent producers of calcium carbonate in the world ocean[10], forms extensive blooms covering large areas in temperate and sub-polar latitudes[9,11].

The response of these two species to CO$_2$-related changes in seawater carbonate chemistry was examined under controlled laboratory conditions. The carbonate system of the growth medium was manipulated by adding acid or base to cover a range from pre-industrial CO$_2$ levels (280 p.p.m.v.) to approximately triple pre-industrial values (about 750 p.p.m.v.). Over this range, *E. huxleyi* and *G. oceanica* experience a slight increase in photosynthetic carbon fixation of 8.5% and 18.6%, respectively (Fig. 1a), and a comparatively larger

Riebesell, U. (2000) Reduced calcification of marine plankton in response to increased atmospheric CO$_2$. *Nature* 407: 364–367.
Reprinted by permission from Macmillan Publishers Limited.

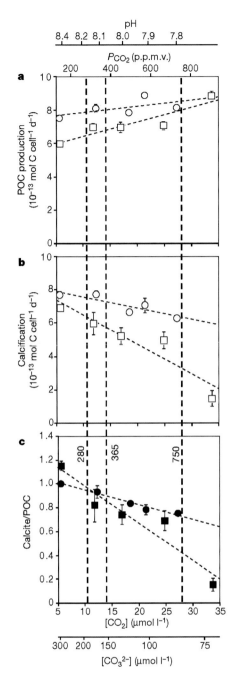

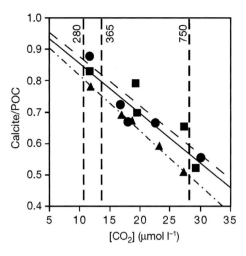

Fig. 2. Ratio of calcification to POC production (calcite/POC) of *Emiliania huxleyi* as a function of CO_2 concentration, $[CO_2]$. Cells were incubated at photon flux densities of 30, 80 and 150 μmol m^{-2}s^{-1} (denoted by circles, squares and triangles and corresponding solid, dashed, dash-dotted regression lines, respectively). Bars denote ±1 s.d. ($n = 3$); lines represent linear regressions. Vertical lines indicate pCO$_2$ values of 280, 365 and 750 p.p.m.v.

Fig. 1. Response of organic and inorganic carbon production to CO_2 concentration in laboratory-cultured coccolithophorids. **a**, Particulate organic carbon (POC) production; **b**, calcification; and **c**, the ratio of calcification to POC production (calcite/POC) of the coccolithophorids *Emiliania huxleyi* (circles) and *Gephyrocapsa oceanica* (squares) as a, function of CO_2 concentration, $[CO_2]$. Bars denote ±1 s.d. ($n = 3$); dotted lines represent linear regressions. Also indicated are corresponding ranges of pH, pCO$_2$, and $\left[CO_3^{2-}\right]$. We note that DIC and total alkalinity differed slightly between experimental sets for the two species; values for pH, pCO$_2$ and $\left[CO_3^{2-}\right]$, therefore, are only approximations. Vertical lines indicate pCO$_2$ values of 280, 365 and 750 p.p.m.v., representing pre-industrial, present day and future concentrations.

decrease in the rate of calcification of 15.7% and 44.7%, respectively (Fig. 1b). The ratio of calcite to organic matter production (calcite/POC) for the two species decreased by 21.0% and 52.5%, respectively, between 280 and 750 p.p.m.v. (Fig. 1c). Since calcite production has been shown to vary with ambient light conditions[12], we have grown *E. huxleyi* under different light/dark cycles and photon flux densities. A similar decrease in the calcite/POC ratio in response to CO_2-related changes in carbonate chemistry was obtained over a fivefold range in photon flux densities (Fig. 2).

Scanning electron microscopy indicated that malformed coccoliths and incomplete coccospheres increased in relative numbers with increasing CO_2 concentrations (Fig. 3). Coccolith under-calcification and malformation is a common phenomenon frequently observed both in natural environments and under laboratory conditions[13]. The systematic trend in the relative abundance of malformed coccoliths and coccospheres observed here, however, suggests a direct effect of seawater carbonate chemistry on the regulatory mechanisms controlling coccolith production inside the cell. Based on light microscopic analysis, no consistent trend was obtained in the number of attached or free coccoliths per coccosphere.

Our laboratory results are consistent with CO_2-related responses of natural plankton assemblages collected in the subarctic north Pacific, a region

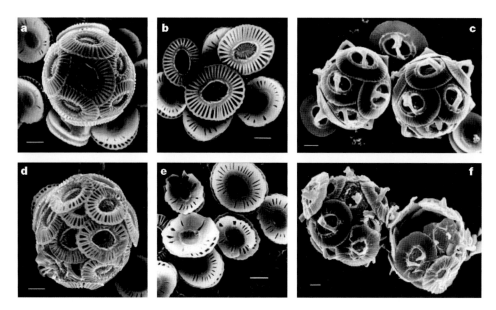

Fig. 3. Scanning electron microscopy (SEM) photographs of coccolithophorids under different CO_2 concentrations, **a**, **b**, **d**, **e**, *Emiliania huxleyi*, and **c**, **f**, *Gephyrocapsa oceanica* collected from cultures incubated at $[CO_2] \approx 12\,\mu mol\,l^{-1}$ (**a–c**) and at $[CO_2] \approx 30$–$33\,\mu mol\,l^{-1}$ (**d–f**), corresponding to pCO_2 levels of about 300 p.p.m.v. and 780–850 p.p.m.v., respectively. Scale bars represent $1\,\mu m$. Note the difference in the coccolith structure (including distinct malformations) and in the degree of calcification of cells grown at normal and elevated CO_2 levels. Pictures are selected from a large set of SEM photographs to depict the general trend in coccolith calcification. As the culture medium was super-saturated with respect to calcium carbonate under all experimental conditions, post-formation calcite dissolution is not expected to have occurred.

where coccolithophorids are major contributors to primary production[14]. After incubation of replicate samples at pCO_2 levels of about 250 p.p.m.v. and about 800 p.p.m.v. for 1.5 to 9 days, the rate of calcification was reduced by 36% to 83% in high-CO_2 relative to low-CO_2 treatments in four independent experiments (Fig. 4). A similar CO_2-dependent response was obtained under reduced light intensities (10% surface irradiance, data not shown). No significant differences were obtained between short and longer-term incubations of the natural assemblages. As short-term incubations are not likely to experience large changes in species composition, the observed response most probably reflects a reduction in carbonate precipitation of the calcifying organisms in the plankton assemblage.

The observed decrease in calcification with increasing pCO_2, if representative of biogenic calcification in the world's ocean, has significant implications for the marine carbon cycle. Owing to its effect on carbonate system equilibria, calcification is a source of CO_2 to the surrounding water[15], whereby the increase in CO_2 concentration, due to calcification is a function of the buffer capacity of sea water. Theoretically, the buffer state of pre-industrial sea water resulted in 0.63

mole CO_2 released per mole $CaCO_3$ precipitated[16] (assuming temperature $T = 15°C$, and salinity $S = 35$). Following the predictions of future atmospheric CO_2 rise, this value will increase to 0.79 in 2100 (assuming Intergovernmental Panel on Climate Change (IPCC) scenario IS92a, ref. 3). At constant global ocean calcification this results in an additional source of CO_2 to the atmosphere. In the case of reduced calcification, this positive feedback is reversed. Assuming a pre-industrial pelagic inorganic carbon production of 0.86 Gt Cy^{-1} (ref. 17) and a CO_2-related decrease in planktonic calcification as observed in our laboratory and field experiments (ranging between 16% and 83%), model calculations yield an additional storage capacity of the surface ocean for CO_2 between 6.2 Gt C and 32.3 Gt C for the period of 1950 to 2100.

Our results indicate that the ratio of calcite to organic matter production in cultured coccolithophorids and in oceanic phytoplankton assemblages is highly sensitive to the seawater pCO_2. Although it is presently not clear what the physiological and ecological role of coccolith formation is[18], we propose that CO_2-dependent changes in calcification may affect cellular processes such as acquisition of inorganic carbon[19] and nutrients[20] as well as trophic

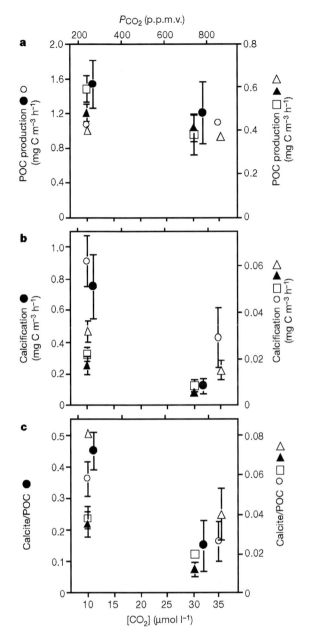

Fig. 4. Effects of CO_2 manipulations on POC production, calcification and the ratio of calcification to POC production (calcite/POC) in subarctic North Pacific phytoplankton assemblages, **a**, POC production; **b**, calcification; and **c**, the calcite/POC ratio. Station P26 (50 °N, 145°W) 1998, 6.8-day CO_2 pre-conditioning (filled circles). Station P26 1999, 2-day CO_2 pre-conditioning (squares). Station P26 1999, 9-day CO_2 pre-conditioning (filled triangles). Station P20 (43°30′ N, 138° 40′W) 1999, 1.5-day CO_2 pre-conditioning (open triangles). Station Z9 (55° N, 145° W) 1999, 1.5-day pre-conditioning (open circles). In all five experiments, POC production did not differ significantly between CO_2 treatments (*t*-test, *p.* ≥ 0.4). The statistical significance of calcification rate differences (*t*-test) is as follows: P26, all experiments and measurements ($p < 0.05$); P20, calcification ($p = 0.056$); calcite/POC ($p = 0.074$); Z9, calcification ($p = 0.135$); calcite/POC ($p = 0.11$). Error bars represent standard errors of means.

interactions, and particle sinking rate[21,22]. These, in turn, may influence the structure and regulation of marine ecosystems in which coccolithophorids are dominant. From a geochemical viewpoint, a decrease in global ocean calcification would enhance CO_2 storage in the upper ocean[3,15,23], thus providing a negative feedback for changes in atmospheric pCO_2. Such a feedback should be taken into account when predicting the role of the ocean in mitigating future anthropogenic CO_2 increases or in reconstructing the relation between ocean productivity and glacial–interglacial variations in PCO_2.

METHODS

Laboratory

Monospecific cultures of the coccolithophorids *Emiliania huxleyi* (strain PML B92/11A) and *Gephyrocapsa oceanica* (strain PC7/1) were grown in dilute batch cultures at 15 °C in filtered (0.2 μm) sea water enriched with nitrate and phosphate to concentrations of 100 and 6.25 μmol l⁻¹, respectively, and with metals and vitamins according to the *f*/2 culture medium (ref. 24). The carbonate system was adjusted through addition of 1 N HCl or 1 N NaOH to the medium. Cells were acclimated to the experimental conditions for 7–9 generations and allowed to grow for about 8 cell divisions during experiments. Cultures were incubated in triplicate at photon flux densities of 150 μmol m⁻² s⁻¹, light/dark (L/D) cycle = 16/8 h (Fig. 1) and of 150, 80 and 30 μmol m⁻² s⁻¹, L/D = 24/0 (Fig. 2). Dissolved inorganic carbon (DIC) was measured coulometrically in duplicate (UIC model 5012)[25]. Alkalinity was determined in duplicate through potentiometric titration[26]. pH, CO_2 and CO_3^{2-} concentrations were calculated from alkalinity, DIC and phosphate concentrations ($T = 15$ °C; $S = 31$) using the dissociation constants of ref. 27. Subsamples for total particulate carbon (TPC) and particulate organic carbon (POC), which in L/D = 16/8 were taken at the end of the dark phase, were filtered onto pre-combusted (12 h, 500 °C) QM-A filters (pore width is about 0.6 μm) and stored at –25 °C. Before analysis, POC filters were fumed for 2 h with saturated HCl solution in order to remove all inorganic carbon. TPC and POC were subsequently measured on a mass spectrometer (ANCA-SL 20-20 Europa Scientific). Particulate inorganic carbon (PIC) was calculated as the difference between TPC and POC. Cell counts obtained with a Coulter Multisizer at the beginning and the end of incubations were used to calculate specific growth rates. PIC and POC production rates were calculated from cellular inorganic and organic carbon contents and specific growth rates.

Field

Ship-board productivity and calcification experiments were conducted at three stations in the subarctic North Pacific Ocean in June of 1998 (one experiment) and September of 1999 (four experiments). Station locations are given in the legend of Fig. 4. Surface seawater (10–20 m) was collected using a trace-metal-clean *in situ* pumping system and dispensed into acid-soaked polycarbonate bottles (3–4 replicate bottles per treatment). Samples were incubated on deck at about 30% surface irradiance levels in a flow-through incubator at *in situ* temperatures (13 ± 1 °C). CO_2 concentrations in samples were manipulated by either bubbling with commercially prepared CO_2/air mixtures (Station P26, 1998/1999) or by the addition of high-purity (trace-metal-clean) HCl/NaOH. Low CO_2 samples contained about 10 µM CO_2 (~250 p.p.m.v.) with a pH of about 8.20 while high CO_2 treatments contained approximately 33 µM CO_2 (~800 p.p.m.v.) with a pH of about 7.75. Total alkalinity (~2180 µEq l^{-1}) was unaffected by CO_2 bubbling while HCl and NaOH additions changed alkalinity by −6.4% and +3.4%, respectively, CO_2 concentrations in samples were calculated from measurements of pH and alkalinity using the algorithm developed by ref. 28. After a pre-conditioning period ranging from 1.5–9 days, samples were incubated with 40 µCi ^{14}C (50 mCi mmol^{-1}) for 6–9 hours, harvested onto 0.4-µm polycarbonate filters, and immediately frozen in scintillation vials at −20 °C. In the laboratory, filter samples were acidified with either 10% HCl or H_3PO_4 to measure acid-stable (organic) carbon. The liberated acid-labile (particulate inorganic) carbon was trapped in NaOH solution contained either in small vials suspended in the primary vials or soaked into small GF/D glass fibre filters stuck onto the caps of the primary vials. Samples were measured on a liquid scintillation counter and corrected for ^{14}C uptake in dark control bottles as well as filtered seawater blanks.

ACKNOWLEDGEMENTS

We thank A. Dauelsberg, B. Höhnisch, A. Terbrüggen, and K.-U. Richter for laboratory assistance, F. Hinz for REM analyses, D. Crawford, M. Lipsen, F. Whitney and C. Mayfield for invaluable help at sea and C. S. Wong for providing space on RV *J.P. Tully*. The *E. huxleyi* strain PML B92/11A was generously supplied by J. Green, Plymouth Marine Laboratory, and the *G. oceanica* strain PC 7/1 by the CODENET Algae collection in Caen. This work was supported by the Netherlands-Bremen Cooperation in Oceanography (NEBROC).

Correspondence and requests for materials should be addressed to U. R. (e-mail: uriebesell@ awi-bremerhaven.de).

NOTES

Received 2 August 1999; accepted 18 July 2000.

1. Milliman, J. D. Production and accumulation of calcium carbonate in the ocean: budget of a nonsteady state. *Glob. Biogeochem. Cycles* **7**, 927–957 (1993).

2. Holligan, P. M. & Robertson, J. E. Significance of ocean carbonate budgets for the global carbon cycle. *Glob. Change Biol.* **2**, 85–95 (1996).

3. Houghton, J. T. *et al.* (eds) *Climate Change 1994: Radiative Forcing of Climate Change and an Evaluation of the IS92 Emission Scenario* (Cambridge Univ. Press, Cambridge, 1995).

4. Wolf-Gladrow, D. A., Riebesell, U., Burkhardt, S. & Bijma, J. Direct effects of CO_2 concentration on growth and isotopic composition of marine plankton. *Tellus* **51B**, 461–476 (1999).

5. Gattuso, J.-P., Frankignoulle, M., Bourge, I., Romaine, S. & Buddemeier, R. W. Effect of calcium carbonate saturation of seawater on coral calcification. *Glob. Planet. Change* **18**, 37–46 (1998).

6. Langdon, C., Takahashi, T., Sweeney, C., Chipman, D., Goddard, J., Marubini, F., Aceves, H., Barnett, H. & Atkinson, M. Effect of calcium carbonate saturation state on the calcification rate of an experimental coral reef. *Global Biogeochem. Cycles* **14**, 639–654 (2000).

7. Westbroek, P., Young, J. R. & Linschooten, K. Coccolith production (biomineralisation) in the marine alga *Emiliania huxleyi*. *J. Protozool.* **36**, 368–373 (1989).

8. Westbroek, P. *et al.* A model system approach to biological climate forcing. The example of *Emiliania huxleyi*. *Glob. Planet. Change* **8**, 27–46 (1993).

9. Winter, A., Jordan, R. W. & Roth, P. H. in *Coccolithophores* (eds Winter, A. & Siesser, W. G.) 161–179 (Cambridge Univ. Press, Cambridge, 1994).

10. Westbroek, P. *et al.* Strategies for the study of climate forcing by calcification. *Bull. Inst. Oceanogr. Monaco* (Spec. Issue) **13**, 37–60 (1994).

11. Holligan, P. M. *et al.* A biogeochemical study of the coccolithophore, *Emiliania huxleyi*, in the North Atlantic. *Glob. Biogeochem. Cycles* **7**, 879–900 (1993).

12. Nielsen, M. V. Growth, dark respiration and photosynthetic parameters of the coccolithophorid *Emiliania huxleyi* (Prymnesiophyceae) acclimated to different day length-irradiance combinations. *J. Phycol.* **33**, 818–822 (1997).

13. Young, J. R. Variation in *Emiliania huxleyi* coccolith morphology in samples from the Norwegian EHUX Experiment, 1992. *Sarsia* **79**, 417–425 (1994).

14. Booth, B. C., Lewin, J. & Postel, J. R. Temporal variation in the structure of autotrophic and heterotrophic communities in the subarctic Pacific. *Prog. Oceanogr.* **32**, 57–99 (1993).

15. Purdie, D. A. & Finch, M. S. Impact of a coccolithophorid bloom on dissolved carbon dioxide in sea water enclosures in a Norwegian fjord. *Sarsia* **79**, 379–387 (1994).

16. Frankignoulle, M., Canon, C. & Gattuso, J.-P. Marine calcification as a source of carbon dioxide: Positive feedback of increasing atmospheric CO_2. *Limnol. Oceanogr.* **39**, 458–462 (1994).

17. Morse, J. W. & Mackenzie, F. T. in *Developments in Sedimentology 48: Geochemistry of Sedimentary Carbonates* (Elsevier, Amsterdam, 1990).

18. Young, J. R. in *Coccolithophores.* (eds Winter, A. & Siesser, W. G.) 63–82 (Cambridge Univ. Press, Cambridge, 1994).

19. McConnaughey, T. A. Calcification, photosynthesis, and global cycles. *Bull. Inst. Océanogr. Monaco* (Spec. Issue) **13**, 137–161 (1994).

20. McConnaughey, T. A. & Whelan, J. F. Calcification generates protons for nutrient and bicarbonate uptake. *Earth Sci. Rev.* **42**, 95–117 (1997).

21. Harris, R. P. Zooplankton grazing on the coccolithophorid *Emiliania huxleyi* and its role in inorganic carbon flux. *Mar. Biol.* **119**, 431–439 (1994).

22. Fritz, J. J. & Balch, W. M. A light-limited continuous culture study of *Emiliania huxleyi*: determination of coccolith detachment and its relevance to cell sinking. *J. Exp. Mar. Biol. Ecol.* **207**, 127–147 (1996).

23. Kheshgi, H. S., Flannery, B. P. & Hoffert, M. I. Marine biota effects on the compositional structure of the world oceans. *J. Geophys. Res.* **96**, 4957–4969 (1991).

24. Guillard, R. R. L. & Ryther, J. H. Studies of marine planktonic diatoms. I. *Cyclothella nana* (Hustedt) and *Detonula confervacea* (Cleve). *Can. J. Microbiol.* **8**, 229–239 (1962).

25. Johnson, K. M., Wills, K. D., Butler, D. B., Johnson, W. K. & Wong, C. S. Coulometric total carbon dioxide analysis for marine studies: maximizing the performance of an automated gas extraction system and coulometric detector. *Mar. Chem.* **44**, 167–187 (1993).

26. Bradshaw, A. L., Brewer, P. G., Shafer, D. K. & Williams, R. T. Measurements of total carbon dioxide and alkalinity by potentiometric titration in the GEOSECS program. *Earth Planet. Sci. Lett.* **55**, 99–115 (1981).

27. Goyet, C. & Poisson, A. New determination of carbonic acid dissociation constants in seawater as a function of temperature and salinity. *Deep-Sea Res.* **36**, 1635–1654 (1989).

28. Lewis, E. & Wallace, D. W. R. Program Developed for CO_2 System Calculations. ORNL/CDIAC-105. (Carbon Dioxide Information Analysis Center, Oak Ridge National Laboratory, US Department of Energy, Oak Ridge, Tennessee, 1998).

18

Tiny Bubbles

Neftel, A., Moor, E., Oeschger, H., & Stauffer, B. (1985). Evidence from polar ice cores for the increase in atmospheric CO_2 in the past two centuries. *Nature*, **315**, 45–47.

Barnola, J.M., Raynaud, D., Korotkevich, Y.S., & Lorius, C. (1987). Vostok ice core provides 160 000-year record of atmospheric CO_2. *Nature*, **329**, 408–414. 7 pages.

Occasionally science gets lucky. Ancient samples of air become trapped in bubbles preserved for hundreds of thousands of years in ice cores, layer upon layer documenting changes in atmospheric composition like a perfect chemical tape recorder, providing information that we can use to test our understanding of the Earth's climate by simulating climates of the past. The oceanographers should be so lucky; a true sample of seawater from the last glacial maximum would revolutionize our understanding of the carbon cycle. It did not have to work out so well. At some depth in the ice, gases become absorbed into the solid ice forming hydrates, but they reemerge as bubbles when the core is brought to the surface and allowed to sit a while. One could also imagine reactive gases like methane combining with the oxygen in the bubbles, but apparently in the dark this is not a problem. Wind-blown dust grains containing $CaCO_3$ have reacted with acids to ruin the CO_2 measurements from the Greenland ice cores, but the gas samples from Antarctica appear to be exquisitely clean and true. Fantastic!

Neftel *et al.* showed some very early ice core CO_2 data that superimposes nicely on the directly measured CO_2 concentrations from Moana Loa (see the two papers by Keeling, this volume). Their measurements from further back in time reveal that CO_2 began rising earlier than had been assumed, back to the year 1750, due to deforestation, the "pioneer effect." They also obtained the preindustrial atmospheric CO_2 concentration of about 280 ppm, a value that had been guessed at, for example, by Bolin (this volume), but never known reliably.

Barnola *et al.* pushed the method farther back in time, finding that atmospheric CO_2 concentrations vary through the glacial/interglacial cycles. The CO_2 concentrations in the ice did not just decrease with age, as if they were just an artifact of some sort, but instead, the concentrations from the last interglacial period look a lot like the preindustrial values from our own interglacial time. More dramatically, the records of CO_2 concentration line up beautifully with estimates of the temperature in Antarctica, derived from isotopes of oxygen and hydrogen in the ice. The glacial climate changes are driven by other factors too, such as changes in the amount of ice on the planet, and ultimately by changes in the Earth's orbit around the sun, but putting the two records together gives a stunning visual impression of the importance of CO_2 in determining the Earth's climate.

Evidence From Polar Ice Cores for the Increase in Atmospheric CO$_2$ in the Past Two Centuries

A. NEFTEL, E. MOOR, H. OESCHGER & B. STAUFFER

Physics Institute, University of Bern, Sidlerstrasse 5, CH-3012 Bern, Switzerland

Precise and continuous measurements of atmospheric CO$_2$ concentration were first begun in 1958 and show a clear increase from 315 parts per million by volume (p.p.m.v.)[1] then to 345 p.p.m.v. now. A detailed knowledge of the CO$_2$ increase since preindustrial time is a prerequisite for understanding several aspects of the role of CO$_2$, such as the contribution of biomass burning to the CO$_2$ increase and the sensitivity of climate to the CO$_2$ concentration in the atmosphere. Estimates of the preindustrial CO$_2$ concentration are in the range 250–290 p.p.m.v. (ref. 2), but the precise level then and the time dependence of the increase to the present levels remain obscure. The most reliable assessment of the ancient atmospheric CO$_2$ concentration is derived from measurements of air occluded in ice cores. An ice core from Siple Station (West Antartica) that allows determination of the enclosed gas concentration with very good time resolution has recently become available. We report here measurements of this core which now allow us to trace the development of the atmospheric CO$_2$ from a period overlapping the Mauna Loa record back over the past two centuries.

Air bubbles are a characteristic feature of natural ice. At locations with mean air temperatures well below the freezing point, ice is formed by sintering of dry firn. At the firn–ice transition, the pore volume becomes separated as isolated bubbles having no further interaction with the atmosphere. The transition process is slow, so that the sampling of air into isolated bubbles spans at least several years, the duration depending mainly on the accumulation rate and the mean annual temperature.

Measurements were made on a 200-m ice core drilled in the Antarctic summer 1983/84 at Siple Station (75°55' S, 83°55' W) by the Polar Ice Coring Office in Nebraska and our institute. The mean annual air temperature is −24 °C and the annual accumulation rate is 500 kg m^{-2}. Only one clearly identifiable melt layer of irregular thickness (2–10 mm) was observed in the entire core at 7 m below surface (m.b.s.). Counting the seasonal variations of the electrical conductivity of the ice core allowed us to date the core over the past 200 yr with an accuracy of ±2 yr (ref. 3). Based on porosity measurements, the time lag between the mean age of the gas and the age of the ice was determined to be 95 yr and the duration of the close-off

process to be 22 yr (ref. 4). These values are, of course, evaluated for one particular core representing the present situation (1983), assuming a homogeneous enclosure process and not taking into account the sealing effect of observed impermeable layers.

The gases from ice samples were extracted by a dry-extraction system, in which bubbles are crushed mechanically to release the trapped gases, and then analysed for CO$_2$ by infrared laser absorption spectroscopy (IRLS) or by gas chromatography (GC). The combination of a needle crusher and IRLS allows us to analyse samples of ice as small as 1 g (ref. 5) (see Fig. 1 legend). During the past 2 yr, a new large dry-extraction device for ice samples up to 1 kg has been built and extensively tested[6]. The ice samples are ground *in vacuo* and the gases collected on a cold finger (20 K) and later analysed by both GC and IRLS. Applying both procedures to neighbouring ice samples gave the same concentration to within 1%. The measurements using the needle crusher, published previously[5,7,8], were performed using a slightly modified procedure and exhibited generally lower CO$_2$ concentrations by 15 p.p.m.v. (see Fig. 1 legend). In 1982 an intercalibration study

Neftel, A. (1985) Evidence from polar ice cores for the increase in atmospheric CO$_2$ in the past two centuries. *Nature* 315: 45–47.
Reprinted by permission from Macmillan Publishers Limited.

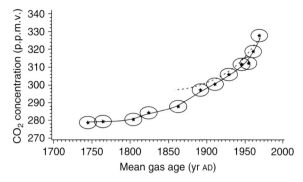

Fig. 1 Measured mean CO_2 concentration plotted against the estimated mean gas age. The horizontal axis of the ellipses indicates the close-off time interval of 22 yr. The uncertainties of the concentration measurements are twice the standard deviation of the mean value, but not lower than the precision of 1% of the measurement. The dotted line represents the model-calculated back extrapolation of the atmospheric CO_2 concentration, assuming only CO_2 input from fossil fuel[2].

Methods. After crushing the ice samples, the gas expands over a cold trap, condensing the water vapour at −80 °C into the absorption cell. Care was taken to minimize the effect of selective transport of CO_2 by water vapour. This effect was discovered while testing the system, when a difference in the measured CO_2 concentration was observed depending on whether the valve between the crusher and the cold trap was open or closed during the crushing process. With the valve closed, the concentrations were systematically higher by ~15 p.p.m.v. Crushing the ice with the valve open causes small ice particles to be transported to warm surfaces, where they immediately evaporate, leading to a considerable amount of water vapour in the system. The part of this water vapour not immediately condensed in the cold trap expands into the absorption cell and then flows slowly back into the cold trap, as indicated by the pressure monitoring in the absorption cell, which leads to a decreased CO_2 concentration in the absorption cell. When the valve is closed, the ice particles created during the crushing process are kept back in the cold crusher and less water vapour reaches the absorption cell after the valve is opened, leading to smaller deviations of the CO_2 concentration. Also, for this slightly modified measuring procedure, the measurements could be controlled by crushing pure gas-free single-crystal ice surrounded by the same amount of standard gas as contained in an average ice sample. With this procedure, small positive deviations in the CO_2 concentration, attributable to water vapour flow from the crusher into the cold trap, can be determined and the measurements corrected accordingly. This correction is below 1% for 10-cm³ samples and below 4% for 1-cm³ samples. All Siple samples were measured using the modified procedure and each fourth measurement was a calibration run.

with the Grenoble laboratory[7] was performed using the small crusher with the older measuring procedure. Based on our new results, the agreement of the inter-calibration must be viewed as a discrepancy, which we will try to resolve in the near future with a new intercalibration series.

Figure 1 shows the mean CO_2 concentration of the 12 depth intervals, measured on the Siple core, plotted against the mean gas age. To extend our data series further back in time, a few samples from the South Pole were also measured (and cross-checked with the large extraction system). These values are listed in Table 1.

Five depth intervals between 82 and 130 m.b.s. of the Siple core were measured with a high spatial resolution. Figure 2a–e shows the individual data series together with the $\delta^{18}O$ results. A seasonal structure could not be identified, indicating that the CO_2 concentration in the bubbles of the Siple core is homogeneously distributed and is independent of the seasonality of the surrounding ice.

Based on earlier studies of ice samples from locations in Greenland, we expected to measure enhanced CO_2 concentration in the extracted gas, especially in summer layers[8,9] from locations with a mean annual air temperature above −25 °C. For comparable mean gas ages, the CO_2 concentration measured on samples from the South Pole (−50 °C) and Siple Station (−24 °C) do not differ within the error limit. Therefore, we conclude that the CO_2 level in the bubbles of the Siple core is not influenced by a temperature effect. This is further supported by the lack of seasonal variations in the five data series measured with high spatial resolution.

Between 68 and 69 m.b.s., the porosity measurements showed two layers with bubble volumes above $80\,cm^3\,kg_{ice}^{-1}$, which correspond to complete enclosure. This is clearly above the $40\,cm^3\,kg_{ice}^{-1}$ expected from a smooth homogeneous enclosure process. These impermeable layers have a sealing effect if they are imbedded in firn with reduced permeability[10]. Because of the layers between 68 and 69 m.b.s., the air below is already completely isolated, about 7 m above the depth obtained, assuming a homogeneous enclosure. Consequently, for this core, the difference between ice and mean gas age is only 80–85 yr instead of 95 yr as estimated previously[4].

The enclosure of air occurs in winter layers at shallower depth than in summer layers, which could lead to a seasonal modulation of the CO_2 concentration in the bubbles during a period of changing atmospheric CO_2 concentration in case of a smooth and homogeneous enclosure process. The abovementioned lack of a seasonal variation in the series from 82 m.b.s. therefore further supports the sealing effect of such layers.

Table 1 Mean CO$_2$ concentration in air extracted from ice samples.

Depth below surface (m)	Samples measured	Siple Station Ice from (yr AD)	Siple Station Air enclosed (yr AD)	CO$_2$ concentration Extracted air (p.p.m.v.)	CO$_2$ concentration Atmosphere (p.p.m.v.)
68.2–68.6	8	1891	1962–1983	328 ± 3.5	328
72.4–72.7	11	1883	1954–1976	318 ± 3.0	321
76.2–76.6	11	1876	1947–1969	312 ± 3.0	315
82.0–83.0	28	1867	1938–1960	311 ± 3.0	
92.0–93.0	25	1850	1921–1943	306 ± 3.0	
102.0–103.0	26	1832	1903–1925	300 ± 3.0	
111.0–112.0	26	1812	1883–1905	297 ± 3.0	
128.0–129.0	47	1782	1842–1864	288 ± 3.0	
147.0–147.2	10	1743	1814–1836	284 ± 3.0	
162.0–162.3	9	1723	1794–1819	280 ± 3.0	
177.0–177.3	10	1683	1754–1776	279 ± 3.0	
187.0–187.3	10	1663	1734–1756	279 ± 3.0	
	IRLS/LES		South Pole	IRLS	LES
139.0–141.0	12/5	820	1660–1880	278 ± 3.0	280 ± 3.0
160.0–162.0	12/5	610	1450–1670	281 ± 3.0	283 ± 3.0
205.0–207.0	8/4	110	950–1170	279 ± 4.0	280 ± 3.0

LES, Large extraction system.

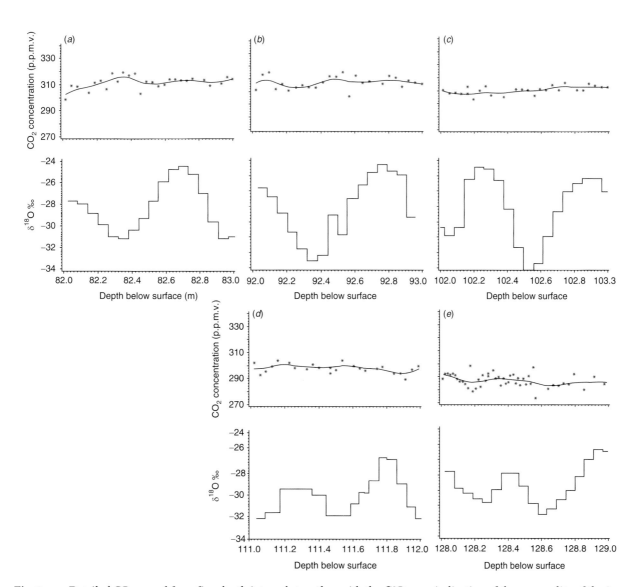

Fig. 2 a–e, Detailed CO$_2$ record from five depth intervals together with the δ^{18}O as an indication of the seasonality of the ice.

If the 328 p.p.m. measured at a depth of 68.5 m.b.s. is matched with the atmospheric South Pole record[11], the mean gas age is 10 yr, corresponding to a difference between mean gas age and ice age of 82 yr, which lies in the above estimated range. This difference is used in calculating the mean gas age for all depths. That the CO_2 concentration measured on the subsequent samples from 72.5 and 76.5 m.b.s. corresponds with the atmospheric South Pole record justifies this age determination and indicates, also, that impermeable layers at a depth of ~68 m.b.s. occur frequently.

There remains the question of the duration of the gas-enclosure time in such an impermeable layer. Assuming, as a first approximation, a constant and continuous enclosure rate ending at the time of core drilling (t_2 = AD1983), the beginning (t_1) can be calculated from the expression

$$\left(t_2 - t_1\right)C_{\text{bubbles}} = \int_{t_1}^{t_2} C_{\text{atm}}(t)\mathrm{d}t, \text{ where } C_{\text{atm}}(t) = a + b$$

$(t - 1958) + c(t - 1958)^2$ (with t in yr AD). The atmospheric concentration, C_{atm}, is taken from a quadratic fit of the measured CO_2 data in the atmosphere. The time t_1 so determined is 1963, so that the duration of the close-off is 20 yr, in good agreement with the estimate of Schwander and Stauffer[4].

We would have more difficulties explaining values lower by 15 p.p.m.v., as obtained, for example by the open-valve method. This would require a 20–40-yr-older mean gas age in the bubbles, depending on the development of the atmospheric CO_2 concentration. A very much prolonged close-off interval, a possibility which we exclude[4], or an incomplete mixing of air in the firn would be required. Also, the atmospheric CO_2 concentration would have to be increased between 1930 and 1960 by ~1 p.p.m.v. yr^{-1}, a similar rate to that of the past 20 yr.

Based on these results, we conclude that the atmospheric CO_2 concentration around 1750 was 280 ± 5 p.p.m.v. and has increased since, essentially because of human factors, by 22.5% to 345 p.p.m.v. in 1984. Back extrapolation of the Mauna Loa curve, on the basis of the fossil fuel input data obtained by Rotty[12] and assuming zero CO_2 input from biomass burning, suggests a preindustrial level of 297 p.p.m.v. (ref. 2). The difference of 17 p.p.m.v. between this calculated value and that measured indicates a significant contribution from biomass burning, even during the past century when the contribution from fossil fuel consumption was still small. The CO_2 increase in the Siple core will be studied in more detail using $\delta^{13}C$ measurements. Preliminary $\delta^{13}C$ results on an ice core from the South Pole are indeed very promising[13]. The CO_2/air $^{13}C/^{12}C$ record on the Siple ice core and the $^{14}C/^{12}C$ Suess effect from tree rings, together with the fossil fuel input-time history, will then enable us to separate the CO_2 inputs from those attributable to fossil fuels and biomass. It will also allow us to determine the effective airborne fraction for the combined input and to check the validity of the present carbon-cycle models.

We thank Drs J. Schwander, W. Bernhard, K. Hänni and R. Walther for assistance with the field work and with the analysis in the laboratory. We also thank D. Raynaud for reviewing the manuscript. This work was supported by the Swiss National Foundation, the Department of Polar Programs of the NSF, the US Department of Energy and the University of Bern.

NOTES

Received 31 December 1984; accepted 15 March 1985.

1. Keeling, C. D. *et al. Tellus* **28**, 538–551 (1976).
2. *WMO Rep.* **10** (World Meteorological Organization, Geneva, 1983).
3. Schwander, J. thesis, Univ. Bern (1984).
4. Schwander, J. & Stauffer, B. *Nature* **311**, 45–47 (1984).
5. Zumbrunn, R., Neftel, A. & Oeschger, H. *Earth planet. Sci. Lett.* **60**, 318–324 (1982).
6. Moor, E. & Stauffer, B. *J. Glaciol.* **30**, 358–361 (1984).
7. Barnola, J. M., Raynaud, D., Neftel, A. & Oeschger, H. *Nature* **303**, 410–412 (1983).
8. Neftel, A., Oeschger, H., Schwander, J., Stauffer, B. & Zumbrunn, R. *Nature* **295**, 220–223 (1982).
9. Stauffer, B. *et al. Ann. Glaciol.* **5**, 160–164 (1984).
10. Stauffer, B. *et al. Ann. Glaciol.* **6**, 108–112 (1985).
11. Keeling, C. D. *et al. Tellus* **28**, 552–564 (1976).
12. Rotty, R. M. in *Carbon Cycle Modelling* Scope 16 (ed. Bolin, B.) 121–126 (Wiley, New York, 1981).
13. Friedli, H. *et al. Geophys. Res. Lett.* **11**, 1145–1148 (1984).

Vostok Ice Core Provides 160,000-Year Record of Atmospheric CO_2

J. M. BARNOLA*, D. RAYNAUD*, Y. S. KOROTKEVICH† & C. LORIUS*

* Laboratoire de Glaciologie et de Géophysique de l'Environnement,
  BP 96, 38402 Saint Martin d'Hères Cedex, France
† Arctic and Antarctic Research Institute, Beringa Street 38, Leningrad 199226, USSR

Direct evidence of past atmospheric CO2 changes has been extended to the past 160,000 years from the Vostok ice core. These changes are most notably an inherent phenomenon of change between glacial and interglacial periods. Besides this major 100,000-year cycle, the CO2 record seems to exhibit a cyclic change with a period of some 21,000 years.

ALTHOUGH the first direct CO_2 measurements in the atmosphere were made in the second half of the nineteenth century, atmospheric CO_2 variations have been monitored in a systematic and reliable manner only since 1958. Fortunately, nature has been taking continuous samples of the atmosphere at the surface of the ice sheets throughout the ages. This natural sampling process takes place when snow is transformed into ice by sintering at the surface of the melt-free zones of the ice sheets, with air becoming isolated from the surrounding atmosphere in the pores of the newly formed ice. After pore closure the gas remains stored in the ice moving within the ice sheet. During this natural air sampling and storage process, different mechanisms could alter the original atmospheric composition[1,2]. But by choosing appropriate sampling sites, ice cores (for example see ref. 1) and experimental methods[3], past CO_2 changes in the atmosphere can be determined with high confidence by analysing the air enclosed in the pores of the ice.

Previous results from ice-core analysis have already provided important reliable information on the 'pre-industrial' CO_2 level and the recent CO_2 increase induced by anthropogenic activities[4–6]. Striking CO_2 changes have also been detected in this way in the ice record covering the last 30–40 kyr[7–10], including the large CO_2 increase associated with the climatic shift from the Last Glacial Maximum (~18 kyr BP) to the Holocene.

Because of the extremely low temperatures at Vostok (present-day mean annual temperature is −55.5 °C) and the good core quality, the 2,083-m-long ice core recovered by the Soviet Antarctic Expeditions at Vostok (East Antarctica) provides a unique opportunity to extend the ice record of atmospheric CO_2 over the last glacial-interglacial cycle back to the penultimate ice age about 160 kyr ago[11]. Over this timescale a high correlation is found between CO_2 concentrations and Antarctic climate, with significant oscillatory behaviour of CO_2 between high levels during interglacial and low levels during glacial periods. The CO_2 record also seems to exhibit a cyclic change with a period of ~21 kyr, that is, around the orbital period corresponding to the precession.

EXPERIMENTAL PROCEDURE

Gas extraction and measurements were performed with the 'Grenoble analytical setup' described by Barnola et al.[12] The method is based on crushing the ice under vacuum without melting, expanding the gas released during the crushing in a pre-evacuated sampling loop, and analysing the CO_2 concentrations by gas chromatography.

The analytical system, except for the stainless steel container in which the ice is crushed, is calibrated for each ice sample measurement with a standard mixture of CO_2 in nitrogen and oxygen. The corresponding accuracy (2σ) is evaluated from the standard deviation of the residuals corresponding to the calibration regression and ranges from 3 to 12 parts per million by volume (p.p.m.v.) for the measurements presented in this article.

Barnola, J. M. (1987) Vostok ice core provides 160,000-year record of atmospheric CO_2. Nature 329: 408–414. Reprinted by permission from Macmillan Publishers Limited.

We recently discovered an additional error due to our experimental system when a significant amount of water vapour is detected by the gas chromatograph. In such a case, selective CO_2 transport by water vapour (similar to that observed by Neftel *et al.*[4]) back to the ice crushing container is suspected of depleting the CO_2 from the air extracted from the ice and injected into the gas chromatograph. Experiments indicate that the presence of water vapour in the analysed air is not systematic but, when it occurs, leads to measured CO_2 concentrations lower by about 13 p.p.m.v. than in its absence. For the 75 Vostok core measurements reported in this paper, we found 22 cases with a water-vapour peak and 37 cases without. The presence of this peak was ambiguous for the 16 remaining measurements, leading in those cases to an additional uncertainty. Another source of uncertainty lies in the fact that two different stainless steel containers were used for ice crushing and that the respective results differ statistically by about 5 p.p.m.v.

For each depth level our 'best estimate' of the CO_2 concentration takes into account the correction of +13 p.p.m.v. for the effect of water vapour where there were unambiguous water-vapour peaks. For consistency with previously published data obtained with the same experimental setup[5,10,12] we added, when evaluating this 'best estimate', 5 p.p.m.v. to the CO_2 concentrations measured with the crushing container giving statistically lower values.

All the identified sources of uncertainty (experimental accuracy, effects of water vapour and crushing container) have been taken into account in plotting the CO_2 envelope shown in Fig. 1 and lead to errors around the 'best estimates' ranging from +3 to +22 p.p.m.v. on the positive side and from −3 to −16 p.p.m.v. on the negative side (Table 1). A single measurement was generally performed on each of the 66 depth levels investigated. Eight levels were nevertheless sampled and measured 2 or 3 times. For these levels the scatter of the multiple measurements is generally much smaller than the width of the envelope.

Although we are confident in the relative CO_2 variations measured, a systematic error could affect the absolute values. A recent comparison, still in progress, between results from the Physics Institute of Bern and our laboratory indicates that the values measured in Bern are systematically higher by about 10 p.p.m.v. The results of the comparison, when completed, will be published jointly by the two groups elsewhere.

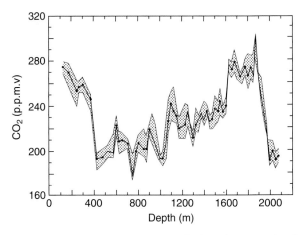

Fig. 1. CO_2 concentrations (p.p.m.v.) plotted against depth in the Vostok ice core. The 'best estimates' of the CO_2 concentrations are indicated by dots and the envelope shown has been plotted taking into account the different uncertainty sources.

VOSTOK CO_2 RECORD

Dating the air found in ice bubbles

Dating the successive ice layers at Vostok has been discussed and an ice chronology established by Lorius *et al.*[11]. This ice chronology cannot, however, be directly applied to the dating of the CO_2 record. Indeed, owing to the air enclosure process in the ice, the air is isolated from the atmosphere in the firn layers well after the precipitation has been deposited at the surface of the ice sheet, and consequently air extracted from the ice is younger than the ice itself.

The following assumptions have been made to date the air in relation to the ice: (1) the air in the firn layers is well mixed with the atmosphere until firn pores close; (2) pore closure occurs roughly in the firn density interval 0.80–0.83 (ref. 13), that is, in the deepest layers of the firn (found between about 86 and 96 m depth at Vostok); and (3) the density–depth profile has remained unchanged over the past 160 kyr.

Based on these assumptions the difference between the age of the ice and the mean age of the air, which is dependent on accumulation rate, is evaluated taking into account the temporal changes of the accumulation rate[11]. This calculated difference varies between about 2,500 yr for the warmest to about 4,300 yr for the coldest periods.

On the other hand, all the air enclosed in an ice sample has not been pinched off from the atmosphere at the same time but rather pore after

Table 1. Mean CO$_2$ concentrations in Vostok ice core (best estimates and corresponding uncertainties, against depth and age).

Depth (m)	Age of the ice (yr BP)	Mean age of the air (yr BP)	CO$_2$ concentration (best estimate) (p.p.m.v.)	Depth (m)	Age of the ice (yr BP)	Mean age of the air (yr BP)	CO$_2$ concentration (best estimate) (p.p.m.v.)
126.4 (Y)*	4,050	1,700	$274.5\,\mp^{5}_{5}$	1,274.2 (Y, N)	87,980	84,700	$218.8\,\mp^{5.5}_{5.5}$
173.1 (N)	5,970	3,530	$270.0\,\mp^{13}_{8}$	1,299.3 (N)	89,940	86,680	$210.0\,\mp^{7}_{7}$
250.3 (N, N)	9,320	6,800	$252.0\,\mp^{13}_{10.5}$	1,322.5 (A)	91,760	88,520	$221.5\,\mp^{4}_{17}$
266.0 (N)	10,040	7,500	$257.0\,\mp^{5}_{5}$	1,349.0 (N)	93,860	90,630	$226.0\,\mp^{10}_{5}$
302.6 (N)	11,870	9,140	$259.0\,\mp^{7}_{7}$	1,374.8 (Y)	95,910	92,700	$234.0\,\mp^{4}_{4}$
375.6 (Y)	16,350	12,930	$245.0\,\mp^{10}_{5}$	1,402.5 (A)	98,130	94,940	$226.5\,\mp^{5}_{18}$
426.4 (N)	20,330	16,250	$193.0\,\mp^{10}_{5}$	1,425.5 (Y)	100,000	96,810	$236.0\,\mp^{5}_{5}$
474.2 (N)	24,280	20,090	$194.5\,\mp^{7}_{7}$	1,451.5 (N)	102,210	98,950	$225.0\,\mp^{6}_{6}$
525.1 (Y)	28,530	24,390	$200.0\,\mp^{5}_{5}$	1,476.1 (N)	104,410	101,040	$229.0\,\mp^{10}_{5}$
576.0 (A)	32,680	29,720	$198.0\,\mp^{4}_{17}$	1,499.6 (Y)	106,610	103,130	$238.5\,\mp^{16}_{11}$
602.3 (Y)	34,770	30,910	$223.0\,\mp^{9}_{9}$	1,526.3 (N)	109,240	105,620	$234.5\,\mp^{9}_{9}$
625.6 (N)	36,600	32,800	$207.0\,\mp^{10}_{10}$	1,547.0 (Y)	111,250	107,650	$244.0\,\mp^{12}_{12}$
651.6 (Y)	38,600	34,870	$210.0\,\mp^{11}_{5}$	1,575.2 (N)	113,850	110,510	$233.5\,\mp^{5}_{5}$
700.3 (N)	42,320	38,660	$207.0\,\mp^{4}_{4}$	1,598.0 (N)	115,850	112,700	$240.0\,\mp^{10}_{7}$
748.3 (A)	45,970	42,310	$178.5\,\mp^{5}_{18}$	1,626.5 (Y)	118,220	115,290	$276.0\,\mp^{6}_{6}$
775.2 (Y)	48,000	44,350	$200.0\,\mp^{10}_{5}$	1,651.0 (Y, N)	120,170	117,410	$271.5\,\mp^{7}_{7}$
800.0 (N, Y, Y)	49,850	46,220	$207.7\,^{+8.4}_{6.6}$	1,676.4 (N)	122,100	119,500	$280.0\,\mp^{10}_{10}$
852.5 (A)	53,770	50,150	$201.0\,\mp^{6}_{19}$	1,700.9 (N)	123,900	121,430	$271.0\,\mp^{8}_{8}$
874.3 (N)	55,450	51,770	$201.0\,\mp^{12}_{12}$	1,726.8 (N, N)	125,730	123,380	$265.3\,\mp^{4}_{4}$
902.2 (N)	57,660	53,860	$219.5\,\mp^{5}_{10}$	1,747.3 (Y, N)	127,150	124,880	$267.0\,\mp^{8}_{5}$
926.8 (A)	59,670	55,780	$214.5\,\mp^{6}_{19}$	1,774.1 (N)	129,020	126,770	$275.0\,\mp^{7}_{7}$
951.9 (A)	61,790	57,800	$206.5\,\mp^{5}_{18}$	1,802.4 (A)	131,030	128,780	$266.5\,\mp^{12}_{10}$
975.7 (Y)	63,880	59,770	$201.0\,\mp^{8}_{8}$	1,825.7 (Y, N)	132,700	130,460	$275.0\,\mp^{10}_{10}$
1,002.5 (N)	66,230	62,080	$192.0\,\mp^{3}_{3}$	1,850.5 (A)	134,510	132,280	$266.0\,\mp^{6}_{19}$
1,023.5 (N)	68,040	63,960	$193.0\,\mp^{7}_{7}$	1,875.9 (Y)	136,450	134,170	$296.5\,\mp^{13}_{8}$
1,052.4 (A)	70,470	66,540	$205.5\,\mp^{12}_{20}$	1,902.0 (Y)	138,660	136,170	$266.0\,\mp^{4}_{4}$
1,074.8 (A)	72,330	68,490	$226.5\,\mp^{11}_{19}$	1,928.0 (Y)	141,170	138,410	$246.5\,\mp^{10}_{18}$
1,101.4 (Y)	74,500	70,770	$243.0\,\mp^{16}_{11}$	1,948.7 (A)	143,440	140,430	$231.0\,\mp^{7}_{20}$
1,124.2 (A)	76,330	72,690	$235.0\,\mp^{9}_{22}$	1,975.3 (N)	146,860	143,370	$217.0\,\mp^{4}_{4}$
1,148.7 (Y)	78,270	74,720	$230.5\,\mp^{7}_{7}$	1,998.0 (A)	150,330	146,340	$191.0\,\mp^{5}_{18}$
1,175.0 (N)	80,320	76,860	$219.5\,\mp^{9}_{9}$	2,025.7 (N)	154,980	150,700	$200.5\,\mp^{8}_{8}$
1,225.7 (A)	84,220	80,900	$222.5\,\mp^{12}_{20}$	2,050.3 (N, N)	159,100	154,970	$191.3\,\mp^{7.5}_{7.5}$
1,251.5 (N)	86,220	82,920	$234.0\,\mp^{12}_{7}$	2,077.5 (N)	163,670	159,690	$195.5\,\mp^{6}_{6}$

* (Y) indicates the presence and (N) the absence of an H$_2$O peak in the chromatographic analysis. Ambiguous cases are denoted (A).

pore over a time interval which, with the above assumptions, lies between 300 and 750 yr. This acts as a low-pass filter and means that each CO$_2$ measurement represents roughly an average value over several hundred years.

All the CO$_2$ ages given in this paper are based on the ice chronology given by Lorius et al.[11], and dating of the air relative to the ice was done as described above. No better quantitative method is currently available to evaluate the age of the air. Nevertheless the two first assumptions made above probably lead to a maximum difference between the two ages (air and ice)[5], whereas the last would lead to an age difference too small (by maybe several hundred of years) for the coldest periods.

Description

Sixty-six depth levels were investigated for CO$_2$ measurements along the 2,083-m core. They are spaced every ~25 m from 850 m depth to the bottom. Because of the presence of fractures in the upper part of the core, the spacing is generally larger above 850 m depth. With this sampling, the mean age difference between two neighbouring levels ranges from ~2,000 to ~4,500 yr.

The record covers the past ~160 kyr and includes the following major climatic periods: the Holocene, the last glaciation, the previous interglacial and the end of the penultimate glaciation[11,14]. Figure 1 shows the results (best estimates and envelope obtained as described above) plotted against depth. Figure 2 shows the CO$_2$ variations with age of the air.

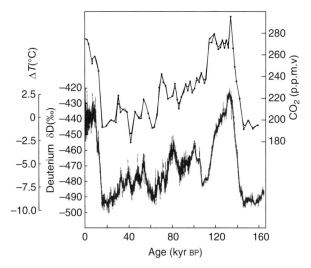

Fig. 2 CO_2 concentrations ('best estimates') and smoothed values (spline function) in p.p.m.v. plotted against age in the Vostok record (upper curves) and atmospheric temperature change derived[14] from the isotopic profile (lower curve). The deuterium scale corresponds to values after correction for deuterium changes of oceanic water[14].

The CO_2 record exhibits two very large changes, one near the most recent part of the record (~400 m depth or 15 kyr BP) and the other near the earliest part of the record (~1,950 m depth or 140 kyr BP), between two levels centred on 190–200 and 260–280 p.p.m.v. roughly typical of the lowest and highest values of the whole record. The high level is comparable with the so-called 'pre-industrial' CO_2 level that prevailed roughly 200 years ago, just before the large anthropogenic disturbance of the CO_2 cycle; the low level ranges among the lowest values of the known geological history of carbon dioxide over the last 10^8 yr (ref. 15). These two large CO_2 changes correspond to the transitions between full glacial conditions (low CO_2) of the last and penultimate glaciations, and the two major warm periods (high CO_2) of the record: the Holocene and the previous interglacial.

Between these two extremities of the record, that is, over the period covering mainly the last glaciation, the CO_2 shows a general decreasing trend with the lowest values found over the last part of the glaciation (between ~1.025 and 425 m depth, or 65 and 15 kyr) and with two marked breaks: a first decrease of ~40 p.p.m.v. at ~1,610 m depth (~114 kyr BP), and a second of ~50 p.p.m.v. at ~1,070 m depth (~68 kyr BP). Other CO_2 changes are superimposed on the general decreasing trend, suggesting an oscillatory behaviour with an apparent period of ~20 kyr, especially between ~80 and 20 kyr BP. These changes are more pronounced in the last

part of the glaciation, but note that some of the variability observed above 850 m depth could be due to the lower ice quality in that part of the core.

The fact that the record does not indicate a continuous long-term trend is an important result in itself regarding possible fractionation of the different gas components due to the interaction with the ice lattice and which should be reflected by such a trend. This adds confidence to the preservation of the atmospheric CO_2 record in the ice over such a long period.

Spectral analysis

The record, as shown in Fig. 2, suggests a cyclic behaviour of the CO_2 changes. Because of the potential link between CO_2 and climate, and consequently between CO_2 and astronomical cyclic forcing of climate, a spectral analysis of the CO_2 record has been done. For comparison with the δD record of Vostok (see below), the same procedure as in ref. 14 has been applied, using values equally spaced every 200 years and obtained by linear interpolation of our best estimates for CO_2 measurements. Because of the short length of the record (relative to the frequencies of interest), we combined different spectral techniques to extract reliable information. Blackman–Tukey (BT) and maximum entropy (ME) methods with and without prewhitening were applied. The BT method has a poor resolution but well defined statistical properties; the use of the ME method increases the frequency resolution but the amplitude estimate has to be interpreted with caution[16]. The prewhitening allows the removal of part of the variance at low frequencies and a more detailed picture in the tilt and precession frequency bands. The ME spectra shown in Fig. 3 were obtained with the Barrodale and Eriksson algorithm[17].

The spectra obtained with the unprewhitened series are dominated by the 100-kyr components (116 kyr for the ME peak). The BT spectrum (not shown in Fig. 3) is relatively flat for higher frequencies but shows a significant bump for periods between 25 and 20 kyr. The ME spectrum indicates a concentration of variance in this frequency band with a well-defined peak at 21 kyr; it also displays a peak at ~55 kyr. Prewhitening, which gives more detailed information for this part of the spectrum, confirms the position of a 21-kyr peak (21.0 kyr for ME). Sensitivity analysis shows the remarkable stability of this peak ($\sigma = 0.1$ kyr) with respect to the lag value (BT) or

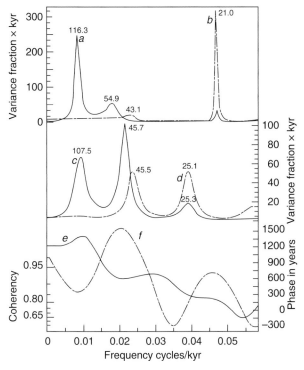

Fig. 3 *a, b,* Variance spectra of the Vostok CO_2 record. The normalized variance density is plotted against frequency in cycles per kyr. The continuous line (*a*) refers to the original series and the dashed line (*b*) to the series after pre-whitening by a first-order difference filter. The spectra were obtained by the ME method applying the Barrodale and Ericksson algorithm with an autoregressive order of 40%. *c, d,* Variance spectra of the Vostok isotope temperature record from ref. 14, obtained with the same procedure, *e, f,* Cross-spectral analysis between CO_2 and temperature for coherency (*e*) and phase spectrum (*f*), obtained by applying the Blackman-Tukey method. The phase (right scale) is positive when CO_2 lags the temperature.

to the autoregressive order (ME). The ME pre-whitened CO_2 spectrum is thus dominated by this component of ~21 kyr and also shows a secondary peak at ~40 kyr (43.1 kyr). As the latter may result from the variance transfer of the 55-kyr peak mentioned above (unprewhitened ME spectrum), it must be interpreted very cautiously.

Because of the relatively small number of measurements with corresponding errors, any interpretation of the spectral analysis needs to be cautious. To check the significance of the spectrum results we have performed some tests with random curves placed within the error band. The results show a rather good stability of the peaks, especially for 21 kyr (20.9 ± 0.2). Some implications of this CO_2 spectrum on the CO_2-climate correlation and on the CO_2 cycle are discussed below.

Comparison with other ice records

Several ice cores from Antarctica and Greenland have already shown CO_2 variations over the past 30–40 kyr (D 10 (ref. 7), Dome C[7,10], Byrd[8] Camp Century[8], Dye 3 (refs 9, 18)). The Vostok profile is in good agreement with the general common features of these records, which show low CO_2 values (~200 p.p.m.v.) during the Last Glacial Maximum (~18 kyr BP) and an increase of the atmospheric CO_2 associated with the glacial-Holocene transition.

Other striking individual CO_2 features are also observed in some of these previous ice records. For instance, abrupt CO_2 changes are recorded in the Dye 3 core between 30 and 40 kyr BP (ref. 9) and in the Dome C core near the end of the glacial–Holocene transition[7,10]. They do not appear in the measured Vostok profile, but our sampling does not allow the search for such rapid changes; furthermore, the low-pass filter effect of the air-trapping process (mentioned above) may prevent any detection in the Vostok core of the changes observed in the Dye 3 core. Note also that the high CO_2 level (> 270 p.p.m.v.) observed in the Dome C core between 23 and 30 kyr BP is not found in the Vostok record. These high CO_2 values could be due to the poor quality of the Dome C ice in this part of the core[19].

Thus, taking into account the resolution of the different records, the Vostok CO_2 results are consistent with the other ice records already published, covering only the most recent part of the time interval revealed by the Vostok core.

Comparison with deep-sea sediment record

$\delta^{13}C$ differences between planktonic and benthic foraminifera found in a deep-sea core (V 19–30) have been used by Shackleton *et al.*[20] to deduce a record of atmospheric CO_2 covering approximately the last climatic cycle and by Shackleton and Pisias[21] to extend this record over the past 340 kyr. The basic idea, following an approach developed by Broecker[22], is that a $\delta^{13}C$ difference between ocean surface and deep waters is associated with the removal of carbon from the surface by photosynthesis and consequently with the atmospheric CO_2 concentration level. The CO_2 record thus obtained agreed fairly well with the available CO_2-ice record, that is, corresponding to the period covering the past ~40 kyr.

We are now able, with the Vostok ice record, to extend the comparison over the past ~160 kyr. For this comparison we use the more recent data by Shackleton and Pisias[21] for the marine record.

Major similarities in the general trends are observed. Both records show rapid and large increases of CO_2 concentrations associated with the two deglaciations and exhibit the highest CO_2 values during the interglacial periods. They also show together a relatively abrupt CO_2 decrease associated with the beginning of the last glaciation.

The comparison is limited in detail because of differences in timescales[14] and in time resolutions between the ice and marine records. There are nevertheless significant differences between the two curves, which include the following observations.

(1) The disparity between previous interglacial and Holocene CO_2 values is much smaller in the Vostok record.

(2) The CO_2 concentrations deduced from the marine record approach Holocene levels several times during the last glaciation, unlike the Vostok record.

(3) The morphology of the Vostok CO_2 curve over the last glaciation (that is, over a large common part of the two records) is definitely not the same as that of the V 19–30 curve. Thus the general CO_2 decrease observed in the Vostok core does not appear in the marine record and the most striking change appears in the Vostok curve as a marked CO_2 decrease (between ~71 and 62 kyr BP in the Vostok timescale) and in the V 19–30 record as a large and sharp CO_2 increase (between ~60 and 55 kyr BP in the V 19–30 timescale).

(4) Spectral analysis has been done[21,23] on the CO_2 record recovered from the marine core study. The V19–30 spectrum exhibits peaks around all the orbital periods (~100 kyr for the Earth's eccentricity, ~41 kyr for the obliquity of the Earth's axis and 23 and 19 kyr for the precession of the equinoxes), whereas the Vostok spectrum shows as mentioned above, aside from the 110 kyr period, a peak at ~21 kyr but nothing clearly corresponding to the obliquity frequency. The comparison is, in fact, not straightforward because the record lengths (~340 kyr for V19–30 and ~160 kyr for Vostok) as well as part of the procedures used for the spectral analysis are different. We have checked the possible influence of these differences by applying equivalent procedures to the Vostok record and to the V19–30 data given in ref. 21, limited to the past 160 kyr. The V19–30 spectrum still shows, in contrast with Vostok, a significant peak around the 40-kyr period.

Thus, although the CO_2 record derived from the carbon isotope differences ($\Delta\delta^{13}C$) between planktonic and benthic foraminifera found in the V19–30 core, as well as the Vostok CO_2 record, indicate that alternations of low and high mean CO_2 concentrations and of glacial and interglacial periods are related, there are significant discrepancies between the changes observed in each of the records. Since the ice record provides *a priori* a more direct and accurate measurement of atmospheric CO_2 changes, these differences suggest that the $\Delta\delta^{13}C$ record of the V19–30 core cannot be interpreted as a pure atmospheric CO_2 curve.

IMPLICATIONS FOR CLIMATE

In this section we compare the CO_2 profile described above with the ice isotopic record obtained on the same core[14] and discuss some climatic implications. A companion paper[24] addresses the potential role of various forcing factors, including CO_2, for producing the temperature record derived from the isotopic composition of the Vostok ice core.

The isotopic record shows the 'dramatic' changes between the two glaciations and the two interglacial periods and provides a detailed description of other climatic variations, especially during the last glacial and the previous interglacial periods.

Our CO_2 profile and the Vostok climatic record (adapted from ref. 14) are plotted together against time in Fig. 2. The climatic significance of the isotopic profile is fully discussed in refs 11 and 14 and high (low) δD values are associated with 'warm' ('cold') temperatures. The temperature scale refers to the atmospheric surface-temperature changes in the Vostok area. The CO_2 and δD records, as obtained from the same ice core, can be directly compared after allowing for the mean age difference between the ice and the air enclosed in the ice (see section on the dating); this age difference has been taken into account in Fig. 2.

The first obvious feature when comparing the two records is the marked similarity in temporal variations. The high correlation ($r^2 = 0.79$) between the linearly interpolated CO_2 curve based on the best estimates and the smoothed δD curve shown in Fig. 2 reflects that generally high (low) CO_2 concentrations correspond to 'warm ('cold') temperatures in the Vostok area. Thus the CO_2 values for the two interglacial periods (mean concentrations of 263 p.p.m.v. for the Holocene and of 272 p.p.m.v. for the previous interglacial period) are much higher than the concentrations measured in the glacial ice (on average 215 p.p.m.v.). Furthermore during the last glaciation, as mentioned above, the mean CO_2 level was higher during the first part (230 p.p.m.v. between 110 and 65 kyr BP) than during the second part (203 p.p.m.v. between 65 and 15 kyr), which also shows colder climatic conditions.

But a more detailed examination of the two records in Fig. 2 indicates differences:

(1) During the previous interglacial period and the transition towards the last glaciation, although both CO_2 and δD seem to peak simultaneously at ~135 kyr BP (note that the CO_2 peak is documented by only one depth level and therefore needs to be confirmed), the CO_2 record shows relatively high and constant values over at least the 10 kyr after the peak (see Fig. 2), whereas the δD curve clearly decreases over the same time interval. A similar effect, although less obvious, seems to appear around 75 kyr BP when the δD curve shows a decreasing trend whereas the average CO_2 apparently remains constant.

(2) The first 'isotopically cold' peak observed during the last glaciation, around 110 kyr BP, has no low-CO_2 companion peak.

The superposition of the CO_2 and δD records also provides information on the comparative timing of the most significant changes in the two signals. The sparse CO_2 sampling and to a smaller extent the uncertainties in the CO_2 measurements and on the age difference between the air trapped in the ice and the ice itself make the determination of a time difference between the two signals of Fig. 2 only possible for differences larger than 3500 yr. Within this uncertainty, the following trends are observed: (1) δD and CO_2 changes appear almost simultaneously when proceeding from glacial to interglacial conditions; (2) as noted above, the situation could be different when proceeding from the previous interglacial to the last glaciation with the CO_2 change clearly lagging behind the δD decrease.

Furthermore information on the comparison between the two time series can be inferred from the spectral analysis. As mentioned above, the CO_2 spectrum reveals a well-defined peak at 21 kyr and, with prewhitening, a relatively weak secondary peak at ~40 kyr that must be interpreted very cautiously. This contrasts (see Fig. 3) with the temperature spectrum deduced from the δD record and given by Jouzel *et al.*[14]. The temperature spectrum does indeed indicate peaks at ~40 and

~20 kyr (that is, which can be associated with the obliquity and precession cycles), with the 40-kyr peak dominating the 20-kyr. The results (Fig. 3) of the cross-spectral analysis for coherence and phase spectrum between CO_2 and temperature (δD) show high coherency in the orbital frequency band and indicate that CO_2 could lag the temperature slightly. This lag is not statistically significant but could reflect, as mentioned above, the temperature signal decreasing before the CO_2 when proceeding from the interglacial to the glacial period.

The above comparison deals with a CO_2 signal that should be of global atmospheric significance and a climatic signal that is *a priori* relevant to Antarctica and consequently to more local conditions. The Vostok δD record shows, in fact, important climatic changes on a worldwide scale[11,14]. In this more general context it is interesting to compare the CO_2 level measured over the previous interglacial with the present atmospheric CO_2 concentrations, because this period of the past has been proposed as an analogue for a future CO_2-warmed world. The Vostok results indicate a mean CO_2 concentration (272 p.p.m.v.) comparable with the 'pre-industrial' CO_2 level (estimated to be 275 ± 10 p.p.m.v. (ref. 25)). Furthermore, the CO_2 peak of about 300 p.p.m.v. found at ~135 kyr BP, if confirmed, is significantly lower than the current level of about 345 p.p.m.v.

The high correlation between atmospheric CO_2 and isotopic temperature found in the Vostok record suggests that the radiative effect of CO_2 coupled with associated feedback mechanisms appears to be a serious candidate as one of the driving forces behind the palaeo-temperature changes recorded in the Vostok core. The relative weights of different types of forcing in the reconstruction of the Vostok temperature variations are discussed in the companion paper[24]. This approach points out the probable importance of CO_2 in influencing the main feature of the climate during the last climatic cycle.

Examining the parts of the CO_2 and δD records containing the most pronounced changes, although the two transitions from glacial to interglacial conditions shown by the record indicate almost simultaneous changes for CO_2 and isotopic temperature, the situation is different when looking at the transition between the previous interglacial period and the last glaciation. As shown in Fig. 2, a large portion, about 7 °C, of the isotopic temperature decrease leading to the glacial conditions takes place between ~132 and 115 kyr, a time during which no significant CO_2 changes are observed. Our results thus indicate that climatic forcings other than CO_2 induced a large part of the temperature change over Antarctica linked with the inter-glacial–glacial transition. The marked CO_2 decrease (from ~275 to 235 p.p.m.v.) appears only during the final phase of the climatic cooling (between ~115 and 110.5 kyr) and is associated with a decrease of ~2 °C in the Vostok isotopic temperature.

Based on the above discussion, we suggest that the interactions between climate and CO_2 could be different depending on whether the climate shifts from glacial to interglacial or from intergla-cial to glacial conditions. In the latter case, orbital and orbitally derived forcing could have influ-enced the marked cooling trend in the Antarctic with a dominant contribution to the temperature decrease, before the potential climatic effect of the change in CO_2 concentrations. Note that the second (in terms of amplitude) cooling trend, found around 75 kyr BP in the Vostok record, exhib-its a similar behaviour of the CO_2–δD comparative variation.

IMPLICATIONS FOR CO_2 CYCLE

As well as providing information on the CO_2–climate interaction, our CO_2 profile allows us to discuss the modifications of the carbon cycle dur-ing the last climatic cycle. Based on previous ice core results over the past 30–40 kyr, several mod-els have been proposed to explain the increase of atmospheric CO_2 associated with the last deglacia-tion. Detailed descriptions of these models can be found in Broecker and Peng[26] or Berger and Keir[27]. These models are based on the idea that the biological productivity of the ocean controls the atmospheric CO_2 but they differ in the mecha-nisms driving these productivity variations.

In one family of models, sea-level variations are responsible for the atmospheric CO_2 changes[22,28]. The general trend of the Vostok CO_2 profile is similar to the benthic $\delta^{18}O$ records, which mainly reflect ice volume or sea level (for example see Fig. 2b, e of the companion paper[24]), making these models attractive. Berger and Keir[27] and Keir and Berger[29] attempt to model the past atmospheric CO_2 modifications due to sea level variations. Choosing appropriate parameters, the morphol-ogy of the derived CO_2 curve is close to that of the Vostok curve and the CO_2 variations lag the sea-level changes by ~1 kyr. To test this time lag, it

is necessary first to evaluate the phase relation-ship between the CO_2 and the Vostok temperature and secondly between the Vostok temperature and the sea level. In deglaciations, as already mentioned, the CO_2 begins to increase almost simultaneously with the Vostok temperature. The precise timing between the temperature and the sea level is difficult to obtain because these two signals are not recorded in the same medium. It is nevertheless reasonable to assume that the sea level cannot increase before the Antarctic tem-perature, as suggested by the interpretation of the marine sediment record, which indicates that changes in Southern Hemisphere sea-surface temperature have generally led the changes in ice volume[30,32]. Although denser sampling would be required for a more definite conclusion, the Vostok results thus suggest that at the end of gla-ciations the CO_2 increased before the sea level and not after as predicted by the models men-tioned. At the beginning of the last glaciation (~120 kyr BP), the CO_2 decreased ~10 kyr after the temperature and thus most probably after the sea-level change. Although this phase relation can-not exclude the influence of the sea level on the CO_2 variations during this period, it is important to note that the CO_2 decrease is very sharp (35 p.p.m.v. in 2 kyr) and seems incompatible with the mechanisms involved in the models proposed. We suggest that this abrupt change of the atmospheric CO_2 between two nearly constant values could be due to an abrupt modification of the deep oceanic circulation possibly linked with the sea-level decrease. Such rather rapid cir-culation changes associated with climatic transi-tions have been shown to occur[33].

In another family of models, changes in oceanic circulation affect the biological productivity of the surface ocean and thus the atmospheric CO_2 concentration. These models stress changes occur-ring in the high latitude deep-water formation areas[34–36] and in the upwelling zones at low lati-tudes[35]. The spectral analysis of the Vostok CO_2 record shows, as well as the 110-kyr component, a very weak and doubtful 41-kyr and a clear 21-kyr peak. This feature can give some indications of the latitudes involved in the forcing of the atmos-pheric CO_2. The latitudinal variation of the insola-tion spectra is complex. Although the presence of an obliquity component (41 kyr) seems to be a characteristic of high latitudes, this component is negligible around the equinoxes and becomes important around the solstices[37]. From marine proxy data, the spectral analysis of the sea-surface temperature of the North Atlantic shows that, apart from the 110-kyr component, the 41-kyr component is dominant for latitudes greater than 45° N (ref. 38), and we can note that the Vostok temperature profile also exhibits a strong 41-kyr period[14]. The quasi-absence of a 41-kyr period in the CO_2 record could thus suggest that the high latitudes do not have a major influence on the CO_2 variations. Nevertheless we cannot exclude an influence of these latitudes during part of the year, in particular around the equinoxes.

At low latitudes, upwelling areas are known to play a key role in the total biological productivity of the oceans and may thus influence the atmospheric CO_2 significantly. Evidence exists from the reconstructed sea-surface temperatures that upwellings were stronger during the Last Glacial Maximum (18 kyr BP)[39]. Similar conclusions are obtained from the palaeoreconstructions of trade-wind strength[40–42], responsible for the upwelling currents. Some of these indicators exhibit similarities with the atmospheric CO_2, for instance at the end of the two last glaciations the upwelling productivity off the NW African coast decreased before the increase in sea level[40], as did the winter trade-wind speed near the Peru coast[41]. Note also that the upwelling productivity off NW Africa was high during only the second part of the glaciation, that is, between ~70 and 15 kyr BP[41], and that the Vostok aluminium concentration profile exhibits three very pronounced spikes suggesting stronger atmospheric circulation at ~150, ~70 and ~20 kyr BP[43]. These spikes correspond to very low CO_2 values (~200 p.p.m.v.).

All these facts suggest that atmospheric circulation, coupled with oceanic circulation, at least through the upwelling currents, could have played an important role in past CO_2 variations. This influence may have been especially important between 70 and 15 kyr BP and thus could be responsible for at least part of the lower CO_2 values observed during this period compared with the values recorded during the previous part of the glaciation (110–70 kyr BP).

In summary, it seems that the atmospheric CO_2 variations recorded in the Vostok core could be explained by two kinds of changes in the oceanic circulation, that is, the deep changes possibly driven by sea level and the surface changes driven by atmospheric circulation. In this case only deep circulation changes could be the cause of the CO_2 variations between the last interglacial and the first part of the glaciation (110–70 kyr), whereas both deep and surface circulation could be responsible for the lower CO_2 values found during the second part of the glaciation (70–15 kyr) relative to the first part. During glacial–interglacial transitions, the CO_2 increase could be initiated by surface circulation changes relayed by the deep circulation changes associated with the sea-level increase.

CONCLUSIONS

An atmospheric CO_2 record over the past 160 kyr has been obtained from the Vostok ice core. This CO_2 record is probably the purest available, covering the last climatic cycle. The CO_2 changes thus revealed, which are of global significance, are well correlated with the Antarctic temperature record derived from the ice isotopic profile measured on the same core. Such a high correlation would be expected if CO_2 plays an important role in forcing the climate.

The results also suggest a different behaviour of the relative timing between CO_2 and Antarctic climate changes depending on whether we proceed from a glacial to an interglacial period or vice versa. This may have implications concerning the relation between cause and effect inside the CO_2–climate system, but more detailed measurements in the transition parts of the record are required before firm conclusions can be drawn. Long-term CO_2 changes are dominated by marked glacial–interglacial oscillations between ~190–200 and 260–280 p.p.m.v. A period of ~20 kyr similar to the orbital precession period, appears in the decreasing CO_2 trend covering most of the last glaciation and is supported by spectral analysis. These CO_2 changes could be linked with oceanic circulation changes but a more precise interpretation would in particular require the determination of the $\delta^{13}C$ of the CO_2 extracted from the air bubbles trapped in the ice and a comparable chronology for marine and ice core records.

We thank all Soviet and French participants in ice drilling and sampling. We acknowledge the efficient logistic support of the Soviet Antarctic Expeditions, the US National Science Foundation (Division of Polar Programs) and the French Polar Expeditions. We thank C. Genthon and J. Jouzel for performing the CO_2 spectral analysis and J. C. Duplessy, C. Genthon and J. Jouzel for helpful discussions. This work was supported in France by the Commission of the European Communities (Climatology Research Programme), CNRS/PIREN, TAAF and in the Soviet Union by Soviet Antarctic Expeditions.

NOTES

Received 26 March; accepted 17 July 1987.

1 Lorius, C. & Raynaud, D. in *Carbon Dioxide: Current Views and Developments in Energy/Climate Research* (eds Bach, W. *et al.*) 145–176 (Reidel, Dordrecht, 1983).

2 Stauffer, B. & Oeschger, H. *Ann. Glaciol.* **7**, 54–59 (1985).

3 Raynaud, D., Delmas, R., Ascencio, J. M. & Legrand, M. *Ann. Glaciol.* **3**, 265–268 (1982).

4 Neftel, A., Moor, E., Oeschger, H. & Stauffer, B. *Nature* **315**, 45–47 (1985).

5 Raynaud, D. & Barnola, J. M. *Nature* **315**, 309–311 (1985).

6 Pearman, G. I., Etheridge, D., De Silva, F. & Fraser, P. J. *Nature* **320**, 248–250 (1986).

7 Delmas, R. J., Ascencio, J. M. & Legrand, M. *Nature* **284**, 155–157 (1980).

8 Neftel, A., Oeschger, H., Schwander, J., Stauffer B. & Zumbrumn, R. *Nature* **295**, 220–223 (1982).

9 Stauffer, B., Hofer, H., Oeschger, H., Schwander, J. & Siegenthaler, U. *Ann. Glaciol.* **5**, 160–164 (1984).

10 Raynaud, D. & Barnola, J. M. in *Current Issues in Climate Research* (eds Ghazi, A. & Fantechi, R.) 240–246 (Reidel, Dordrecht, 1985).

11 Lorius, C. *et al. Nature* **316**, 591–596 (1985).

12 Barnola, J. M., Raynaud, D., Neftel, A. & Oeschger, H. *Nature* **303**, 410–412 (1983).

13 Schwander, J. & Stauffer, B. *Nature* **311**, 45–47 (1984).

14 Jouzel, J. *et al. Nature* **329**, 403–408 (1987).

15 Gammon, R. H., Sundquist, E. T. & Fraser, P. J. in *Atmospheric Carbon Dioxide and the Global Carbon Cycle* (DOE Report ER-0239) 25–62 (1985).

16 Pestiaux, P. & Berger, A. L. in *Milankovitch and Climate* (eds Berger, A. *et al.*) 417–445 (Reidel, Dordrecht, 1984).

17 Barrodale, I., & Ericksson, R. E. *Geophys* **45**, 420–432 (1980).

18 Stauffer, B., Neftel, A., Oeschger, H. & Schwander, J. in *Greenland ice core: Geophysics, Geochemistry and the Environment* (eds Langway, C. C. *et al.*) (Geophysical Monograph 33) 85–89 (American Geophysical Union, Washington DC, 1985).

19 Barnola, J. M. thesis, Université Scientifique et Médicale de Grenoble (1984).

20 Shackleton, N. J., Hall, M. A., Line, J. & Cang, Shuxi *Nature* **306**, 319–322 (1983).

21 Shackleton, N. J. & Pisias, N. G. in *The Carbon Cycle and Atmospheric CO$_2$: Natural Variations Archean to Present* (eds Sundquist, E. T. & Broecker, W. S.) (Geophysical Monograph 32) 303–317 (American Geophysical Union, Washington, DC, 1985).

22 Broecker, W. S. *Progr. Oceanogr.* **11**, 151–197 (1982).

23 Pisias, N. G. & Shackleton, N. J. *Nature* **310**, 757–759 (1984).

24 Genthon, C. *et al. Nature* **329**, 414–418 (1987).

25 Scientific Committee on Problems of the Environment in *Scope 29: The Greenhouse Effect, Climatic Change and Ecosystems* (eds Bolin, B. *et al.*) xxv–xxxi (Wiley, Chichester, 1986).

26 Broecker, W. S. & Peng, T. H. *Radiocarbon* **28**, 309–327 (1986).

27 Berger, W. H. & Keir, R. S. in *Climate Processes and climate sensitivity* (eds Hansen, J. E. & Takahashi, T.) (Geophysical Monograph 29) 337–350 (American Geophysical Union, Washington, DC, 1984).

28 Berger, W. H. *Naturwissenschaften* **69**, 87–88 (1982).

29 Keir, R. S. & Berger, W. H. *J. geophys. Res.* **88**, 6027–6038 (1983).

30 Hays, J. D., Imbrie, J. & Shackleton, N. J. *Science* **194**, 1121–1132 (1976).

31 CLIMAP project members *Quat. Res.* **21**, 123–224 (1984).

32 Labeyrie, L. D. *et al. Nature* **322**, 701–706 (1986).

33 Duplessy, J. C. & Shackleton, N. J. *Nature* **316**, 500–507 (1985).

34 Knox, F. & McElroy, M. B. *J. geophys. Res.* **89**, 4629–4637 (1984).

35 Siegenthaler, U. & Wenk, T. *Nature* **308**, 624–626 (1984).

36 Sarmiento, J. L. & Toggweiler, J. R. *Nature* **308**, 621–624 (1984).

37 Berger, A. & Pestiaux, P. in *Milankovitch and Climate* (eds Berger, A. *et al.*) 83–111 (Reidel, Dordrecht, 1984).

38 Ruddiman, W. F. & McIntyre, A. *Bull. geol. Soc. Am.* **95**, 381–391 (1984).

39 CLIMAP project members *Science* **191**, 1131–1136 (1976).

40 Sarntheim, M., Winn, K. & Zahn, R. in *Biviers Symposium on Abrupt Climatic Changes* (in the press).

41 Molina Crutz, A. *Quat. Res.* **8**, 324–328 (1977).

42 Boyle, E. *J. geophys. Res.* **88**, 7667–7680 (1983).

43 DeAngelis, M., Barkov, N. I. & Petrov, V. N. *Nature* **325**, 318–321 (1987).

Index